CONCISE
CALCULUS

CONCISE CALCULUS

Sheng Gong

University of Science & Technology of China, China

translated by Youhong Gong

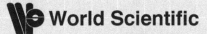

World Scientific

NEW JERSEY · LONDON · SINGAPORE · BEIJING · SHANGHAI · HONG KONG · TAIPEI · CHENNAI

Published by

World Scientific Publishing Co. Pte. Ltd.

5 Toh Tuck Link, Singapore 596224

USA office: 27 Warren Street, Suite 401-402, Hackensack, NJ 07601

UK office: 57 Shelton Street, Covent Garden, London WC2H 9HE

Library of Congress Cataloging-in-Publication Data
Names: Gong, Sheng, 1930– | Gong, Youhong, translator.
Title: Concise calculus / by Sheng Gong (University of Science and Technology of China, China) ;
    translated by Youhong Gong (University of Science and Technology of China, China).
Description: New Jersey : World Scientific, 2016. | Includes index.
Identifiers: LCCN 2016052209| ISBN 9789814291484 (hardcover : alk. paper) |
    ISBN 9789814291491 (pbk : alk. paper)
Subjects: LCSH: Calculus--Textbooks.
Classification: LCC QA303.2 .G6637 2016 | DDC 515--dc23
LC record available at https://lccn.loc.gov/2016052209

British Library Cataloguing-in-Publication Data
A catalogue record for this book is available from the British Library.

Typeset by Stallion Press
Email: enquiries@stallionpress.com

Printed in Singapore

To the students with a passion for mathematics

Preface

This book grew out of the collected lecture notes of the calculus course for the Department of Modern Physics, at the University of Science and Technology of China (USTC), in the academic year of 1965.

In order to help students majoring in physics complete their mathematics prerequisite more quickly, Professor Sheng Gong taught calculus a different way.

In Chapter 1, he brings the Fundamental Theorem of calculus to the student's attention with only a brief preparation. And then, following this main theme of interplay between differential and integral calculus, he presents the subsequent topics: the calculation of derivatives and integrals in Chapter 2, applications of calculus in Chapter 3, and ordinary differential equations in Chapter 4. In other words, he presents the main structure of the theory of calculus first, and then lets the students see how the other topics are positioned within the structure, and how they are related to each other. This effectively makes the students understand the entire course much more clearly. There are plenty of examples of how to solve physics problems using calculus as a tool.

After introducing the students to vector spaces and analytic geometry in Chapter 5, and partial derivatives and multiple integrals in Chapter 6, he adopts the notion of exterior differential forms in Chapter 7 to unify Green's Theorem, Gauss's Theorem, and Stokes' Theorem into the Stokes' Theorem in exterior differential form:

$$\int_{\partial D} \omega = \int_D d\omega.$$

He points out that this is the Fundamental Theorem of calculus in higher dimensional spaces. Taylor expansions of multi-variable functions and several physics topics are studied as applications of calculus in Chapter 8.

Concise but not simplistic, this text consists of the entire contents of the theory of calculus, including the precise definition of limits and continuity in Chapter 9. Chapter 10 studies not only infinite series and infinite integrals, but also the connections between them, followed by Fourier series and Fourier integrals in Chapter 11.

In the past several decades, this book has been used as the textbook for non-mathematics majors at USTC with excellent results.

In 2002, Professor Shiing-Shen Chern invited Professor Sheng Gong to give a lecture series (total of ten hours) based on the main ideas of this textbook in the Chern Institute of Mathematics at Nankai University in Tianjin, China.

In 2003, Professor Shing-Tung Yau invited Professor Sheng Gong to give a lecture series (total of twenty hours) based on the main ideas of this textbook in the Center of Mathematical Sciences at Zhejiang University in Hangzhou, China.

In 2004, Professor Sheng Gong gave a lecture series (total of twenty hours) based on the main ideas of this textbook in USTC in Hefei, China.

All of these lectures earned high praise from the students and professors in the audience.

Professor Wenjun Wu especially wrote a forward for this book (Fourth Edition in Chinese).

When Professor Sheng Gong started teaching this course at USTC in the 1960s, Professor Shenglei Zhang was a colleague of his. He provided all of the exercises, gave important suggestions and made necessary changes to the original lecture notes as well as to the first edition of this textbook. The success of this textbook could not have been achieved without his considerable efforts.

In the beginning of this project, Professor Sheng Gong asked me to add some more material about applications of calculus to the English edition of this book. When I actually began work on it, I found that the original content formed a consecutive and integrated whole. This fast-paced yet thorough treatment of the material is the highlight of this book. Therefore, aside from the necessary changes due to translation from Chinese to English, I have made no alterations, in order to preserve the original intentions of the author. There is no additional material compared to the original Chinese edition.

Thanks to Dr. Weiqi Gao for the many ways in which he helped make the English edition of this book possible. Also, thanks to Angela Gao for being the English consultant with her busy schedule.

I would especially like to express my appreciation to Zhang Ji, Jessie Tan and Tan Rok Ting of World Scientific Publishing, for their patience and cooperation during the preparation of this edition.

Special thanks to Dr. Suqi Pan for his detailed editing of the book.

Youhong Gong
November 16, 2015
St. Louis, U.S.A.

Contents

Preface   vii

1. Basic Concepts   1

    1.1   Functions and Limits   1
          1.1.1   Limits of Sequences and Functions   1
          1.1.2   Continuous Functions   2
    1.2   Definite Integrals   7
          1.2.1   Calculation of Areas   7
          1.2.2   Definition of Definite Integrals   11
          1.2.3   Logarithmic Function   18
    1.3   Derivatives and Differentials   24
          1.3.1   Tangent Lines of a Curve   24
          1.3.2   Velocity and Density   25
          1.3.3   Definition of Derivatives   27
          1.3.4   Differentials   31
          1.3.5   Mean-Value Theorem   33
    1.4   Fundamental Theorem of Calculus   38

2. Calculations of Derivatives and Integrals   43

    2.1   Differentiation   43
          2.1.1   Calculations of Derivatives and Differentials . . .   43
          2.1.2   Derivatives and Differentials of Higher Order . . .   54
          2.1.3   Approximate Calculation by Derivatives   58
    2.2   Integration   67
          2.2.1   Indefinite Integrals   67
          2.2.2   Definite Integrals   86

2.2.3   Approximate Calculations of Definite Integrals . .   93

3. Some Applications of Differentiation and Integration   101

3.1   Areas, Volumes, Arc Lengths   101
       3.1.1   Areas   101
       3.1.2   Volumes   103
       3.1.3   Arc Lengths   106
3.2   Techniques for Graphing Functions   111
       3.2.1   Increasing and Decreasing Functions   112
       3.2.2   Concavity   114
       3.2.3   Asymptotes   115
       3.2.4   Examples of Graphing Functions   118
       3.2.5   Curvatures   120
3.3   Taylor Expansions and Extreme Value Problems   126
       3.3.1   Taylor Expansions   126
       3.3.2   Extreme Value Problems   131
3.4   Examples in Physics   141

4. Ordinary Differential Equations   149

4.1   First Order Differential Equations   149
       4.1.1   Concepts   149
       4.1.2   Separation of Variables   152
       4.1.3   Linear Differential Equations   161
4.2   Second Order Differential Equations   167
       4.2.1   Reducible Differential Equations   167
       4.2.2   Second Order Linear Differential Equations   171
       4.2.3   Linear Differential Equations with Constant
                Coefficients   181
       4.2.4   Mechanical Vibration   190
       4.2.5   General Linear Differential Equations and Systems
                of Linear Equations   196

5. Vector Algebra and Analytic Geometry in Three-
    Dimensional Space   209

5.1   Coordinate System of Three-Dimensional Space and
       Concept of Vectors   209
       5.1.1   Rectangular Coordinate System   209
       5.1.2   Addition and Scalar Multiplication of Vectors . .   211

5.2		Products of Vectors	217
	5.2.1	Inner Products of Vectors	217
	5.2.2	Cross Products of Vectors	220
	5.2.3	Scalar Triple Products of Vectors	222
5.3		Planes and Lines	226
	5.3.1	Equations of Planes	226
	5.3.2	Equations of Lines	229
5.4		Quadric Surfaces	234
	5.4.1	Cylindrical Surfaces	234
	5.4.2	Surfaces of Revolution	236
	5.4.3	Conical Surfaces	238
	5.4.4	Ellipsoid	239
	5.4.5	Hyperbolic Paraboloid	241
	5.4.6	Hyperboloid of One Sheet	242
	5.4.7	Hyperboloid of Two Sheets	242
	5.4.8	Elliptic Paraboloid	242
5.5		Transformations of Coordinates	244
	5.5.1	Translation of Axes	244
	5.5.2	Rotation of Axes	246

6. Multiple Integrals and Partial Derivatives — 251

6.1		Multiple Integrals	251
	6.1.1	Limits and Continuity of Functions of Several Variables	251
	6.1.2	Multiple Integration	253
	6.1.3	Calculation of Multiple Integrals	257
6.2		Partial Derivatives	270
	6.2.1	Partial Derivatives and Total Differentials	270
	6.2.2	Derivatives of Implicit Functions	278
6.3		Jacobian Determinants, Area Elements, Volume Elements	295
	6.3.1	Properties of Jacobian Determinant	295
	6.3.2	Area Elements and Volume Elements	296

7. Line Integrals, Surface Integrals and Exterior Differential Forms — 317

7.1		Scalar Fields and Vector Fields	317
	7.1.1	Contour Surfaces and Gradient of a Scalar Field	317
	7.1.2	Streamlines of Vector Fields	321
7.2		Line Integrals	327

7.2.1   Line Integrals of the First Kind   327
7.2.2   Applications of Line Integrals of the First Kind
        (Areas of Surfaces of Revolution)   330
7.2.3   Line Integrals of the Second Kind   331
7.2.4   Calculation of Line Integrals of the Second Kind .   335
7.2.5   Relation Between Line Integrals of the First Kind
        and the Second Kind   338
7.2.6   Circulations of Vector Fields and Line Integrals of
        Vectors   339
7.3   Surface Integrals   345
7.3.1   Surface Integrals of the First Kind   345
7.3.2   Flux of Vector Fields, Surface Integrals of the Sec-
        ond Kind (Integral with respect to the projections
        of the area element)   348
7.3.3   Calculation of Surface Integrals of the Second Kind   350
7.4   Stokes Theorem   357
7.4.1   Green's Theorem   357
7.4.2   Gauss's Theorem, Divergence   361
7.4.3   Stokes' Theorem, and The Curl of a Vector Field   367
7.5   Total Differentials and Line Integrals   377
7.5.1   Line Integrals that are Independent of Paths . . .   377
7.5.2   Potential Fields   381
7.5.3   Solenoidal Vector Fields   383
7.6   Exterior Differential Forms   387
7.6.1   Exterior Products, Exterior Differential Forms . .   387
7.6.2   Exterior Differentiation, Poincaré Lemma and its
        Inverse   394
7.6.3   Mathematical Meaning of Gradient, Divergence
        and Curl   399
7.6.4   Fundamental Theorem of Calculus in Several
        Variables (Stokes' Theorem)   401

8. Some Applications of Calculus in Several Variables   405

8.1   Taylor Expansions and Extremal Problems   405
8.1.1   Taylor Expansions of Functions in Several
        Variables   405
8.1.2   Extremal Problems of Functions in Several
        Variables   406
8.1.3   Conditional Extremum Problems   411

8.2   Examples of Applications in Physics   417
       8.2.1   Barycenter, Moment of Inertia and Gravitational
               Force   417
       8.2.2   Complete System of Equations of Fluid Dynamics   423
       8.2.3   Propagation of Sound   426
       8.2.4   Heat Exchange   427

9. The ε-δ Definitions of Limits   433

9.1   The ε-N Definition of Limits of Number Sequences   433
       9.1.1   Definition of Limits of Number Sequences   433
       9.1.2   Properties of Limits of Number Sequences   435
       9.1.3   Criteria for the Existence of Limits   438
9.2   The ε-δ Definition of Continuity of Functions   447
       9.2.1   Limits of Functions   447
       9.2.2   Definition of Continuous Functions   454
       9.2.3   Properties of Continuous Functions   457
       9.2.4   Uniform Continuity of Functions   460
9.3   Existence of Definite Integrals   466
       9.3.1   Darboux Sums   466
       9.3.2   Integrability of Continuous Functions   468
       9.3.3   Generalization of the Concept of Definite Integrals
               (Improper Integrals)   475

10. Infinite Series and Infinite Integrals   485

10.1   Number Series   485
        10.1.1 Basic Concepts   485
        10.1.2 Some Convergence Criteria   487
        10.1.3 Conditionally Convergent Series   493
10.2   Function Series   502
        10.2.1 Infinite Sums   502
        10.2.2 Uniformly Convergent Sequences of Functions . .   504
        10.2.3 Uniformly Convergent Function Series   508
        10.2.4 Existence Theorem of Implicit Functions   512
        10.2.5 Existence and Uniqueness Theorem of the Solution
                of Ordinary Differential Equations   516
10.3   Power Series and Taylor Series   524
        10.3.1 Convergence Radius of Power Series   524
        10.3.2 Properties of Power Series   527

10.3.3 Taylor Series 532
10.3.4 Applications of Power Series 539
10.4 Infinite Integrals and Integrals with Parameters 553
10.4.1 Convergence Criteria for Infinite Integrals 553
10.4.2 Integrals with Parameters 565
10.4.3 Infinite Integrals with Parameters 570
10.4.4 Several Important Infinite Integrals 583

11. Fourier Series and Fourier Integrals   597

11.1 Fourier Series 597
11.1.1 Orthogonality of the System of Trigonometric
        Functions 597
11.1.2 Bessel Inequality 607
11.1.3 Convergence Criterion for Fourier Series 611
11.2 Fourier Integrals 616
11.2.1 Fourier Integrals 616
11.2.2 Fourier Transforms 619
11.2.3 Applications of Fourier Transforms 624
11.2.4 Higher-Dimensional Fourier Transforms 625

Answers   627

Index   671

Chapter 1

Basic Concepts

1.1 Functions and Limits

1.1.1 *Limits of Sequences and Functions*

If the term a_n in an infinite sequence

$$a_1, a_2, \cdots, a_n, \cdots$$

approaches a fixed number A as the ordinal number n of its term increases without bound, we say that the sequence has *limit A*. We denote this as

$$\lim_{n \to \infty} a_n = A.$$

This concept can also be stated as: for any arbitrarily small positive number, we can always find a term in the sequence, such that the absolute value of the difference between any term after the term we found and the number A is less than the given small positive number.

For two sequences with limits, the sequences formed by the sums, differences, products and quotients of the terms of the given sequences, have limits which are the sum, difference, product and quotient of the limits of the given sequences (the limit that is the divisor cannot be zero).

Symbolically expressed, if $\lim_{n \to \infty} a_n = A$, $\lim_{n \to \infty} b_n = B$, then

$$\lim_{n \to \infty} (a_n \pm b_n) = A \pm B, \quad \lim_{n \to \infty} (a_n \cdot b_n) = A \cdot B,$$

$$\lim_{n \to \infty} \frac{a_n}{b_n} = \frac{A}{B}, \quad (B \neq 0).$$

Furthermore, for any constant k

$$\lim_{n \to \infty} k \cdot a_n = k \cdot A.$$

If there is a third sequence with $\lim\limits_{n\to\infty} c_n = C$ and the terms of the three sequences satisfy the relation $a_n \leqslant b_n \leqslant c_n$, then

$$A \leqslant B \leqslant C.$$

For a function $y = f(x)$, if the value of $f(x)$ approaches a fixed number A when the independent variable x approaches a point a, we say that the function has a limit A as x approaches a. We denote this as

$$\lim_{x\to a} f(x) = A.$$

The value a can be a finite number or an infinity.

Similar to the sequences case, for functions $f(x)$, $g(x)$ and $h(x)$ we have the following properties.

If $\lim\limits_{x\to a} f(x)$ and $\lim\limits_{x\to a} g(x)$ exist and are finite, then

$$\lim_{x\to a}(f(x) \pm g(x)) = \lim_{x\to a} f(x) \pm \lim_{x\to a} g(x),$$

$$\lim_{x\to a}(f(x) \cdot g(x)) = \lim_{x\to a} f(x) \cdot \lim_{x\to a} g(x),$$

$$\lim_{x\to a} \frac{f(x)}{g(x)} = \frac{\lim\limits_{x\to a} f(x)}{\lim\limits_{x\to a} g(x)},$$

where $\lim\limits_{x\to a} g(x) \neq 0$.

Moreover, if

$$f(x) \leqslant h(x) \leqslant g(x)$$

when x is close to a, and

$$\lim_{x\to a} f(x) = \lim_{x\to a} g(x) = A$$

then $\lim\limits_{x\to a} h(x) = A$.

More detailed discussion about the concept of limits can be found in Chapter 9.

1.1.2 *Continuous Functions*

In practical applications, the value of one quantity often depends on that of another. For example, the length L of a metal rod depends on the temperature T since the metal rod will expand when heated and contract when cooled. The relation between L and T is

$$L = L_0(1 + \alpha T),$$

where L_0 is the length of the metal rod at $0°C$, and α is the coefficient of linear expansion.

This is a *function* relation between the variable L and the variable T.

Sometimes, there are numerical boundaries for the variables. For instance, $-273.15°C$, the absolute zero on the Kelvin scale, is lower than the temperature of any matter.

For a simple function relation, the region of change of a variable is usually an interval. The notation $[a, b]$ denotes all real numbers x that satisfy the inequality

$$a \leqslant x \leqslant b,$$

and it is called a *closed interval*. The notation (a, b) denotes all real numbers x that satisfy the inequality

$$a < x < b,$$

and it is called an *open interval*. The notation $[a, b)$ denotes all real numbers x that satisfy the inequality

$$a \leqslant x < b.$$

Notations $(a, b]$, $[a, +\infty)$, $(-\infty, b]$, $(-\infty, b)$, $(a, +\infty)$, and $(-\infty, +\infty)$ can be defined in similar ways.

Some variables can be thought of as constants since the changes of those variables are very limited under certain conditions. For instance, the gravitational acceleration g near the surface of the earth can be considered as a constant within the city limit of Beijing. Also, there are constants representing the relations between two variables. For instance, when the diameter of a circle changes, so does its circumference, but the ratio of the circumference to the diameter is a constant $\pi = 3.14159265\cdots$.

Example 1. When an object falls from a high position, the distance s between the object and the point where it falls from depends on the time t since the object starts to fall:

$$s = \frac{1}{2}gt^2 ,$$

where g is the gravitational acceleration.

Example 2. Within a 24-hour period, the outdoor temperature θ changes with time t. The curve in Figure 1.1 represents the relation between these two variables in one particular day. The temperature θ_0 of a specific time t_0 can be identified by the curve.

Example 3. If you mail a parcel through the post office, the postage depends on the weight of the parcel. It is 80 cents for parcels under 20 grams, and the postage will increase 80 cents for every 20 grams weight increased. Thus the postage is a function of the weight, as depicted in Figure 1.2.

$$y = \begin{cases} 80, & \text{if } 0 \leqslant x \leqslant 20; \\ 160, & \text{if } 20 < x \leqslant 40; \\ 240, & \text{if } 40 < x \leqslant 60; \\ \vdots \end{cases}$$

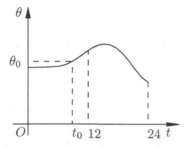
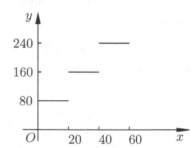

Fig. 1.1   Fig. 1.2

From these examples, we can see that a function is a correspondence between two variables. From this correspondence, the value of one variable can be determined by the value of another variable. This correspondence is often denoted as $y = f(x)$. That is, y is a function of x. The letter f represents the correspondence. The variable x is called the *independent variable*, and the variable y is called the *dependent variable*. The region of change of independent variables is called the *domain* of the function. The value of a function $f(x)$ at $x = a$ is denoted as $f(a)$. The region of change of dependent variables is called the *range* of the function.

Many function relations are "continuous". The position x of a moving particle is a continuous function of time t, $x = f(t)$. If a particle moves from one position to another position, it passes every point on a path connecting the two positions. The graph of a continuous function is a continuous curve, as depicted in Figure 1.3.

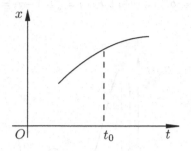
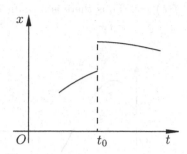

Fig. 1.3   Fig. 1.4

Consider a particular time $t = t_0$, if t has an increment Δt, then $f(t)$ also has an increment $\Delta x = f(t + \Delta t) - f(t)$. The increment of Δx approaches zero when Δt approaches to zero. That is $f(t) \to f(t_0)$ as $t \to t_0$. If $\Delta x = f(t + \Delta t) - f(t)$ does not approach zero when $\Delta t \to 0$, the curve is said to be *discontinuous* at $t = t_0$, as depicted in Figure 1.4.

In other words, a function $f(t)$ is *continuous* at a point $t = t_0$ if the equation

$$\lim_{t \to t_0} f(t) = f(t_0)$$

holds. Otherwise $f(t)$ is discontinuous at t_0. It is easy to see that if $f(t)$ is continuous at $t = t_0$, and $f(t_0) > 0$, then $f(t) > 0$ when t is close enough to t_0.

If $x = f(t)$ is continuous at every point of a closed interval $a \leqslant x \leqslant b$ (or an open interval $a < x < b$), then we say that $f(t)$ is a *continuous function* on $[a, b]$ (or on (a, b)).

Suppose $f(t)$ is a continuous function on $[a, b]$, and $f(a)$, $f(b)$ carry different signs. That is, the points $(a, f(a))$ and $(b, f(b))$ locate at different sides of the t-axis. Since $x = f(t)$ is a continuous function, the graph of f is a continuous curve that connects these two points. The curve must pass through the t-axis (see Figure 1.5.) Therefore, there exists a point c in the open interval (a, b), such that $f(c) = 0$.

It follows that, if $f(t)$ is continuous on $[a, b]$, and r is a value between $f(a)$ and $f(b)$, then there exists a point c between a and b, such that $f(c) = r$.

Indeed, consider the function $F(t) = f(t) - r$. It is still a continuous function on $[a, b]$, and $F(a) = f(a) - r$ has a different sign from $F(b) =$

$f(b) - r$. Thus there is a point c, $a < c < b$, such that $F(c) = f(c) - r = 0$, and $f(c) = r$.

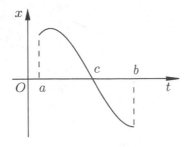

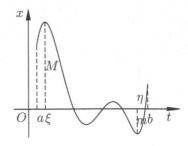

Fig. 1.5   Fig. 1.6

A continuous function on a closed interval reaches its maximum value M and minimum value m, as depicted in Figure 1.6. In other words, there exist two points ξ and η in $[a, b]$ such that $f(\xi) = M$ and $f(\eta) = m$, and the inequality $m \leqslant f(t) \leqslant M$ holds for all t in $[a, b]$. This conclusion is not necessarily true if the domain of the function is an open interval. All elementary functions are continuous in their domains. Detailed discussions of continuous functions can be found in Section 9.2.

Exercises 1.1

1. Suppose $f(x) = x2^x$. Find $f(0)$, $f(1)$, $f(-2)$, $f\left(\dfrac{1}{x}\right)$, $\dfrac{1}{f(x)}$, $f(x) + 1$, $f(x + 1)$, $|f(-a)|$, $f(a^2)$, $[f(a)]^2$.

2. Suppose
$$f(x) = \begin{cases} x, & \text{if } x \neq 0; \\ 1, & \text{if } x = 0. \end{cases}$$
Is this a function? How many functions are defined in this expression? one or two? What is the domain of $f(x)$? Sketch the graph of $f(x)$. Find $f(1)$, $f(-1)$, $f(0)$, $f(t)$. At which points is $f(x)$ continuous, and at which points is $f(x)$ discontinuous?

3. Suppose the distance between two electric charges q_1 and q_2 is r. Find the function relation of the electric force f between the two point charges and their distance r.

4. Cut away a sector of degree θ from a round iron sheet of radius R. Make a cone out of the remaining part of the sheet. Find the function relation of its volume with respect to θ.

5. On an 8-meter beam, a 30kg weight is distributed evenly between 2 meters and 4 meters from the left, and a 10kg weight is applied at the 5 meter point. Find the weight borne by the portion of the beam from the left end to the point x meters from the left end expressed in terms of x.

6. Suppose $\varphi(x) = \dfrac{f(x+h) - f(x)}{h}$. Calculate $\varphi(x)$ for
   (1) $f(x) = ax + b$;   (2) $f(x) = x^2$;
   (3) $f(x) = a^x$.

7. Study the continuity of the following functions.
   (1) $y = x^2$;   (2) $y = \sin x$;   (3) $y = x^2 \sin x$;
   (4) $y = \dfrac{1 - x^2}{2 + x}$;   (5) $y = \dfrac{1}{\sin \pi x}$;

   (6) $y = \begin{cases} \dfrac{|x|}{x}, & \text{if } x \neq 0; \\ 0, & \text{if } x = 0. \end{cases}$ (7) $y = \begin{cases} x, & \text{if } 0 \leqslant x < 1; \\ 4x - 2, & \text{if } 1 < x < 3; \\ 13 - x, & \text{if } 3 \leqslant x < \infty. \end{cases}$

8. Show that the following equations have solutions on the given interval.
   (1) $x^3 - 3x = 1$ on $[1, 2]$;
   (2) $x2^x = 1$ on $[0, 1]$.

1.2   Definite Integrals

1.2.1   *Calculation of Areas*

We have learned methods for calculating areas of shapes bounded by straight lines and of circular discs in middle school. To find the areas of more complicated shapes, we need to develop better techniques. One of the ideas that we will discuss next is to use the sum of areas of small regular shapes to approach the areas of irregular shapes.

We introduce this idea by way of an example. Consider the parabola $y = x^2$ in Figure 1.7. Let C be a point on the x-axis such that the length of the line segment OC is equal to 1. Draw a vertical line from C that intersects the parabola at point A. Now we calculate the area of the triangle OAC with one curved side.

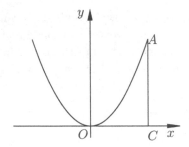
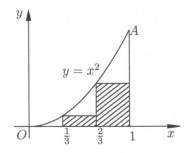

Fig. 1.7   Fig. 1.8

Partition the interval $[0, 1]$ to three equal sections. The area of the two shaded rectangles under the parabola in Figure 1.8 is

$$S_3 = \frac{1}{3}\left[\left(\frac{1}{3}\right)^2 + \left(\frac{2}{3}\right)^2\right] = \frac{1}{3^3}\left[1^2 + 2^2\right] = \frac{5}{27}.$$

This number can be considered as a rough estimate of the area we want.

If we partition $[0, 1]$ to four equal parts and calculate the area of the three shaded rectangles in Figure 1.9, we have

$$S_4 = \frac{1}{4}\left[\left(\frac{1}{4}\right)^2 + \left(\frac{2}{4}\right)^2 + \left(\frac{3}{4}\right)^2\right] = \frac{1}{4^3}[1^2 + 2^2 + 3^2] = \frac{7}{32}.$$

This seems a better approximation compared with S_3.

Continuing the same process, partitioning $[0, 1]$ to n equal parts and calculating the area of the $n - 1$ shaded rectangles in Figure 1.10, we have

$$S_n = \frac{1}{n}\left[\left(\frac{1}{n}\right)^2 + \left(\frac{2}{n}\right)^2 + \cdots + \left(\frac{n-1}{n}\right)^2\right]$$

$$= \frac{1}{n^3}[1^2 + 2^2 + \cdots + (n-1)^2] = \frac{1}{n^3}\frac{(n-1)n(2n-1)}{6}$$

$$= \frac{1}{6}\left(1 - \frac{1}{n}\right)\left(2 - \frac{1}{n}\right) = \frac{1}{3} - \frac{1}{2n} + \frac{1}{6n^2}.$$

The area of the stair shaped figure becomes closer to the area of the triangle with a curved side as the number of sections n of $[0, 1]$ becomes greater. This process gives a sequence of numbers

$$\{S_n\} : S_3, S_4, \cdots, S_n, \cdots$$

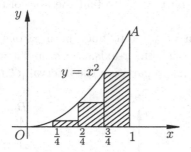

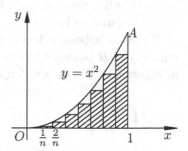

Fig. 1.9   Fig. 1.10

with values

$$\frac{5}{27}, \frac{7}{32}, \cdots, \frac{1}{6}\left(1 - \frac{1}{n}\right)\left(2 - \frac{1}{n}\right), \cdots.$$

Since $\frac{1}{n}$ and $\frac{1}{n^2}$ approach zero as n becomes greater, S_n approaches a fixed number $\frac{1}{3}$. In other words, the limit of the sequence $\{S_n\}$ is $\frac{1}{3}$.

We can also see that, if the length of OC is not 1, but an arbitrary real number a, then by the same method, we can find that the area of the triangle OAC with a curved side is $\frac{a^3}{3}$. Moreover, the area of the shaded region $ACDB$ in Fig. 1.11 is $\frac{b^2 - a^2}{3}$, where $OC = a$ and $OD = b$.

Next, we consider a slightly more complicated example. We want to find the area of the right triangle having a curve that is described by the function $y = x^3$ as hypotenuse. Performing the same process as we did with the $y = x^2$ case, we can arrive at the formula for the area. In this process, we need to use the equation

$$1^3 + 2^3 + \cdots + n^3 = (1 + 2 + \cdots + n)^2$$

which can be proved by mathematical induction.

If the hypotenuse of the triangle is the curve $y = x^4$, then we need to find the sum

$$1^4 + 2^4 + \cdots + n^4$$

if we still use the same process to find the area of the right triangle. More generally, the sum

$$1^m + 2^m + \cdots + n^m$$

is needed when we work with the function $y = x^m$ to find the area of the triangle defined as before.

To avoid tedious calculations, when we want to find the area of the region $ACDB$ bounded by the curve $y = x^4$, the x-axis, $x = a$, and $x = b$ in Figure 1.12, we use a different technique to partition the interval $[C, D]$.

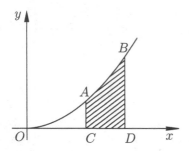
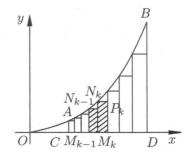

Fig. 1.11   Fig. 1.12

Suppose the points on the line segment CD which separate each section of the partition are

$$(C =)M_0, M_1, M_2, \cdots, M_{n-1}, M_n(= D).$$

If $OC = a$ and $OD = b$, then we choose $M_0, M_1, M_2, \cdots, M_{n-1}, M_n$ such that

$$OM_1 = aq, \ OM_2 = aq^2, \ \cdots, \ OM_n = aq^n$$

where $q = \sqrt[n]{\dfrac{b}{a}}$. Thus,

$$M_1C = a(q - 1), \quad M_2M_1 = aq(q - 1), \quad M_3M_2 = aq^2(q - 1), \cdots,$$
$$M_kM_{k-1} = aq^{k-1}(q - 1), \cdots, \quad M_nM_{n-1} = aq^{n-1}(q - 1).$$

In this way, the length of each section of the partition is not the same. It is shorter for the sections close to C, and longer for the sections close to D, but still, the length of each section will approach zero as the partition gets finer.

Under this partition, the area of the rectangle $M_{k-1}N_{k-1}P_kM_k$ is

$$aq^{k-1}(q - 1)(aq^{k-1})^4 = (q - 1)(aq^{k-1})^5.$$

The sum of all these areas of such rectangles is an approximation of the area of the right triangle with a curved hypotenuse

$$S_n = (q-1)(aq^{1-1})^5 + (q-1)(aq^{2-1})^5 + \cdots + (q-1)(aq^{n-1})^5$$
$$= (q-1)a^5(1 + q^5 + \cdots + q^{5(n-1)})$$
$$= (q-1)a^5 \frac{q^{5n}-1}{q^5-1}.$$

Since $q^n = \dfrac{b}{a}$, we have

$$S_n = a^5 \frac{\left(\dfrac{b}{a}\right)^5 - 1}{\dfrac{q^5-1}{q-1}}.$$

The limit of the value of q is 1 as n goes to infinity. Since

$$\frac{q^5-1}{q-1} = q^4 + q^3 + q^2 + q + 1,$$

and it approaches 5 as n goes to infinity, we have

$$S_n \to \frac{b^5 - a^5}{5}$$

as $n \to \infty$.

With the same method, we can calculate the area under the curve $y = x^m$ and between $x = a$, $x = b$,

$$S = \frac{b^{m+1} - a^{m+1}}{m+1},$$

where the number m is an arbitrary integer, and $m \neq -1$.

If $m = -1$, then the function becomes $y = \dfrac{1}{x}$. The above method does not apply to this function. The detailed discussion about finding areas under this function can be found in Section 1.2.3.

1.2.2   Definition of Definite Integrals

We studied how to find the area under some special functions in the previous section. Similar ideas can be applied to any continuous functions. Without loss of generosity, assume that $f(x) \geqslant 0$ on $[a, b]$. Partition $[a, b]$ into n subintervals (the length of each subinterval is not necessarily equal), and denote the set of dividing points as

$$a = x_0 < x_1 < \cdots < x_{n-1} < x_n = b.$$

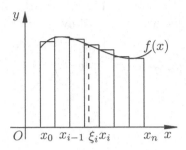

Fig. 1.13

Draw a rectangle with the subinterval $[x_{i-1}, x_i]$ as the base and $f(\xi_i)$ as the hight, where ξ_i is an arbitrary point in the subinterval $[x_{i-1}, x_i]$. The area of this rectangle is $f(\xi_i)(x_i - x_{i-1})$. The sum of the areas of all such rectangles is

$$f(\xi_1)(x_1 - x_0) + f(\xi_2)(x_2 - x_1) + \cdots + f(\xi_n)(x_n - x_{n-1})$$

and is denoted as

$$\sum_{i=1}^{n} f(\xi_i)(x_i - x_{i-1}). \tag{1}$$

This is an approximation of the area S that we want to find. The finer the partition of the interval $[a, b]$, the smaller the difference between the value of this sum and S.

Let

$$\lambda = \max_{1 \leqslant i \leqslant n} (x_i - x_{i-1})$$

be the length of the longest interval among $(x_1 - x_0), \cdots, (x_n - x_{n-1})$, then the area of the region under the curve $f(x)$ can be defined by

$$S = \lim_{\lambda \to 0} \sum_{i=1}^{n} f(\xi_i)(x_i - x_{i-1}). \tag{2}$$

Therefore, finding the area under a curve becomes calculating a limit of this type.

Let's look at some other examples where limits of this type occur.

If an object moves along a straight line with a constant velocity v, then the distance travelled by the object between times t_0 and T is:

$$s = v(T - t_0).$$

What is the distance between time t_0 and T if the velocity is not a constant?

We partition the interval $[t_0, T]$ to n sections:

$$t_0 < t_1 < t_2 < \cdots < t_n = T,$$

in such a way that the length of the interval $[t_{i-1}, t_i]$ becomes very small when n is very large. Thus, the change of the velocity $v = f(t)$ in each interval is also very small. Hence, at any specific time $t = \xi_i$ in $[t_{i-1}, t_i]$, $v = f(\xi_i)$ can be considered the velocity for every t in $[t_{i-1}, t_i]$. Therefore, the distance the object traveled in the time interval $[t_{i-1}, t_i]$ is approximately equal to

$$f(\xi_i)(t_i - t_{i-1}).$$

The total distance in the time interval $[t_0, T]$ is approximately equal to

$$s \approx \sum_{i=1}^{n} f(\xi_i)(t_i - t_{i-1}). \tag{3}$$

Let the length of the longest time interval in the partition be

$$\lambda = \max_{1 \leqslant i \leqslant n} (t_i - t_{i-1}).$$

It is obvious that the value of approximate distance will be more accurate if the number n of small intervals in $[t_0, T]$ gets larger.

Let n approach infinity, in other words, take the limit as λ approaches zero. Then

$$s = \lim_{\lambda \to 0} \sum_{i=1}^{n} f(\xi_i)(t_i - t_{i-1}) \tag{4}$$

is the distance that we want to find.

In physics, work A is the product of force F and displacement s:

$$A = F \cdot s.$$

Suppose an object moves along a straight line. If the force F acting on the object is not a constant, its value changes when the position of the object changes, that is, F is a function of s. Then what is the work done by the force in the displacement interval $[s_0, s]$?

Partition the interval $[s_0, s]$ to n sections

$$s_0 < s_1 < s_2 < \cdots < s_n = s,$$

in such a way that the length of the interval $[s_{i-1}, s_i]$ becomes smaller when n becomes larger. Thus, we can assume that the change of $F = F(s)$

is very small in $[s_{i-1}, s_i]$, and it can be considered as a constant. The force $F = F(\eta_i)$ at a point $s = \eta_i$ in $[s_{i-1}, s_i]$ can be considered as the force on every point of $[s_{i-1}, s_i]$. Hence, the work on this path segment is approximately equal to

$$F(\eta_i)(s_i - s_{i-1}).$$

The force on the entire path $[s_0, s]$ is approximately equal to

$$\sum_{i=1}^{n} F(\eta_i)(s_i - s_{i-1}). \tag{5}$$

Let the length of the longest displacement interval in the partition be

$$\lambda = \max_{1 \leqslant i \leqslant n} (s_i - s_{i-1}).$$

Let n approach infinity, in other words, take the limit as λ approaches zero. Then

$$A = \lim_{\lambda \to 0} \sum_{i=1}^{n} F(\eta_i)(s_i - s_{i-1}) \tag{6}$$

is the work that we want to find.

From these examples we can see that, although the subjects of the problems are different, the calculation methods are the same, that is, to evaluate the limit

$$\lim_{\lambda \to 0} \sum_{i=1}^{n} f(\xi_i)(x_i - x_{i-1})$$

where ξ_i is a point in $[x_{i-1}, x_i]$.

Suppose $f(x)$ is a continuous function on $[a, b]$, T is an arbitrary partition of $[a, b]$ with dividing points

$$a = x_0 < x_1 < \cdots < x_{n-1} < x_n = b.$$

Consider the sum

$$\sigma = \sum_{i=1}^{n} f(\xi_i)\Delta x_i , \tag{7}$$

where $\Delta x_i = x_i - x_{i-1}$, and ξ_i is a point in $[x_{i-1}, x_i]$. It depends on both the partition T and the choice of ξ_is. It is called a *partial sum* of the function $f(x)$ on the interval $[a, b]$. Let $\lambda(T) = \max_{1 \leqslant i \leqslant n} \Delta x_i$, the length of the longest interval among $[x_{i-1}, x_i]$, $(i = 1, 2, \cdots, n)$. If the partial sum σ approaches a value I as $\lambda(T)$ approaches zero regardless of the difference in partition and the choices of ξ_is, we call I the *definite integral* of the function $f(x)$ on the interval $[a, b]$, and denote it as

$$I = \int_a^b f(x)\, dx.$$

The symbol $\int$ is the integral sign which can be thought of as an elongated letter S standing for sum. The function $f(x)$ is called the *integrand*, x is the *integral variable* and a, b are called the *lower limit* and the *upper limit* respectively.

The existence theorem of definite integrals will be discussed in Chapter 9 Section 9.3.

From the definition, it is clear that the definite integral of a function $f(x)$ on an interval $[a, b]$ is a global property of the function and the interval. That is, its value depends on the function's values on every part of the interval.

It is also clear from the definition that the value of the definite integral has nothing to do with the name of the integration variable, i.e.,

$$\int_a^b f(x)\, dx = \int_a^b f(t)\, dt = \int_a^b f(\xi)\, d\xi = \cdots .$$

By the definition of definite integrals, we can write what we calculated in Section 1.2.1 as follows.

(1) $\int_0^a x^2\, dx = \dfrac{a^3}{3},$ (2) $\int_a^b x^2\, dx = \dfrac{b^3 - a^3}{3},$

(3) $\int_a^b x^4\, dx = \dfrac{b^5 - a^5}{5},$

(4) $\int_a^b x^m\, dx = \dfrac{b^{m+1} - a^{m+1}}{m+1},$ where m is an integer and $m \neq -1$.

Thus, to find the area of a region bounded by the curve $y = f(x)$ and straight lines $x = a$, $x = b$, $y = 0$, is to calculate the definite integral

$$\int_a^b f(x)\, dx.$$

To find the distance (displacement) of an object moving along a straight line from time t_0 to time T is to evaluate the definite integral

$$\int_{t_0}^T v(t)\, dt,$$

where $v(t)$ is the velocity function of time t. To find the work of a force on an object moving along a straight line from point s_0 to point s is to evaluate the definite integral

$$\int_{s_0}^{s} F(s)\, ds,$$

where $F(s)$ is the force function of distance s.

In the definition of the definite integral $\int_a^b f(x)\, dx$, we assumed that $a < b$. What does the integral $\int_a^b f(x)\, dx$ mean if $a \geqslant b$? We define that

$$\int_b^a f(x)\, dx = -\int_a^b f(x)\, dx$$

where $a < b$, and

$$\int_a^a f(x)\, dx = 0.$$

The following properties can be derived directly from the definition of definite integrals.

(1) $\displaystyle\int_a^b (f(x) \pm g(x))\, dx = \int_a^b f(x)\, dx \pm \int_a^b g(x)\, dx.$

(2) $\displaystyle\int_a^b cf(x)\, dx = c\int_a^b f(x)\, dx,$ where c is a constant.

(3) $\displaystyle\int_a^b f(x)\, dx = \int_a^c f(x)\, dx + \int_c^b f(x)\, dx$ regardless the numerical order of a, b and c.

(4) $\displaystyle\int_a^b f(x)\, dx \leqslant \int_a^b g(x)\, dx,$ if $f(x) \leqslant g(x)$ on $[a, b]$.

We leave the verification of above properties to our readers.

(5) $\displaystyle\left|\int_a^b f(x)\, dx\right| \leqslant \int_a^b |f(x)|\, dx.$

In fact, by the inequality

$$-|f(x)| \leqslant f(x) \leqslant |f(x)|,$$

we have

$$-\int_a^b |f(x)|\, dx \leqslant \int_a^b f(x)\, dx \leqslant \int_a^b |f(x)|\, dx.$$

It follows that

$$\left| \int_a^b f(x)\,dx \right| \leqslant \int_a^b |f(x)|\,dx.$$

This property can be viewed as another version of the inequality

$$|a + b| \leqslant |a| + |b|.$$

(6) For any continuous function $f(x)$ defined on an interval $[a, b]$, there exists a point ξ in $[a, b]$, such that

$$\int_a^b f(x)\,dx = f(\xi)(b - a).$$

In fact, if m and M are the minimum and the maximum values of $f(x)$ on $[a, b]$, then $m \leqslant f(x) \leqslant M$ for all x in $[a, b]$. By (4), we have

$$m(b - a) \leqslant \int_a^b f(x)\,dx \leqslant M(b - a).$$

It follows that

$$m \leqslant \frac{\displaystyle\int_a^b f(x)\,dx}{b - a} \leqslant M.$$

Let

$$\mu = \frac{1}{b - a} \int_a^b f(x)\,dx,$$

then $m \leqslant \mu \leqslant M$. By the continuity of the function, there exists a point ξ in $[a, b]$ such that $f(\xi) = \mu$. Hence

$$\frac{1}{b - a} \int_a^b f(x)\,dx = f(\xi),$$

and

$$\int_a^b f(x)\,dx = f(\xi)(b - a).$$

This is called the **Mean-Value Theorem for Integrals**. The geometric meaning of this theorem is obvious: For any trapezoid with a curved edge as shown in Figure 1.14, there exists a rectangle with width $b - a$ and length $|f(\xi)|$ (where ξ is a point between a and b.), such that the area of the trapezoid is equal to the area of this rectangle.

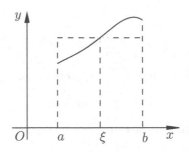

Fig. 1.14

1.2.3   *Logarithmic Function*

Now we pick up the problem left over in Section 1.2.1: finding the area of the region under the curve $y = x^m$ and between vertical lines $x = a$, $x = b$ for $m = -1$.

Let's look at an example: Consider a cylindrical tube containing ideal gas. When the gas expands, it pushes a piston. Suppose the gas expands from volume V_0 to V_1 under a constant temperature. Find the work done by the pressure of the gas on the piston (Figure 1.15).

Suppose the area of the piston is A, and take the positive direction of x-axis as the direction of the piston movement. Thus, the piston moves from $s_0 = \dfrac{V_0}{A}$ to $s = \dfrac{V_1}{A}$ when the ideal gas expands from V_0 to V_1. If at $x = s_i$, the piston has a movement Δs_i to $x = s_i + \Delta s_i$, then the work at this time is approximately equal to

$$Ap(s_i)\Delta s_i,$$

where $p(s_i)$ is the pressure of the piston at $x = s_i$.

If the piston moves from s_0 to s, by the definition of definite integral, the work is

$$\int_{s_0}^{s} Ap(s)\,ds = A\int_{s_0}^{s} p(s)\,ds.$$

The ideal gas law states

$$pV = RT,$$

where V is volume, T is temperature and R is a constant. Thus, we have

$$p = RT\frac{1}{V}.$$

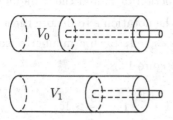
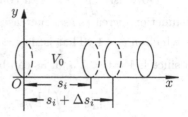

Fig. 1.15   Fig. 1.16

Since $V = As$, we have $p = \dfrac{RT}{A} \cdot \dfrac{1}{s}$, and the work is

$$A \int_{s_0}^{s} \frac{RT}{A} \cdot \frac{1}{s}\, ds = RT \int_{s_0}^{s} \frac{1}{s}\, ds.$$

Therefore, the problem becomes to evaluate the integral of type $\displaystyle\int_{a}^{b} \frac{1}{s}\, ds$.

The graph of the function $r = \dfrac{1}{s}$ is a hyperbola. We denote the area of the region bounded by $r = \dfrac{1}{s}$, $s = 1$ and $s = x$ by $\ln x$,

$$\int_{1}^{x} \frac{1}{s}\, ds = \ln x,$$

where $x > 0$.

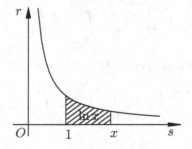

Fig. 1.17

Obviously, $\int_1^x \frac{1}{s}\,ds$ is a continuous function of x, and the value of the function increases as x increases. We use the notation e to denote the x for which $\ln x = 1$. That is, $\ln e = 1$ or $\int_1^e \frac{1}{s}\,ds = 1$. Such a number e exists since $\ln 1 = 0 < 1$ and, as can be seen in Figure 1.18,

$$\ln 4 = \int_1^4 \frac{1}{s}\,ds > \frac{1}{2}\cdot 1 + \frac{1}{3}\cdot 1 + \frac{1}{4}\cdot 1 = \frac{13}{12} > 1.$$

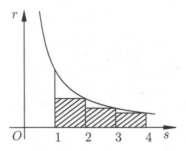

Fig. 1.18

Next, we study the function $\ln x$.

For three points on the Os-axis, $x_1 > 0$, $x_2 > 0$ and $x_1 x_2$, the equation

$$\ln x_1 x_2 = \ln x_1 + \ln x_2$$

is always true. This is because

$$\ln x_1 x_2 = \int_1^{x_1 x_2} \frac{1}{s}\,ds$$

and by definite integral property (3) in Section 1.2.2, we have

$$\int_1^{x_1 x_2} \frac{1}{s}\,ds = \int_1^{x_1} \frac{1}{s}\,ds + \int_{x_1}^{x_1 x_2} \frac{1}{s}\,ds.$$

To calculate $\int_{x_1}^{x_1 x_2} \frac{1}{s}\,ds$, partition the interval $[x_1, x_1 x_2]$ to n subintervals with the same length (assume $x_2 > 1$, otherwise consider the interval

$[x_1x_2, x_1])$, then we have

$$\int_1^{x_1x_2} \frac{1}{s}\, ds = \lim_{n\to\infty} \left[\frac{1}{x_1 + \dfrac{x_1x_2 - x_1}{n}} + \frac{1}{x_1 + 2\dfrac{x_1x_2 - x_1}{n}} + \cdots \right.$$

$$\left. + \frac{1}{x_1 + \dfrac{x_1x_2 - x_1}{n}} \right] \frac{x_1x_2 - x_1}{n}$$

$$= \lim_{n\to\infty} \left[\frac{1}{1 + \dfrac{x_2 - 1}{n}} + \frac{1}{1 + 2\dfrac{x_2 - 1}{n}} + \cdots + \frac{1}{1 + n\dfrac{x_2 - 1}{n}} \right] \frac{x_2 - 1}{n}$$

$$= \int_1^{x_2} \frac{1}{s}\, ds.$$

Thus,

$$\int_1^{x_1x_2} \frac{1}{s}\, ds = \int_1^{x_1} \frac{1}{s}\, ds + \int_1^{x_2} \frac{1}{s}\, ds.$$

It follows that

$$\ln x_1x_2 = \ln x_1 + \ln x_2.$$

Let $x_2 = \dfrac{1}{x_1}$, then

$$\ln 1 = \ln x_1 + \ln x_2 = \ln x_1 + \ln \frac{1}{x_1}.$$

Since

$$\ln 1 = \int_1^1 \frac{1}{s}\, ds = 0,$$

we have

$$\ln \frac{1}{x_1} = -\ln x_1.$$

Let $x_2 = x_1 = x$, then

$$\ln x^2 = \ln x + \ln x = 2\ln x.$$

Continue the process, we have

$$\ln x^3 = \ln(x^2 \cdot x) = \ln x^2 + \ln x = 2\ln x + \ln x = 3\ln x$$

$$\cdots$$

$$\ln x^n = \ln(x^{n-1} \cdot x) = \ln x^{n-1} + \ln x = (n-1)\ln x + \ln x = n\ln x.$$

Therefore, for any positive integer n, we have

$$\ln x^n = n \ln x.$$

For any negative integer $k = -n$, we have

$$\ln x^k = \ln \frac{1}{x^n} = -\ln x^n = -n \ln x = k \ln x.$$

Thus, for any integer β, the equation

$$\ln x^\beta = \beta \ln x$$

is always true.

Let $x_1 = x_0^{\frac{1}{m}}$, where m is an integer. Then $x_0 = x_1^m$. Thus,

$$\ln x_0 = \ln x_1^m = m \ln x_1 = m \ln x_0^{\frac{1}{m}}.$$

That is,

$$\ln x_0^{\frac{1}{m}} = \frac{1}{m} \ln x_0.$$

Let $x_0 = x^n$ in above equation, where n is a positive integer, then

$$\ln x^{\frac{n}{m}} = \frac{1}{m} \ln x^n = \frac{n}{m} \ln x.$$

Thus, for any rational number μ, we have

$$\ln x^\mu = \mu \ln x.$$

For any irrational number α, we can always find two sequences of rational numbers

$$\alpha_1, \alpha_2, \cdots$$

and

$$\alpha_1', \alpha_2', \cdots$$

such that $\alpha_n < \alpha < \alpha_n'$, and

$$\lim_{n \to \infty} \alpha_n = \lim_{n \to \infty} \alpha_n' = \alpha.$$

If $x > 1$, then by the definition of $\ln x$, we have

$$\ln x^{\alpha_n} < \ln x^\alpha < \ln x^{\alpha_n'}.$$

It follows that

$$\alpha_n \ln x < \alpha \ln x < \alpha_n' \ln x.$$

Take the limit on the left and the right side of the above inequality, we have

$$\ln x^{\alpha} = \alpha \ln x.$$

For $x = e$, we get

$$\ln e^{\alpha} = \alpha \ln e = \alpha.$$

Hence, $\ln x$ is the logarithmic function with base e,

$$\ln x = \int_1^x \frac{1}{s}\, ds = \log_e x.$$

A similar proof can show that this equation is also true for $0 < x \leqslant 1$. It can be proved that e is an irrational number with value $2.71828\cdots$.

The function $\ln x$ is called the *natural logarithmic function*. We use this function more often than the logarithmic function with base 10 in our text.

A more advanced discussion about the theory of definite integrals can be found in Chapter 9 Section 9.3.

Exercises 1.2

1. The velocity of a moving particle is $v = t + 4\, m/s$. Find the distance the particle traveled in the first 10 seconds.

2. Compare (without calculation) the values of the following pair of definite integrals. Which one is greater?

   (1) $\displaystyle\int_0^{\frac{\pi}{2}} \sin^{10} x\, dx$ and $\displaystyle\int_0^{\frac{\pi}{2}} \sin^2 x\, dx$;

   (2) $\displaystyle\int_{11}^{15} \log x\, dx$ and $\displaystyle\int_{11}^{15} (\log x)^2\, dx$;

   (3) $\displaystyle\int_{\frac{1}{2}}^{1} \ln x\, dx$ and $\displaystyle\int_{\frac{1}{2}}^{1} (\ln x)^2\, dx$.

3. Prove that $\displaystyle\int_a^b x^3\, dx = \frac{b^4 - a^4}{4}$.

4. Prove that $\displaystyle\int_a^b x^{-2}\, dx = \frac{b^{-1} - a^{-1}}{-1}$.

5. Prove that the equation

   $$\int_a^b x^m\, dx = \frac{b^{m+1} - a^{m+1}}{m+1}$$

   holds for any integer $m \neq -1$ and interval $[a, b]$ that does not contain the origin.

6. The decay rate v of radioactive material is a function of time, $v = v(t)$. Find the expression of the mass decayed from time T to time T_1 as
   (1) an approximate value by a partial sum;
   (2) an accurate value by a definite integral.

7. Prove that

$$\int_0^{10} \frac{1}{x^3 + 16}\, dx \leqslant \frac{5}{8}.$$

8. The force acting upon a particle moving along a straight line is

$$F = s^2 + 1.$$

Find the work done by this force as the particle moves from $s = 1$ to 10.

1.3   Derivatives and Differentials

1.3.1   *Tangent Lines of a Curve*

The high school definition of a tangent line to a circle is a line that intersects with the circle at only one point. Obviously this definition is not suitable for the parabola $y = x^2$, since both the x-axis and the y-axis intersect with the curve at one point but y-axis is obviously not a tangent line. We need to find another way to define tangent lines of curves.

Suppose the coordinates of a point P on a curve C (Figure 1.19) is $(x_0, f(x_0))$, and the coordinates of another point P' on the same curve is $(x_0 + \Delta x, f(x_0 + \Delta x))$.

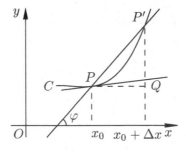

Fig. 1.19

The line segment which connects P and P' is a secant line, and its slope is

$$\tan \varphi = \frac{P'Q}{PQ} = \frac{f(x_0 + \Delta x) - f(x_0)}{\Delta x}.$$

As the point P' moves towards the point P along the curve C, the secant line PP' turns about P, and Δx approaches zero. It follows that, the slope of the secant line approaches the limit

$$\lim_{\Delta x \to 0} \tan \varphi = \lim_{\Delta x \to 0} \frac{f(x_0 + \Delta x) - f(x_0)}{\Delta x}.$$

In other words, the secant line PP' approaches a limit position as P' approaches P, and the line at this limit position is called the *tangent line* to the curve C at point P.

If we denote the limit of the slope of the secant line as k, then the equation of the tangent line of the curve $y = f(x)$ at a point (x_0, y_0) can be written as

$$y - y_0 = k(x - x_0).$$

Example: Find the equation of the tangent line to the parabola $y = x^2$ at the point $(2, 4)$.

Solution: The slope of the tangent line of $y = x^2$ at the point $(2, 4)$ is

$$\lim_{\Delta x \to 0} \frac{f(2 + \Delta x) - f(2)}{\Delta x} = \lim_{\Delta x \to 0} \frac{(2 + \Delta x)^2 - 4}{\Delta x} = \lim_{\Delta x \to 0} (4 + \Delta x) = 4.$$

The equation of the tangent line is

$$y - 4 = 4(x - 2)$$

or

$$4x - y - 4 = 0.$$

To find the tangent line of a curve $y = f(x)$ at a point x is to find the limit

$$\lim_{\Delta x \to 0} \frac{f(x + \Delta x) - f(x)}{\Delta x}. \tag{1}$$

1.3.2 *Velocity and Density*

Let's study two examples in physics.

A particle moves along a straight line. Suppose the function relation between the distance s and the time t is

$$s = s(t).$$

The distance that the particle traveled between time t_0 and time t is

$$s(t) - s(t_0).$$

The ratio

$$\bar{v}(t) = \frac{s(t) - s(t_0)}{t - t_0}$$

is the mean velocity of the particle in the time interval $[t_0, t]$. It is a function of the independent variable t.

In different time intervals of the same length, the distances the particle traveled may be different. So $\bar{v}(t)$ can not represent the velocity of the partial at the time t_0. Obviously, as the time interval $[t_0, t]$ gets shorter, the value of $\bar{v}(t)$ will get closer to the velocity at t_0. The limit

$$\lim_{t \to t_0} \frac{s(t) - s(t_0)}{t - t_0} \tag{2}$$

is the instantaneous velocity of the particle at t_0.

For the next example, suppose we have a rod as in Figure 1.20. Let x be the length from the end point A to the point M. The mass m of the rod AM is a function of x: $m = m(x)$.

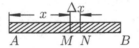

Fig. 1.20

How can we determine the linear density of the rod at point M if the density of the rod is not even?

Take a point N near M, and let $MN = \Delta x$, the mass of the rod from M to N is

$$m(x + \Delta x) - m(x).$$

The ratio

$$\frac{m(x + \Delta x) - m(x)}{\Delta x}$$

is the mean linear density of the segment MN. The mean linear density becomes closer to the linear density at M when Δx becomes sufficiently small. The limit

$$\lim_{\Delta x \to 0} \frac{m(x + \Delta x) - m(x)}{\Delta x} \tag{3}$$

is the linear density of the rod at M.

The methods of solving these three different problems are the same, that is, calculating the limit

$$\lim_{\Delta x \to 0} \frac{f(x + \Delta x) - f(x)}{\Delta x}.$$

1.3.3 *Definition of Derivatives*

Suppose $y = f(x)$ is a function defined on the closed interval $[a, b]$, and x_0 is a point in the interval. If the limit

$$\lim_{\Delta x \to 0} \frac{f(x_0 + \Delta x) - f(x_0)}{\Delta x}$$

exists, then we say that the function $f(x)$ is differentiable at x_0, and this limit is called the derivative of $f(x)$ at x_0, denoted as $f'(x_0)$ or $\left.\dfrac{dy}{dx}\right|_{x=x_0}$.

Obviously, the value of $f'(x_0)$ is related to the point x_0. It changes when x_0 changes between a and b. Therefore, if the function $f(x)$ is differentiable at every point of $[a, b]$, then $f'(x)$ is a new function of x on the same interval, called the *derivative* of $f(x)$.

Let Δx be a small change of x such that $x + \Delta x$ also lies in the same interval that contains x, and $\Delta y = f(x + \Delta x) - f(x)$. Then the rate of change of f from x to $x + \Delta x$ is $\dfrac{\Delta y}{\Delta x}$, and the derivative

$$\frac{dy}{dx} = \lim_{\Delta x \to 0} \frac{f(x + \Delta x) - f(x)}{\Delta x} = \lim_{\Delta x \to 0} \frac{\Delta y}{\Delta x}.$$

is the limit of the rate of change of f as $\Delta x \to 0$.

The notation $\dfrac{dy}{dx}$ represents the derivative of the function $f(x)$ at the point x. We do not consider it as a ratio until the actual meaning is defined individually for dy and dx.

From the definition of derivative we can see that, the slope of a tangent line to a curve at a point is the derivative of the function of the curve at the point; the instantaneous velocity of a moving particle at a time t_1 is the derivative of $s(t)$ at t_1. The leaner density at a point of a rod is the derivative of the mass function $m(x)$ with respect to x.

The derivative of a function is the rate of change of the dependent variable of the function with respect to the independent variable.

Differentiability is a "local" property of a function since it only depends on the nature of the function nearby the point at which the derivative of the function needs to be found.

If a function $y = f(x)$ is differentiable at a point $x = x_0$, then it is continuous at the point. In fact, taking the limit on both sides of the equation

$$f(x + \Delta x) - f(x) = \frac{f(x + \Delta x) - f(x)}{\Delta x} \cdot \Delta x,$$

as $\Delta x \to 0$. Since the limit

$$\lim_{\Delta x \to 0} \frac{f(x + \Delta x) - f(x)}{\Delta x},$$

exists, we have

$$\lim_{\Delta x \to 0} [f(x + \Delta x) - f(x)] = 0.$$

Therefore, $f(x)$ is continuous at x_0. Obviously, continuity is also a "local" property.

Example 1. The derivative of any constant is equal to zero, that is

$$\frac{dC}{dx} = 0.$$

where C is a constant.

Example 2. Find the derivative of $y = x^n$ (n is a positive integer) at the point $x = x_0$.

Solution: By the definition of derivative we have

$$\lim_{\Delta x \to 0} \frac{(x_0 + \Delta x)^n - x_0^n}{\Delta x} = \lim_{\Delta x \to 0} \frac{x_0^n + nx_0^{n-1}\Delta x + \cdots + (\Delta x)^n - x_0^n}{\Delta x}$$

$$= \lim_{\Delta x \to 0} \left[nx_0^{n-1} + \frac{n(n-1)}{2} x_0^{n-2} \Delta x + \cdots + (\Delta x)^{n-1} \right]$$

$$= nx_0^{n-1}.$$

Thus

$$\frac{dx^n}{dx} = nx^{n-1}.$$

Example 3. Find the derivative of $y = \sin x$ at the point $x = x_0$.

Solution: By the definition of derivative we have

$$\lim_{\Delta x \to 0} \frac{\sin(x_0 + \Delta x) - \sin x_0}{\Delta x} = \lim_{\Delta x \to 0} \frac{(2 \sin \frac{\Delta x}{2} \cos \frac{2x_0 + \Delta x}{2})}{\Delta x}.$$

To complete the solution, we need to show that

$$\lim_{\alpha \to 0} \frac{\sin \alpha}{\alpha} = 1.$$

Consider a circle with radius 1 centered at the origin. Suppose α ($0 < \alpha < \frac{\pi}{2}$) is an angle between OA (the positive side of x-axis) and the straight line OB (Figure 1.21). Let C be the intersection of the circle and OB, and CD be a straight line that is perpendicular to OA. Also, let BA be a tangent line of the circle at A.

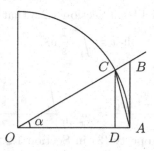

Fig. 1.21

Obviously, the area of the triangle $\triangle OAB$ is greater than the area of the sector OAC, and the area of the sector OAC is greater than the area of the triangle $\triangle ODC$.

The area of the triangle $\triangle OAB$ is $\frac{1}{2}\tan\alpha$. The area of the sector OAC is $\frac{1}{2}\alpha$. The area of $\triangle OAC$ is $\frac{1}{2}\sin\alpha$. Thus, we have

$$\frac{1}{2}\tan\alpha > \frac{1}{2}\alpha > \frac{1}{2}\sin\alpha.$$

It follows that

$$\frac{1}{\sin\alpha} > \frac{1}{\alpha} > \frac{1}{\tan\alpha}$$

and

$$1 > \frac{\sin\alpha}{\alpha} > \cos\alpha.$$

The right side of this inequality approaches 1 as α approaches zero, so $\frac{\sin\alpha}{\alpha} \to 1$ as $\alpha \to 0$. The same proof can be applied for the case where $\alpha < 0$, and $\alpha \to 0$.

Using this result, we have that

$$\lim_{\Delta x \to 0} \frac{2\sin\dfrac{\Delta x}{2}\cos\dfrac{2x_0+\Delta x}{2}}{\Delta x} = \lim_{\Delta x \to 0} \frac{\sin\dfrac{\Delta x}{2}}{\dfrac{\Delta x}{2}}\cos\left(x_0+\frac{\Delta x}{2}\right) = \cos x_0.$$

Therefore,

$$\frac{d\sin x}{dx} = \cos x.$$

We can show that $\dfrac{d\cos x}{dx} = -\sin x$ in a similar way.

Example 4. Find the derivative of $y = \ln x$ at $x = x_0$.

Solution: We have learned from Section 1.2.3 that

$$\ln x = \int_1^x \frac{1}{s}\, ds,$$

so

$$\lim_{\Delta x \to 0} \frac{\ln(x_0 + \Delta x) - \ln x_0}{\Delta x} = \lim_{\Delta x \to 0} \frac{\displaystyle\int_1^{x_0 + \Delta x} \frac{1}{s}\, ds - \int_1^{x_0} \frac{1}{s}\, ds}{\Delta x}.$$

By definite integral property (3) in Section 1.2.2, the right side of the above equation is

$$\lim_{\Delta x \to 0} \frac{1}{\Delta x} \int_{x_0}^{x_0 + \Delta x} \frac{1}{s}\, ds.$$

By definite integral property (6) in Section 1.2.2, there exists a point ξ in $[x_0, x_0 + \Delta x]$, such that

$$\int_{x_0}^{x_0 + \Delta x} \frac{1}{s}\, ds = \frac{1}{\xi}(x_0 + \Delta x - x_0) = \frac{1}{\xi}\Delta x.$$

Hence,

$$\lim_{\Delta x \to 0} \frac{1}{\Delta x} \int_{x_0}^{x_0 + \Delta x} \frac{1}{s}\, ds = \lim_{\Delta x \to 0} \frac{1}{\Delta x} \frac{1}{\xi} \Delta x = \lim_{\Delta x \to 0} \frac{1}{\xi} = \frac{1}{x_0}.$$

Therefore,

$$\frac{d\ln x}{dx} = \frac{1}{x}.$$

In Section 1.2.3, we proved that $\ln x$ is the logarithmic function with base e, $\ln x = \log_e x$. By the definition of derivative,

$$\frac{d\ln x}{dx} = \frac{d\log_e x}{dx} = \lim_{\Delta x \to 0} \frac{\log_e(x + \Delta x) - \log_e x}{\Delta x}$$

$$= \lim_{\Delta x \to 0} \log_e \left(1 + \frac{\Delta x}{x}\right)^{\frac{1}{\Delta x}}.$$

We just showed that

$$\frac{d\ln x}{dx} = \frac{1}{x}.$$

Therefore,

$$\lim_{\Delta x \to 0} \log_e \left(1 + \frac{\Delta x}{x}\right)^{\frac{1}{\Delta x}} = \frac{1}{x}.$$

Let $x = 1$ and $\Delta x = \frac{1}{n}$. By the continuity of the logarithm function, since

$$\lim_{n \to \infty} \log_e \left(1 + \frac{1}{n}\right)^n = 1,$$

we have

$$\lim_{n \to \infty} \left(1 + \frac{1}{n}\right)^n = e.$$

This equation can be used to calculate the approximate value of the number e (see Example 3 in Chapter 9 Section 9.1.3).

1.3.4 *Differentials*

A concept closely related to derivative is differential. Suppose a function $y = f(x)$ has the derivative $f'(x)$ at a point x. Then the product of an increment of x and the derivative $f'(x)\Delta x$ is called the *differential* of the function $y = f(x)$ at point x, and denoted as

$$dy = f'(x)\Delta x. \tag{4}$$

We now explain the meaning of the differential. In some real problems, the change of a function value needs to be found when the independent variable value of the function changes. For instance, we may need to find the increment of the volume of a ball when its radius r has an increase Δr. The volume of the ball is $V = \frac{4}{3}\pi r^3$. The increment of the volume as r changes to $r + \Delta r$ is

$$\Delta V = \frac{4}{3}\pi(r + \Delta r)^3 - \frac{4}{3}\pi r^3 = \frac{4}{3}\pi(3r^2\Delta r + 3r\Delta r^2 + \Delta r^3)$$

$$= 4\pi r^2 \Delta r + 4\pi r(\Delta r)^2 + \frac{4}{3}\pi(\Delta r)^3.$$

The increment ΔV can be approximated by the first term of the right hand side of the above equation $4\pi r^2 \Delta r$, when the change Δr is small. The term $4\pi r^2 \Delta r$ is the differential of V.

In general, by the definition of derivative,

$$\lim_{\Delta x \to 0} \frac{\Delta y}{\Delta x} = \lim_{\Delta x \to 0} \frac{f(x_0 + \Delta x) - f(x_0)}{\Delta x} = f'(x_0).$$

Let

$$\alpha = \frac{\Delta y}{\Delta x} - f'(x_0).$$

Then we have $\alpha \to 0$ as $\Delta x \to 0$, that is $\lim\limits_{\Delta x \to 0} \alpha = 0$. By equation (4), we have

$$\Delta y = f'(x_0)\Delta x + \alpha \Delta x.$$

In other words, for a change Δx of independent veritable x, the change of dependent variable Δy, also called the difference, consists two parts: the differential $f'(x_0)\Delta x$, and $\alpha \Delta x$. Since $\alpha \to 0$ as $\Delta x \to 0$, $\alpha \Delta x$ approaches zero faster than Δx. And if $f'(x_0) \neq 0$, and Δx is small, the differential is the dominant part of Δy. Thus the deferential part can be used as the approximate value of Δy when Δx is sufficiently small.

We now study the geometric meaning of differential. Suppose $M(x, y)$ is a point on a curve $y = f(x)$, and $M'(x + \Delta x, y + \Delta y)$ is another point on the curve near M.

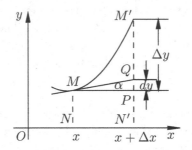

Fig. 1.22

Let vertical lines MN and $M'N'$ be perpendicular to x-axis. Also, let the tangent line to the curve at M intersect $M'N'$ at Q. Suppose α is the angle between the tangent line and the x-axis. Then we have

$$MP = \Delta x;$$
$$M'P = M'N' - PN' = f(x + \Delta x) - f(x) = \Delta y;$$
$$PQ = MP \tan \alpha = f'(x)\Delta x = dy.$$

Thus, Δy is the change of the curve on the vertical axis, and dy is the change of the tangent on the vertical axis. Also,

$$\Delta y \overset{.}{-} dy = M'Q.$$

This difference also tends to zero as Δx tends to zero. It tends to zero at a faster rate than Δx.

Consider the function $y = x$. We have $f'(x) = x' = 1$. Thus, the differential of this function is

$$dx = 1 \cdot \Delta x = \Delta x.$$

This implies that, the differential of the independent variable is the change of the variable. Hence we can rewrite (4) as follows

$$dy = f'(x)dx.$$

Here dx and dy are the differentials of the independent variable and the dependent variable respectively. The derivative of the function is therefore the ratio of these two differentials:

$$\frac{dy}{dx} = f'(x).$$

The differential is also a "local" property of a function. It only depends on the nature of the function near the point x where the differential is to be found.

Example 1. The differential of any constant is equal to zero.

Example 2. If $y = x^n$ (n is a positive integer), then $dy = nx^{n-1}\,dx$.

Example 3. If $y = \sin x$, then $dy = \cos x\,dx$.

Example 4. If $y = \ln x$, then $dy = \dfrac{1}{x}\,dx$.

1.3.5 *Mean-Value Theorem*

Corresponding to the Mean-Value Theorem for Integrals, we have the **Mean-Value Theorem for Derivatives**: If a function $f(x)$ is continuous on a closed interval $[a, b]$, and differentiable on the open interval (a, b), then there exists a point ξ in (a, b) such that

$$f(b) - f(a) = f'(\xi)(b - a).$$

The geometric meaning of this theorem is clear: when the above equation is rewritten as

$$\frac{f(b) - f(a)}{b - a} = f'(\xi),$$

the left hand side, $\dfrac{f(b) - f(a)}{b - a}$, is the slope of the secant line which connects the points $(a, f(a))$ and $(b, f(b))$. The theorem tells us that we can find a

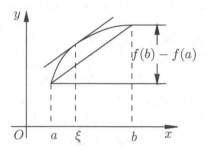

Fig. 1.23

point ξ in (a, b) such that the tangent line to the curve at the point $(\xi, f(\xi))$ is parallel to the secant line.

Before we prove the theorem, we introduce the concept of extreme values as preparation.

If the value of $f(x)$ at a point x_0 is greater than all the values of $f(x)$ at x near x_0, then we say that $f(x_0)$ is a *maximum value* of $f(x)$.

In other words, if there is a positive number δ, such that for any point x in the interval $(x_0 - \delta, x_0 + \delta)$, the inequality

$$f(x) \leqslant f(x_0)$$

holds, then $f(x_0)$ is a maximum value. Similarly, if for any point x in the interval $(x_0 - \delta, x_0 + \delta)$, the inequality

$$f(x) \geqslant f(x_0)$$

holds, then $f(x_0)$ is a *minimum value*.

A maximum or a minimum of a function is called an *extremum*. A point where a function has an extremum is called an *extremum point*.

If a function $f(x)$ is differentiable at a point x_0, and $f(x_0)$ is an extremum of f, then $f'(x_0) = 0$.

Indeed, let $f(x_0)$ be a maximum value of f, then there exists an open interval $(x_0 - \delta, x_0 + \delta)$ with $\delta > 0$, such that for any x in this interval

$$f(x) \leqslant f(x_0).$$

In other words, for any real number h which satisfies $0 < h < \delta$,

$$f(x_0 + h) \leqslant f(x_0) \quad \text{and} \quad f(x_0 - h) \leqslant f(x_0).$$

Thus,

$$\frac{f(x_0 + h) - f(x_0)}{h} \leqslant 0,$$

$$\frac{f(x_0 - h) - f(x_0)}{-h} \geqslant 0.$$

Taking the limit on the left side of both equations as $h \to 0$, we have

$$f'(x_0) \leqslant 0, \quad f'(x_0) \geqslant 0.$$

Therefore,

$$f'(x_0) = 0.$$

The same result can be obtained in a similar way if $f(x_0)$ is a minimum value of f.

This result is called the **Fermat Theorem**.

More detailed discussion about extreme values of a function can be found in Chapter 3 Section 3.3.2.

The next result follows from the above theorem and is called **Rolle's Theorem**: If a function $f(x)$ is continuous on a closed interval $[a, b]$ and is differentiable on the open interval (a, b), and $f(a) = f(b)$, then there exists a point ξ in (a, b), such that

$$f'(\xi) = 0.$$

Proof: Suppose m and M are minimum and maximum values of $f(x)$ on $[a, b]$ respectively. then there exist ξ and η in $[a, b]$, such that

$$f(\xi) = M, \quad \text{and} \quad f(\eta) = m.$$

If $M = m$, then $f(x)$ is a constant function on $[a, b]$, and $f'(\xi) = 0$ for all ξ in $[a, b]$.

If $M > m$, then ether M or m is different from $f(a)$ (and $f(b)$). Assume that

$$M = f(\xi) > f(a) = f(b),$$

by Fermat theorem,

$$f'(\xi) = 0.$$

If $f(a) = f(b) > f(\eta) = m$, then $f'(\eta) = 0$.

With these preparations, we are ready to prove The Mean-Value Theorem for Derivatives.

Consider the function

$$\varphi(x) = f(x) - f(a) - \frac{f(b) - f(a)}{b - a}(x - a),$$

for x in $[a, b]$.

Since $f(x)$ is continuous on $[a, b]$ and differentiable on (a, b), the function $\varphi(x)$ is also continuous on $[a, b]$ and differentiable on (a, b). Moreover,

$$\varphi(a) = \varphi(b).$$

By Rolle's Theorem, there exists a point ξ in (a, b), such that $\varphi'(\xi) = 0$. Taking the derivative of $\varphi(x)$, we have

$$f'(\xi) = \frac{f(b) - f(a)}{b - a}.$$

Remark: Although the Mean-Value Theorem for Derivatives has a different form and a different geometric meaning from the Mean-Value Theorem for Integrals, they represent the same principle. Indeed, for the purpose of explanation, we state the Mean-Value Theorem for Derivatives in another notation:

If $F(x)$ is continuous on $[a, b]$ and differentiable on (a, b), then there exists a point ξ in (a, b), such that

$$\frac{F(b) - F(a)}{b - a} = F'(\xi).$$

Let $F(x) = \displaystyle\int_a^x f(t)\, dt$ in above equation. Then it becomes the equation in the Mean-Value Theorem for Integrals. The only difference is that the ξ in the theorem might be equal to a or b.

Differentiation and integration are two parts of calculus. They are two aspects of the same theory. Therefore there is a correspondence between theorems and formulas in these two parts.

The Mean-Value Theorem for Derivatives that we proved is also called **Lagrange's Mean-Value Theorem**. One generalization of Lagrange's Mean-Value Theorem is **Cauchy's Mean-Value Theorem**: Suppose that functions $f(x)$ and $g(x)$ are continuous on $[a, b]$ and are differentiable on (a, b). Also suppose that $g'(x) \neq 0$ for any point x in (a, b). Then there exists a point ξ in (a, b) such that

$$\frac{f(b) - f(a)}{g(b) - g(a)} = \frac{f'(\xi)}{g'(\xi)}.$$

Proof: Since $g'(x) \neq 0$ for all x in (a, b), by Rolle's Theorem, we have $g(a) \neq g(b)$. Let

$$F(x) = f(x) - \frac{f(b) - f(a)}{g(b) - g(a)}(g(x) - g(a)) - f(a).$$

Then it is easy to see that

$$F(b) - F(a) = 0.$$

By Rolle's Theorem, there exists a point ξ in (a, b), such that

$$F'(\xi) = f'(\xi) - \frac{f(b) - f(a)}{g(b) - g(a)} g'(x) = 0,$$

and the theorem follows.

Obviously, Cauchy's Mean-Value Theorem reduces to Lagrange's Mean-Value Theorem if $g(x) = x$.

Exercises 1.3

1. Find the equation of the tangent line to $y = x^3$ at $(1, 1)$.

2. A particle moves along a straight line. The function between its displacement and the time is $s = 5t^2 + 6$.

   (1) Find the mean velocity in the time interval $2 \leqslant t \leqslant 2 + \Delta t$, for $\Delta t = 1$, $\Delta t = 0.1$, $\Delta t = 0.01$, and $\Delta t = 0.001$.

   (2) Estimate the instantaneous velocity at $t = 2$ using the results from (1).

   (3) Find instantaneous velocity at $t = 2$ using the definition.

3. A rod AB with uneven mass have a length of 20 cm. Let M be a point on the rod between the two end points A and B. The mass of the part of the rod from A to M is directly proportional to the square of the length of the rod from A to M. Find the following values if the length of AM is 2 cm and the mass of AM is 8g.

   (1) The mean linear density of AM.

   (2) The mean linear density of AB.

   (3) The linear density at point A.

   (4) The linear density at an arbitrary point M.

4. A rod AB with uneven mass have a length of 30 cm. Let l be the length from the end point A. The function relation between the mass m and the length l is $m = 3l^2 + 5l$. Find the following values.

   (1) The mean linear density of the rod.

   (2) The linear density at the point 5cm from the end point A.

   (3) The linear density at the end point B.

5. Find the equation of the tangent line to

$$y = x - \frac{1}{x}$$

   at the x intercepts.

6. Find the equation of the tangent line to $y = x^2$ which is parallel to the secant line connecting the two points on the parabola with x coordinates $x = 1$ and $x = 3$ respectively.

7. Find the equation(s) of the tangent line(s) to $y = x^2$ that makes a $45°$ angle with the line $3x - y + 1 = 0$.

8. Prove that the area of the triangle bounded by the coordinate axes and the tangent line to the hyperbola $xy = a^2$ is equal to the constant $2a^2$.

9. Find the angle between the curves $y = \sin x$ and $y = \cos x$. That is, find the angle between the two tangent lines to the curves at their intersection point.

10. Find the increment Δy of the function $y = x^2$ when the independent variable x changes from 1 to 1.02. Also, find the differential dy.

11. Suppose $y = x^2 + x$. Find the increment of the function Δy and the differential dy at $x = 1$ for $\Delta x = 10$, 1, 0.1, and 0.01. Observe the decrease of $\Delta y - dy$ as the value of Δx gets smaller.

12. Consider a square with the length of each side 8cm. Find the increase of its area when the length of each side increases by 1cm (0.5cm, 0.1cm). What is the dominant part of this increase?

13. Find the differential of $y = \sin 2\varphi$ when φ changes from $\dfrac{\pi}{6}$ to $\dfrac{61\pi}{360}$.

14. Suppose $f(x) = (x-2)(x-3)(x-4)(x-5)$. Without taking the derivative, find how many real roots does the equation $f'(x) = 0$ have? What are the intervals that contain these roots?

15. Suppose $P(x)$ is an nth degree polynomial with real coefficients. Prove that all the roots of $P'(x)$ are real if all the roots of $P(x)$ are real.

16. Suppose a function $f(x)$ is continuous on $[a, b]$ and differentiable on (a, b), where $b > a > 0$. Prove that there exists a point ξ in (a, b) such that

$$2\xi(f(b) - f(a)) = (b^2 - a^2)f'(\xi).$$

1.4   Fundamental Theorem of Calculus

We start this section with an example. Suppose $v(t)$ is the velocity of a particle moving along a straight line. Then the distance traveled $s = s(t)$ of the particle from the beginning to time t is the definite integral $\displaystyle\int_0^t v(\tau)\, d\tau$.

For simplicity, assume $s(0) = 0$, then

$$\int_0^t v(\tau)\, d\tau = s(t) \quad (= s(t) - s(0)).$$

From Section 1.3 we know that $s'(t) = v(t)$. That is

$$ds(t) = v(t)\, dt.$$

The differential of $s(t)$ is $v(t)dt$, the integral of $v(t)$ is $s(t)$. Thus, the part $v(\tau)d\tau$ in the integral $\int_0^t v(\tau)d\tau$ is the differential of $s(\tau)$. In other words, the distance $\int_0^t v(\tau)d\tau$ is an "accumulation" of the product of the velocity at each point and infinitesimal displacement $d\tau$.

Generally, we have the following theorem.

The Fundamental Theorem of Calculus (Differential Form): Suppose a function $f(x)$ is continuous on a closed interval $[a, b]$, and x is a fixed point in $[a, b]$. Let

$$\Phi(x) = \int_a^x f(t)\, dt \quad (a \leqslant x \leqslant b).$$

Then $\Phi(x)$ is differentiable on $[a, b]$, and

$$\Phi'(x) = f(x) \quad (a \leqslant x \leqslant b).$$

That is

$$d\Phi(x) = f(x)\, dx.$$

In other words, if the integral of $f(x)$ is $\Phi(x)$, then the differential of $\Phi(x)$ is $f(x)\, dx$. The differential of the integral of $f(t)$ is $f(t)$ multiply by dt.

Proof: We only need to show that $\Phi'(x) = f(x)$, that is

$$\lim_{\Delta x \to 0} \frac{\Phi(x + \Delta x) - \Phi(x)}{\Delta x} = f(x).$$

From the definition of $\Phi(x)$, we have

$$\Phi(x + \Delta x) - \Phi(x) = \int_a^{x+\Delta x} f(t)\, dt - \int_a^x f(t)\, dt = \int_x^{x+\Delta x} f(t)\, dt.$$

Since $f(t)$ is continuous on $[a, b]$, by the Mean-Value Theorem for Integral, there exists a point ξ in the interval $[x, x + \Delta x]$ such that

$$\int_x^{x+\Delta x} f(t)\, dt = f(\xi)(x + \Delta x - x) = f(\xi)\Delta x.$$

Thus,

$$\frac{\Phi(x + \Delta x) - \Phi(x)}{\Delta x} = f(\xi).$$

Since ξ is in the interval $[x, x + \Delta x]$, we have $\xi \to x$ as $\Delta x \to 0$. And $\lim\limits_{\xi \to x} f(\xi) = f(x)$ since $f(t)$ is continuous at x. Therefore,

$$\lim_{\Delta x \to 0} \frac{\Phi(x + \Delta x) - \Phi(x)}{\Delta x} = \lim_{\xi \to x} f(\xi) = f(x).$$

From the discussion above, we can obtain another form of this theorem.

The Fundamental Theorem of Calculus (Integral Form): Suppose a function $\Phi(x)$ is differentiable on $[a, b]$, and

$$\frac{d\Phi(x)}{dx} = f(x)$$

where $f(x)$ continuous. Then

$$\int_a^x f(t)\, dt = \Phi(x) - \Phi(a) \quad (a \leqslant x \leqslant b).$$

In other words, if the differential of $\Phi(x)$ is $f(x)dx$, then the integral of $f(x)$ is $\Phi(x)$, the integral of the derivative of $\Phi(x)$ is $\Phi(x)$ itself (up to a constant).

Proof: Suppose

$$G(x) = \int_a^x f(t)\, dt.$$

Then $\dfrac{dG(x)}{dx} = f(x)$. Since $\dfrac{d\Phi(x)}{dx} = f(x)$, we have

$$\frac{dG(x)}{dx} - \frac{d\Phi(x)}{dx} = 0.$$

That is

$$\frac{d(\Phi(x) - G(x))}{dx} = 0.$$

Let $F(x) = \Phi(x) - G(x)$. Then $\dfrac{dF(x)}{dx} = 0$. We know that the geometric meaning of $\dfrac{dF(x)}{dx}$ is the slope of the tangent line to the curve $y = F(x)$ at the point x. This slope is zero for every point in $[a, b]$ implies that all the tangent line of F are parallel to the x-axis. The only possibility is

$$y = F(x) = C \quad \text{(a constant)}.$$

It follows that $F(x) = \Phi(x) - G(x) = C$ (constant), and $\Phi(x) = G(x) + C$. Let $x = a$. Then we have

$$\Phi(a) = G(a) + C.$$

By the definition of $G(x)$, we have $G(a) = 0$, so $\Phi(x) - G(x) = \Phi(a)$ and $G(x) = \Phi(x) - \Phi(a)$, and the theorem follows.

The Fundamental Theorem of Calculus is also called the **Newton–Leibniz Formula**.

From the above two theorems we can obtain the following corollary:

Suppose a function $f(x)$ is continuous on an interval $[a, b]$, and there exists a function $H(x)$ such that $\dfrac{dH}{dx} = f(x)$ on $[a, b]$. Then

$$\int_a^b f(x)\, dx = H(b) - H(a).$$

Such a function $H(x)$ is called a *antiderivative* of $f(x)$. Any two antiderivatives of $f(x)$ differ by a constant. The entire set of antiderivative of $f(x)$ is called the *indefinite integral* of $f(x)$, denoted by $\displaystyle\int f(x)\, dx$.

Obviously, indefinite integral and derivative are inverse operations of each other.

By this corollary, the calculation of definite integrals becomes finding indefinite integrals or antiderivatives. Here are some examples:

$$\int_0^1 x^2\, dx = \left.\frac{x^3}{3}\right|_0^1 = \frac{1}{3}.$$

$$\int_0^{\frac{\pi}{2}} \cos x\, dx = \left.\sin x\right|_0^{\frac{\pi}{2}} = 1.$$

$$\int_1^2 \frac{-1}{x^2}\, dx = \left.\frac{1}{x}\right|_1^2 = \frac{1}{2} - 1 = -\frac{1}{2}.$$

$$\int_2^4 \frac{1}{x}\, dx = \left.\ln x\right|_2^4 = \ln 4 - \ln 2 = \ln 2.$$

We will use this method to calculate some definite integrals in the next chapter.

Chapter 2

Calculations of Derivatives and Integrals

2.1 Differentiation

2.1.1 *Calculations of Derivatives and Differentials*

In Section 3 of Chapter 1, we calculated the derivatives of the following functions using the definition of derivative:

(i) $\dfrac{dC}{dx} = 0$, where C is a constant;

(ii) $\dfrac{dx^n}{dx} = nx^{n-1}$, where n is a positive integer;

(iii) $\dfrac{d\sin x}{dx} = \cos x$, $\dfrac{d\cos x}{dx} = -\sin x$;

(iv) $\dfrac{d\ln x}{dx} = \dfrac{1}{x}$.

Suppose $u(x)$, $v(x)$ are differentiable functions (assume that all functions in this chapter are differentiable without further declaration). Then their sum, difference, product and quotient (for non-zero denominator functions) are also differentiable, and can be obtained using the following formulas:

(1) $[u(x) + v(x)]' = u'(x) + v'(x)$;

(2) $[u(x) - v(x)]' = u'(x) - v'(x)$;

(3) $[u(x)v(x)]' = u'(x)v(x) + u(x)v'(x)$;

(4) $\left[\dfrac{u(x)}{v(x)}\right]' = \dfrac{u'(x)v(x) - v'(x)u(x)}{[v(x)]^2}$.

The first two equations above can be derived easily from the definition

43

of derivative. The third property is based on the fact that

$$[u(x)v(x)]' = \lim_{\Delta x \to 0} \frac{u(x + \Delta x)v(x + \Delta x) - u(x)v(x)}{\Delta x}$$

$$= \lim_{\Delta x \to 0} \left[\frac{u(x + \Delta x)v(x + \Delta x) - u(x)v(x + \Delta x)}{\Delta x} \right.$$

$$\left. + \frac{u(x)v(x + \Delta x) - u(x)v(x)}{\Delta x} \right]$$

$$= \lim_{\Delta x \to 0} \frac{u(x + \Delta x) - u(x)}{\Delta x} v(x + \Delta x)$$

$$+ \lim_{\Delta x \to 0} u(x) \frac{v(x + \Delta x) - v(x)}{\Delta x}$$

$$= u'(x)v(x) + u(x)v'(x).$$

The continuity of the function $v(x)$ is applied in the proof.

Let $u(x) = C$ be a constant function in (3). Then a fifth equation follows:

(5) $[Cv(x)]' = Cv'(x)$.

We leave the verification of equation (4) to our readers. Now we can calculate the derivatives of many functions.

Example 1. Find the derivative of the function $y = \log_a x$.

Solution: Since $a^y = x$, we have $\ln x = y \ln a$. That is

$$y = \log_a x = \frac{\ln x}{\ln a}.$$

Thus

$$y' = \left(\frac{\ln x}{\ln a} \right)' = \frac{1}{x \ln a}$$

by (5).

Example 2. Find the derivative of the function $y = \tan x$.

Solution: By (4)

$$y' = \left(\frac{\sin x}{\cos x} \right)' = \frac{(\sin x)' \cos x - (\cos x)' \sin x}{\cos^2 x}$$

$$= \frac{\cos^2 x + \sin^2 x}{\cos^2 x} = \sec^2 x.$$

Example 3. Find the derivative of the function $y = \sec x$.

Solution: By (4)

$$y' = \left(\frac{1}{\cos x}\right)' = \frac{(1)'\cos x - (\cos x)'1}{\cos^2 x}$$

$$= \frac{\sin x}{\cos^2 x} = \sec x \tan x.$$

Suppose y is a function of u, $y = f(u)$, and u is a function of x, $u = \varphi(x)$. Suppose both functions are differentiable. The function

$$y = f(\varphi(x))$$

is called the composition of f and φ. The variable u is an intermediate variable.

A change of x, Δx, induces a change in u,

$$\Delta u = \varphi(x + \Delta x) - \varphi(x)$$

which in turn induces a change in y,

$$\Delta y = f(u + \Delta u) - f(u).$$

Since

$$\frac{\Delta y}{\Delta x} = \frac{\Delta y}{\Delta u} \cdot \frac{\Delta u}{\Delta x},$$

and $\Delta u \to 0$ as $\Delta x \to 0$, we have

$$y' = \lim_{\Delta x \to 0} \frac{\Delta y}{\Delta x} = \lim_{\Delta x \to 0}\left(\frac{\Delta y}{\Delta u} \cdot \frac{\Delta u}{\Delta x}\right)$$

$$= \lim_{\Delta x \to 0}\frac{\Delta y}{\Delta u} \cdot \lim_{\Delta x \to 0}\frac{\Delta u}{\Delta x} = \lim_{\Delta u \to 0}\frac{\Delta y}{\Delta u} \cdot \lim_{\Delta x \to 0}\frac{\Delta u}{\Delta x}$$

$$= \frac{dy}{du} \cdot \frac{du}{dx} = f'(u)\varphi'(x).$$

Thus, we obtain equation (6) as follows:

(6) If $y = f(u)$ and $u = \varphi(x)$ are differentiable functions, then the derivative of the composite function $y = f(\varphi(x))$ is

$$\frac{dy}{dx} = f'(u)\varphi'(x).$$

Example 4. Find the derivative of the function $y = \sin x^3$.

Solution: The function $y = \sin x^3$ can be considered as the composition of functions

$$y = \sin u, \quad u = x^3.$$

By (6), we have

$$y' = \cos u \cdot 3x^2 = 3x^2 \cos x^3.$$

Example 5. Find the derivative of the function

$$y = \left(x + \frac{1}{x}\right)^{100}.$$

Solution: The function

$$y = \left(x + \frac{1}{x}\right)^{100}$$

can be considered as the composition of functions

$$y = u^{100}, \quad u = x + \frac{1}{x}.$$

By (6), we have

$$y' = 100u^{99}\left(1 - \frac{1}{x^2}\right) = 100\left(x + \frac{1}{x}\right)^{99}\left(1 - \frac{1}{x^2}\right).$$

Example 6. Suppose $s(t) = \left(\dfrac{1+t}{1-t}\right)^3$. Evaluate $s'(0)$.

Solution: Consider the function $s(t) = \left(\dfrac{1+t}{1-t}\right)^3$ as the composition of functions

$$s = u^3, \quad u = \frac{1+t}{1-t}.$$

By (6), we have

$$s'(t) = 3u^2\frac{2}{(1-t)^2} = \frac{6(1+t)^2}{(1-t)^4}.$$

Thus

$$s'(0) = 6.$$

Equation (6) is called the *chain rule*, and it can be generalized to compositions of finitely many functions.

If a function is the composition of

$$y = f(u), \quad u = \varphi(v), \quad v = \psi(x),$$

that is,

$$y = f[\varphi(\psi(x))],$$

then the derivative of y with respect to x can be calculated by the formula

$$\frac{dy}{dx} = f'(u)\varphi'(v)\psi'(x).$$

Example 7. Find the derivative of the function $y = \sin^3(x + x^2)$.

Solution: Consider the function $y = \sin^3(x + x^2)$ as the composition of functions

$$y = u^3, \quad u = \sin v, \quad v = x + x^2.$$

Then

$$y' = 3u^2 \cos v(1 + 2x) = 3(1 + 2x)\sin^2(x + x^2)\cos(x + x^2).$$

If y is an one-to-one function of x, $y = f(x)$, then x is also a function of y, $x = g(y)$. It is called the inverse of f, and the equation $y = f(g(y))$ holds.

Suppose $f(x)$ and $g(y)$ are monotonic and differentiable functions. By the formula of derivative of composite functions, we have

$$1 = f'(x)g'(y).$$

That is

$$g'(y) = \frac{1}{f'(x)}$$

if $f'(x) \neq 0$. Thus, we obtain

(7) If $x = g(y)$ is the inverse function of $y = f(x)$, and $f'(x) \neq 0$, then

$$g'(y) = \frac{1}{f'(x)}.$$

There is a simple geometric explanation of the above equation. In Figure 2.1, equations $y = f(x)$ and $x = g(y)$ represent the same curve C. Also

$$\tan \alpha = f'(x), \quad \tan \beta = g'(x).$$

Since

$$\alpha + \beta = \frac{\pi}{2},$$

we have

$$\tan \alpha = \frac{1}{\tan \beta},$$

and this is

$$g'(y) = \frac{1}{f'(x)}.$$

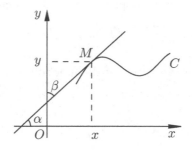

Fig. 2.1

Example 8. Show that $(e^x)' = e^x$.

Proof: Since $y = e^x$ is the inverse function of $x = \ln y$, by (7), we have

$$(e^x)' = \frac{1}{(\ln y)'} = \frac{1}{\dfrac{1}{y}} = y = e^x.$$

The equation

$$(a^x)' = a^x \ln a$$

can be obtained in a similar way.

Example 9. Show that $(\arcsin x)' = \dfrac{1}{\sqrt{1-x^2}}$ and $(\arctan x)' = \dfrac{1}{1+x^2}$.

Proof: Since $y = \arcsin x$ is the inverse function of $x = \sin y$, by (7), we have

$$(\arcsin x)' = \frac{1}{(\sin y)'} = \frac{1}{\cos y} = \frac{1}{\sqrt{1-\sin^2 y}} = \frac{1}{\sqrt{1-x^2}}.$$

Since y is between $-\dfrac{\pi}{2}$ and $\dfrac{\pi}{2}$, we have $\cos y > 0$. Hence we take the positive square root.

Since $y = \arctan x$ is the inverse function of $x = \tan y$, we have

$$(\arctan x)' = \frac{1}{(\tan y)'} = \frac{1}{\sec^2 y} = \frac{1}{1+\tan^2 y} = \frac{1}{1+x^2}.$$

The derivatives

$$(\arccos x)' = \frac{-1}{\sqrt{1-x^2}}, \quad (\text{arccot}\, x)' = \frac{-1}{1+x^2}$$

can be obtained in a similar way.

We have already known that the equation $(x^n)' = nx^{n-1}$ holds for any positive integers. For any real number μ, we also have the same result.

Example 10. The equation $(x^\mu)' = \mu x^{\mu-1}$ is true for any real number μ.

Proof: Since $x^\mu = e^{\mu \ln x}$, by the formula of derivative of composite functions, we have

$$(x^\mu)' = (e^{\mu \ln x})' = e^{\mu \ln x}(\mu \ln x)' = x^\mu \frac{\mu}{x} = \mu x^{\mu-1}.$$

We gather the derivatives we calculated so far in the following list:

(1) $(C)' = 0$, (C is a constant);

(2) $(x^\mu)' = \mu x^{\mu-1}$ (μ is any real number);

(3) $(\sin x)' = \cos x$;

(4) $(\cos x)' = -\sin x$;

(5) $(\tan x)' = \sec^2 x$;

(6) $(\cot x)' = -\csc^2 x$;

(7) $(\sec x)' = \sec x \tan x$;

(8) $(\csc x)' = -\csc x \cot x$;

(9) $(\ln x)' = \dfrac{1}{x}$;

(10) $(\log_a x)' = \dfrac{1}{x \ln a}$;

(11) $(e^x)' = e^x$;

(12) $(a^x)' = a^x \ln a$;

(13) $(\arcsin x)' = \dfrac{1}{\sqrt{1 - x^2}}$;

(14) $(\arccos x)' = \dfrac{-1}{\sqrt{1 - x^2}}$;

(15) $(\arctan x)' = \dfrac{1}{1 + x^2}$;

(16) $(\operatorname{arccot} x)' = \dfrac{-1}{1 + x^2}$.

These formulas together with the seven rules for derivatives allow us to calculate most of the derivatives we may encounter. Among the rules, (1), (3) and (6) are more important than others.

(1) $[u(x) + v(x)]' = u'(x) + v'(x)$;

(3) $[u(x)v(x)]' = u'(x)v(x) + u(x)v'(x)$;

(6) If $y = f(u)$ and $u = \varphi(x)$, then

$$\frac{dy}{dx} = f'(u)\varphi'(x).$$

This is because (2) can be derived from (1), and (4) can be derived from (3) by the fact that if we take derivative on both side of the equation

$$\left(\frac{u(x)}{v(x)}\right) v(x) = u(x),$$

then (4) follows.

Moreover, (5) is a special case of (3), and (7) is a special case of (6). Also, all the other formulas in this list can be derived from (2), (3) and

(11):

$$(x^\mu)' = \mu x^{\mu-1}, \quad (\sin x)' = \cos x, \quad (e^x)' = e^x.$$

The functions in the above list, as well as their finite arithmetic combinations and compositions are called *elementary functions*.

For instance,

$$y = \ln(\sqrt{3^{2x} + 1} + \cos(x^2 + 1)),$$

$$y = \left(\frac{a^x + \sin x}{\ln x}\right)^{\frac{1}{2}},$$

$$y = x^x$$

are elementary functions.

With these basic formulas and rules, we can calculate the derivatives of all elementary functions.

Example 11. Find the derivative of the power exponential function $y = x^x$.

Solution: Since $x^x = e^{x \ln x}$, we have

$$y' = (e^{x \ln x})' = e^{x \ln x} \cdot (x \ln x)' = x^x (\ln x + 1).$$

Similarly, the derivative of the general power exponential function $y = u(x)^{v(x)}$ is

$$y' = [u(x)^{v(x)}]' = u(x)^{v(x)} \left[v'(x) \ln u(x) + v(x) \frac{u'(x)}{u(x)}\right].$$

Example 12. Find the derivative of the function $y = x^{a^x}$, $(a > 0)$.

Solution:

$$y' = (e^{a^x \ln x})' = e^{a^x \ln x} \left(a^x \ln a \ln x + \frac{a^x}{x}\right)$$

$$= a^x x^{a^x} \left(\ln a \ln x + \frac{1}{x}\right).$$

Example 13. Find the derivative of the function $y = \ln(x + \sqrt{x^2 + a^2})$.

Solution:

$$y' = \frac{1}{x + \sqrt{x^2 + a^2}} \left[1 + \frac{1}{2\sqrt{x^2 + a^2}} \cdot 2x\right]$$

$$= \frac{1}{x + \sqrt{x^2 + a^2}} \cdot \frac{\sqrt{x^2 + a^2} + x}{\sqrt{x^2 + a^2}} = \frac{1}{\sqrt{x^2 + a^2}}.$$

Example 14. Show that $(\ln|x|)' = \dfrac{1}{x}$ regardless of whether $x > 0$ or $x < 0$.

Solution: In fact, $\ln|x| = \ln x$ when $x > 0$, so it is obvious that $(\ln|x|)' = \dfrac{1}{x}$. When $x < 0$, we have $\ln|x| = \ln(-x)$, thus

$$(\ln|x|)' = [\ln(-x)]' = \frac{1}{-x}(-1) = \frac{1}{x}.$$

Example 15. Compute $\dfrac{d\rho}{d\varphi}$ if $\rho(\varphi) = \dfrac{1}{2}\arctan\dfrac{2\varphi}{1-\varphi^2}$.

Solution:

$$\frac{d\rho}{d\varphi} = \frac{1}{2}\frac{1}{1+(\frac{2\varphi}{1-\varphi^2})^2}\frac{2(1+\varphi^2)}{(1-\varphi^2)^2} = \frac{1}{1+\varphi^2}.$$

Example 16. A container of conical shape has hight h cm. The radius of its base is R cm. Fill the container with water from the top at a rate of A cm^3 per second. Find the rate at which the water surface is rising when the water surface is at half of the hight of the container.

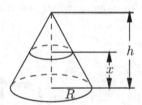

Fig. 2.2

Solution: Let x be the hight of the water surface. Then x is a function of time t:

$$x = x(t).$$

We need to find the value of $\dfrac{dx}{dt}$ when $x = \dfrac{h}{2}$. The water in the container forms a frustum with hight x and radius of the top surface $\dfrac{h-x}{h}R$. The volume of the frustum is

$$V = \frac{\pi R^2}{3h^2}[h^3 - (h - x)^3].$$

The rate of increasing of the volume of water is the derivative of V with respect to t

$$\frac{dV}{dt} = \frac{\pi R^2}{h^2}(h - x)^2 \frac{dx}{dt}.$$

On the other hand, the rate of increasing of the volume of water is the same as the rate of pouring water, that is, $A\text{cm}^3$ per second, thus we have

$$\frac{dx}{dt} = \frac{Ah^2}{\pi R^2(h - x)^2}.$$

Therefore,

$$\frac{dx}{dt} = \frac{4A}{\pi R^2}$$

when $x = \dfrac{h}{2}$.

Example 17. In order to perform aerial photography, an airplane is flying at an altitude of 2km with velocity of 200km/h over a target. Find the angular velocity $\dfrac{d\theta}{dt}$ of the camera when the airplane is directly over the target (Figure 2.3).

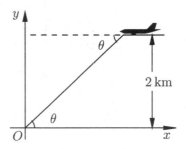

Fig. 2.3

Solution: Consider the target as the origin of the coordinate system. Let x be the horizontal distance between the target and the point on the ground that is directly below the airplane.

As in Figure 2.3, $\tan\theta = \dfrac{2}{x}$, so $\theta = \arctan\dfrac{2}{x}$. Thus,

$$\frac{d\theta}{dt} = \frac{-2}{x^2 + 4}\frac{dx}{dt}.$$

Since $\dfrac{dx}{dt} = -200$, we have

$$\frac{d\theta}{dt} = \frac{400}{x^2 + 4}.$$

Therefore, when the plane is directly over the target,

$$\frac{d\theta}{dt} = 100 \text{ radian/h} = \frac{5}{\pi} \text{ degree/s}.$$

We studied in Chapter 1 that if the derivative of a function $y = f(x)$ at a point x exists, so does the differential

$$dy = f'(x)dx.$$

For instance, the differential of the function $y = \arctan x$ is

$$dy = \frac{1}{1 + x^2}dx$$

since $(\arctan x)' = \dfrac{1}{1 + x^2}$.

The differential of the function $y = \ln(x + \sqrt{x^2 + 1})$ is

$$dy = \frac{1}{\sqrt{x^2 + 1}}dx$$

since $y' = [\ln(x + \sqrt{x^2 + 1})]' = \dfrac{1}{\sqrt{x^2 + 1}}$. Thus, it is easy to get the rules for differentials of functions from the corresponding rules for derivatives:

$$d(u \pm v) = du \pm dv, \quad d(uv) = u\,dv + v\,du,$$

$$d\left(\frac{u}{v}\right) = \frac{v\,du - u\,dv}{v^2},$$

where u and v are functions of x, and $v \neq 0$ for the quotient case.

Actually, these equations can be derived directly from the definition of differential. For instance,

$$d(uv) = (uv)'\,dx = uv'\,dx + vu'\,dx = u\,dv + v\,du.$$

We now study the differential of composite functions. Suppose $y = f(x)$, $x = \varphi(t)$ are differentiable functions. We know the derivative of the composite function $y = f(\varphi(t))$ is

$$y' = f'(x)\varphi'(t).$$

Thus,

$$dy = f'(x)\varphi'(t)dt.$$

Since

$$dx = \varphi'(t)dt,$$

we have

$$dy = f'(x)dx$$

This implies that the form of the differential $dy = f'(x)\,dx$ of the function $y = f(x)$ does not change regardless of whether x represents an independent variable or an intermediate variable. This is called the *invariance of differential forms*.

2.1.2   *Derivatives and Differentials of Higher Order*

The derivative $f'(x)$ of a function $f(x)$ is still a function with the same independent variable x. So it would make sense to consider the existence of the derivative of $f'(x)$. If the derivative of $f'(x)$ exists, it is called the *second order derivative* of $f(x)$, denoted as $f''(x)$ or $\dfrac{d^2 f(x)}{dx^2}$.

From this definition, the second derivative of the function $f(x)$ at a point x is the limit

$$f''(x) = \lim_{\Delta x \to 0} \frac{f'(x + \Delta x) - f'(x)}{\Delta x}.$$

Suppose the distance travelled by a moving particle has the relation $s = f(t)$ with time t. We know that the velocity of the particle is the first derivative of this function. And from the above definition we see that the second derivative is the rate of change of the velocity, i.e., the acceleration of the particle. A geometric interpretation of the second derivative of a function will be given later.

The second derivative $f''(x)$ of a function $f(x)$ is still a function of x. If the derivative of the function $f''(x)$ exists, then it is the third derivative of $f(x)$, denoted as $f'''(x)$ or $\dfrac{d^3 f(x)}{dx^3}$. In general, if the derivative of the function $f^{(n-1)}(x)$ exists, then it is called the nth derivative of the function $f(x)$, denoted as $f^{(n)}(x)$ or $\dfrac{d^n f(x)}{dx^n}$.

Obviously, it is not necessary to develop new methods to calculate higher order derivatives. We only need to take the derivative repeatedly for the function $f(x)$.

Similarly, the differential of the first order differential is called the second order differential, that is

$$d^2y = d(dy).$$

Suppose $y = f(x)$ and $y' = f'(x)$. If the derivative of $f'(x)$ exists, then

$$d^2y = d(dy) = d[f'(x)dx] = f''(x)dx^2.$$

If the derivative of $f''(x)$ exists, then we can calculate the third order differential

$$d^3y = d(d^2y) = d[f''(x)dx] = f'''(x)dx^3.$$

In general, If the derivative of $f^{(n-1)}(x)$ exists, then we can calculate the nth order differential

$$d^ny = d(d^{(n-1)}y) = d[f^{(n-1)}(x)dx^{n-1}] = f^{(n)}(x)dx^n.$$

We will not discuss the properties of higher order differentials separately since they can be derived from the properties of higher order derivatives.

Example 1. The derivative of a polynomial

$$y = a_nx^n + a_{n-1}x^{n-1} + \cdots + a_1x + a_0$$

is

$$y = a_n nx^{n-1} + a_{n-1}(n-1)x^{n-2} + \cdots + a_1.$$

It is still a polynomial, only with one less degree. By this, we see that the nth derivative of the polynomial is a constant

$$y^{(n)} = n!a_n.$$

Hence, the derivatives of the polynomial of order higher than n are all equal to zero

$$y^{(n+1)} = y^{(n+2)} = \cdots = 0.$$

Example 2. Consider the exponential function $y = e^{ax}$. Since

$$y' = ae^{ax}, \quad y'' = a^2e^{ax}, \quad y''' = a^3e^{ax}, \cdots,$$

we have

$$y^{(n)} = a^ne^{ax}.$$

Example 3. Consider the logarithmic function $y = \ln(1 + x)$. Since

$$y' = \frac{1}{1+x}, \quad y'' = \frac{-1}{(1+x)^2}, \quad y''' = \frac{1 \cdot 2}{(1+x)^3}, \cdots,$$

we have

$$y^{(n)} = (-1)^{(n-1)} \frac{(n-1)!}{(1+x)^n}.$$

Example 4. The nth derivatives of trigonometric functions $y = \sin x$, $y = \cos x$ are

$$(\sin x)^{(n)} = \sin\left(x + \frac{n\pi}{2}\right);$$

$$(\cos x)^{(n)} = \cos\left(x + \frac{n\pi}{2}\right).$$

These two equations can be verified by induction. In fact

$$(\sin x)' = \cos x = \sin\left(x + \frac{\pi}{2}\right),$$

so the equation holds for $n = 1$. Assume that the equation holds for $n = k$. Then for $n = k + 1$, we have

$$(\sin x)^{(k+1)} = [(\sin x)^{(k)}]' = \left[\sin\left(x + \frac{k\pi}{2}\right)\right]'$$

$$= \cos\left(x + \frac{k\pi}{2}\right) = \sin\left(x + \frac{(k+1)\pi}{2}\right).$$

Thus the first equation is true. The second equation can be verified in a similar way.

If the nth derivatives exist for both functions u and v, then it is easy to see that

$$(Cu)^{(n)} = Cu^{(n)},$$

where C is a constant; and

$$(u \pm v)^{(n)} = u^{(n)} \pm v^{(n)}.$$

The formula for the nth derivative of the product of two functions is not so straightforward. We have,

$$(uv)' = u'v + uv', \quad (uv)'' = u''v + 2u'v' + uv'',$$

$$(uv)''' = u'''v + 3u''v' + 3u'v'' + uv'''.$$

In these equations, the coefficient of each term on the right-hand side is the same as the coefficient of each term in the corresponding binomial expansions. Therefore, the generalized formula is

$$(uv)^{(n)} = u^{(n)}v + C_n^1 u^{(n-1)}v' + C_n^2 u^{(n-2)}v'' + \cdots$$
$$+ C_n^r u^{(n-r)}v^r + \cdots + uv^{(n)},$$

where

$$C_n^r = \frac{n(n-1)\cdots(n-r+1)}{r!}.$$

This can be obtained by induction, and we leave this work to our readers.

This formula of calculating higher order derivatives of the product of two functions can be written in a simplified form:

$$(uv)^{(n)} = \sum_{r=0}^{n} C_n^r u^{(n-r)}v^{(r)}, \qquad (1)$$

where $u^{(0)} = u$ and $v^{(0)} = v$.

This is called **Leibniz formula**. It is similar in form to the binomial expansion formula.

$$(u+v)^n = \sum_{r=0}^{n} C_n^r u^{n-r}v^r.$$

The difference is that in each term $C_n^r u^{(n-r)}v^{(r)}$ in (1), the numbers $n-r$ and r represent the order of derivatives, and that in each term $C_n^r u^{n-r}v^r$ of the binomial expansion, the number $n-r$ and r represent the power. The differential form of this equation is

$$d^n(uv) = \sum_{r=0}^{n} C_n^r d^{n-r}u\, d^r v.$$

Example 5. Find the 50th derivative of the function

$$y = x^2 \cos ax.$$

Solution: Let $u = \cos ax$, $v = x^2$. Since

$$u^{(n)} = a^n \cos\left(ax + \frac{n\pi}{2}\right),$$
$$v' = 2x, \quad v'' = 2, \quad v''' = v^{(4)} = \cdots = 0,$$

we have

$$y^{(50)} = a^{50}x^2 \cos(ax + 25\pi) + 50a^{49}(2x)\cos\left(ax + \frac{49\pi}{2}\right)$$
$$+ \frac{50 \cdot 49}{2} \cdot 2a^{48} \cos(ax + 24\pi)$$
$$= -a^{50}x^2 \cos ax - 100a^{49}x \sin ax + 2450a^{48} \cos ax$$
$$= a^{48}[(2450 - a^2x^2)\cos ax - 100ax \sin ax].$$

Now we study the derivatives of functions defined parametrically. Suppose $x = \varphi(t)$, $y = \psi(t)$, and $x = \varphi(t)$ has the inverse function $t = \omega(x)$. Then we have $y = \psi(\omega(x))$. By the rule of derivative for composite functions,

$$\frac{dy}{dx} = \frac{dy}{dt}\frac{dt}{dx} = \psi'(t)\frac{1}{\varphi'(t)} = \frac{\psi'(t)}{\varphi'(t)}.$$

Here we assume the existence of $\varphi'(t)$ and $\psi'(t)$, also $\varphi'(t) \neq 0$.

Furthermore, if $\varphi''(t)$ and $\psi''(t)$ exist, then again by the rule we have

$$\frac{d^2y}{dx^2} = \frac{d}{dx}\left(\frac{\psi'(t)}{\varphi'(t)}\right) = \frac{d}{dt}\left(\frac{\psi'(t)}{\varphi'(t)}\right)\frac{dt}{dx}$$

$$= \frac{\psi''(t)\varphi'(t) - \varphi''(t)\psi'(t)}{[\varphi'(t)]^2}\frac{1}{\varphi'(t)} = \frac{\psi''(t)\varphi'(t) - \varphi''(t)\psi'(t)}{[\varphi'(t)]^3}.$$

Example 6. Find the equation of the tangent line to the ellipse $x = a\cos t$, $y = b\sin t$ at $t = \dfrac{\pi}{4}$.

Solution: The point on the ellipse corresponding to $t = \dfrac{\pi}{4}$ is $\left(\dfrac{a}{\sqrt{2}}, \dfrac{b}{\sqrt{2}}\right)$. Since

$$\varphi'(t)|_{t=\frac{\pi}{4}} = -a\sin t|_{t=\frac{\pi}{4}} = -\frac{a}{\sqrt{2}},$$

$$\psi'(t)|_{t=\frac{\pi}{4}} = b\cos t|_{t=\frac{\pi}{4}} = \frac{b}{\sqrt{2}},$$

the equation of the tangent line is

$$y - \frac{b}{\sqrt{2}} = \frac{\dfrac{b}{\sqrt{2}}}{-\dfrac{a}{\sqrt{2}}}\left(x - \frac{a}{\sqrt{2}}\right),$$

or

$$bx + ay = \sqrt{2}ab.$$

2.1.3 *Approximate Calculation by Derivatives*

We know that the change Δy of a function $y = f(x)$ is often represented approximately by the differential dy:

$$\Delta y \approx dy,$$

or

$$f(x_0 + \Delta x) - f(x_0) \approx f'(x_0)\Delta x.$$

This equation can be rewritten as

$$f(x_0 + \Delta x) \approx f(x_0) + f'(x_0)\Delta x. \tag{2}$$

The value of $f(x)$ at the point $x_0 + \Delta x$ can be approximated by its value at the x_0, $f(x_0)$, and its derivative at x_0, $f'(x_0)$. In general, the smaller the value of $|\Delta x|$, the smaller the difference is between Δy and dy. This equation is often used in approximate calculations.

Example 1. Show that

$$(1 + x)^\alpha \approx 1 + \alpha x \tag{3}$$

when $|x|$ is sufficiently small, where α is an arbitrary real number.

Proof: Let $f(x) = (1 + x)^\alpha$, $x_0 = 0$ and $\Delta x = x$. Then

$$f(x_0 + \Delta x) = (1 + x)^\alpha, \quad f(x_0) = 1, \quad f'(x_0) = \alpha.$$

Substitute into equation (2), we have

$$(1 + x)^\alpha \approx 1 + \alpha x.$$

We can calculate some approximate values using this equation. For example, we can find an approximate value of $\sqrt[5]{245}$. Since

$$\sqrt[5]{245} = (243 + 2)^{\frac{1}{5}} = \left[243\left(1 + \frac{2}{243}\right)\right]^{\frac{1}{5}} = 3\left(1 + \frac{2}{243}\right)^{\frac{1}{5}},$$

let $x = \dfrac{2}{243}$ and substitute it into equation (3), we have

$$\sqrt[5]{245} = 3\left(1 + \frac{1}{5}\frac{2}{243}\right) = 3.0049.$$

Example 2. Find an approximate value of $\sin 30°13'$.

Solution:

$$\sin 30°13' = \sin\left(\frac{\pi}{6} + \frac{13\pi}{60 \times 180}\right) = \sin\left(\frac{\pi}{6} + \frac{13\pi}{10800}\right).$$

Let $f(x) = \sin x$, $x_0 = \dfrac{\pi}{6}$, $\Delta x = \dfrac{13\pi}{10800}$, and substitute them into equation (2) we have

$$\sin 30°13' = \sin\left(\frac{\pi}{6} + \frac{13\pi}{10800}\right) \approx \sin\frac{\pi}{6} + \cos\frac{\pi}{6} \cdot \frac{13\pi}{10800} = 0.5033.$$

Example 3. The pendulum of a wall clock contracts 0.01cm in the winter. If the period of the pendulum is 1 second, By how much does the clock run fast?

Solution: The period of pendulum is

$$T = 2\pi\sqrt{\frac{l}{g}},$$

where l is the length of pendulum, g is the gravitational acceleration. Since the period is 1 second, the original length of the pendulum is

$$l = \frac{g}{(2\pi)^2}.$$

To find how much the clock runs fast is to find the change of the period. By the formula above we have

$$dT = \frac{\pi}{\sqrt{g}}\frac{1}{\sqrt{l}}\,dl = \frac{\pi}{\sqrt{g}}\frac{1}{\sqrt{l}}\Delta l.$$

Since

$$\Delta l = -0.01, \quad l = \frac{g}{(2\pi)^2},$$

we have the change of the period

$$\Delta T \approx dT = \frac{2\pi^2}{g}(-0.01) = -0.0002 \text{ seconds}.$$

This implies that the clock run fast 0.0002 seconds per period. Therefore, the clock run fast

$$86400 \times 0.0002 = 17.28 \text{ seconds}$$

every day.

Another application of differentials is to calculate the approximate values of solutions of equations of the form $f(x) = 0$.

Suppose an approximate value of the solution α is x_0. Let $\Delta x = \alpha - x_0$. Then we have

$$f(x_0) + f'(x_0)\Delta x \approx f(x_0 + \Delta x) = 0.$$

Thus

$$\Delta x \approx -\frac{f(x_0)}{f'(x_0)}.$$

Hence

$$\alpha \approx x_1 = x_0 - \frac{f(x_0)}{f'(x_0)}.$$

The value x_1 is an improvement of the initial approximate value x_0. And it can also be improved in the same way:

$$x_2 = x_1 - \frac{f(x_1)}{f'(x_1)}.$$

This process can be continued until a satisfactory approximation is reached. And in general, we can approach the solution α of the equation rapidly if the initial approximation x_0 was properly chosen.

Since the number x_1 is the x intercept of the tangent line of the curve $f(x)$ at the point $(x_0, f(x_0))$, this method is called the *tangent method*.

The initial approximate value x_0 can be found by observing the graph of the function or by the following method:

Suppose the signs of values of $f(x)$ at points a and b are different. By the properties of continuous functions from the last chapter, there is a solution of the equation $f(x) = 0$ between a and b. It is easy to see that the x intercept of the line segment connecting the points $(a, f(a))$ and $(b, f(b))$ is

$$x_0 = \frac{af(b) - bf(a)}{f(b) - f(a)}.$$

We can consider this number as the initial approximate value of the solution.

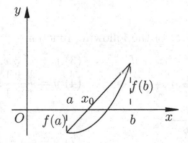

Fig. 2.4

If this initial value is not close enough, then we can find another value in a similar method:

Connect points $(x_0, f(x_0))$ and $(a, f(a))$ (or $(b, f(b))$) by a line segment depending on whether the sign of $f(x_0)$ is different from the sign of $f(a)$ (or $f(b)$). We use the x intercept of the line segment as a better initial value.

This process can be continued if a better yet initial value is still needed.

Example 4. Find the real solution of the equation $f(x) = x^3 + 2x^2 - 3x - 7 = 0$.

Solution: Since $f(1.8) = -0.088 < 0$, $f(1.9) = 1.379 > 0$, there exists a solution of the equation between 1.8 and 1.9.

Let $x_0 = 1.8$, then

$$x_1 = 1.8 - \frac{f(1.8)}{f'(1.8)} = 1.8 + \frac{0.088}{13.92} = 1.806.$$

Substitute it into the original equation, we have

$$f(1.806) = -0.0042 < 0.$$

Thus the solution should locate between 1.806 and 1.9. The second approximate value is

$$x_2 = 1.806 - \frac{f(1.806)}{f'(1.806)} = 1.806 + \frac{0.0042}{13.9348} = 1.8063.$$

Since $f(1.8063) < 0$, the solution is between 1.8063 and 1.9. Since $f(1.807) > 0$, the solution is between 1.8063 and 1.807. Thus the value $x_2 = 1.8063$ can be considered as the approximation of the real solution of the equation with an error less than 0.001.

The other two solutions of this equation are complex numbers.

Exercises 2.1

1. Find the derivatives of the following functions.

(1) $y = ax^3 + bx^2 + cx + d$;

(2) $y = \sqrt[5]{x} + \dfrac{a}{\sqrt[3]{x}} + \dfrac{b}{\sqrt[5]{x}}$;

(3) $y = x^3 \sqrt[5]{x}$;

(4) $y = \dfrac{x+1}{x-1}$;

(5) $y = \dfrac{3x^2 + 9x - 2}{5x + 8}$;

(6) $y = \sin x \cos^2 x$;

(7) $y = \sin x \tan x + \cot x$;

(8) $y = \sqrt[3]{\dfrac{x+1}{x-1}}$;

(9) $y = \dfrac{0.3x^5 + a \sin x}{(a+b)\cos x}$;

(10) $y = \dfrac{\ln x}{x^a}$;

(11) $y = a^x \ln x$;

(12) $y = \dfrac{1 - 10^x}{1 + 10^x}$;

(13) $y = \dfrac{t^3 \arctan t}{e^t}$;

(14) $y = (a^2 + b^2)x^a e^x \arctan x$;

(15) $y = \dfrac{1 - \ln x}{1 + \ln x}$;

(16) $y = x^2 \log_3 x$;

(17) $y = \sin^\alpha x \ln^\beta x$;

(18) $y = \sqrt[3]{1 + \ln^2 x}$;

(19) $y = \dfrac{\arcsin x}{\sqrt{x + x^2}} + \ln \sqrt{\dfrac{1 - x}{1 + x}}$;

(20) $y = \sqrt{x + \sqrt{x + \sqrt{x}}}$;

(21) $y = \sqrt[3]{1 + \sqrt[3]{1 + \sqrt[3]{x}}}$;

(22) $y = \sin(\cos^2 x)\cos(\sin^2 x)$;

(23) $y = \sin(\sin(\sin x))$;

(24) $y = e^{\sqrt{x^2+1}}$;

(25) $y = \sin[\cos^5(\arctan x^3)]$;

(26) $y = \sec^2 \dfrac{x}{a^2} + \tan^2 \dfrac{x}{b^2}$;

(27) $y = a^x e^{\arctan x}$;

(28) $y = \cos \dfrac{1}{x^2} e^{\cos \frac{1}{x^2}}$;

(29) $y = \arcsin \dfrac{1 - x^2}{1 + x^2}$;

(30) $y = \ln[\ln^2(\ln^3 x)]$;

(31) $y = \ln \tan \left(\dfrac{x}{2} + \dfrac{\pi}{4} \right)$;

(32) $y = \arctan \dfrac{1 + x}{1 - x}$;

(33) $y = \sqrt[x]{x}$;

(34) $y = x^x$;

(35) $y = e^x + e^{e^x} + e^{e^{e^x}}$;

(36) $y = x^{x^x}$;

(37) $y = (\sin x)^{\cos x}$;

(38) $y = x^{a^x}$;

(39) $y = (\ln x)^x x^{\ln x}$;

(40) $y = \arccos(\sin^2 x - \cos^2 x)$;

(41) $y = \dfrac{1}{\arccos^2(x^2)}$;

(42) $r = \arctan \left(\dfrac{\sin \theta + \cos \theta}{\sin \theta - \cos \theta} \right)$;

(43) $\rho = \dfrac{2}{\sqrt{a^2 - b^2}} \arctan \left(\sqrt{\dfrac{a - b}{a + b}} \tan \dfrac{\varphi}{2} \right)$;

(44) $y = \dfrac{x}{2}\sqrt{a^2 - x^2} + \dfrac{a^2}{2} \arcsin \dfrac{x}{a}$;

(45) $y = \arctan(x + \sqrt{1 + x^2})$;

(46) $y = \ln(e^x + \sqrt{1 + e^{2x}})$;

(47) $y = \ln(\cos^2 x + \sqrt{1 + \cos^4 x})$;

(48) $s = \dfrac{t^6}{1+t^{12}} - \operatorname{arccot} t^6$;   (49) $y = \arctan e^x - \ln \sqrt{\dfrac{e^{2x}}{e^{2x}+1}}$;

(50) $y = \ln \cos \arctan \dfrac{e^x - e^{-x}}{2}$.

2. Let $f(x) = \dfrac{1}{3}x^3 + \dfrac{1}{2}x^2 - 2x$, find x if

   (1) $f'(x) = 0$;   (2) $f'(x) = -2$;   (3) $f'(x) = 10$.

3. Find the derivatives of the following functions at the given points.

   (1) $f(x) = \dfrac{3x^3 - 3x + \sqrt{x} - 1}{x}$, $x = \dfrac{1}{4}$;

   (2) $f(x) = \dfrac{p}{1 + e\cos x}$, $x = \dfrac{1}{2}$.

4. Find the derivative of the function
$$y = \frac{1}{x^2 - 5}$$
   by the following two different methods and compare the two.
   (1) Use the rule of derivatives of quotients.
   (2) Rewrite the function as $y = (x^2 - 5)^{-1}$, then use the rule of derivatives of composite functions.

5. Find the derivative of the function
$$y = \ln \sqrt[3]{\frac{(x+2)(x+3)}{x+1}}$$
   by the following two different methods and compare the two.
   (1) Use the rule of derivatives of composite functions.
   (2) Rewrite the function as
$$y = \frac{1}{3}[\ln(x+2) + \ln(x+3) - \ln(x+1)],$$
   then find the derivative.

6. Find the derivative of the function
$$y = \frac{2x+1}{2x-1}$$
   by the following two different methods and compare the two.
   (1) Use the rule of derivatives of quotients.
   (2) Rewrite the function as
$$y = 1 + \frac{2}{2x-1},$$
   then find the derivative.

7. Suppose $\varphi(x)$, $\psi(x)$ are differentiable functions. Find $\dfrac{dy}{dx}$.

(1) $y = \sqrt{\varphi^2(x) + \psi^2(x)}$; (2) $y = \sqrt[\varphi(x)]{\psi(x)}$;

(3) $y = \left[\dfrac{\varphi(x)}{\psi(x)}\right]^{\ln\frac{\varphi(x)}{\psi(x)}}$;

(4) $y = \arctan[1 + \varphi(x) + \varphi(x)^{\psi(x)}]$.

8. Find $\dfrac{dx}{dy}$ for the following functions.

(1) $y = xe^x$; (3) $y = 2e^{-x} - e^{-2x}$.

(2) $y = \arctan\dfrac{1}{x}$;

9. Find the second derivative of the following functions.

(1) $y = e^{-x^2}$; (3) $y = x^2 a^x$;

(2) $y = \dfrac{\arcsin x}{\sqrt{1-x^2}}$; (4) $y = \sin x \arctan\dfrac{x}{a}$;

(5) $y = (1+x^2)\arctan x$; (6) $y = x[\sin(\ln x) + \cos(\ln x)]$.

10. Suppose that the functions $u = \varphi(x)$, $v = \psi(x)$ have second derivatives. Find $\dfrac{d^2y}{dx^2}$ for the following compositions.

(1) $y = \ln\dfrac{u}{v}$; (2) $y = \sqrt{u^2 + v^2}$.

11. Suppose that the function $f(x)$ has third derivative. Find $\dfrac{d^2y}{dx^2}$ and $\dfrac{d^3y}{dx^3}$ for the following compositions.

(1) $y = f(x^2)$; (2) $y = f(e^x + x)$.

12. Find the following higher-order derivatives.

(1) $(x^2 e^x)^{(50)}$; (2) $[\ln(1+x)^x]^{(30)}$;

(3) $[(x^2+1)\sin x]^{(20)}$;

(4) $\left(\dfrac{1+x}{\sqrt{1-x}}\right)^{(100)}$; (5) $\left(\dfrac{1}{x^2-3x+2}\right)^{(n)}$;

(6) $\left(\dfrac{e^x}{x}\right)^{(n)}$; (7) $\left(\dfrac{1-x}{1+x}\right)^{(n)}$.

13. Verify that the function
$$y = e^{-x} + \frac{1}{2}(\cos x + \sin x)$$
satisfies the equation $y' + y = \cos x$.

14. Verify that the function $y = A\sin(\omega t + \varphi)$ satisfies the equation
$$\frac{d^2y}{dt^2} + \omega^2 y = 0.$$

15. If r_1 and r_2 are the solutions of the equation $r^2 + pr + q = 0$, show that the function $y = c_1 e^{r_1 x} + c_2 e^{r_2 x}$ satisfies the equation $y'' + py' + qy = 0$, where c_1 and c_2 are constants.

16. Find the differential of the following functions

(1) $y = \ln\left(\dfrac{\pi}{2} - \dfrac{x}{4}\right)$;

(2) $y = a^{\ln \tan x}$;

(3) $y = \sin x - x \cos x$;

(4) $y = \dfrac{1}{2a} \ln \dfrac{x - a}{x + a}$.

17. The approximation $f(x) \approx f(0) + f'(0)x$ holds for small x. Show that

(1) $\sin x \approx x$;

(2) $\ln(1 + x) \approx x$;

(3) $\tan x \approx x$;

(4) $\sqrt[n]{a^n + x} \approx a + \dfrac{x}{na^{n-1}}$, $(a > 0)$.

18. Find the approximate value of the following:

(1) $\sqrt[5]{1.01}$;

(2) $e^{1.01}$;

(3) $\sqrt[10]{1000}$;

(4) $\sin 29°$;

(5) $\ln 1.03$;

(6) $\tan 45°4'$.

19. Find the approximate value of the solution of the equation $x^3 - 3x + 1 = 0$ in $[1, 2]$ to within two decimal points.

20. Suppose $x = \sqrt{1 + t}$, $y = \sqrt{1 - t}$. Verify that

$$\frac{dy}{dx} = \frac{-x}{y}, \quad \frac{d^2 y}{dx^2} = -\frac{2}{y^3}.$$

21. Suppose $x = \ln(1 + t^2)$, $y = t - \arctan t$. Find $\dfrac{dy}{dx}, \dfrac{d^2 y}{dx^2}$.

22. Suppose $x = a(t - \sin t)$, $y = a(1 - \cos t)$. Find $\dfrac{dy}{dx}, \dfrac{d^2 y}{dx^2}$.

23. Find the equations of the tangent and the normal lines of the following curves at the given point.

(1) $x = a(t - \sin t)$, $y = a(1 - \cos t)$ at $t = \pi$;

(2) $x = \sin t$, $y = \cos 2t$ at $t = \dfrac{\pi}{6}$;

(3) $x = \dfrac{3at}{1 + t^2}$, $y = \dfrac{3at^2}{1 + t^2}$ at $t = 2$.

24. A particle moves along a straight line. Starting from point A, the distance between the particle and point A after t seconds is $s = \dfrac{1}{4}t^4 - 4t^3 + 16t^2$. Find the time when the velocity is zero.

25. When air is blown into a balloon, the rate of increase of the radius is 10 cm/s. Find the rates of increase of the volume and the surface area of the balloon when the radius is 10 cm.

26. The radius of a wheel of a car is R. When the car starts, the angular velocity of the wheel is $\theta(t) = at^3$. Find the velocity of the car.

27. A balloon lifts up vertically from a point that is 500 m away from an observer. The velocity of the balloon is 140 m/min. Find the rate of increase of the angle between the observer's line of sight and the ground when the hight of balloon is 500 m.

28. A conical funnel has hight 18 cm. The radius of the base of the conical funnel is 6 cm. A cylindrical tube has diameter 10 cm. Water is poured from the funnel into the tube. The rate of decrease of the depth of water in the funnel is 1 cm/min when the depth of water in the funnel is 12 cm. Find the rate of increase of the depth of water in the cylindrical tube at this time.

29. A ladder with length 5 m leans on a vertical wall. If the foot of the ladder is slipping away from the wall at a rate of 3 m/s, find the rate at which the top of the ladder is sliding down along the wall when the foot of the ladder is 1.4 m away from the wall.

2.2   Integration

2.2.1   *Indefinite Integrals*

The Fundamental Theorem of Calculus tells us that the calculation of definite integrals can be a reduced to the calculation of indefinite integrals followed by the evaluation at boundary points.

Differentiation and integration are inverse operations. So corresponding to the formulas of derivatives, there is a list of formulas for integrals.

1. $\displaystyle\int 0\,dx = C$, ($C$ is a constant);   2. $\displaystyle\int 1\,dx = x + C$;

3. $\displaystyle\int x^n\,dx = \frac{x^{n+1}}{n+1} + C$, ($n \neq -1$);   4. $\displaystyle\int \frac{1}{x}\,dx = \ln|x| + C$;

5. $\displaystyle\int \cos x\,dx = \sin x + C$;   6. $\displaystyle\int \sin x\,dx = -\cos x + C$;

7. $\displaystyle\int \sec^2 x\,dx = \tan x + C$;   8. $\displaystyle\int \csc^2 x\,dx = -\cot x + C$;

9. $\displaystyle\int e^x\,dx = e^x + C$;

10. $\displaystyle\int a^x\,dx = \frac{a^x}{\ln a} + C \quad (a > 0,\ a \neq 1);$

11. $\displaystyle\int \frac{1}{1+x^2}\,dx = \arctan x + C;$

12. $\displaystyle\int \frac{1}{\sqrt{1-x^2}}\,dx = \arcsin x + C.$

It is easy to find the correspondence between this list and the list of formulas of derivatives. For instance, the integration formula

$$\int \cos x\,dx = \sin x + C$$

corresponds to the derivative formula

$$(\sin x)' = \cos x.$$

Similarly, from the rules for the calculation of derivatives and differentials, we can derive corresponding rules for the calculation of integrals. The equations

$$(u + v)' = u' + v'$$

and

$$(cv)' = cv'$$

are (1) and (5) in the list of the rules of derivatives. The corresponding rules for integrals are

$$\int [f(x) + g(x)]\,dx = \int f(x)dx + \int g(x)\,dx \tag{1'}$$

and

$$\int cf(x)\,dx = c\int f(x)\,dx$$

where c is a constant.

In the last section we mentioned that the following three rules are fundamental, from which the others can be derived.

(1) $(u + v)' = u' + v'$,

(3) $(uv)' = u'v + uv'$,

(6) If $y = f(u)$, $u = g(x)$, then

$$\frac{dy}{dx} = f'(u)g'(x).$$

What are the corresponding rules in the list of rules for integrals? We have seen the corresponding integral rule (1') for rule (1) above. The integration by parts and integration by substitution rules that we are going to discuss later in this section, are the corresponding rules for (3) and (6). Let's look at two examples of the application of rule (1').

Example 1. Find $\int \left(3x^2 + 5x + 6 + \dfrac{1}{x} + \dfrac{2}{x^2} \right) dx$.

Solution:

$$\int \left(3x^2 + 5x + 6 + \frac{1}{x} + \frac{2}{x^2} \right) dx$$

$$= 3 \int x^2 \, dx + 5 \int x \, dx + 6 \int dx + \int \frac{1}{x} \, dx + 2 \int \frac{1}{x^2} \, dx$$

$$= x^3 + \frac{5}{2} x^2 + 6x + \ln |x| - \frac{2}{x} + C.$$

Example 2. Find $\int \dfrac{x^2 + 1}{\sqrt{x}} \, dx$.

Solution:

$$\int \frac{x^2 + 1}{\sqrt{x}} \, dx = \int x^{\frac{3}{2}} \, dx + \int x^{-\frac{1}{2}} \, dx = \frac{2}{5} x^{\frac{5}{2}} + 2x^{\frac{1}{2}} + C.$$

Now, we introduce two basic methods of integration. The first one is *integration by substitution*:

Let $x = \varphi(t)$ be a differentiable function with $\varphi'(t)$ continuous. Then

$$\int f(x) \, dx = \int f[\varphi(t)] \varphi'(t) \, dt.$$

This is called the formula of integration by substitution.

In fact, If the antiderivative of $f(x)$ is the function $F(x)$, then

$$\int f(x) dx = F(x) + C.$$

By the rule of derivatives of composite functions, we have

$$dF[\varphi(t)] = \frac{dF}{dx} \cdot \frac{dx}{dt} dt = f[\varphi(t)] \varphi'(t) \, dt.$$

Integrate the above equation, we have

$$\int f[\varphi(t)] \varphi'(t) \, dt = \int dF = F[\varphi(t)] + C$$

$$= F(x) + C = \int f(x) \, dx.$$

This is the formula we want.

Using this formula, to calculate the antiderivative of a function $g(t)$, we can try to write it as $f[\varphi(t)] \varphi'(t)$ such that the antiderivative of the function $f(x)$ is relatively easy to get.

Example 3. Find $\displaystyle\int \frac{1}{ax+b}\, dx$.

Solution:

$$\int \frac{1}{ax+b}\, dx = \frac{1}{a}\int \frac{1}{ax+b}\, d(ax+b).$$

Let $t = ax + b$, then the right hand side of the above equation becomes

$$\frac{1}{a}\int \frac{1}{t}\, dt = \frac{1}{a}\ln|t| + C = \frac{1}{a}\ln|ax+b| + C.$$

Similarly, we can find that

$$\int (ax+b)^n\, dx = \frac{(ax+b)^{n+1}}{a(n+1)} + C,$$

where $n \neq -1$.

Example 4. Find $\displaystyle\int \frac{1}{a^2+x^2}\, dx$.

Solution:

$$\int \frac{1}{a^2+x^2}\, dx = \frac{1}{a^2}\int \frac{1}{1+\left(\frac{x}{a}\right)^2}\, dx$$

$$= \frac{1}{a}\int \frac{1}{1+\left(\frac{x}{a}\right)^2}\, d\left(\frac{x}{a}\right) = \frac{1}{a}\arctan\frac{x}{a} + C.$$

Example 5. Find $\displaystyle\int \frac{1}{x^2-a^2}\, dx$.

Solution:

$$\int \frac{1}{x^2-a^2}\, dx = \frac{1}{2a}\int \left(\frac{1}{x-a} - \frac{1}{x+a}\right)\, dx$$

$$= \frac{1}{2a}\int \frac{1}{x-a}\, d(x-a) - \frac{1}{2a}\int \frac{1}{x+a}\, d(x+a)$$

$$= \frac{1}{2a}\left(\ln|x-a| - \ln|x+a|\right) + C = \frac{1}{2a}\ln\left|\frac{x-a}{x+a}\right| + C.$$

Example 6. Find $\displaystyle\int \frac{1}{\sqrt{a^2-x^2}}\, dx$.

Solution:

$$\int \frac{1}{\sqrt{a^2-x^2}}\, dx = \frac{1}{a}\int \frac{1}{\sqrt{1-\left(\frac{x}{a}\right)^2}}\, dx$$

$$= \int \frac{1}{\sqrt{1-\left(\frac{x}{a}\right)^2}}\, d\left(\frac{x}{a}\right) = \arcsin\frac{x}{a} + C.$$

Example 7. Find $\int \dfrac{x}{1+x^2}\, dx$.

Solution:

$$\int \frac{x}{1+x^2}\, dx = \frac{1}{2} \int \frac{2x}{1+x^2}\, dx$$

$$= \frac{1}{2} \int \frac{1}{1+x^2}\, d(1+x^2) = \frac{1}{2}\ln(1+x^2) + C.$$

Example 8. Find $\int \dfrac{\ln x}{x}\, dx$.

Solution:

$$\int \frac{\ln x}{x}\, dx = \int \ln x\, d(\ln x) = \frac{1}{2}\ln^2 x + C.$$

Example 9. Find $\int \sin^5 x\, dx$.

Solution:

$$\int \sin^5 x dx = \int \sin^4 x \sin x dx = -\int (1-\cos^2 x)^2\, d(\cos x)$$

$$= -\int (1 - 2\cos^2 x + \cos^4 x)\, d(\cos x)$$

$$= -\left(\cos x - \frac{2}{3}\cos^3 x + \frac{1}{5}\cos^5 x\right) + C.$$

Example 10. Find $\int \sec x\, dx$.

Solution:

$$\int \sec x\, dx = \int \frac{1}{\cos x}\, dx = \int \frac{\cos x}{\cos^2 x}\, dx = \int \frac{1}{1-\sin^2 x}\, d(\sin x).$$

Let $t = \sin x$, then the right hand side becomes

$$\int \frac{1}{1-t^2}\, dt = \frac{1}{2}\int \left(\frac{1}{1-t} + \frac{1}{1+t}\right) dt = \frac{1}{2}\ln\left|\frac{1+t}{1-t}\right| + C.$$

Substituting $t = \sin x$ back, we get

$$\int \sec x\, dx = \frac{1}{2}\ln\left|\frac{1+\sin x}{1-\sin x}\right| + C = \ln|\sec x + \tan x| + C.$$

Sometimes, the substation formula can be used in the other direction, that is, by the substitution $x = \varphi(x)$, we can turn the integral $\int f(x)\, dx$ into

$$\int f(x)\, dx = \int f(\varphi(t))\varphi'(t)\, dt = \int g(t)\, dt,$$

where the antiderivative $G(t)$ of $g(t)$ is relatively easy to find. Then we have

$$\int f(x)\, dx = G(\psi(x)) + C,$$

where $t = \psi(x)$ is the inverse function of $x = \varphi(t)$.

Example 11. Find $\int \sqrt{a^2 - x^2}\, dx$.

Solution: We must have $|x| \leqslant a$. So assume that $x = a\sin t$, then $dx = a\cos t\, dt$.

$$\int \sqrt{a^2 - x^2}\, dx = a^2 \int \cos^2 t\, dt = a^2 \int \frac{1 + \cos 2t}{2}\, dt$$

$$= \frac{a^2}{2}\left(t + \frac{\sin 2t}{2}\right) + C = \frac{a^2}{2}(t + \sin t \cos t) + C$$

$$= \frac{a^2}{2}t + \frac{a}{2}\sin t\sqrt{a^2 - (a\sin t)^2} + C$$

$$= \frac{a^2}{2}\arcsin\frac{x}{a} + \frac{x}{2}\sqrt{a^2 - x^2} + C.$$

Example 12. Find $\int \frac{1}{\sqrt{x^2 + 1}}\, dx$.

Solution: Since the domain of the integrand is all real numbers, we can let $x = \tan t$, then $dx = \sec^2 t\, dt$.

$$\int \frac{1}{\sqrt{x^2 + 1}}\, dx = \int \frac{1}{\sqrt{\tan^2 t + 1}}\sec^2 t\, dt = \int \sec t\, dt$$

$$= \ln(\sec t + \tan t) + C = \ln(\tan t + \sqrt{1 + \tan^2 t}) + C$$

$$= \ln(x + \sqrt{1 + x^2}) + C.$$

Example 13. Find $\int \frac{1}{\sqrt{x^2 + 4x + 5}}\, dx$.

Solution: Since $x^2 + 4x + 5 = (x + 2)^2 + 1$, let $t = x + 2$, we have

$$\int \frac{1}{\sqrt{x^2 + 4x + 5}}\, dx = \int \frac{1}{\sqrt{t^2 + 1}}\, dt = \ln(t + \sqrt{t^2 + 1}) + C$$

$$= \ln(x + 2 + \sqrt{x^2 + 4x + 5}) + C.$$

Example 14. Find $\int \frac{1}{\sqrt{x^2 - 1}}\, dx$.

Solution: Since we must have $|x| > 1$, let $x = \sec t$, then $dx = \sec t \tan t \, dt$

$$\int \frac{1}{\sqrt{x^2 - 1}} dx = \int \frac{1}{\sqrt{\sec^2 t - 1}} \sec t \tan t \, dt$$

$$= \int \sec t \, dt = \ln(\sec t + \tan t) + C$$

$$= \ln(\sec t + \sqrt{\sec^2 t - 1}) + C$$

$$= \ln(x + \sqrt{x^2 - 1}) + C.$$

The other important method of integration is *integration by parts*. This corresponds to rule (3) of differentials $d(uv) = u \, dv + v \, du$. In fact, integrating both sides of this equation we have

$$uv = \int u \, dv + \int v \, du.$$

That is

$$\int u \, dv = uv - \int v \, du.$$

It follows that

$$\int uv' \, dx = uv - \int vu' \, dx.$$

This is called the formula of integration by parts.

Integration by parts is useful if the integrand can be written as uv', in such a way that both the antiderivative of v' and the antiderivative of $u'v$ are relatively easy to find.

Example 15. Find $\int x^3 \ln x \, dx$.

Solution: Let $u = \ln x$, $dv = x^3 \, dx$, then

$$u' = \frac{1}{x}, \quad v = \frac{1}{4}x^4.$$

Thus

$$\int x^3 \ln x \, dx = \frac{1}{4}x^4 \ln x - \frac{1}{4} \int x^3 \, dx$$

$$= \frac{1}{4}x^4 \ln x - \frac{1}{16}x^4 + C.$$

Example 16. Find $\int \arctan x \, dx$.

Solution: Let $u = \arctan x$, $dv = dx$, then

$$u' = \frac{1}{1+x^2}, \quad v = x.$$

Thus

$$\int \arctan x \, dx = x \arctan x - \int \frac{x}{1+x^2} \, dx$$

$$= x \arctan x - \frac{1}{2} \ln(1+x^2) + C.$$

Example 17. Find $\displaystyle\int e^{ax} \cos bx \, dx$ and $\displaystyle\int e^{ax} \sin bx \, dx$.

Solution: Let $u = \cos bx$, $dv = e^{ax} dx$ and $u = \sin bx$, $dv = e^{ax} dx$ in these two integrals respectively, then we have that

$$\int e^{ax} \cos bx \, dx = \frac{1}{a} e^{ax} \cos bx + \frac{b}{a} \int e^{ax} \sin bx \, dx,$$

$$\int e^{ax} \sin bx \, dx = \frac{1}{a} e^{ax} \sin bx - \frac{b}{a} \int e^{ax} \cos bx \, dx.$$

Solving the system of equations for the two integrals, we get

$$\int e^{ax} \cos bx \, dx = \frac{b \sin bx + a \cos bx}{a^2 + b^2} e^{ax} + C,$$

$$\int e^{ax} \sin bx \, dx = \frac{a \sin bx - b \cos bx}{a^2 + b^2} e^{ax} + C.$$

Sometimes, we may need to use integration by parts more than once in the evaluation of integrals.

Example 18. Find $\displaystyle\int x^2 e^x \, dx$.

Solution: Let $u = x^2$, $dv = e^x \, dx$, then

$$\int x^2 e^x dx = x^2 e^x - 2 \int x e^x \, dx.$$

Apply integration by parts again for the integral on the right hand side. Let $u = x$, $dv = e^x dx$. Then

$$\int x^2 e^x \, dx = x^2 e^x - 2\left(x e^x - \int e^x \, dx\right) = (x^2 - 2x + 2)e^x + C.$$

Example 19. Find $\displaystyle\int x^2 \sin x \, dx$.

Solution: Let $u = x^2$, $dv = \sin x \, dx$, then

$$\int x^2 \sin x \, dx = -x^2 \cos x + 2 \int x \cos x \, dx.$$

Let $u = x$, $dv = \cos x \, dx$ in the integral on the right hand side. Then

$$\int x^2 \sin x \, dx = -x^2 \cos x + 2 \left(x \sin x - \int \sin x \, dx \right)$$

$$= -x^2 \cos x + 2x \sin x + 2 \cos x + C.$$

Example 20. Find $\int \sqrt{x^2 + a^2} \, dx$.

Solution: Let $u = \sqrt{x^2 + a^2}$, $dv = dx$, then

$$\int \sqrt{x^2 + a^2} \, dx = x\sqrt{x^2 + a^2} - \int \frac{x^2}{\sqrt{x^2 + a^2}} \, dx$$

$$= x\sqrt{x^2 + a^2} - \int \frac{(x^2 + a^2) - a^2}{\sqrt{x^2 + a^2}} \, dx$$

$$= x\sqrt{x^2 + a^2} - \int \sqrt{x^2 + a^2} \, dx + a^2 \int \frac{1}{\sqrt{x^2 + a^2}} \, dx$$

$$= x\sqrt{x^2 + a^2} - \int \sqrt{x^2 + a^2} \, dx + a^2 \ln(x + \sqrt{x^2 + a^2}) + C.$$

Solving for the integral, we get

$$\int \sqrt{x^2 + a^2} \, dx = \frac{x}{2} \sqrt{x^2 + a^2} + \frac{a^2}{2} \ln(x + \sqrt{x^2 + a^2}) + C.$$

Example 21. Find $J_n = \int \dfrac{1}{(x^2 + a^2)^n} \, dx$, $a \neq 0$.

Solution: Let $u = \dfrac{1}{(x^2 + a^2)^n}$, $dv = dx$, then

$$J_n = \frac{x}{(x^2 + a^2)^n} + 2n \int \frac{x^2}{(x^2 + a^2)^{n+1}} \, dx.$$

Since

$$\int \frac{x^2}{(x^2 + a^2)^{n+1}} \, dx = \int \frac{x^2 + a^2 - a^2}{(x^2 + a^2)^{n+1}} \, dx$$

$$= \int \frac{1}{(x^2 + a^2)^n} \, dx - a^2 \int \frac{1}{(x^2 + a^2)^{n+1}} \, dx$$

$$= J_n - a^2 J_{n+1},$$

we have

$$J_n = \frac{x}{(x^2 + a^2)^n} + 2n(J_n - a^2 J_{n+1}).$$

That is

$$J_{n+1} = \frac{1}{2na^2} \frac{x}{(x^2 + a^2)^n} + \frac{2n-1}{2n} \frac{1}{a^2} J_n.$$

This is the recursive formula of the integral J_n. Since

$$J_1 = \frac{1}{a} \arctan \frac{x}{a} + C,$$

from the formula, we have

$$J_2 = \frac{1}{2a^2} \frac{x}{x^2 + a^2} + \frac{1}{2a^3} \arctan \frac{x}{a} + C,$$

$$J_3 = \frac{1}{4a^2} \frac{x}{(x^2 + a^2)^2} + \frac{3}{8a^4} \frac{x}{x^2 + a^2} + \frac{3}{8a^5} \arctan \frac{x}{a} + C.$$

We have derived the calculation rules of integrals from the corresponding rules of derivatives of sums, products and composition of functions. Since the calculation rules of derivatives of differences, quotients and inverse functions can be derived from the above rules, there is no need to derive the corresponding integration rules for them separately.

Now we study the integration of rational functions.

If the integrand is a rational fraction, and the degree of the numerator is higher than the degree of the denominator, then it can be transformed to the sum of a polynomial and a proper rational fraction by a long division. It is easy to evaluate the integral of the polynomial, so we need to evaluate the integral of the proper rational fraction. Let's look at some examples of integrals of proper rational fractions first:

Example 22. Find $\int \frac{x+1}{x^2 + x + 1} dx$.

Solution: Complete the square for the denominator of the integrand, we have

$$\int \frac{x+1}{x^2 + x + 1} dx = \int \frac{x+1}{\left(x + \frac{1}{2}\right)^2 + \frac{3}{4}} dx.$$

Let $t = x + \dfrac{1}{2}$, then

$$\int \frac{x+1}{x^2+x+1}\,dx = \int \frac{t+\dfrac{1}{2}}{t^2+\dfrac{3}{4}}\,dx = \int \frac{t}{t^2+\dfrac{3}{4}}\,dt + \frac{1}{2}\int \frac{1}{t^2+\dfrac{3}{4}}\,dt$$

$$= \frac{1}{2}\int \frac{1}{t^2+\dfrac{3}{4}}\,d\left(t^2+\frac{3}{4}\right) + \frac{\sqrt{3}}{3}\int \frac{1}{1+\left(\dfrac{2t}{\sqrt{3}}\right)^2}\,d\left(\frac{2t}{\sqrt{3}}\right)$$

$$= \frac{1}{2}\ln\left(t^2+\frac{3}{4}\right) + \frac{\sqrt{3}}{3}\arctan\frac{2t}{\sqrt{3}} + C$$

$$= \frac{1}{2}\ln(x^2+x+1) + \frac{\sqrt{3}}{3}\arctan\frac{2x+1}{\sqrt{3}} + C.$$

Example 23. Find $\displaystyle\int \frac{5x+6}{x^2+3x+1}\,dx$.

Solution: Complete the square for the denominator of the integrand, we have

$$\int \frac{5x+6}{x^2+3x+1}\,dx = \int \frac{5x+6}{\left(x+\dfrac{3}{2}\right)^2-\dfrac{5}{4}}\,dx.$$

Let $t = x + \dfrac{3}{2}$, then

$$\int \frac{5x+6}{x^2+3x+1}\,dx = \int \frac{5t-\dfrac{3}{2}}{t^2-\dfrac{5}{4}}\,dt = 5\int \frac{t}{t^2-\dfrac{5}{4}}\,dt - \frac{3}{2}\int \frac{1}{t^2-\dfrac{5}{4}}\,dt$$

$$= \frac{5}{2}\ln\left|t^2-\frac{5}{4}\right| - \frac{3}{2\sqrt{5}}\ln\left|\frac{2t-\sqrt{5}}{2t+\sqrt{5}}\right| + C$$

$$= \frac{5}{2}\ln(x^2+3x+1) - \frac{3}{2\sqrt{5}}\ln\left|\frac{2x+3-\sqrt{5}}{2x+3+\sqrt{5}}\right| + C.$$

It is not hard to get the basic idea of how to evaluate the integral of rational functions with the general form

$$\int \frac{mx+n}{x^2+px+q}\,dx$$

from the above two examples. The discriminant of quadratic expression in the denominator of the integrand is less than zero ($p^2 - 4q < 0$) in Example 22, and is greater than zero ($p^2 - 4q > 0$) in Example 23.

Example 24. Find $\displaystyle\int \frac{5x - 6}{(x^2 + x + 2)^2}\, dx$.

Solution: Complete the square for the denominator of the integrand, we have

$$\int \frac{5x - 6}{(x^2 + x + 2)^2}\, dx = \int \frac{5x - 6}{\left[\left(x + \dfrac{1}{2}\right)^2 + \dfrac{7}{4}\right]^2}\, dx.$$

Let $t = x + \dfrac{1}{2}$, then

$$\int \frac{5x - 6}{(x^2 + x + 2)^2}\, dx = \int \frac{5t + \dfrac{7}{2}}{\left(t^2 + \dfrac{7}{4}\right)^2}\, dt$$

$$= 5 \int \frac{t}{\left(t^2 + \dfrac{7}{4}\right)^2}\, dt + \frac{7}{2} \int \frac{1}{\left(t^2 + \dfrac{7}{4}\right)^2}\, dt.$$

It is easy to evaluate the first integral on the right hand side of the above equation,

$$\int \frac{t}{\left(t^2 + \dfrac{7}{4}\right)^2}\, dt = -\frac{1}{2}\frac{1}{t^2 + \dfrac{7}{4}} + C.$$

The second integral can be evaluated by using the result of Example 21 of integration by parts.

$$\int \frac{1}{\left(t^2 + \dfrac{7}{4}\right)^2}\, dt = \frac{2}{7}\frac{t}{t^2 + \dfrac{7}{4}} + \frac{4}{7\sqrt{7}} \arctan \frac{2t}{\sqrt{7}} + C.$$

Thus

$$\int \frac{5x - 6}{(x^2 + x + 2)^2}\, dx = -\frac{5}{2}\frac{1}{x^2 + x + 2} + \frac{1}{2}\frac{2x + 1}{x^2 + x + 2}$$

$$+ \frac{2}{\sqrt{7}} \arctan \frac{2x + 1}{\sqrt{7}} + C.$$

Generalizing this example, it is not hard to find the integrals with the general form

$$\int \frac{mx + n}{(x^2 + px + q)^k}\, dx, \quad (k = 2, 3, \cdots).$$

The next example will introduce a method with which a proper rational fraction in general form can be transformed to the sum of two kinds of simple forms

$$\frac{A}{(x-a)^k},$$

and

$$\frac{mx+n}{(x^2+px+q)^k}.$$

Example 25. Find $\int \dfrac{x}{(x^2+1)(x-1)}\, dx$.

Solution: Suppose the integrand can be written as the sum of two rational proper fractions

$$\frac{x}{(x^2+1)(x-1)} = \frac{Ax+B}{(x^2+1)} + \frac{C}{(x-1)},$$

where A, B, C are unknown numbers.

Rewrite the two rational fractions on the right hand side with the common denominator and compare the coefficients of the terms with the same degree in both numerators, we have the following system of equations

$$\begin{cases} A+C = 0, \\ -A+B = 1, \\ -B+C = 0. \end{cases}$$

Solving for A, B, C, we get

$$A = -\frac{1}{2}, \quad B = \frac{1}{2}, \quad C = \frac{1}{2}.$$

Thus

$$\int \frac{x}{(x^2+1)(x-1)}\, dx = -\frac{1}{2}\int \frac{x-1}{x^2+1}\, dx + \frac{1}{2}\int \frac{1}{x-1}\, dx$$

$$= -\frac{1}{4}\ln(1+x^2) + \frac{1}{2}\arctan x + \frac{1}{2}\ln|x-1| + C.$$

Example 26. Find $\int \dfrac{x^3+1}{x^4-3x^3+3x^2-x}\, dx$.

Solution: Factorize the denominator of the integrand,

$$x^4 - 3x^3 + 3x^2 - x = x(x^3 - 3x^2 + 3x - 1) = x(x-1)^3.$$

Suppose

$$\frac{x^3+1}{x^4-3x^3+3x^2-x} = \frac{A}{x} + \frac{B}{(x-1)^3} + \frac{C}{(x-1)^2} + \frac{D}{x-1}.$$

Rewrite the four rational fractions on the right side of the equation with the common denominator and compare the coefficients of the terms with the same degree in their numerators, we have the following system of equations

$$\begin{cases} A+D=1, \\ -3A+C-2D=0, \\ 3A+B-C+D=0, \\ -A=1. \end{cases}$$

Solving for A, B, C, D, we get

$$A=-1, \quad B=2, \quad C=1, \quad D=2.$$

Thus

$$\int \frac{x^3+1}{x^4-3x^3+3x^2-x}\,dx$$

$$= -\int \frac{1}{x}\,dx + 2\int \frac{1}{(x-1)^3}\,dx + \int \frac{1}{(x-1)^2}\,dx + 2\int \frac{1}{x-1}\,dx$$

$$= -\ln|x| - \frac{1}{(x-1)^2} - \frac{1}{x-1} + 2\ln|x-1| + C$$

$$= -\frac{x}{(x-1)^2} + \ln\frac{(x-1)^2}{|x|} + C.$$

Some integrals can be calculated by transforming the integrand to a rational fraction with substitutions of variables. Here are several examples.

Example 27. Find $\displaystyle\int \frac{\sqrt{x+1}+2}{(x+1)^2 - \sqrt{x+1}}\,dx.$

Solution: Let $t = \sqrt{x+1}$, then $dx = 2t\,dt$. Thus

$$\int \frac{\sqrt{x+1}+2}{(x+1)^2 - \sqrt{x+1}}\,dx = 2\int \frac{t+2}{t^3-1}\,dt$$

$$= \int \left(\frac{2}{t-1} - \frac{2t+2}{t^2+t+1} \right) dt$$

$$= \ln\left| \frac{(t-1)^2}{t^2+t+1} \right| - \frac{2}{\sqrt{3}} \arctan \frac{2t+1}{\sqrt{3}} + C$$

$$= \ln\left| \frac{(\sqrt{x+1}-1)^2}{x+\sqrt{x+1}+2} \right| - \frac{2}{\sqrt{3}} \arctan \frac{2\sqrt{x+1}+1}{\sqrt{3}} + C.$$

Example 28. Find $\displaystyle\int \frac{1}{\sqrt[3]{(x+1)^2(x-1)}}\, dx$.

Solution: The integrand can be rewrite as

$$\frac{1}{\sqrt[3]{(x+1)^2(x-1)}} = \sqrt[3]{\frac{x+1}{x-1}}\,\frac{1}{x+1}.$$

Let $t = \sqrt[3]{\dfrac{x+1}{x-1}}$. Then $x = \dfrac{t^3+1}{t^3-1}$, $dx = \dfrac{-6t^2}{(t^3-1)^2}\,dt$. Thus

$$\int \frac{1}{\sqrt[3]{(x+1)^2(x-1)}}\,dx = -3\int \frac{1}{t^3-1}\,dt = \int\left(-\frac{1}{t-1} + \frac{t+2}{t^2+t+1}\right)dt$$

$$= \frac{1}{2}\ln\left|\frac{t^2+t+1}{(t-1)^2}\right| + \sqrt{3}\arctan\frac{2t+1}{\sqrt{3}} + C$$

$$= \frac{1}{2}\ln\frac{\left|\sqrt[3]{(x+1)^2} + \sqrt[3]{x^2-1} + \sqrt[3]{(x-1)^2}\right|}{(\sqrt[3]{x+1} - \sqrt[3]{x-1})^2}$$

$$+ \sqrt{3}\arctan\frac{2\sqrt[3]{x+1} + \sqrt[3]{x-1}}{\sqrt{3}\sqrt[3]{x-1}} + C.$$

Example 29. Find $\displaystyle\int \frac{1}{5+4\sin x}\, dx$.

Solution: Let $t = \tan\dfrac{x}{2}$. Then

$$\sin x = \frac{2\tan\dfrac{x}{2}}{1+\tan^2\dfrac{x}{2}} = \frac{2t}{1+t^2}, \qquad \cos x = \frac{1-\tan^2\dfrac{x}{2}}{1+\tan^2\dfrac{x}{2}} = \frac{1-t^2}{1+t^2}.$$

Since $x = 2\arctan t$, $dx = \dfrac{2}{1+t^2}\,dt$, we have

$$\int \frac{1}{5+4\sin x}\,dx = \int \frac{\dfrac{2}{1+t^2}}{5+\dfrac{8t}{1+t^2}}\,dt = \int \frac{2}{5t^2+8t+5}\,dt$$

$$= \frac{2}{5}\int \frac{1}{(t+\dfrac{4}{5})^2 + \dfrac{9}{25}}\,dt = \frac{2}{5}\frac{1}{\dfrac{3}{5}}\arctan\left(\frac{t+\dfrac{4}{5}}{\dfrac{3}{5}}\right) + C$$

$$= \frac{2}{3}\arctan\left(\frac{5t+4}{3}\right) + C$$

$$= \frac{2}{3}\arctan\left(\frac{5\tan\dfrac{x}{2}+4}{3}\right) + C.$$

Example 30. Find $\displaystyle\int \frac{1}{(1+\cos^2 x)(2+\sin^2 x)}\, dx.$

Solution: Rewrite the integrand as

$$\frac{1}{(1+\cos^2 x)(2+\sin^2 x)} = \frac{1}{4}\left(\frac{1}{1+\cos^2 x} + \frac{1}{2+\sin^2 x}\right).$$

We have

$$\int \frac{1}{(1+\cos^2 x)(2+\sin^2 x)}\, dx = \frac{1}{4}\int \frac{1}{1+\cos^2 x}\, dx + \frac{1}{4}\int \frac{1}{2+\sin^2 x}\, dx.$$

Let $t = \tan x$ in the first integral. Then

$$\cos^2 x = \frac{1}{1+t^2}, \qquad dx = \frac{1}{1+t^2}\, dt.$$

Hence

$$\int \frac{1}{1+\cos^2 x}\, dx = \int \frac{1}{2+t^2}\, dt = \frac{1}{\sqrt{2}}\arctan\frac{t}{\sqrt{2}} + C_1$$

$$= \frac{1}{\sqrt{2}}\arctan\left(\frac{1}{\sqrt{2}}\tan x\right) + C_1.$$

Similarly, we find that

$$\int \frac{1}{2+\sin^2 x}\, dx = \frac{1}{\sqrt{6}}\arctan\left(\sqrt{\frac{3}{2}}\tan x\right) + C_2.$$

Finally

$$\int \frac{1}{(1+\cos^2 x)(2+\sin^2 x)}\, dx = \frac{1}{4\sqrt{2}}\arctan\left(\frac{1}{\sqrt{2}}\tan x\right)$$

$$+ \frac{1}{4\sqrt{6}}\arctan\left(\sqrt{\frac{3}{2}}\tan x\right) + C.$$

The purpose of introducing these basic methods of evaluating indefinite integrals is to show our readers that some complicated integrands can be transformed to functions we are already familiar with, and the process of finding the indefinite integrals can be simplified.

Some integrals of elementary functions can not be expressed as elementary functions, such as

$$\int e^{-x^2}\, dx, \quad \int \sin x^2\, dx, \quad \int \cos x^2\, dx,$$

$$\int \frac{\sin x}{x}\, dx, \quad \int \frac{\cos x}{x}\, dx, \quad \int \frac{1}{\ln x}\, dx,$$

$$\int \frac{1}{\sqrt{1-k^2\sin^2 x}}\, dx, \quad \int \sqrt{1-k^2\sin^2 x}\, dx.$$

We do not discuss these integrals here as they need special preparation.

Exercises 2.2.1

1. Find the following differentials.

   (1) $d(x^2 + a)$;

   (2) $d(\sqrt{x})$;

   (3) $d\left(\dfrac{1}{x}\right)$;

   (4) $d(\sqrt{1 + x^2})$;

   (5) $d(\ln(x + \sqrt{x^2 + 1}))$;

   (6) $d(\cos x + 1)$;

   (7) $d(e^{2x})$;

   (8) $d\left(\arctan \dfrac{x}{a}\right)$.

2. Finish the right hand side of the following equations.

   (1) $x \, dx = d?$

   (2) $\dfrac{1}{x} \, dx = d?$

   (3) $\dfrac{1}{x^2} \, dx = d?$

   (4) $\dfrac{1}{\sqrt{x}} \, dx = d?$

   (5) $\sin x \, dx = d?$

   (7) $e^{-x} \, dx = d?$

   (6) $\dfrac{1}{\cos^2 x} \, dx = d?$

   (8) $\dfrac{1}{\sqrt{x + 2}} \, dx = d?$

   (9) $\dfrac{x}{\sqrt{x^2 + a^2}} \, dx = d?$

   (10) $\dfrac{1}{\sqrt{1 - x^2}} \, dx = d?$

   (11) $\dfrac{1}{\sqrt{1 + x^2}} \, dx = d?$

   (12) $\dfrac{1}{1 + x^2} \, dx = d?$

3. Find the following integrals.

   (1) $\displaystyle\int \dfrac{x + 1}{\sqrt{x}} \, dx$;

   (2) $\displaystyle\int \sqrt[m]{x^n} \, dx$;

   (3) $\displaystyle\int (\sqrt{x} + 1)(x - \sqrt{x} + 1) \, dx$;

   (4) $\displaystyle\int \dfrac{\sqrt{x} - x^3 e^x + x^2}{x^3} \, dx$;

   (5) $\displaystyle\int \left(1 - \dfrac{1}{x^2}\right) \sqrt{x - \sqrt{x}} \, dx$;

   (6) $\displaystyle\int \left(\dfrac{1 - x}{x}\right)^2 \, dx$;

   (7) $\displaystyle\int \dfrac{\sqrt{x^3} + 1}{\sqrt{x} + 1} \, dx$;

   (8) $\displaystyle\int \dfrac{e^{3x} + 1}{e^x + 1} \, dx$;

   (9) $\displaystyle\int \dfrac{\sqrt{1 + x^2} + \sqrt{1 - x^2}}{\sqrt{1 - x^4}} \, dx$;

   (10) $\displaystyle\int (2^x + 3^x)^2 \, dx$.

4. Find the following integrals of the type $\displaystyle\int \dfrac{f'(x)}{f(x)} \, dx$.

(1) $\displaystyle \int \frac{x}{1+x^2}\, dx;$

(2) $\displaystyle \int \frac{2x-3}{x^2-3x+8}\, dx;$

(3) $\displaystyle \int \tan x\, dx;$

(4) $\displaystyle \int \frac{\sin x}{1+\cos x}\, dx;$

(5) $\displaystyle \int \frac{e^{2x}}{1+e^{2x}}\, dx.$

5. Find the following integrals using integration by substitution.

(1) $\displaystyle \int (2x-3)^{100}\, dx;$

(2) $\displaystyle \int \frac{\arctan x}{1+x^2}\, dx;$

(3) $\displaystyle \int \frac{1}{\sin^2(2x+\frac{\pi}{4})}\, dx;$

(4) $\displaystyle \int e^{\sin x}\cos x\, dx;$

(6) $\displaystyle \int x^2\sqrt[3]{1+x^3}\, dx;$

(5) $\displaystyle \int \frac{x}{\sqrt{1-x^2}}\, dx;$

(7) $\displaystyle \int \frac{6x-5}{\sqrt{3x^2-5x+6}}\, dx;$

(8) $\displaystyle \int \frac{\sin 2x}{\sqrt{2-\sin^4 x}}\, dx;$

(9) $\displaystyle \int \frac{\sqrt{\tan x}}{\cos^2 x}\, dx;$

(10) $\displaystyle \int \frac{\arctan \frac{x}{2}}{\sqrt{4-x^2}}\, dx;$

(11) $\displaystyle \int \frac{\arctan \sqrt{x}}{\sqrt{x}(1+x)}\, dx;$

(12) $\displaystyle \int \cot \frac{x}{b-a}\, dx;$

(13) $\displaystyle \int \frac{1}{x\ln x\ln(\ln x)}\, dx;$

(14) $\displaystyle \int (e^x+1)^3 e^x\, dx;$

(15) $\displaystyle \int \sqrt{\frac{\ln(x+\sqrt{x^2+1})}{1+x^2}}\, dx.$

6. Find the following integrals using integration by parts.

(1) $\displaystyle \int x\sin 2x\, dx;$

(2) $\displaystyle \int xe^{-x}\, dx;$

(3) $\displaystyle \int x^2 a^x\, dx;$

(4) $\displaystyle \int \arcsin x\, dx;$

(5) $\displaystyle \int x\arctan x\, dx;$

(6) $\displaystyle \int x^2\ln(1+x)\, dx;$

(7) $\displaystyle \int x^n\ln x\, dx,\ n\neq -1;$

(8) $\int \dfrac{x}{\cos^2 x}\, dx;$

(9) $\int \dfrac{x}{\sqrt{1+2x}}\, dx;$

(10) $\int x^2 \sin x \cos x\, dx;$

(12) $\int \ln(x + \sqrt{1+x^2})\, dx;$

(11) $\int x \ln \dfrac{1+x}{1-x}\, dx;$

(13) $\int \dfrac{x^2}{(1+x^2)^2}\, dx;$

(14) $\int \dfrac{x^2 e^x}{(x+2)^2}\, dx;$

(15) $\int \dfrac{x\cos x}{\sin^3 x}\, dx.$

7. Find the following integrals of rational functions.

(1) $\int \dfrac{1}{x^2 + x - 2}\, dx;$

(2) $\int \dfrac{1}{(x^2+1)(x^2+2)}\, dx;$

(3) $\int \dfrac{x^5}{x+1}\, dx;$

(4) $\int \dfrac{x^2}{x^4 + 3x^2 + 2}\, dx;$

(5) $\int \dfrac{2x^2 - 5}{x^4 - 5x^2 + 6}\, dx;$

(6) $\int \dfrac{4x - 3}{(x-2)^2}\, dx;$

(7) $\int \dfrac{x^3 + 1}{x^3 - x^2}\, dx;$

(8) $\int \dfrac{1}{(x+1)^2(x^2+1)}\, dx;$

(9) $\int \dfrac{1}{(x+1)(x+2)(x+3)}\, dx;$

(10) $\int \dfrac{1}{(x^2 + 4x + 6)^2}\, dx.$

8. Find the following integrals.

(1) $\int \dfrac{1}{\sin x + \cos x}\, dx;$

(2) $\int \dfrac{1}{1 + \sin x}\, dx;$

(3) $\int \dfrac{1}{a \sin x + b \cos x}\, dx;$

(4) $\int \dfrac{1}{\sin^4 x \cos^4 x}\, dx;$

(5) $\int \dfrac{\sin x \cos x}{1 + \sin^4 x}\, dx;$

(6) $\int \dfrac{1}{(2 + \cos x)\sin x}\, dx;$

(7) $\int \dfrac{1}{1 + \sqrt{1+x}}\, dx;$

(8) $\int \dfrac{x}{\sqrt{2 + 4x}}\, dx;$

(9) $\int \dfrac{\sqrt{x}}{\sqrt{x} - \sqrt[3]{x}}\, dx;$

(10) $\int \sqrt{\dfrac{1-x}{1+x}}\, dx;$

(11) $\int \dfrac{1}{x\sqrt{x^2 - 1}}\, dx;$

(12) $\int \dfrac{x^2}{\sqrt{a^2 - x^2}}\, dx;$

(13) $\int \sqrt{3 + 4x - 4x^2}\, dx;$

(14) $\int e^{\sqrt{x}}\, dx;$

(15) $\displaystyle\int \frac{1}{\cos^4 x}\, dx;$ (16) $\displaystyle\int \frac{1-\tan x}{1+\tan x}\, dx;$

(17) $\displaystyle\int \frac{e^x - e^{-x}}{e^x + e^{-x}}\, dx;$ (19) $\displaystyle\int \frac{1}{\sqrt{1-2x-x^2}}\, dx;$

(18) $\displaystyle\int \sqrt{1+\sin x}\, dx;$ (20) $\displaystyle\int \sqrt{x^3 + x^4}\, dx;$

(21) $\displaystyle\int \frac{1}{\sqrt{x-1}-\sqrt{x-2}}\, dx;$ (22) $\displaystyle\int \frac{x}{x-\sqrt{x^2-1}}\, dx;$

(23) $\displaystyle\int \sqrt{x}\ln^2 x\, dx;$ (25) $\displaystyle\int \cos x \cos 2x \cos 3x\, dx;$

(24) $\displaystyle\int \frac{1}{1+x(\ln^2 x)}\, dx;$

(26) $\displaystyle\int x^5 e^{x^3}\, dx;$ (27) $\displaystyle\int (\tan^2 x + \tan^4 x)\, dx;$

(28) $\displaystyle\int \frac{1}{(\tan x + 1)\sin^2 x}\, dx;$ (29) $\displaystyle\int \frac{1}{\sin^2 x \cos^2 x}\, dx;$

(30) $\displaystyle\int \frac{\arcsin x}{\sqrt{(1-x^2)^3}}\, dx;$ (31) $\displaystyle\int \frac{2^x}{1-4^x}\, dx;$

(32) $\displaystyle\int \frac{\ln(1+x)}{\sqrt{1+x}}\, dx;$ (34) $\displaystyle\int \frac{x^2}{\sqrt{1-x^2}}\, dx;$

(33) $\displaystyle\int \ln(1+x^2)\, dx;$

(35) $\displaystyle\int \frac{1}{x^2\sqrt{4-x^2}}\, dx;$ (36) $\displaystyle\int \frac{1}{(\sin x + \cos x)^2}\, dx;$

(37) $\displaystyle\int \frac{\arcsin \sqrt{x}}{\sqrt{1-x}}\, dx;$ (38) $\displaystyle\int \frac{\sin 2x}{\sqrt{1+\cos^4 x}}\, dx;$

(39) $\displaystyle\int \frac{x^2 \arctan x}{1+x^2}\, dx;$ (40) $\displaystyle\int \frac{1-x^2}{1+x^4}\, dx.$

2.2.2 Definite Integrals

Evaluating definite integrals becomes easier when we know how to find the antiderivatives of the integrand functions, since the value of a definite integral is the difference of the antiderivative function values at the upper limit and the lower limit.

Corresponding to the rules for indefinite integrals

$$\int [f(x) + g(x)] \, dx = \int f(x) \, dx + \int g(x) \, dx,$$

$$\int Cg(x) \, dx = \int Cg(x) \, dx,$$

where C is a constance, the rules for definite integrals are

$$\int_a^b [f(x) + g(x)] \, dx = \int_a^b f(x) \, dx + \int_a^b g(x) \, dx,$$

$$\int_a^b Cg(x) \, dx = \int_a^b Cg(x) \, dx.$$

We verified these two equations in Section 2 of Chapter 1. So it is also reasonable to consider the rules for indefinite integrals as consequence of the corresponding rules for definite integrals by varying the upper limits of the definite integrals.

The two methods of integration we introduced earlier, integration by substitution and integration by parts, in definite integrals have no major differences with the corresponding formulas for indefinite integrals.

1. *Integration by Substitution*: Suppose $f(x)$ is a continuous function on a closed interval $[a, b]$, and $\varphi(t)$ is a differentiable function on $[\alpha, \beta]$ that satisfies the following conditions (i) $\varphi'(t)$ is a continuous function; (ii) all values of $\varphi(t)$ are in the interval $[a, b]$; (iii) $\varphi(\alpha) = a$, $\varphi(\beta) = b$. Then

$$\int_a^b f(x) \, dx = \int_\alpha^\beta f[\varphi(t)] \varphi'(t) \, dt.$$

This is called the formula of *integration by substitution for definite integrals*. In fact, if $F(x)$ is an antiderivative of $f(x)$ on $[a, b]$, then by the Fundamental Theorem of Calculus (integral form), we have

$$\int_a^b f(x) \, dx = F(b) - F(a).$$

Applying the rule of derivatives of composite functions, we have

$$\frac{d}{dt} F[\varphi(t)] = \frac{d}{dx} F(x) \frac{dx}{dt} = f(x) \varphi'(t) = f[\varphi(t)] \varphi'(t).$$

Thus, the function $F[\varphi(t)]$ is an antiderivative of $f[\varphi(t)] \varphi'(t)$ on $[\alpha, \beta]$. It follows that

$$\int_\alpha^\beta f[\varphi(t)] \varphi'(t) \, dt = F[\varphi(t)] \Big|_\alpha^\beta = F[\varphi(\beta)] - F[\varphi(\alpha)] = F(b) - F(a).$$

(The notation $H(x)|_\eta^\xi$ represents $H(\xi) - H(\eta)$.)

Example 1. Evaluate the definite integral $\int_0^1 \sqrt{1 - x^2}\, dx$.

Solution: Let $x = \sin t$. Then when $x = 0$, $t = 0$ since $\sin 0 = 0$, and when $x = 1$, $t = \dfrac{\pi}{2}$ since $\sin \dfrac{\pi}{2} = 1$. Thus

$$\int_0^1 \sqrt{1 - x^2}\, dx = \int_0^{\frac{\pi}{2}} \sqrt{1 - \sin^2 t}\, d\sin t$$

$$= \int_0^{\frac{\pi}{2}} \cos^2 t\, dt = \frac{1}{2} \cdot \frac{\pi}{2} = \frac{\pi}{4}.$$

Example 2. Evaluate the definite integral $\int_0^a \dfrac{1}{(x^2 + a^2)^{\frac{3}{2}}}\, dx$, $(a > 0)$.

Solution: Let $x = a \tan t$. Then when $x = 0$, $t = 0$ since $\tan 0 = 0$, and when $x = a$, $t = \dfrac{\pi}{4}$ since $\tan \dfrac{\pi}{4} = 1$. And $dx = a \sec^2 t\, dt$. Thus

$$\int_0^a \frac{1}{(x^2 + a^2)^{\frac{3}{2}}}\, dx = \int_0^{\frac{\pi}{4}} \frac{a \sec^2 t}{a^3 \sec^3 t}\, dt = \frac{1}{a^2} \int_0^{\frac{\pi}{4}} \cos t\, dt = \frac{\sqrt{2}}{2a^2}.$$

Example 3. Evaluate the definite integral $\int_0^1 \dfrac{\ln(1 + x)}{1 + x^2}\, dx$.

Solution: Let $x = \dfrac{1 - t}{1 + t}$. Then when $x = 0$, $t = 1$, and when $x = 1$, $t = 0$. Thus

$$I = -\int_1^0 \frac{\ln 2 - \ln(1 + t)}{1 + t^2}\, dt = \int_0^1 \frac{\ln 2}{1 + t^2}\, dt - \int_0^1 \frac{\ln(1 + t)}{1 + t^2}\, dt$$

$$= \frac{\pi}{4} \ln 2 - I.$$

Solving for I we get

$$I = \frac{\pi}{8} \ln 2.$$

Example 4. If $f(x)$ is an odd function, i.e., $f(-x) = -f(x)$, then

$$\int_{-l}^l f(x)\, dx = 0.$$

If $f(x)$ is an even function, i.e., $f(-x) = f(x)$, then

$$\int_{-l}^l f(x)\, dx = 2 \int_0^l f(x)\, dx.$$

In fact, let $x = -t$, then

$$\int_{-l}^{0} f(x)\,dx = -\int_{l}^{0} f(-t)\,dt = \int_{0}^{l} f(-t)\,dt.$$

The equation

$$\int_{-l}^{0} f(x)\,dx = -\int_{0}^{l} f(t)\,dt$$

holds if $f(x)$ is an odd function. The equation

$$\int_{-l}^{0} f(x)\,dx = \int_{0}^{l} f(t)\,dt$$

holds if $f(x)$ is an even function.

Since

$$\int_{-l}^{l} f(x)\,dx = \int_{-l}^{0} f(x)\,dx + \int_{0}^{l} f(x)\,dx,$$

the conclusion follows.

2. *Integration by Parts*: If $u'(x)$ and $v'(x)$ are continuous functions on $[a, b]$, then

$$\int_{a}^{b} u(x)v'(x)\,dx = u(x)v(x)\Big|_{a}^{b} - \int_{a}^{b} v(x)u'(x)\,dx$$

or

$$\int_{a}^{b} u(x)\,dv(x) = u(x)v(x)\Big|_{a}^{b} - \int_{a}^{b} v(x)\,du(x).$$

This is called the formula of *integration by parts for definite integrals*. In fact, since

$$\left[\int u(x)v'(x)\,dx\right]' = u(x)v'(x)$$

we have

$$\int_{a}^{b} u(x)v'(x)\,dx = \left[\int u(x)v'(x)\,dx\right]\Big|_{a}^{b}$$

$$= \left[u(x)v(x) - \int v(x)u'(x)\,dx\right]\Big|_{a}^{b}$$

$$= u(x)v(x)\Big|_{a}^{b} - \int_{a}^{b} v(x)u'(x)\,dx.$$

Example 5. Evaluate the definite integral $\displaystyle\int_{0}^{\sqrt{3}} x \arctan x\,dx.$

Solution:

$$\int_0^{\sqrt{3}} x \arctan x \, dx = \frac{1}{2} \int_0^{\sqrt{3}} \arctan x \, d(x^2)$$

$$= \frac{1}{2} x^2 \arctan x \Big|_0^{\sqrt{3}} - \frac{1}{2} \int_0^{\sqrt{3}} x^2 \frac{1}{1+x^2} \, dx$$

$$= \frac{2\pi}{3} - \frac{\sqrt{3}}{2}.$$

Example 6. Evaluate the definite integrals

$$J_m = \int_0^{\frac{\pi}{2}} \sin^m x \, dx,$$

and

$$J'_m = \int_0^{\frac{\pi}{2}} \cos^m x \, dx,$$

where m is a positive integer.

Solution:

$$J_m = \int_0^{\frac{\pi}{2}} \sin^m x \, dx = \int_0^{\frac{\pi}{2}} \sin^{m-1} x \, d(-\cos x)$$

$$= -\sin^{m-1} x \cos x \Big|_0^{\frac{\pi}{2}} + (m-1) \int_0^{\frac{\pi}{2}} \cos^2 x \sin^{m-2} x \, dx$$

$$= (m-1) \int_0^{\frac{\pi}{2}} \sin^{m-2} x (1 - \sin^2 x) \, dx$$

$$= (m-1) \int_0^{\frac{\pi}{2}} \sin^{m-2} x \, dx - (m-1) \int_0^{\frac{\pi}{2}} \sin^m x \, dx.$$

Thus

$$J_m = (m-1) J_{m-2} - (m-1) J_m.$$

Therefore, we have the induction formula

$$J_m = \frac{m-1}{m} J_{m-2}.$$

Since

$$J_0 = \int_0^{\frac{\pi}{2}} dx = \frac{\pi}{2},$$

$$J_1 = \int_0^{\frac{\pi}{2}} \sin x \, dx = 1,$$

the general formulas of calculating J_m are

$$J_{2n} = \int_0^{\frac{\pi}{2}} \sin^{2n} x \, dx = \frac{(2n-1)(2n-3)\cdots 3 \cdot 1}{2n(2n-2)\cdots 4 \cdot 2} \cdot \frac{\pi}{2}$$

$$= \frac{(2n-1)!!}{(2n)!!} \cdot \frac{\pi}{2},$$

$$J_{2n+1} = \int_0^{\frac{\pi}{2}} \sin^{2n+1} x \, dx = \frac{2n(2n-2)\cdots 4 \cdot 2}{(2n+1)(2n-1)\cdots 3 \cdot 1}$$

$$= \frac{(2n)!!}{(2n+1)!!}.$$

The same result can be obtained for J_m'.

Summarizing the above conclusions, we get

$$\int_0^{\frac{\pi}{2}} \sin^m x \, dx = \int_0^{\frac{\pi}{2}} \cos^m x \, dx = \begin{cases} \dfrac{(m-1)!!}{m!!} \dfrac{\pi}{2}, & \text{if } m \text{ is even;} \\ \dfrac{(m-1)!!}{m!!}, & \text{if } m \text{ is odd.} \end{cases}$$

Example 7. Suppose $f''(x)$ is a continuous function. Then

$$\int_0^1 f(x) \, dx = \frac{f(0)+f(1)}{2} - \frac{1}{2}\int_0^1 x(1-x)f''(x) \, dx.$$

Solution:

$$\int_0^1 x(1-x)f''(x) \, dx = x(1-x)f'(x)\Big|_0^1 - \int_0^1 (1-2x)f'(x) \, dx$$

$$= \int_0^1 (2x-1)f'(x) \, dx = (2x-1)f(x)\Big|_0^1 - 2\int_0^1 f(x) \, dx$$

$$= f(1) + f(0) - 2\int_0^1 f(x) \, dx.$$

Solve for $\int_0^1 f(x) \, dx$ and the result follows.

Exercises 2.2.2

1. Evaluate the following definite integrals.

(1) $\displaystyle\int_1^e \frac{\ln x}{x} \, dx;$

(2) $\displaystyle\int_{\frac{1}{\pi}}^{\frac{2}{\pi}} \frac{\sin \frac{1}{x}}{x^2} \, dx;$

(3) $\displaystyle\int_0^{\ln 2} x e^{-x}\, dx$;

(4) $\displaystyle\int_0^{\pi} x^2 \sin^2 x\, dx$;

(5) $\displaystyle\int_2^3 \frac{1}{2x^2 + 3x - 2}\, dx$;

(7) $\displaystyle\int_0^4 \frac{1}{1 + \sqrt{x}}\, dx$;

(6) $\displaystyle\int_0^2 |1 - x|\, dx$;

(8) $\displaystyle\int_0^{\ln 2} \sqrt{e^x - 1}\, dx$;

(9) $\displaystyle\int_0^{\frac{\pi}{2}} \sin^3 x \cos^3 x\, dx$;

(11) $\displaystyle\int_0^{\frac{\pi}{2}} e^{2x} \sin x\, dx$;

(10) $\displaystyle\int_0^{\frac{\pi}{4}} \frac{1}{1 + a^2 \cos^2 x}\, dx$;

(12) $\displaystyle\int_0^1 \frac{\sqrt{e^x}}{\sqrt{e^x} + e^{-x}}\, dx$;

(13) $\displaystyle\int_0^a \ln(t + \sqrt{t^2 + a^2})\, dt,\ (a > 0)$;

(14) $\displaystyle\int_0^a \frac{x^2}{\sqrt{x^2 + a^2}}\, dx,\ (a > 0)$;

(16) $\displaystyle\int_0^{\frac{2\pi}{3}} \frac{1}{5 + 4\cos\theta}\, d\theta$;

(15) $\displaystyle\int_1^e \ln^3 x\, dx$;

(17) $\displaystyle\int_0^{\frac{\pi}{2}} \sin x \sin 2x \sin 3x\, dx$;

(18) $\displaystyle\int_{\pi}^{2\pi} \sqrt{1 + \sin x}\, dx$;

(19) $\displaystyle\int_{-1}^1 \frac{1}{(1 + x^2)^2}\, dx$;

(20) $\displaystyle\int_{-\frac{1}{2}}^{\frac{1}{2}} (1 - 3x^2) \ln \frac{1 + x}{1 - x}\, dx$.

2. Show that the following equations hold if m, n are positive integers.

(1) $\displaystyle\int_0^{2\pi} \sin mx \cos nx\, dx = 0$;

(2) $\displaystyle\int_0^{2\pi} \cos mx \cos nx\, dx = \int_0^{2\pi} \sin mx \sin nx\, dx$

$$= \begin{cases} 0, & \text{if } m \neq n; \\ \pi, & \text{if } m = n. \end{cases}$$

3. Show that the following equations hold if n is a positive integer.

(1) $\displaystyle\int_0^{\pi} \sin^n x\, dx = 2 \int_0^{\frac{\pi}{2}} \sin^n x\, dx$;

(2) $\displaystyle\int_0^{\pi} \cos^{2n} x\, dx = 2 \int_0^{\frac{\pi}{2}} \cos^{2n} x\, dx$.

4. Evaluate the following definite integrals by using the induction formulas

for $\displaystyle I_n = \int_0^{\frac{\pi}{2}} \sin^n x\, dx = \int_0^{\frac{\pi}{2}} \cos^n x\, dx$.

(1) $\displaystyle\int_0^{\frac{\pi}{2}} \sin^5 x\,dx$;

(2) $\displaystyle\int_0^{\pi} \cos^8 x\,dx$;

(3) $\displaystyle\int_0^{\pi} \sin^{11} x\,dx$;

(4) $\displaystyle\int_0^{\frac{\pi}{4}} \cos^7 2x\,dx$;

(5) $\displaystyle\int_0^{\pi} \sin^6 \frac{x}{2}\,dx$;

(6) $\displaystyle\int_0^1 \sqrt{(1-x^2)^3}\,dx$;

(7) $\displaystyle\int_0^1 (1-x^2)^n\,dx$;

(8) $\displaystyle\int_0^a x^2 \sqrt{a^2-x^2}\,dx$.

5. Suppose $f(x)$ is a periodical function with period T. Show that
$$\int_a^{a+T} f(x)\,dx = \int_0^T f(x)\,dx.$$
Explain this equation geometrically.

6. Show that $\displaystyle\int_0^x f(x)\,dx$ is an even function if $f(x)$ is an odd function. Is $\displaystyle\int_0^x f(x)\,dx$ an odd function if $f(x)$ is an even function?

7. Suppose
$$f(x) = a_1 \cos x + b_1 \sin x + a_2 \cos 2x + b_2 \sin 2x,$$
where a_1, b_1, a_2, b_2 are constants. Show that
$$\int_{-\pi}^{\pi} f^2(x)\,dx = \pi(a_1^2 + b_1^2 + a_2^2 + b_2^2).$$

8. Suppose
$$f(x) = a_1 \cos x + b_1 \sin x + a_2 \cos 2x + b_2 \sin 2x,$$
$$g(x) = c_1 \cos x + d_1 \sin x + c_2 \cos 2x + d_2 \sin 2x,$$
where $a_1, b_1, a_2, b_2, c_1, d_1, c_2, d_2$ are constants. Show that
$$\left[\int_{-\pi}^{\pi} f(x)g(x)\,dx\right]^2 \leqslant \int_{-\pi}^{\pi} f^2(x)\,dx \cdot \int_{-\pi}^{\pi} g^2(x)\,dx.$$

2.2.3 *Approximate Calculations of Definite Integrals*

Not all antiderivatives can be expressed as elementary functions. In such cases, sometimes it makes sense to calculate an approximate value of the definite integral. Next, we study some methods to estimate the values of definite integrals.

1. *Trapezoidal Method*: In Section 2 of Chapter 1, we studied that, geometrically, the value of integral

$$I = \int_a^b f(x)\,dx$$

is the area of the region bounded by the curve $y = f(x)$, and the straight lines $x = a$, $x = b$ and $y = 0$ as in Figure 2.5.

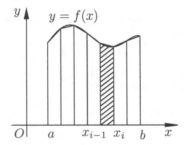

Fig. 2.5

Partition the interval $[a, b]$ into n equal sections, $a = x_0 < x_1 < \cdots < x_{i-1} < x_i < \cdots x_n = b$. Denote $y_i = f(x_i)$. Then the line segment that connects $(x_i, 0)$ and (x_i, y_i) is parallel to y-axis for $i = 0, 1, \cdots, n$. When the number of sections n in the partition is large enough, each part of the curve $f(x)$ between two line segments described above can be approximated by a straight line with end points (x_{i-1}, y_{i-1}) and (x_i, y_i), $i = 0, 1, \cdots, n$. Thus, the area of the shaded quadrilateral in Figure 2.5 is equal to

$$\Delta \bar{I}_i = \frac{f(x_i) + f(x_{i-1})}{2}(x_i - x_{i-1})$$
$$= \frac{b - a}{n}\frac{y_i + y_{i-1}}{2}.$$

The sum for all quadrilaterals under the curve $f(x)$ is

$$\bar{I} = \sum_{i=1}^n \frac{b - a}{n}\frac{y_i + y_{i-1}}{2}$$
$$= \frac{b - a}{n}\left(\frac{y_0}{2} + y_1 + y_2 + \cdots + y_{n-1} + \frac{y_n}{2}\right).$$

This is an approximation of the value of the integral. The approximation

$$\int_a^b f(x)\, dx \approx \frac{b-a}{n}\left(\frac{y_0}{2} + y_1 + y_2 + \cdots + y_{n-1} + \frac{y_n}{2}\right)$$

is called the *trapezoidal rule*.

To estimate the error of the value of the integral by using this method, consider the integral of $f(x)$ on the ith interval

$$\Delta I_i = \int_{x_{i-1}}^{x_i} f(x)\, dx.$$

Let $x = x_{i-1} + t\Delta x_i$. Then by Example 7 in Section 2.2.3 we have

$$\Delta I_i = \Delta x_i \int_0^1 f(x_{i-1} + t\Delta x_i)\, dt$$

$$= \frac{\Delta x_i}{2}[f(x_{i-1}) + f(x_i)] - \frac{(\Delta x_i)^3}{2}\int_0^1 t(1-t)f''(x_{i-1} + t\Delta x_i)\, dt.$$

Thus, if $|f''(x)| \leqslant M_2$ on the interval $[a,b]$, then

$$|\Delta I_i - \Delta \bar{I}_i| = \frac{(\Delta x_i)^3}{2}\left|\int_0^1 t(1-t)f''(x_{i-1} + t\Delta x_i)\, dt\right|$$

$$\leqslant \frac{(\Delta x_i)^3}{2}\int_0^1 t(1-t)|f''(x_{i-1} + t\Delta x_i)|\, dt$$

$$\leqslant \frac{(\Delta x_i)^3}{2}M_2 \int_0^1 t(1-t)\, dt \leqslant \frac{(b-a)^3 M_2}{12n^3}.$$

Therefore

$$|I - \bar{I}| \leqslant \sum_{i=1}^n |\Delta I_i - \Delta \bar{I}_i| \leqslant \frac{(b-a)^3}{12n^2}M_2.$$

This implies that the error is an infinitesimal of order not lower than $\frac{1}{n^2}$.

Example 1. Find the approximate value of $\ln 5$ by the trapezoidal method.

Solution: We calculate the approximate value of the integral $\int_1^5 \frac{1}{x}\, dx$. Partition the interval $[1,5]$ into 8 sections. The length of each section is $\frac{b-a}{n} = 0.5$. The points of division are

$$x = 1,\ 1.5,\ 2,\ 2.5,\ 3,\ 3.5,\ 4,\ 4.5,\ 5.$$

The corresponding y values on $y = \frac{1}{x}$ are

$$y = 1,\ \frac{2}{3},\ \frac{1}{2},\ \frac{2}{5},\ \frac{1}{3},\ \frac{2}{7},\ \frac{1}{4},\ \frac{2}{9},\ \frac{1}{5}.$$

By trapezoidal rule, we have

$$\ln 5 = \int_1^5 \frac{1}{x}\, dx$$

$$\approx 0.5 \left(\frac{1 + \dfrac{1}{5}}{2} + \frac{2}{3} + \frac{1}{2} + \frac{2}{5} + \frac{1}{3} + \frac{2}{7} + \frac{1}{4} + \frac{2}{9} \right) = 1.63.$$

Since $f''(x) = \dfrac{2}{x^3}$, the maximum value of $|f''(x)|$ on $[1,5]$ is $M_2 = 2$. Therefore the error does not exceed

$$\frac{(b-a)^3}{12n^2} M_2 = \frac{4^3}{12 \cdot 8^2} \cdot 2 = \frac{1}{6} \approx 0.166.$$

2. *Parabolic Method*: To get a more accurate value of the approximation, consider substituting $f(x)$ by a second degree polynomial instead of a straight line on the small interval $[x_{i-1}, x_i]$.

Partition the interval $[a, b]$ into $2n$ equal sections, and let $y_k = f(x_k)$. The interval $[x_{2k-2}, x_{2k}]$ consists of two small sections (x_{2k-2}, x_{2k-1}) and (x_{2k-1}, x_{2k}). The curve $f(x)$ can be approximated by the parabola $y = \alpha x^2 + \beta x + \gamma$ passing through the three points M_{2k-2}, M_{2k-1} and M_{2k} as in Figure 2.6. (The coefficients α, β and γ can be easily determined from the given conditions.)

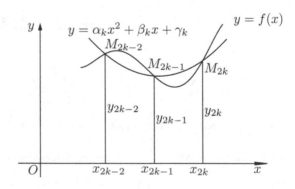

Fig. 2.6

The values of integral of $f(x)$ on every interval $[x_{2k-2}, x_{2k}]$ ($k = 1, 2, \cdots, n$) are approximated by the area under the parabola on the same interval:

$$\int_{x_{2k-2}}^{x_{2k}} f(x)\, dx \approx \int_{x_{2k-2}}^{x_{2k}} (\alpha x^2 + \beta x + \gamma)\, dx = \bar{I}_k.$$

We can transform the coordinate system so that $x_{2k-1} = 0$. Denote the width of each section of the partition as $\dfrac{b-a}{2n} = h$, then $x_{2k-2} = -h$ and $x_{2k} = h$. Thus,

$$y_{2k-2} = \alpha h^2 - \beta h + \gamma, \quad y_{2k-1} = \gamma,$$
$$y_{2k} = \alpha h^2 + \beta h + \gamma.$$

The area under the curve remains the same under this coordinate transformation, and it becomes

$$\bar{I}_k = \int_{-h}^{h} (\alpha x^2 + \beta x + \gamma)\,dx = \frac{h}{3}(2\alpha h^2 + 6\gamma)$$
$$= \frac{h}{3}[(\alpha h^2 + \beta h + \gamma) + 4\gamma + (\alpha h^2 - \beta h + \gamma)]$$
$$= \frac{b-a}{6n}(y_{2k} + 4y_{2k-1} + y_{2k-2}).$$

Thus

$$\int_a^b f(x)\,dx = \sum_{k=1}^{n} \int_{x_{2k-2}}^{x_{2k}} f(x)\,dx$$
$$\approx \frac{b-a}{6n} \sum_{k=1}^{n}(y_{2k} + 4y_{2k-1} + y_{2k-2})$$
$$= \frac{b-a}{6n}\left(y_0 + y_{2n} + 4\sum_{k=1}^{n} y_{2k-1} + 2\sum_{k=1}^{n-1} y_{2k}\right).$$

This is called parabolic formula (Simpson's rule), and it can be shown that the error of the value of approximation will not exceed

$$\frac{(b-a)^5}{180(2n)^4} M_4$$

where M_4 is the maximum value of $|f^{(4)}(x)|$ on $[a, b]$.

The error is an infinitesimal of the same order as $\dfrac{1}{n^4}$. This implies that the approximation of the integral by the parabolic formula is more accurate than the approximation of the integral by the trapezoidal rule.

Example 2. Evaluate the integral $\displaystyle\int_0^1 e^{-x^2}\,dx$ to the accuracy of 0.00001 using the parabolic formula.

Solution: The absolute value of the fourth derivative of e^{-x^2} does not exceed 12. Its error does not exceed $\dfrac{12}{180(2n)^4}$. The value of $\dfrac{12}{180(2n)^4}$ is less than 0.00001 for $n = 5$. The terms of the parabolic formula is calculated as follows.

$$x_0 = 0, \quad y_0 = 1.00000,$$
$$x_{10} = 1, \quad y_{10} = 0.36788.$$

The term is 1.36788.

$$x_2 = 0.2, \quad y_2 = 0.96079,$$
$$x_4 = 0.4, \quad y_4 = 0.85214,$$
$$x_6 = 0.6, \quad y_6 = 0.69768,$$
$$x_8 = 0.8, \quad y_8 = 0.52729.$$

The term is $(3.03790)(2) = 6.07580$.

$$x_1 = 0.1, y_1 = 0.99005,$$
$$x_3 = 0.3, y_3 = 0.91393,$$
$$x_5 = 0.5, y_5 = 0.77880,$$
$$x_7 = 0.7, y_7 = 0.61263,$$
$$x_9 = 0.9, y_9 = 0.44486.$$

The term is $(3.74027)(4) = 14.96108$.
Therefore,

$$\int_0^1 e^{-x^2}\, dx \approx \frac{1}{30}(1.36788 + 6.07580 + 14.96108) = 0.746825.$$

Example 3. The width of a river is 20m. The following list of numbers give the depth of the river in every 2m from one bank to the other:

$$0.2, \ 0.5, \ 0.9, \ 1.1, \ 1.3, \ 1.7, \ 2.1, \ 1.5, \ 1.1, \ 0.6, \ 0.2.$$

Find the cross-sectional area A of the river.

Solution: Suppose the curve of the cross-sectional area of the river bottom is $y = f(x)$. From the statement of the problem, we have $A = \int_0^{20} f(x)\, dx$ and

$$y_0 = 0.2, \ y_1 = 0.5, \ y_2 = 0.9, \ y_3 = 1.1, \ y_4 = 1.3, \ y_5 = 1.7,$$
$$y_6 = 2.1, \ y_7 = 1.5, \ y_8 = 1.1, \ y_9 = 0.6, \ y_{10} = 0.2.$$

By parabolic formula, we have

$$A = \int_0^{20} f(x)\, dx \approx \frac{20}{30}[0.2 + 0.2 + 2(0.9 + 1.3 + 2.1 + 1.1)$$
$$+ 4(0.5 + 1.1 + 1.7 + 1.5 + 0.6)] = 21.9m^2.$$

Exercises 2.2.3

1. Evaluate the value of the approximation of the integral

$$\int_0^1 \frac{1}{1+x^2}\,dx$$

by trapezoidal rule and parabolic formula. Partition the interval $[0, 1]$ into 10 equal sections. The error will not exceed 10^{-3}.

2. Evaluate the value of the approximation of the integral

$$\int_0^{\frac{\pi}{2}} \frac{\sin x}{x}\,dx$$

by parabolic formula. Partition the interval $[0, \frac{\pi}{2}]$ into 10 equal sections.

Chapter 3

Some Applications of Differentiation and Integration

3.1 Areas, Volumes, Arc Lengths

3.1.1 *Areas*

We learned in Chapter 1 that the area bounded by the curve $y = f(x)(\geqslant 0)$ and the lines $x = a$, $x = b$, $y = 0$ is (Figure 3.1)

$$A = \int_a^b f(x)\,dx = \int_a^b y\,dx. \tag{1}$$

If $y = f(x)$ changes signs on $[a, b]$ (Figure 3.2), then

$$A = \int_a^b |f(x)|\,dx. \tag{2}$$

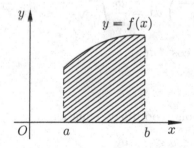

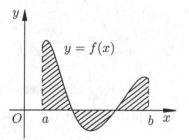

Fig. 3.1   Fig. 3.2

If the area is bounded by closed curves (Figure 3.3), then

$$A = \int_a^b |f_1(x) - f_2(x)|\,dx. \tag{3}$$

101

Example 1. Find the area bounded by $y = x^2$ and $y = \sqrt{x}$ (Figure 3.4).

Solution: Since the intersect points of two curves are $(0,0)$ and $(1,1)$, by equation (3) we have

$$A = \int_0^1 (\sqrt{x} - x^2)\, dx = \left(\frac{2}{3}x^{\frac{3}{2}} - \frac{1}{3}x^3\right)\Big|_0^1 = \frac{1}{3}.$$

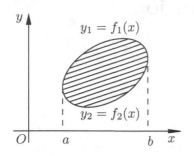

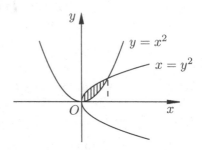

Fig. 3.3   Fig. 3.4

Example 2. Find the area inside the ellipse $\dfrac{x^2}{a^2} + \dfrac{y^2}{b^2} = 1$ (Figure 3.5).

Solution: The area A is

$$A = \int_{-a}^a \left(\frac{b}{a}\sqrt{a^2 - x^2} - \frac{-b}{a}\sqrt{a^2 - x^2}\right) dx = \frac{2b}{a}\int_{-a}^a \sqrt{a^2 - x^2}\, dx$$

$$= \frac{4b}{a}\int_0^a \sqrt{a^2 - x^2}\, dx = 4ab \int_0^{\frac{\pi}{2}} \cos^2 t\, dt$$

$$= 4ab \cdot \frac{1}{2} \cdot \frac{\pi}{2} = \pi ab.$$

Suppose a curve is expressed by a polar coordinate equation

$$\rho = f(\theta), \quad \alpha \leqslant \theta \leqslant \beta.$$

We want to evaluate the area bounded by the curve $\rho = f(\theta)$ and the two rays $\theta = \alpha$, $\theta = \beta$. The area of a sector with radius r and central angle θ is $\dfrac{1}{2}r^2\theta$.

Let

$$\alpha = \theta_0 < \theta_1 < \cdots < \theta_n = \beta$$

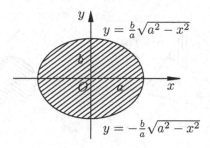

Fig. 3.5

be a partition of the angle between α and β ($\alpha \leqslant \theta \leqslant \beta$) (Figure 3.6), then the area of the sector bounded by $\rho = f(x)$ and rays $\theta = \theta_{i-1}$, $\theta = \theta_i$ is approximately equal to

$$\frac{1}{2}f^2(\xi_i)\Delta\theta_i, \quad (\theta_{i-1} \leqslant \xi_i \leqslant \theta_i).$$

Therefore, the total area is approximately equal to

$$\sum_{i=1}^{n}\frac{1}{2}f^2(\xi_i)\Delta\theta_i.$$

Let $d = \max\limits_{1\leqslant i\leqslant n}|\Delta\theta_i|$, then

$$A = \lim_{d\to 0}\sum_{i=1}^{n}\frac{1}{2}f^2(\xi_i)\Delta\theta_i = \frac{1}{2}\int_{\alpha}^{\beta}f^2(\theta)d\theta.$$

Example 3. Find the area inside of the Lemniscate $\rho^2 = a^2\cos 2\theta$ (Figure 3.7).

Solution: Since the graph of the curve is symmetrical in both x-axis and y-axis, we have

$$A = 4\cdot\frac{1}{2}\int_{0}^{\frac{\pi}{4}}\rho^2 d\theta = 2\int_{0}^{\frac{\pi}{4}}a^2\cos 2\theta d\theta = 2a^2\left(\frac{1}{2}\sin 2\theta\right)\Big|_{0}^{\frac{\pi}{4}} = a^2.$$

3.1.2   Volumes

Suppose a solid is bounded between two parallel planes perpendicular to the x-axis at $x = a$ and $x = b$, $(a < b)$. Suppose $S(x)$ is the area of the cross section of the solid at $x(a \leqslant x \leqslant b)$. (Figure 3.8)

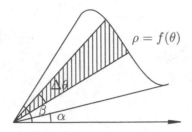

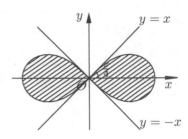

Fig. 3.6   Fig. 3.7

To find the volume of the solid, make a partition of $[a, b]$, $a = x_0 < x_1 < \cdots < x_n = b$. The planes $x = x_i$, which are perpendicular to the x-axis, cut the solid into thin slices of thickness of $\Delta x_i = x_i - x_{i-1}$.

Let ξ_i be a point between x_{i-1} and x_i. Then the volume of the part of the solid between planes $x = x_{i-1}$ and $x = x_i$ is approximately equal to

$$\Delta V_i = S(\xi_i)\Delta x_i.$$

Therefore, the volume of the solid is

$$V = \int_a^b S(x)\, dx.$$

Consider a solid generated by revolving the curve $y = f(x)$, $(a \leqslant x \leqslant b)$, about the x-axis (Figure 3.9). The cross section of this solid by any plane perpendicular to the x-axis at x, $(a \leqslant x \leqslant b)$, is a disc with radius y. Thus, the area of this disc is $S(x) = \pi y^2$.

Therefore, the volume of this solid of revolution is

$$V = \pi \int_a^b y^2\, dx = \pi \int_a^b (f(x))^2\, dx.$$

Example 1. Find the volume of solid obtained by rotating the ellipse

$$\frac{x^2}{a^2} + \frac{y^2}{b^2} = 1$$

about the x-axis.

Solution: By the equation above, the volume is

$$V = \pi \int_{-a}^a \frac{b^2}{a^2}(a^2 - x^2)\, dx = 2\pi \int_0^a \frac{b^2}{a^2}(a^2 - x^2)\, dx = \frac{4}{3}\pi a b^2.$$

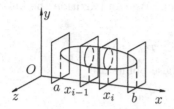

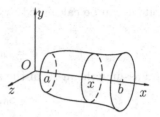

Fig. 3.8 Fig. 3.9

Example 2. A circular cylinder with radius a is cut by a plane having an angle α ($\alpha < \frac{\pi}{2}$) with the base and passing through a diameter AB. Find the volume of the part of the cylinder between the cutting plane and the base (Figure 3.10).

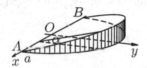

Fig. 3.10

Solution: Let the diameter AB be on the x-axis and the center of the base of the circular cylinder be at the origin. Then all the sections of the solid by planes perpendicular to the x-axis are right triangles with an acute angle α, and the leg adjacent to α is of the length $\sqrt{a^2 - x^2}$. Thus, the area of the section can be expressed as

$$S(x) = \frac{1}{2}\sqrt{a^2 - x^2}\sqrt{a^2 - x^2}\tan\alpha = \frac{1}{2}(a^2 - x^2)\tan\alpha.$$

Therefore, the volume is

$$V = \int_{-a}^{a} \frac{1}{2}(a^2 - x^2)\tan\alpha\, dx = \frac{2}{3}a^3\tan\alpha.$$

If the function of the area of sections $S(x)$ is not known exactly, then the volume can be calculated by approximation. We can partition the interval $[a, b]$ as

$$a = x_0, \ x_1, \ x_2, \ \cdots, \ x_{2n} = b$$

and assume the areas of the sections at these points are

$$S_0, \ S_1, \ S_2, \ \cdots, \ S_{2n}.$$

Then by the parabolic formula, the volume is approximately equal to

$$\frac{b-a}{6n} \left(S_0 + S_{2n} + 4 \sum_{k=1}^{n} S_{2k-1} + 2 \sum_{k=1}^{n-1} S_{2k} \right).$$

3.1.3 Arc Lengths

We learned how to find the length of a line segment and the length of an arc of a circle in elementary geometry. Now we study how to find the length of a curve.

Suppose a curve is expressed by the parametric equations

$$x = \varphi(t), \quad y = \psi(t),$$

where $\varphi(t)$ and $\psi(t)$ are functions on $[\alpha, \beta]$ with $\varphi'(t)$ and $\psi'(t)$ continuous (Figure 3.11).

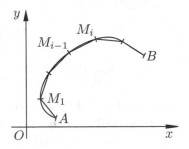

Fig. 3.11

Partition the curve from one end A to the other end B by

$$A = M_0, \ M_1, \ \cdots, \ M_{i-1}, \ M_i, \ \cdots, \ M_n = B.$$

Connect each pair of adjacent points by a line segment. We get an inscribed broken line from A to B with a total length approximately equal to the length of the curve. The finer the partition, the closer the two lengths. We define the length of the curve to be the limit of the length of the broken line as the partition points become infinitely dense.

Suppose the parametric coordinates of the point M_i on the curve is $(\varphi(t_i), \psi(t_i))$, and $t_0 = \alpha$, $t_n = \beta$. Then the length of the line segment $M_{i-1}M_i$ is equal to

$$\sqrt{[\varphi(t_i) - \varphi(t_{i-1})]^2 + [\psi(t_i) - \psi(t_{i-1})]^2}.$$

Thus, the length of the entire inscribed broken line is

$$\sum_{i=1}^{n} \sqrt{[\varphi(t_i) - \varphi(t_{i-1})]^2 + [\psi(t_i) - \psi(t_{i-1})]^2}.$$

Since

$$\varphi(t_i) - \varphi(t_{i-1}) \approx \varphi'(t_{i-1})\Delta t_i,$$
$$\psi(t_i) - \psi(t_{i-1}) \approx \psi'(t_{i-1})\Delta t_i,$$

the length of the curve s is approximately equal to

$$\sum_{i=1}^{n} \sqrt{[\varphi'(t_{i-1})]^2 + [\psi'(t_{i-1})]^2}\Delta t_i.$$

Let $\lambda = \max\limits_{1 \leqslant i \leqslant n} |\Delta t_i|$. Then the length of the curve s is

$$s = \lim_{\lambda \to 0} \sum_{i=1}^{n} \sqrt{[\varphi'(t_{i-1})]^2 + [\psi'(t_{i-1})]^2}\Delta t_i.$$

By the continuity of $\varphi'(t)$ and $\psi'(t)$,

$$s = \int_{\alpha}^{\beta} \sqrt{[\varphi'(t)]^2 + [\psi'(t)]^2}\, dt. \tag{4}$$

For any point $M = (\varphi(t), \psi(t))$ on the curve, the length of the arc from A to M is

$$s(t) = \int_{\alpha}^{t} \sqrt{[\varphi'(t)]^2 + [\psi'(t)]^2}\, dt.$$

Taking the derivative of $s(t)$ with respect to t, we have

$$\frac{ds}{dt} = \sqrt{[\varphi'(t)]^2 + [\psi'(t)]^2}, \tag{5}$$

and

$$ds = \sqrt{[\varphi'(t)]^2 + [\psi'(t)]^2}\, dt, \tag{6}$$

or,

$$ds^2 = dx^2 + dy^2.$$

There is a geometric explanation of this equation.

Let $M(x, y)$ be an arbitrary point on the curve and $M'(x + \Delta x, y + \Delta y)$ be a point nearby, $\Delta s = \widehat{MM'}$ be the length of the arc from M to M'. Draw a tangent line MT to the curve at M. As we can see from Figure 3.12, $NP = dy$ is the rise of the slope of MT and $MN = \Delta x = dx$ is the run.

Equation (6) states that the differential of the length of arc is equal to the hypotenuse of the right triangle $\triangle MNP$, which is the segment MP of the tangent line. That is, $ds = MP$.

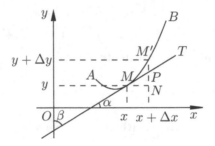

Fig. 3.12

If we choose the tangent vector to be in the direction of increasing curve length then

$$\frac{dx}{ds} = \cos \alpha, \quad \frac{dy}{ds} = \cos \beta.$$

where α, β are the angles between the tangent line MT and the x-, and y-axis respectively.

Example 1. A disc of radius a is rolled along a straight line. A point P on the rim of the disc draws a curve with arches. Find the length of one of the arch (Figure 3.13).

Solution: Let the straight line on which the disc is rolling on be the x-axis, and the start point of one arch be the origin. Draw a vertical line PM perpendicular to x-axis, and draw a horizontal line PQ perpendicular

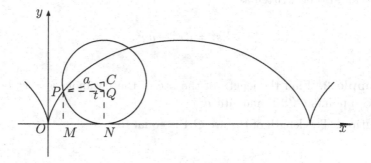

Fig. 3.13

to the vertical radius CN of the disc. Let $t = \angle PCN$ and consider t as a parameter. Then

$$ON = \widehat{PN} = at.$$

And the parametric equations of the arch is

$$x = OM = ON - MN = at - a\sin t = a(t - \sin t),$$
$$y = PM = NC - QC = a - a\cos t = a(1 - \cos t),$$

where $0 \leqslant t \leqslant 2\pi$.

By equation (4), we have

$$s = \int_0^{2\pi} \sqrt{\left(\frac{dx}{dt}\right)^2 + \left(\frac{dy}{dt}\right)^2}\, dt = \int_0^{2\pi} \sqrt{a^2(1 - \cos t)^2 + a^2 \sin^2 t}\, dt$$
$$= a\int_0^{2\pi} \sqrt{2 - 2\cos t}\, dt = 2a\int_0^{2\pi} \sin\frac{t}{2}\, dt = 8a.$$

If the equation of the curve is given by

$$y = f(x), \quad (a \leqslant x \leqslant b),$$

then it can be considered as parametric equations

$$x = x, \quad y = f(x),$$

where x is the parameter. By equation (4), we have

$$s = \int_a^b \sqrt{1 + [f'(x)]^2}\, dx,$$
$$ds = \sqrt{1 + [f'(x)]^2}\, dx.$$

This can also be written as

$$ds^2 = dx^2 + dy^2.$$

Example 2. Find the length of the arc of the parabola $y = x^2$ between the two points $O(0,0)$ and $A(a, a^2)$.

Solution: The length of the arc OA is equal to

$$\int_0^a \sqrt{1 + y'^2}\, dx = \int_0^a \sqrt{1 + (2x)^2}\, dx = \frac{1}{2} \int_0^{2a} \sqrt{1 + t^2}\, dx$$

$$= \frac{1}{4}[2a\sqrt{1 + 4a^2} + \ln(2a + \sqrt{1 + 4a^2})].$$

Exercises 3.1

1. Find the areas bounded by the following curves.

   (1) $y = x^2$, $x + y = 2$;　　　　　　(4) $\rho = 2a\cos\theta$, $(a > 0)$;

   (2) $y^2 + 8x = 16$, $y^2 - 24x = 48$;　(5) $\rho = a(1 + \cos\theta)$, $(a > 0)$.

   (3) $y^2 = x^2(a^2 - x^2)$, $(a > 0)$;

   (6) $\begin{cases} x = a\cos t; \\ y = a\sin t; \end{cases}$　　　　$\begin{cases} x = a\cos^3 t; \\ y = a\sin^3 t; \end{cases}$

   where $0 \leqslant t \leqslant 2\pi$, $(a > 0)$.

2. Find the volumes of the following solids of revolution:

   (1) revolving $y = \sin x$, $0 \leqslant x \leqslant \pi$ about the x-axis;

   　　revolving $y = \sin x$, $0 \leqslant x \leqslant \pi$ about the y-axis;

   (2) revolving $x^2 + (y - b)^2 = a^2$, $(0 < a < b)$ about the x-axis;

   (3) revolving the shape bound by the curves $y^2 = 4x$ and $x = 1$, about the x-axis;

   (4) revolving the shape bounded by the curves $x = a\cos^3 t$, $y = a\sin^3 t$ about the x-axis;

   　　revolving the shape bounded by the curves $x = a\cos^3 t$, $y = a\sin^3 t$ about the y-axis; where $0 \leqslant t \leqslant 2\pi$, $(a > 0)$.

3. Find the volume of the solid bounded by the plane $z = 1$ and the paraboloid of revolution $x^2 + y^2 = 2z$.

4. Show that the volume of the solid of revolution generated by revolving the shape bounded by $0 < a \leqslant x \leqslant b$, $0 \leqslant y \leqslant f(x)$, about the y-axis is

$$V = 2\pi \int_a^b x f(x) \, dx.$$

5. Show that the volume of a cone with base radius a and hight h is

$$V = \frac{1}{3} Sh,$$

   where $S = \pi a^2$.

6. Find the length of the following arcs.
   (1) $x^2 + y^2 = 2x$.
   (2) $y = a \left(\dfrac{e^{\frac{x}{a}} + e^{-\frac{x}{a}}}{2} \right)$, $\quad 0 \leqslant x \leqslant b$, $\quad (a > 0)$.
   (3) $x = a \cos^3 t$, $\quad y = a \sin^3 t$, $0 \leqslant t \leqslant 2\pi$.

7. Show that the length s of the curve $\rho = f(\theta)$, $\alpha \leqslant \theta \leqslant \beta$, is

$$s = \int_\alpha^\beta \sqrt{\rho^2 + \rho'^2} \, d\theta.$$

8. Find the length of the arc of the following curves by using the formula in Exercise 7.
   (1) $\rho = a(1 + \cos\theta)$, $\quad (a > 0)$.
   (2) $\rho = a\theta$, $\quad 0 \leqslant \theta \leqslant 2\pi$.

3.2　Techniques for Graphing Functions

One method of graphing a function is choosing a group of independent variables $x_1, x_2, \cdots, x_n$, then calculate the corresponding function values $y_1, y_2, \cdots, y_n$, plotting the points $(x_1, y_1), (x_2, y_2), \cdots, (x_n, y_n)$ on a graphing paper, then connecting the points by a smooth curve. The curve is considered as the graph of the function. There is a drawback with this method. If there is a big change of the function values between two chosen points, like the points P and Q in the Figure 3.14, then simply connecting these two points by a smooth curve will make the graph inaccurate.

To remedy this problem, we leverage our knowledge of derivatives to learn how the function values on the entire domain changes.

With the tool of derivatives, the graph of a function can be plotted more precisely.

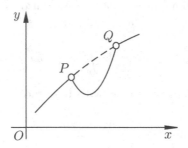

Fig. 3.14

3.2.1 *Increasing and Decreasing Functions*

If for any two points x_1, x_2 in an interval (a, b) in the domain of the function $y = f(x)$, $f(x_2) > f(x_1)$ whenever $x_2 > x_1$, then we say that $f(x)$ is increasing in (a, b). If $f(x_2) < f(x_1)$ whenever $x_2 > x_1$, then we say that $f(x)$ is decreasing in (a, b).

The graph of the curve goes up in (a, b) if $f(x)$ is increasing in (a, b), and the graph of the curve goes down in (a, b) if $f(x)$ is decreasing in (a, b).

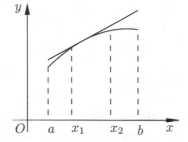 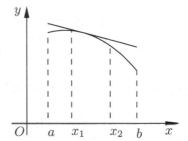

Fig. 3.15   Fig. 3.16

As we can see from Figure 3.15, if the slope of the tangent line at any point of the curve is positive, $f'(x) > 0$, then the curve goes up (the value of the function increases) as x increases. Similarly in Figure 3.16, if the slope of the tangent line at any point of the curve is negative, $f'(x) < 0$, then the curve goes down (the value of the function decreases) as x increases.

Therefore, whether a function increases or decreases on an interval can be determined by the sign of its first derivative on this interval.

If $f'(x) > 0$ (or $f'(x) < 0$) on an interval (a, b), then $f(x)$ increases (or decreases) on the interval.

In fact, let x_1, x_2 be any two points in (a, b) and $x_1 < x_2$. By the Mean-Value Theorem,

$$f(x_2) - f(x_1) = f'(\xi)(x_2 - x_1), \quad x_1 < \xi < x_2.$$

Since $f'(x) > 0$ on (a, b), we have $f'(\xi) > 0$. Hence,

$$f(x_2) > f(x_1).$$

This shows that $f(x)$ is increasing in (a, b). Similarly, we can prove that if $f'(x) < 0$ on (a, b), then $f(x)$ is decreasing.

Example: Find the intervals of the domain of the function $f(x) = (x - 1)^2(x - 2)^3$ where it increases or decreases.

Solution: Since

$$f'(x) = 2(x - 1)(x - 2)^3 + 3(x - 2)^2(x - 1)^2$$
$$= (x - 1)(5x - 7)(x - 2)^2,$$

we have $f'(x) = 0$ at $x_1 = 1$, $x_2 = \dfrac{7}{5}$, and $x_3 = 2$.

These three points divide the entire x-axis into four intervals:

$$(-\infty, 1), \ \left(1, \frac{7}{5}\right), \ \left(\frac{7}{5}, 2\right), \ (2, +\infty).$$

By looking at the sign of $f'(x)$ on each interval, we can determine if the function is increasing or decreasing on the interval. We list the results as follows:

x	$(-\infty, 1)$	$\left(1, \dfrac{7}{5}\right)$	$\left(\dfrac{7}{5}, 2\right)$	$(2, +\infty)$
$f'(x)$	$+$	$-$	$+$	$+$
$f(x)$	increasing	decreasing	increasing	increasing

Now that the directions of changes of the function on its domain are clear, the function can be graphed according to these information (Figure 3.17).

3.2.2 *Concavity*

From the graph of the last example we can see that the function increases
on the intervals $(-\infty, 1)$ and $(2, +\infty)$. But on the former interval the curve
"bends" downward and on the latter interval the curve "bends" upward.
We need more detailed description of this kind of "behavior" of functions.

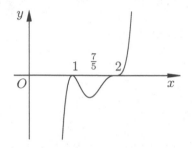

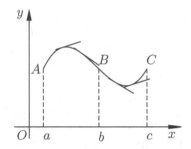

Fig. 3.17   Fig. 3.18

In this section, we study the concavity of functions. In Figure 3.18,
the part of the curve $\widehat{AB}$ is a concave arc, and $\widehat{BC}$ is a convex arc. The
definition of these two concepts are based on the position of the tangent
lines of the curve. The tangent lines of the curve at each point of the arc
$\widehat{AB}$ lie above the curve, and the tangent lines at each point of $\widehat{BC}$ lie under
the curve. If the tangent lines at each point of a curve lie above the curve,
then the curve is *concave down* (*concave*). If the tangent lines at each point
of a curve lie under the curve, then the curve is *concave up* (*convex*).

In Figure 3.18, the slope of the tangent lines of the arc $\widehat{AB}$ gets smaller
when x changes from a to b. This implies that $f'(x)$ is a decreasing function.
The slope of the tangent lines of the arc $\widehat{BC}$ gets greater when x changes
from b to c. This implies that $f'(x)$ is an increasing function.

Let $y = f(x)$ be a function on (a, b). If $f''(x) > 0$ on (a, b), then $f'(x)$
is increasing and the curve is concave up. If $f''(x) < 0$, then $f'(x)$ is
decreasing, and the curve is concave down.

Thus, the sign of the second derivative of a function on an interval of
its domain can be used to determine the concavity of the function on the
interval.

In Figure 3.18, the point b on the x-axis is the place where the curve
changes its concavity. It is called an infection point of the function $f(x)$.

Obviously, $f''(b) = 0$. A point in the domain of a function $f(x)$ is called an *inflection point* of f if f changes from concave down to concave up or from concave up to concave down at this point.

We need to mention that $f''(x) = 0$ is not a sufficient condition for an inflection point of a function. For instance, consider the function $y = x^4$, although

$$f''(0) = 12x^2|_{x=0} = 0,$$

$x = 0$ is not an inflection point of the function.

To determine whether a point x_0 is an inflection point of a function, we need to observe whether the sign of the second derivative of the function changes when the variable x moves from the left side of x_0 to the right side of x_0. If the sign changes, then x_0 is an inflection point. If not, then x_0 is not an inflection point.

All of the discussion above is based on the existence of the second derivative of functions. There are situations where the second derivative of a function does not exist at a certain point, but the point is an inflection point of the function.

Let's look at the function

$$y = f(x) = (x - 2)^{\frac{5}{3}}.$$

Its second derivative is

$$f''(x) = \frac{10}{9}(x - 2)^{-\frac{1}{3}}$$

for $x \neq 2$, and it does not exist for $x = 2$.

The value of $f''(x)$ is less than zero on the left side of $x = 2$ and greater than zero on the right side of $x = 2$. Since the function is continuous at $x = 2$, the curve changes its concavity at the point $M_0(2,0)$. Therefore, $x = 2$ is an inflection point of f.

The sign of the first derivative of a function determines the increasing or decreasing on certain intervals of the domain, and the sign of the second derivative determines the concavity of the function. With the help of these information, we now can make much better graphs of functions. In the next section, we study the nature of a function as its graph extends to infinity.

3.2.3 *Asymptotes*

A hyperbola has two asymptotes. When the hyperbola extends to infinity, it gets closer and closer to its asymptotes. Therefore, the shape of the hyperbola as it tends to infinity can be described by its asymptotes. Finding the asymptotes of a function if it has them, will help us obtain a more accurate graph of the function.

An *asymptote* can be defined as: If the distance of a point of a curve and a fixed line approaches to zero when the point moves along the curve to infinity, then the fixed line is called an asymptote of the curve.

The simplest asymptotes are horizontal asymptotes and vertical asymptotes.

Suppose a function $y = f(x)$ satisfies

$$\lim_{x \to a} f(x) = \infty.$$

Then $x = a$ is a vertical asymptote of the function (Figure 3.19).

If

$$\lim_{x \to \infty} f(x) = b,$$

where b is a finite number, then $y = b$ is a horizontal asymptote of the function.

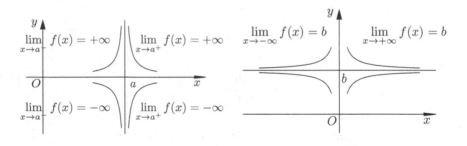

Fig. 3.19   Fig. 3.20

To find oblique asymptote, we assume that the equation of the asymptote is

$$y = ax + b \tag{1}$$

where $a \neq 0$.

In Figure 3.21, we can see that there is a relation

$$MK = MK' \cos \alpha$$

between the distance MK of a moving point M on the curve to the line (1) and the vertical line segment MK', where α is the angle between the line (1) and x-axis. Thus, the distance MK approaching zero is equivalent to the length of MK' approaching zero. Therefore, we have

$$\lim_{x\to\infty} MK' = 0$$

as the point M move along the curve to infinity. It follows that

$$\lim_{x\to\infty} (f(x) - ax - b) = 0. \tag{2}$$

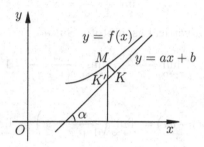

Fig. 3.21

Any straight line that satisfies this equation is an *oblique asymptote*.

Hence, we can conclude that: The necessary and sufficient condition for a line $y = ax + b$ to be an oblique asymptote of a function $f(x)$ is

$$\lim_{x\to\infty} (f(x) - ax - b) = 0.$$

Now we determine a and b in this condition. Equation (2) implies the equation

$$\lim_{x\to\infty} \frac{f(x) - ax - b}{x} = 0,$$

and that is,

$$\lim_{x\to\infty} \left(\frac{f(x)}{x} - a \right) = 0,$$

or

$$\lim_{x\to\infty} \frac{f(x)}{x} = a. \tag{3}$$

The value of a can be calculated from this equation.

The equation (2) also implies

$$\lim_{x\to\infty} (f(x) - ax) = b. \tag{4}$$

So b can be found, and the equation of the asymptote

$$y = ax + b$$

follows.

Example: Find the asymptotes of the curve $y = \dfrac{x^2}{1+x}$.

Solution: Since

$$\lim_{x \to -1} \frac{x^2}{1+x} = \infty,$$

the line $x = -1$ is a vertical asymptote. Since

$$\lim_{x \to \infty} \frac{f(x)}{x} = \lim_{x \to \infty} \frac{x^2}{x(1+x)} = 1$$

we have $a = 1$ and

$$b = \lim_{x \to \infty} (f(x) - ax) = \lim_{x \to \infty} \left(\frac{x^2}{1+x} - x \right) = \lim_{x \to \infty} \frac{-x}{1+x} = -1,$$

and the other asymptote is $y = x - 1$.

3.2.4   *Examples of Graphing Functions*

Now we are able to obtain more accurate graphs of functions by what we just learned.

Example 1. Sketch the graph of the function $f(x) = \dfrac{(x-1)^3}{(x+1)^2}$.

Solution: The function is well defined for all x, $-\infty < x < \infty$, except at the point $x = -1$. Since

$$\lim_{x \to -1} f(x) = -\infty,$$

the line $x = -1$ is a vertical asymptote of $f(x)$. Since

$$\lim_{x \to \infty} \frac{f(x)}{x} = \lim_{x \to \infty} \frac{(x-1)^3}{x(x+1)^2} = 1$$

and

$$\lim_{x \to \infty} (f(x) - x) = \lim_{x \to \infty} \frac{(x-1)^3 - x(x+1)^2}{(x-1)^2}$$

$$= \lim_{x \to \infty} \frac{-5x^2 + 2x - 1}{(x+1)^2} = -5,$$

$y = x - 5$ is an oblique asymptote of $f(x)$.

It is easy to see that $f(x) = 0$ when $x = 1$, so the graph intersects the x-axis at the point $(1, 0)$. Since

$$f'(x) = \frac{(x-1)^2(x+5)}{(x+1)^3},$$

we have

$$f'(1) = f'(-5) = 0.$$

Since

$$f''(x) = \frac{24(x-1)}{(x+1)^4},$$

we have

$$f''(1) = 0.$$

These special points divide the x-axis into several intervals. We list the changes of the function on these intervals as follows:

x	$(-\infty, -5)$	$(-5, -1)$	$(-1, 1)$	$(1, +\infty)$
$f'(x)$	$+$	$-$	$+$	$+$
$f''(x)$	$-$	$-$	$-$	$+$
$f(x)$	concave down, increasing	concave down, decreasing	concave down, increasing	concave up, increasing

Obviously, $(1, 0)$ is the inflation point since $f''(x)$ changes its sign as x moves from the left of $x = 1$ to the right.

By this list and the asymptotes $x = -1$, $y = x - 5$, we can sketch the graph as in Figure 3.22.

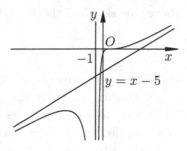

Fig. 3.22

Example 2. Sketch the graph of the function $f(x) = \dfrac{(x-3)^2}{4(x-1)}$.

Solution: The domain of the function is $-\infty < x < 1$, $1 < x < \infty$. Since

$$f'(x) = \frac{(x-3)(x+1)}{4(x-1)^2}$$

and

$$f''(x) = \frac{2}{(x-1)^3},$$

we have $f'(3) = 0$, $f'(-1) = 0$. Since

$$\lim_{x \to 1} f(x) = \lim_{x \to 1} \frac{(x-3)^2}{4(x-1)} = \infty,$$

the line $x = 1$ is a vertical asymptote. Since

$$a = \lim_{x \to \infty} \frac{f(x)}{x} = \lim_{x \to \infty} \frac{(x-3)^2}{4x(x-1)} = \frac{1}{4},$$

$$b = \lim_{x \to \infty} (f(x) - ax) = \lim_{x \to \infty} \left[\frac{(x-3)^2}{4(x-1)} - \frac{x}{4} \right] = -\frac{5}{4},$$

the oblique asymptote is $y = \dfrac{1}{4}x - \dfrac{5}{4}$.

The changes of the function on the intervals divided by the special points are as follows:

x	$(-\infty, -1)$	$(-1, 1)$	$(1, 3)$	$(3, +\infty)$
$f'(x)$	$+$	$-$	$-$	$+$
$f''(x)$	$-$	$-$	$+$	$+$
$f(x)$	concave down, increasing	concave down, decreasing	concave up, decreasing	concave up, increasing

From the analysis above, the sketch of the graph is Figure 3.23.

3.2.5 *Curvatures*

In the last section, we studied the concavity of curves, that is, the direction of a curve's bends. In this section, we study the rate of change of the direction of a curve with respect to a change in its length. This concept is important in some other subjects such as engineering and differential geometry.

The degree of the curvedness of a graph of function is related to the change of direction of its tangent lines. In other words, it is related to the

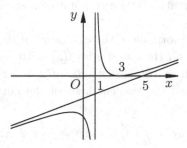

Fig. 3.23

angle between the tangent lines at two end points of a curve. It is also related to the length of the curve (Figure 3.24). For the same amount of change in the direction of tangent lines, the longer the curve, the less curved it is.

Let $\Delta\varphi$ be the angle between the tangent lines at two end points of a curve, Δs be the length of the curve. Then the ratio $\dfrac{\Delta\varphi}{\Delta s}$ describes the average degree of curvedness of the curve. To define the degree of curvedness, the curvature, at a point A of the curve, we consider a nearby point B. We then calculate the average curvature of the arc $\overset{\frown}{AB}$, $\dfrac{\Delta\varphi}{\Delta s}$. The *curvature* of the curve at point A is defined to be the limit of this fraction as B tends to A, that is, as Δs approaches zero:

$$k = \lim_{\Delta s \to 0} \frac{\Delta\varphi}{\Delta s}.$$

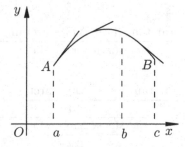

Fig. 3.24

Obviously the curvature at every point of a straight line is zero since $\Delta\varphi = 0$.

Now we derive the formula of the curvature of the curve $y = f(x)$.

Let A be a point on the curve $y = f(x)$ with coordinate $(x, f(x))$, B be another point with coordinate $(x + \Delta x, f(x + \Delta x))$, α, β be the angles between the x-axis and the tangent lines of the curve at point A and B respectively, then

$$\tan\alpha = f'(x), \quad \tan\beta = f'(x + \Delta x).$$

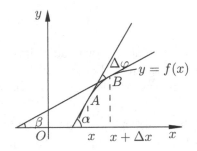

Fig. 3.25

The angle between these two tangent lines is

$$\Delta\varphi = |\beta - \alpha| = |\arctan f'(x + \Delta x) - \arctan f'(x)|.$$

On the other hand, let $s(x)$ be a function of arc length of the curve, then the length of the arc AB is

$$\Delta s = s(x + \Delta x) - s(x).$$

The *average curvature* is

$$\frac{\Delta\varphi}{\Delta s} = \frac{|\arctan f'(x + \Delta x) - \arctan f'(x)|}{s(x + \Delta x) - s(x)}.$$

That is

$$\frac{\Delta\varphi}{\Delta s} = \frac{\left|\dfrac{\arctan f'(x + \Delta x) - \arctan f'(x)|}{\Delta x}\right|}{\dfrac{s(x + \Delta x) - s(x)}{\Delta x}}.$$

(Translator's note: This is true only in the $\Delta x > 0$ case. However, the result reached here, formula (5) is true also for the $\Delta x < 0$ case.)

Obviously, when Δs approaches zero, Δx also approaches zero. The numerator of the above equation tends to

$$\left| \frac{d}{dx}(\arctan f'(x)) \right| = \left| \frac{d}{dx} \arctan y' \right| = \frac{|y''|}{1+y'^2}$$

as Δx tends to zero.

The limit of the denominator is

$$\frac{ds(x)}{dx} = \sqrt{1+y'^2}.$$

Thus,

$$k = \lim_{\Delta s \to 0} \frac{\Delta \varphi}{\Delta s} = \frac{|y''|}{(1+y'^2)^{\frac{3}{2}}}, \tag{5}$$

and this is the formula of curvature that we want to find.

Suppose a curve is given parametrically by

$$x = \varphi(t), y = \psi(t), \quad (\alpha \leqslant t \leqslant \beta).$$

Since

$$y' = \frac{\psi'(t)}{\varphi'(t)}$$

and

$$y'' = \frac{\psi''(t)\varphi'(t) - \varphi''(t)\psi'(t)}{[\varphi'(t)]^3},$$

Substituting these two equations into (5), we have

$$k = \frac{|\psi''(t)\varphi'(t) - \varphi''(t)\psi'(t)|}{[\varphi'^2(t) + \psi'^2(t)]^{\frac{3}{2}}}. \tag{6}$$

Example: The parametric equation of a circle with radius R is

$$x = R\cos t, \quad y = R\sin t, \quad (0 \leqslant t \leqslant 2\pi).$$

By equation (6), we have

$$k = \frac{1}{R}.$$

The curvature of a circle at every point is equal to the reciprocal of its radius. The larger the radius the smaller the curvature.

Suppose the curvature of a curve $y = f(x)$ at a point A is $k(\neq 0)$. Draw a normal line of the curve at A, and let C be a point on the part of the normal line that lies on the concave side of the curve, such that the length of the line segment AC is equal to $\frac{1}{k}$. Draw a circle with radius AC centered

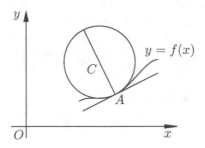

Fig. 3.26

at C (Figure 3.26). Since the point C is on the normal, the circle is tangent to the curve at A. Moreover, the circle has the same curvature as the curve at A. Thus, in a neighborhood of A, the shape of curve is closer to this circle than any other circle that passes the point A. This circle is called the *circle of curvature* of the curve at point A, and its radius is called the *radius of curvature*. The point C is called the *center of curvature*.

Since the circle of curvature has these properties, when solving problems with respect to concavity or curvature of curves, we can use circles of curvatures instead of the curves to simplify the problem. For instance, when studying the movement of a particle along a curve, if we consider the circle of curvature instead of the arc of the curve at a neighborhood of a point, then the knowledge of circular motion can be applied to study the curvilinear motion. Suppose a particle with mass m moves along a curve, and the tangential velocity at a point on the curve is v. If the radius of curvature of the curve at this point is R, then the centripetal acceleration of the particle at this point is $\dfrac{v^2}{R}$ and the centripetal force at this point is $\dfrac{mv^2}{R}$.

Exercises 3.2

1. Show that the function $y = e^{-x^2}$ is increasing on $(-\infty, 0)$ and is decreasing on $(0, +\infty)$.

2. Show the following.
   (1) The inequality
   $$nb^{n-1}(a-b) \leqslant a^n - b^n \leqslant na^{n-1}(a-b)$$

holds for $a > b > 0$ and $n > 1$.

(2) The inequality $e^x > 1 + x$ holds for $x \neq 0$.

(3) The inequality $x > \ln(1 + x)$ holds for $x > 0$.

3. Show that if $f'(x) > g'(x)$ for $x > 0$, and $f(0) = g(0)$, then $f(x) > g(x)$.

4. Find the inflection points and the intervals where the curve is concave up or concave down for the following functions.

(1) $y = x^3 - 3x^2 - 9x + 9$;   (3) $y = x + \sin x$;

(2) $y = (1 + x^2)e^x$;   (4) $y = \ln(1 + x^2)$.

5. Find the values of a and b such that the point $(1, 3)$ is an inflection point of the function $y = ax^3 + bx^2$.

6. Find the asymptotes of the following curves.

(1) $y = \dfrac{1}{x^2 - 3x + 2}$;   (3) $\dfrac{x^2}{a^2} - \dfrac{y^2}{b^2} = 1$;

(2) $y = c + \dfrac{a^3}{(x - b)^2}$;   (4) $y = xe^{\frac{2}{x}} + 1$.

7. Sketch the graphs of the following functions.

(1) $y = \dfrac{8a^2}{x^2 + a^2}$;

(2) $y = x - 2\arctan x$;

(3) $y = x - \dfrac{1}{x}$;   (4) $y = \dfrac{x^3}{3 - x^2}$;

(5) $y = \dfrac{\ln x}{x}$;   (6) $y = \sqrt{\dfrac{x - 1}{x + 1}}$;

(7) $y = e^{-x^2}$;   (8) $y = e^{-x}\sin x$, $(0 \leqslant x \leqslant \pi)$.

8. Find the curvature of the curve at the given point.

(1) $xy = 4$ at $(2, 2)$;   (3) $y = x^3$ at any point;

(2) $y = \ln x$ at $(1, 0)$;   (4) $x = 3t^2$, $y = t - t^3$ at $t = 1$;

(5) $x = a\cos t$, $y = b\sin t$ at $t = \dfrac{\pi}{4}$, $(a > 0, b > 0)$.

9. Prove that the radius of curvature at any point P of the catenary

$$y = a\left(\frac{e^{\frac{x}{a}} + e^{-\frac{x}{a}}}{2}\right), \ (a > 0)$$

is equal to the length of the segment of the normal line at P between the point P and the x-axis.

3.3 Taylor Expansions and Extreme Value Problems

3.3.1 *Taylor Expansions*

In Chapter 1 Section 1.3, we mentioned that derivatives (and differentials) are "local" properties of functions. Higher order derivatives (and higher order differentials) are also "local" properties of functions. Taylor expansions of a function describes the properties or characters of a function at a neighborhood of a point in the domain of the function by its derivatives (and higher order derivatives).

Given a function $y = f(x)$ with nth derivative at a point $x = x_0$. We want to find an nth degree polynomial $P_n(x)$ such that

$$P_n(x_0) = f(x_0),\ P_n'(x_0) = f'(x_0),\ P_n''(x_0) = f''(x_0),$$

$$\cdots$$

$$P_n^{(n)}(x_0) = f^{(n)}(x_0).$$

The polynomial

$$P_n(x) = f(x_0) + f'(x_0)(x - x_0) + \frac{f''(x_0)}{2!}(x - x_0)^2 + \cdots$$
$$+ \frac{f^{(n)}(x_0)}{n!}(x - x_0)^n.$$

obviously satisfies these requirements.

(How did we get this polynomial?)

To study the difference between $f(x)$ and $P_n(x)$, we assume $f(x)$ has $(n+1)$th derivative. Denote the difference between $f(x)$ and $P_n(x)$ as

$$f(x) - P_n(x) = \frac{(x - x_0)^{n+1}}{(n + 1)!} Q_n.$$

Consider a function ϕ with respect to the points where f has its $(n+1)$th derivative

$$\phi(z) = f(x) - f(z) - \frac{x - z}{1!} f'(z) - \frac{(x - z)^2}{2!} f''(z) - \cdots$$
$$- \frac{(x - z)^n}{n!} f^{(n)}(z) - \frac{(x - z)^{n+1}}{(n + 1)!} Q_n.$$

It is easy to see that $\phi(x_0) = 0$ and $\phi(x) = 0$. By the Mean-Value Theorem, there exists a point $\xi = x_0 + \theta(x - x_0)$, where $0 < \theta < 1$, such that $\phi'(\xi) = 0$.

On the other hand

$$\phi'(z) = \frac{(x - z)^n}{n!} [Q_n - f^{(n+1)}(z)],$$

so we have

$$Q_n = f^{(n+1)}(\xi).$$

Thus,

$$f(x) = f(x_0) + f'(x_0)(x - x_0) + \frac{f''(x_0)}{2!}(x - x_0)^2 + \cdots$$
$$+ \frac{f^{(n)}(x_0)}{n!}(x - x_0)^n + \frac{f^{(n+1)}(\xi)}{(n+1)!}(x - x_0)^{n+1}$$

where $\xi = x_0 + \theta(x - x_0)$.

This is called the *Taylor expansion* of the function $f(x)$ at the point x_0. The difference between $f(x)$ and $P_n(x)$

$$\frac{f^{(n+1)}(\xi)}{(n+1)!}(x - x_0)^{n+1}$$

is called the *remainder* of the expansion.

The distance between x and x_0 becomes very small as x gets closer to x_0, so the difference between $P_n(x)$ and $f(x)$ has the same order of magnitude as $(x - x_0)^{n+1}$. That means

$$\lim_{x \to x_0} \frac{f(x) - P_n(x)}{(x - x_0)^{n+1}} = C,$$

where C constant.

If two functions $g(x)$ and $h(x)$ satisfy

$$\lim_{x \to x_0} \frac{g(x)}{h(x)} = C,$$

then we use the notation $y(x) = O[h(x)]$ to denote this relation between them. Hence, we can write

$$f(x) - P_n(x) = O[(x - x_0)^{n+1}].$$

Example 1. Find the Taylor expansion of $f(x) = e^x$ at $x = 0$.

Solution: Since

$$f(x) = f'(x) = f''(x) = \cdots = f^{(n)}(x) = e^x,$$

we have

$$f(0) = f'(0) = f''(0) = \cdots = f^{(n)}(0) = 1.$$

Thus

$$e^x = 1 + \frac{x}{1!} + \cdots + \frac{x^n}{n!} + \frac{x^{n+1}}{(n+1)!}e^{\theta x},$$

where $0 < \theta < 1$.

Example 2. Find the Taylor expansion of $f(x) = \sin x$ at $x = 0$.

Solution: Since
$$f'(x) = \sin\left(x + \frac{\pi}{2}\right), \cdots, f^{(k)}(x) = \sin\left(x + k\frac{\pi}{2}\right),$$
we have
$$f(0) = 0, \ f'(0) = 1, \ f''(0) = 0, \ f'''(0) = -1, \cdots.$$

In general,
$$f^{(2m)}(0) = 0, \ f^{(2m+1)}(0) = (-1)^m$$
for any positive integer m. Thus
$$\sin x = x - \frac{x^3}{3!} + \frac{x^5}{5!} + \cdots + (-1)^n \frac{x^{2n+1}}{(2n+1)!}$$
$$+ \frac{x^{n+3}}{(2n+3)!} \sin\left(\theta x + \frac{2n+3}{2}\pi\right).$$

The partial sums of this Taylor expansion can be used as approximation functions of $\sin x$:
$$\sin x \approx x, \ \sin x \approx x - \frac{x^3}{3!}, \ \sin x \approx x - \frac{x^3}{3!} + \frac{x^5}{5!}, \cdots.$$

Moreover,
$$|\sin x - x| < \frac{x^3}{6}, \ \left|\sin x - x + \frac{x^3}{3!}\right| < \frac{x^5}{120},$$
$$\left|\sin x - x + \frac{x^3}{3!} - \frac{x^5}{5!}\right| < \frac{x^7}{5040}, \cdots.$$

The difference between x and $\sin x$ is less than 0.001 if
$$\frac{x^3}{3!} < 0.001.$$

This implies that $x < 0.1817$ (about $10°$). In other words, if x is less than 10 degrees, the error of using x as an approximate value of $\sin x$ is less than 0.001.

Example 3. Find the Taylor expansion of $f(x) = (1+x)^\alpha$ at $x = 0$, where $x > -1$ and α is a real number.

Solution: Since
$$f'(x) = \alpha(1+x)^{\alpha-1}, \ f''(x) = \alpha(\alpha-1)(1+x)^{\alpha-2},$$
$$\cdots,$$
$$f^{(k)}(x) = \alpha(\alpha-1)\cdots(\alpha-k+1)(1+x)^{\alpha-k},$$

we have

$$f(0) = 1, \ f'(0) = \alpha,$$

$$\cdots,$$

$$f^{(k)}(0) = \alpha(\alpha - 1) \cdots (\alpha - k + 1).$$

So the Taylor expansion is

$$(1 + x)^\alpha = 1 + \alpha x + \frac{\alpha(\alpha - 1)}{2!}x^2 + \cdots$$

$$+ \frac{\alpha(\alpha - 1) \cdots (\alpha - n + 1)}{n!}x^n$$

$$+ \frac{\alpha(\alpha - 1) \cdots (\alpha - n)}{(n + 1)!}(1 + \theta x)^{\alpha - n - 1}x^{n+1}.$$

More detailed discussion about Taylor expansion can be found in Chapter 10.

The concept of Taylor expansion can be interpreted from another point of view.

The properties of some functions at the neighborhood of a point are complicated, so it is natural to use elementary functions to approach these function in such a neighborhood.

Elementary functions are composed of by three categories of functions:

(1) The power function x^μ, where μ is a real number;

(2) Trigonometric functions and their inverses, such as $\sin x$, $\cos x$, $\arcsin x$, $\arccos x$, $\cdots$;

(3) The exponential function and its inverse e^x, $\ln x$.

Taylor expansions approximate a function with a polynomial of power functions $x, x^2, \cdots, x^n$. Taylor series expresses a function as an infinite series of power functions. We study Taylor series in Chapter 10. Fourier expansions approximate a function with a polynomial of trigonometric functions. Fourier series expresses a function as an infinite series of trigonometric functions. We study Fourier expansions and Fourier series in Chapter 11. By the Euler formula $e^{ix} = \cos x + i \sin x$ (where $i = \sqrt{-1}$) in the complex number field, a polynomial of exponential functions is equivalent to a polynomial of trigonometric functions.

The Mean-Value Theorem in Chapter 1 Section 1.3 is a special case of Taylor expansion ($n = 0$). Let's look at an application of the Taylor expansion:

Suppose $f(x)$, $g(x)$ have second derivatives at $x = x_0$, with $g'(x_0) \neq 0$, and

$$\lim_{x \to x_0} f(x) = 0, \quad \lim_{x \to x_0} g(x) = 0.$$

Then
$$\lim_{x \to x_0} \frac{f(x)}{g(x)} = \frac{f'(x_0)}{g'(x_0)}.$$

This equation is called the *l'Hôpital's rule*.

The proof is not hard: Since the Taylor expansions of $f(x)$ and $g(x)$ at $x = x_0$ are

$$f(x) = f(x_0) + f'(x_0)(x - x_0) + O[(x - x_0)^2],$$
$$g(x) = g(x_0) + g'(x_0)(x - x_0) + O[(x - x_0)^2],$$

respectively, and $f(x_0) = g(x_0) = 0$, we have

$$\lim_{x \to x_0} \frac{f(x)}{g(x)} = \lim_{x \to x_0} \frac{f'(x_0)(x - x_0) + O[(x - x_0)^2]}{g'(x)(x - x_0) + O[(x - x_0)^2]}$$
$$= \lim_{x \to x_0} \frac{f'(x_0) + O[(x - x_0)]}{g'(x_0) + O[(x - x_0)]} = \frac{f'(x_0)}{g'(x_0)}.$$

Furthermore, if $f(x)$, $g(x)$ have nth derivatives at the same point $x = x_0$ with $g^{(n)}(x_0) \neq 0$, and

$$\lim_{x \to x_0} f(x) = \lim_{x \to x_0} f'(x) = \cdots = \lim_{x \to x_0} f^{(n-1)}(x) = 0,$$

$$\lim_{x \to x_0} g(x) = \lim_{x \to x_0} g'(x) = \cdots = \lim_{x \to x_0} g^{(n-1)}(x) = 0.$$

Then
$$\lim_{x \to x_0} \frac{f(x)}{g(x)} = \frac{f^{(n)}(x_0)}{g^{(n)}(x_0)}.$$

Some problems can be transformed to the above situations. For instance, to find

$$\lim_{x \to x_0} \frac{f(x)}{g(x)}$$

if $\lim_{x \to x_0} f(x) = \infty$ and $\lim_{x \to x_0} g(x) = \infty$, or to find

$$\lim_{x \to x_0} f(x)g(x)$$

if $\lim_{x \to x_0} f(x) = 0$ and $\lim_{x \to x_0} g(x) = \infty$.

Example 4. Evaluate $\lim\limits_{x \to 0} \dfrac{1 - \cos x}{x^2}$.

Solution: Since

$$\lim_{x \to 0} (1 - \cos x) = 0, \quad \lim_{x \to 0} x^2 = 0,$$

we have

$$\lim_{x \to 0} \frac{1 - \cos x}{x^2} = \lim_{x \to 0} \frac{\sin x}{2x} = \frac{1}{2}.$$

Example 5. Evaluate $\lim_{x \to 0} \dfrac{\tan x - \sin x}{x^3}$.

Solution: Since

$$\lim_{x \to 0} (\tan x - \sin x) = 0, \quad \lim_{x \to 0} x^3 = 0,$$

and

$$\lim_{x \to 0} (\tan x - \sin x)' = \lim_{x \to 0} (\sec^2 x - \cos x) = 0,$$

$$\lim_{x \to 0} (x^3)' = \lim_{x \to 0} 3x^2 = 0,$$

$$\lim_{x \to 0} (\sec^2 x - \cos x)' = \lim_{x \to 0} (2 \sec^2 x \tan x + \sin x) = 0,$$

$$\lim_{x \to 0} (3x^2)' = \lim_{x \to 0} 6x = 0,$$

$$\lim_{x \to 0} (2 \sec^2 x \tan x + \sin x)'$$

$$= \lim_{x \to 0} 2(\sec^4 x + 4 \sec^2 x \tan^2 x + \cos x) = 3,$$

$$\lim_{x \to 0} (6x)' = 6,$$

we have

$$\lim_{x \to 0} \frac{\tan x - \sin x}{x^3} = \frac{3}{6} = \frac{1}{2}.$$

3.3.2 *Extreme Value Problems*

The extreme value problem can be studied from the point of view of Taylor expansions. Cut a sector from a round iron sheet. Make a container in the shape of a cone with the rest of the sheet. What is the angle of the sector to cut when the volume of the container is the maximum?

Suppose the central angle of the remainder sheet is x. The circumference of the base circle of the cone made by the sheet is Rx (Figure 3.27). The radius of the base is

$$r = \frac{Rx}{2\pi}.$$

The hight of the cone is

$$h = \sqrt{R^2 - r^2} = \sqrt{R^2 - \left(\frac{Rx}{2\pi}\right)^2} = \frac{R}{2\pi} \sqrt{4\pi^2 - x^2}.$$

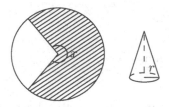

Fig. 3.27

Thus, the volume of the cone is

$$V = \frac{1}{3}\pi r^2 h = \frac{1}{3}\pi \left(\frac{Rx}{2\pi}\right)^2 \frac{R}{2\pi}\sqrt{4\pi^2 - x^2}$$
$$= \frac{R^3 x^2}{24\pi^2}\sqrt{4\pi^2 - x^2}.$$

Therefore, the problem becomes: at what value of x does the function

$$f(x) = \frac{R^3 x^2}{24\pi}\sqrt{4\pi^2 - x^2}, \quad (0 \leqslant x \leqslant 2\pi)$$

have its maximum?

Let's look at another example:

A ship is anchored at a point A, 9 km away from the coast (Figure 3.28), ($AO = 9$ km). A person needs to deliver a letter to a place B on the coast. The distance between B and O is 15 km. If the walking velocity of the person is 5 km/h and the velocity of rowing a boat is 4 km/h, then at which point should the person disembark to make the delivery in the shortest time?

Suppose the point that the person disembarks is C and the distance of this point from O is x km. We have

$$AC = \sqrt{9^2 + x^2}, \quad CB = 15 - x.$$

Thus, the time this person needs is

$$t = \frac{\sqrt{9^2 + x^2}}{4} + \frac{15 - x}{5}, \quad (0 \leqslant x \leqslant 15).$$

Therefore, the problem becomes: at what value of x does the function

$$f(x) = \frac{\sqrt{9^2 + x^2}}{4} + \frac{15 - x}{5}$$

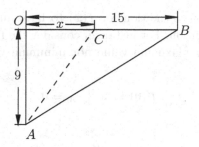

Fig. 3.28

have its minimum.

The problems can be different but the principle is the same: for a given function $f(x)$ that is defined on $[a, b]$, find a value of x such that the function reaches its minimum or maximum at this point.

We have studied the concepts of relative minimum values, relative maximum values and extremes of a function in Chapter 1 Section 1.3.5.

For a curve as in Figure 3.29, the function has relative minimum values at x_1 and x_3; the function has relative maximum values at x_0 and x_2.

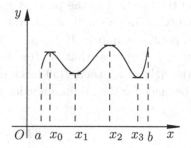

Fig. 3.29

The extreme maximum (minimum) value of a function $f(x)$ on a closed interval $[a, b]$ is the maximum (minimum) value among all relative maximum (minimum) values and the values of $f(x)$ at the end points.

The problem becomes how to find extreme points of the function.

If a function is differentiable at x_0, and has an extrema at this point, then $f'(x_0) = 0$ (Figure 3.29). This result was proved in Chapter 1 Section

1.3.5.

This is a necessary condition for $f(x)$ to have an extrema at x_0, but not a sufficient condition. For instance, consider the function $y = x^3$. The function has neither maximum value nor minimum value at point $x = 0$ (Figure 3.30), but

$$f'(0) = 3x^2|_{x=0} = 0.$$

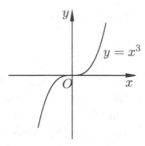

Fig. 3.30

How to determine whether a point is an extrema point of a function when the derivative of the function at the point is zero?

Suppose $f'(x_0) = 0$. Consider a small neighborhood of x_0, $(x_0 - \delta, x_0 + \delta)$, where δ is a small positive number. If $f'(x_0) > 0$ for $x_0 - \delta < x < x_0$ and $f'(x_0) < 0$ for $x_0 < x < x_0 + \delta$, that is, $f(x)$ increases on the left of x_0 and decreases on the right of x_0, then $f(x)$ has a maximum value at x_0. Similar criterion can be made for the minimum value of $f(x)$. (Figure 3.31)

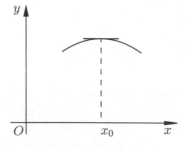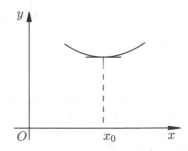

Fig. 3.31

If $f'(x)$ is constantly greater than zero (or constantly less than zero) in a neighborhood of x_0, then x_0 is not an extrema point of $f(x)$.

Example 1. Find extrema points of $f(x) = (x-1)^2(x-2)^3$ on $[0,2]$.

Solution:

$$f'(x) = 2(x-1)(x-2)^3 + 3(x-2)^2(x-1)^2$$
$$= (x-1)(x-2)^2(5x-7).$$

Let $f'(x) = 0$. We have $x = 1$, $x = \dfrac{7}{5}$ and $x = 2$, where $x = 1$, $x = \dfrac{7}{5}$ are possible extrema points.

Since

$$f'(x) = (x-1)(x-2)^2(5x-7),$$

we have $f'(x) > 0$ on the left side of $x = 1$, and $f'(x) < 0$ on the right side of $x = 1$. Thus, $f(x)$ has a maximum value at $x = 1$.

Similarly, we can verify that $f(x)$ has a minimum value at $x = \dfrac{7}{5}$. And the values of the function at the two points are

$$f(1) = 0, \quad f\left(\frac{7}{5}\right) = -0.035.$$

If $f''(x_0) > 0$, since

$$f''(x_0) = \lim_{x \to x_0} \frac{f'(x) - f'(x_0)}{x - x_0},$$

we have

$$\frac{f'(x) - f'(x_0)}{x - x_0} > 0$$

on a neighborhood of x_0.

Since $f'(x_0) = 0$, the above inequality becomes

$$\frac{f'(x)}{x - x_0} > 0.$$

This implies that $f'(x) < 0$ when $x < x_0$, and $f'(x) > 0$ when $x > x_0$. Thus $f(x)$ has a minimum value at x_0. Similarly, if $f''(x) < 0$, then $f(x)$ has a maximum value at x_0. Thus, we obtain another criterion for extreme values: Suppose $f'(x_0) = 0$. The function $f(x)$ has a minimum value at x_0 if $f''(x_0) > 0$; $f(x)$ has a maximum value at x_0 if $f''(x_0) < 0$.

If $f''(x) = 0$, then we can only determine the extreme values by the previous criterion or by Taylor expansion. The details are left as an exercise for the reader.

We have assumed that $f(x)$ is differentiable at x_0 in the above study. The points on which $f'(x)$ does not exist can also be the extrema points. For instance, for

$$y = f(x) = 1 + (x-2)^{\frac{2}{3}},$$

$$f'(x) = \frac{2}{3}(x-2)^{-\frac{1}{3}}, \quad (x \neq 2),$$

the function $f'(x)$ does not exist at $x = 2$. But $f'(x) < 0$ on the left side of $x = 2$ and $f'(x) > 0$ on the right side of $x = 2$. Therefore, $f(x)$ has a minimum value at $x = 2$, and the minimum value is 1.

Example 2. The problem in the example of the conical container is to find the maximum value of the function

$$f(x) = \frac{R^3}{24\pi^2} x^2 \sqrt{4\pi^2 - x^2}$$

on $[0, 2\pi]$.

We need to find the extrema points of the function

$$f(x) = x^2 \sqrt{4\pi^2 - x^2}.$$

Since

$$f'(x) = \frac{8\pi^2 x - 3x^3}{\sqrt{4\pi^2 - x^2}},$$

let $f'(x) = 0$ and solve $8\pi^2 x - 3x^3 = 0$, we get

$$x_1 = 0, \quad x_2 = -2\pi\sqrt{\frac{2}{3}}, \quad x_3 = 2\pi\sqrt{\frac{2}{3}}.$$

Obviously,

$$x_1 = 0, \quad x_2 = -2\pi\sqrt{\frac{2}{3}}$$

are meaningless for this problem. Since

$$f''(x_3) < 0,$$

$f(x)$ has a maximum value at x_3. This is the only maximum value of $f(x)$ on $[0, 2\pi]$, so the container has the maximum volume when the central angle of the cut sector is

$$2\pi - 2\pi\sqrt{\frac{2}{3}} = \frac{2}{3}\pi(3 - \sqrt{6}).$$

Example 3. The problem in the example of the ship is to find the minimum value of the function

$$f(x) = \frac{\sqrt{9^2 + x^2}}{4} + \frac{15 - x}{5}$$

on $[0, 15]$.

Since

$$f'(x) = \frac{1}{4} \frac{x}{\sqrt{9^2 + x^2}} - \frac{1}{5},$$

let $f'(x) = 0$ and solve $5x = 4\sqrt{9^2 + x^2}$, we get $x = 12$. Also

$$f''(x) = \frac{1}{4} \frac{\sqrt{9^2 + x^2} - \dfrac{x^2}{\sqrt{9^2 + x^2}}}{9^2 + x^2} = \frac{1}{4} \frac{9^2}{(9^2 + x^2)^{\frac{3}{2}}}.$$

Therefore,

$$f''(12) = \frac{1}{4} \frac{9^2}{(9^2 + 12^2)^{\frac{3}{2}}} > 0.$$

This implies that $f(x)$ has a minimum value at $x = 12$. Since this is the only minimum value of $f(x)$ on $[0, 15]$, the person should disembark at the point 12 km away from the point O.

Example 4. An object with weight G is placed on a horizontal plane. It is moved by an acting force F, overcoming friction. Find the angle between the force and the horizontal plane such that a minimum value of the force is needed to make the object move (Assume the friction coefficient is μ.)
Solution: Suppose the angle between the force and the horizontal plane is θ. From physics, we know that the friction force is

$$f = \mu(G - F_y) = \mu(G - F \sin \theta).$$

The acting force is equal to the friction force when the object start to move, that is $f = F \cos \theta$, or $\mu(G - F \sin \theta) = F \cos \theta$. Thus,

$$F = \frac{\mu G}{\cos \theta + \mu \sin \theta}, \quad \left(0 \leqslant \theta < \frac{\pi}{2}\right).$$

The smallest value of F is obtained when the denominator

$$g(\theta) = \cos \theta + \mu \sin \theta$$

is the largest.

Solving

$$g'(\theta) = -\sin \theta + \mu \cos \theta = 0,$$

we get $\tan\theta = \mu$ or $\theta = \arctan\mu$.

Since

$$g''(\theta) = -\cos\theta - \mu\sin\theta = -\cos\theta(1 + \mu\tan\theta)$$

$$= -\frac{1}{\sqrt{1+\tan^2\theta}}(1 + \mu\tan\theta),$$

we have

$$g''(\arctan\mu) = -\sqrt{1+\mu^2} < 0.$$

Therefore, $g(\theta)$ has a maximum value at $\theta = \arctan\mu$. It follows that F has a minimum value at $\theta = \arctan\mu$.

Exercises 3.3

1. Find the Taylor expansions and the remainders of the following functions at $x = 0$.

   (1) $\cos x$;   (2) $\ln(1+x)$;   (3) $\ln(1-x)$;

   (4) xe^x;   (5) $\arctan x$;   (6) $\arcsin x$.

2. Approximate the following values to five decimals.

   (1) $\sin 10°$;   (2) $\sqrt[5]{1000}$.

3. Find the following limits.

   (1) $\displaystyle\lim_{x\to 0}\frac{e^x - e^{-x}}{\sin x}$;

   (2) $\displaystyle\lim_{x\to 0}\frac{\tan x - x}{x - \sin x}$;

   (3) $\displaystyle\lim_{x\to 0}\frac{e^{\alpha x} - \cos\alpha x}{e^{\beta x} - \cos\beta x}$;

   (4) $\displaystyle\lim_{x\to\frac{\pi}{4}}\frac{\tan x - 1}{\sin 4x}$;

   (5) $\displaystyle\lim_{x\to 0}\frac{(e^{x^2} - 1)\sin x^2}{x^2(1 - \cos x)}$;

   (7) $\displaystyle\lim_{x\to 0}\frac{x(e^x + 1) - 2(e^x - 1)}{x^3}$;

   (6) $\displaystyle\lim_{x\to 0}\frac{x\cot x - 1}{x^2}$;

   (8) $\displaystyle\lim_{x\to 0}\frac{1 - \cos x^2}{x^2\sin x^2}$;

   (9) $\displaystyle\lim_{x\to 0}\frac{\ln\tan 7x}{\ln\tan 2x}$;

   (10) $\displaystyle\lim_{x\to\frac{\pi}{2}}\frac{\tan x}{\tan 3x}$;

   (11) $\displaystyle\lim_{x\to 0}\frac{e^x - e^{-x}}{\sin x\cos x}$;

   (12) $\displaystyle\lim_{x\to 0}\frac{e^{a\sqrt{x}} - 1}{\sqrt{\sin bx}}$;

   (13) $\displaystyle\lim_{x\to 0}\frac{e^{\tan x} - e^x}{\tan x - x}$;

   (14) $\displaystyle\lim_{x\to 0}\frac{(a+x)^x - a^x}{x^2}$, $(a > 0)$;

(15) $\lim\limits_{x \to 0} \dfrac{e^x \sin x - x(1 + x)}{x^3}$;

(16) $\lim\limits_{x \to 0} \dfrac{(1 + x)^{\frac{1}{x}} - e}{x}$;

(17) $\lim\limits_{x \to \infty} x \left(e^{-\frac{1}{x}} - 1\right)$;

(19) $\lim\limits_{x \to \frac{\pi}{2}} (\cos x)^{\frac{\pi}{2} - x}$;

(18) $\lim\limits_{x \to 1} \left(\dfrac{x}{x - 1} - \dfrac{1}{\ln x}\right)$;

(20) $\lim\limits_{x \to 1} \left(\tan \dfrac{\pi x}{4}\right)^{\tan \frac{\pi x}{2}}$;

(21) $\lim\limits_{x \to \infty} \left(\cos \dfrac{1}{x}\right)^x$;

(23) $\lim\limits_{t \to 0} \left(\dfrac{\sin t}{t}\right)^{\frac{1}{t^2}}$;

(22) $\lim\limits_{x \to \frac{\pi}{2}} (\tan x)^{2x - \pi}$;

(24) $\lim\limits_{x \to \infty} x \sin \dfrac{k}{x}$;

(25) $\lim\limits_{x \to \infty} \left(1 + \dfrac{1}{x^2}\right)^x$;

(26) $\lim\limits_{x \to 0} \left(\dfrac{\cot x}{x} - \dfrac{1}{x^2}\right)$;

(27) $\lim\limits_{x \to 0} \left(\dfrac{2}{\pi} \arccos x\right)^{\frac{1}{x}}$;

(28) $\lim\limits_{x \to 0} \left(\cot x - \dfrac{1}{x}\right)$;

(30) $\lim\limits_{x \to 0} \left[\dfrac{(1 + x)^{\frac{1}{x}}}{e}\right]^{\frac{1}{x}}$.

(29) $\lim\limits_{x \to 1^-} \ln x \cdot \ln(1 - x)$;

4. Find the extreme values of the following functions.

(1) $y = 2x^3 - 3x^2$;

(3) $y = \sqrt[3]{(x^2 - a^2)^2}$;

(2) $y = \dfrac{3x^2 + 4x + 4}{x^2 + x + 1}$;

(4) $y = x - \ln(1 + x)$;

(5) $y = x^2 e^{-x^2}$;

(6) $y = \cos x + \sin x, \ \left(-\dfrac{\pi}{2} \leqslant x \leqslant \dfrac{\pi}{2}\right)$;

(7) $y = x - \sin x$;

(8) $y = x^2 (x - 1)^3$.

5. Find the extreme values of the following functions on the given interval.

(1) $y = x^4 - 2x^2 + 5$ on $[-2, 2]$;

(2) $y = \sqrt{100 - x^2}$ on $[-6, 8]$;

(3) $y = \dfrac{x - 1}{x + 1}$ on $[0, 4]$;

(4) $y = \sin 2x - x$ on $\left[-\dfrac{\pi}{2}, \dfrac{\pi}{2}\right]$.

6. Show that if $p > 1$, then

$$x^p + (1 - x)^p \geqslant \dfrac{1}{2^{p-1}}$$

for any x in $[0, 1]$.

7. Given a rectangular piece of cardboard with length 8 cm and width 5 cm, cut off four squares from the corners such that the remaining part of the cardboard can be folded into a box without a cover. Find the length of the side of the small squares such that the box has maximum volume.

8. Find the height of a conical funnel such that the funnel has maximum volume. Assume that the generator of the cone is 20 cm.

9. Prove that the amount of material needed to build a conical tent is least when the height is $\sqrt{2}$ times of the radius of the base of the conical tent.

10. Given a round table with radius a, find the height of a lamp hanging above the center of the table such that the brightness at the edge of the table has a maximum value.

    Hint: Brightness $I = k\dfrac{\sin\varphi}{r^2}$, where φ is the angle of light, r is the distance between the light source and the table, and k is the intensity of the light source.

11. Three points A, B and C are not on the same straight line, and $\angle ABC = 60°$. A car travels from A to B with velocity 80 km/h. A train travels at the same time from B to C with velocity 50 km/h. If $AB = 200$ km, then how many hours later is the distance between the car and the train the shortest?

12. Let $x_1, x_2, \cdots, x_n$ be n numbers. Find x, such that the value of the sum

$$(x_1 - x)^2 + (x_2 - x)^2 + \cdots + (x_n - x)^2$$

    is the smallest. If $x_1, x_2, \cdots, x_n$ are the measured results of a quantity A, then x is usually taken as the value of A.

13. The volume of a cylinder-shaped bucket without a cover is $4.71 \left(\approx \dfrac{3}{2}\pi\right)$ m². The metal used to make the base of the bucket costs \$3.00 per square meter. The metal used to make the side of the bucket costs \$2.00 per square meter. How do we minimize the cost of this bucket?

14. City A is located on the bank of a river. City B is located 40 km away from the bank. (Assume the bank is a straight line and city A is viewed as a point on the line.) The perpendicular line from city B to the bank meets the bank at point C, which is 50 km away from A. A water plant needs to be built for both cities. The cost of the water pipes from the water plant to city A is \$500 per kilometer. The cost of the water pipes

from the water plant to city B is $700 per kilometer. Where should the water plant be built on the bank to minimize cost of water pipes?

15. A railway running from south to north connects city A and city B. The distance between the two cities is 15 km. A factory is located 2 km west of city B. A shipment needs to be made from city A to the factory. The railway fare is $3.00 per kilometer. The highway fare is $5.00 per kilometer. From which point on the railway should we choose to build a highway to the factory so that the total fare is the lowest?

16. A picture with height 1.4 m is hanging on a wall. The base of the picture is 1.8 m higher than the eye level of an observer. What is the distance between the observer and the wall such that the visual angle is the largest?

17. One canal with width a is perpendicular to another canal with width b. Find the length of the longest ship that can travel in the canals.

3.4 Examples in Physics

In this section, we are not going to repeat the definition of the definite integrals when we do the calculations. We go directly from $\Delta Q = q(x)\Delta x$ to integration.

Example 1. A disc with non-uniform density has radius R. The areal density is given by

$$\rho = ar + b,$$

where r is the distance from a point on the disc to the center of the disc, and a and b are constants. Find the mass of the disc.

Solution: Consider the figure bounded by and containing the area between two concentric circles with radius r and $r + dr$ (Figure 3.32). The mass is approximately equal to

$$\Delta m \approx 2\pi r(ar + b)\,dr.$$

Thus,

$$m = 2\pi \int_0^R r(ar + b)\,dr = \frac{1}{3}\pi R^2(2aR + 3b).$$

Example 2. A half circular sluice gate has radius 1 m. Find the pressure on the sluice gate when water is full.

Solution: Let the horizontal axis be the y-axis, and the positive direction of the x-axis be downward vertically (Figure 3.33.) Consider the pressure ΔP on the shaded region as in Figure 3.33. The area of this region is approximately equal to

$$2\sqrt{1 - x^2}\, dx.$$

Thus,

$$\Delta P = 2gx\sqrt{1 - x^2}\, dx,$$

where g is the acceleration of gravity.

Hence, the pressure on the entire sluice gate is

$$P = 2g \int_1^0 x\sqrt{1 - x^2}\, dx = \frac{2}{3}g.$$

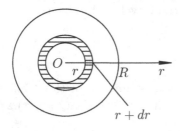

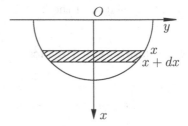

Fig. 3.32   Fig. 3.33

Example 3. A rod with uniform density ρ is bent into a half circle of radius a (Figure 3.34.) Find the moment of inertia when the rod revolves about its diameter (x-axis.)

Solution: The equation of the half circle is

$$y = \sqrt{a^2 - x^2}.$$

The moment of inertia of a small arc ds is

$$dI_x = y^2 \rho\, ds = \rho y^2 \sqrt{1 + y'^2}\, dx.$$

Thus, the moment of inertia of the rod is

$$I_x = \int_{-a}^{a} \rho y^2 \sqrt{1 + y'^2}\, dx = 2\rho \int_0^a (a^2 - x^2)\frac{a}{\sqrt{a^2 - x^2}}\, dx$$

$$= \frac{1}{2}\pi \rho a^3.$$

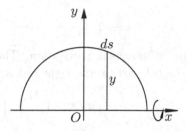

Fig. 3.34

Example 4. Two rods with uniform density are placed on a straight line. The distance between them is l. Find the gravitational force between them (Figure 3.35.) (Assume the length of both rods is l and the mass of both rods is M.)

Solution: Let the given straight line be the x-axis, and the left end point of the rod on the left be the origin of the coordinates. (Figure 3.35)

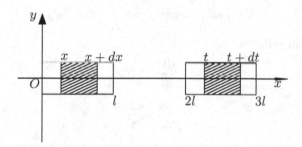

Fig. 3.35

The mass of the shaded part on the right rod is

$$\Delta m_2 = \frac{M}{l} \, dt.$$

The mass of the shaded part on the left rod is

$$\Delta m_1 = \frac{M}{l} \, dx.$$

By the law of universal gravitation, the gravity between the two shaded parts is

$$\Delta F = k \left(\frac{M}{l} \right)^2 \frac{1}{(t-x)^2} \, dx \, dt,$$

where k is the constant of universal gravitation. Thus, the gravity of the entire left rod to the shaded part on the right rod is

$$F_1 = k \left(\frac{M}{l} \right)^2 dt \int_0^l \frac{1}{(t-x)^2} \, dx = k \left(\frac{M}{l} \right)^2 \left[\frac{1}{t-l} - \frac{1}{t} \right] dt.$$

Integrate again we get the gravity of the entire left rod to the entire right rod

$$F = k \left(\frac{M}{l} \right)^2 \int_{2l}^{3l} \left[\frac{1}{t-l} - \frac{1}{t} \right] dt = k \left(\frac{M}{l} \right)^2 \ln \frac{4}{3}.$$

Example 5. The shape of a container is the result of the curve $x = f(y)$ revolving about the y-axis (Figure 3.36.) Water is poured into the container at a rate of $2t \text{cm}^3/\text{s}$. Determine the function $f(y)$ such that the rate of rise of the water surface is $\dfrac{2}{\pi}$ cm/s.

Solution: The volume of the revolving solid is

$$V(y) = \int_0^y \pi x^2 \, dy = \pi \int_0^y [f(y)]^2 \, dy.$$

By the assumption, we have

$$\frac{dV}{dt} = 2t, \tag{1}$$

$$\frac{dy}{dt} = \frac{2}{\pi}. \tag{2}$$

Since

$$\frac{dV}{dt} = \frac{dV}{dy} \frac{dy}{dt} = \pi [f(y)]^2 \frac{2}{\pi} = 2[f(y)]^2,$$

by (1), we have

$$f(y) = \sqrt{t}.$$

Also, by (2), we have

$$y = \frac{2}{\pi} t.$$

Therefore,

$$x = f(y) = \sqrt{\frac{\pi}{2} y}.$$

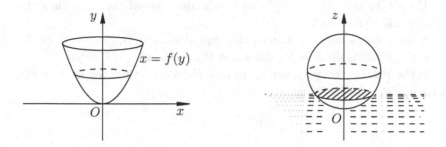

Fig. 3.36 Fig. 3.37

Example 6. A ball with uniform density 0.1 kg/m^3 is placed in water. The radius of the ball is 1 m. Find the height of the part of the ball under water. If a force is applied to the ball, and it presses half of the ball into water, find the work done by the force against the buoyancy (assume the density of water is 1×10^3 kg/m^3).

Solution: Suppose the surface of the water is $z = z$. Then the buoyancy to the ball is

$$b(z) = \int_0^z \pi[1^2 - (1-\tau)^2]\,d\tau = \pi\left(z^2 - \frac{1}{3}z^3\right).$$

If the ball sits still at $z = h$, that is, the weight of the ball is equal to the buoyancy, then

$$\pi\left(h^2 - \frac{1}{3}h^3\right) = \frac{4}{3}\pi \times 1^3(0.1).$$

Simplify this equation, we get

$$f(h) = h^3 - 3h^2 + 0.4 = 0.$$

Now we need to solve this equation.

Since $f(0.40) = -0.016 < 0$, and $f(0.39) = 0.003 > 0$, there exists a root of the equation between 0.39 and 0.40. Let $h_0 = 0.39$, we have

$$h_1 = 0.39 - \frac{f(0.39)}{f'(0.39)} = 0.39 + \frac{0.0030}{1.8873} = 0.3916.$$

Substitute this value into the original equation, we have

$$f(0.3916) = -0.0001 < 0.$$

This implies that the root is between 0.39 and 0.3916. Also, by calculation, $f(0.3915) > 0$, and this implies that the root is between 0.3915 and 0.3916.

If we take $h = 0.3916$ as the approximate value of the root, then the error is less than 0.0001.

Since there is only one root of this equation in $[0, 1]$, the height of the part of the ball sinking under the water is 0.3916 m.

If the ball be pressed down dz m into the water, then the work of the force against the buoyancy is

$$dW = f(z)\, dz.$$

Thus, the work of the pressure that presses down half of the ball is

$$W = \int_1^{0.3916} \pi \left(z^2 - \frac{1}{3}z^3 \right) dz = \pi \left[\frac{1}{3}z^3 - \frac{1}{12}z^4 \right]\Big|_{0.3916}^1$$
$$= \pi \left[\left(\frac{1}{3} - \frac{1}{12} \right) - \left(\frac{1}{3}(0.3916)^3 - \frac{1}{12}(0.3916)^4 \right) \right]$$
$$\approx 0.7285.$$

Exercises 3.4

1. A particle moves along a straight line. The acceleration at time t is $t^2 - 1$, and the velocity is $\frac{1}{3}$ at time $t = 1$. Find the function of the distance s with respect to time t (assume that $s(0) = 0$).

2. A particle moves along a straight line. Because of resistance, the decrease of the velocity per second is 2 m/s. If the velocity at the beginning is 25 m/s, how far can the particle travel?

3. A rocket is launched upright to the sky with the initial velocity zero. The acceleration is the function $\dfrac{A}{a - bt}$ with positive values (where $a > 0$ and $b > 0$). Find the velocity and the height from the ground at time t_1 (where $t_1 < \dfrac{a}{b}$).

4. A particle moves along a circle of radius 1 m. The initial velocity is 3 m/s. The angular acceleration is $6(t - 1)$ m/s^2. How much time will it need to travel 9 m.

5. Find the work done by a force that lifts an object with mass m from the ground to height h.

6. The density of clod is 1.5 t/m^3. Find the work that needs to be done to dig an impounding reservoir with the shape of a half ball (assume the reservoir is 20 m deep).

7. A total of 24.5 N force is needed to stretch a spring 0.025 m from its original length of 1 m. Find the work that is needed to stretch this spring from 1.1 m to 1.2 m.

8. A rod has length 30 cm and linear density 30 g/cm. Fix one end of the rod and let the other end turn in a horizontal plane. If the angular velocity is 10 π/s, find the kinetic energy of the rod.

9. A board in the shape of an isosceles trapezoid with top 6 m, base 10 m and height 5 m is placed 20 m under the surface of the water vertically. Find the pressure of the water to the board.

10. The decomposition rate of a substance is directly proportional to the time of the chemical reaction. If the ratio is 2, how much of the substance will remain after 3 seconds? Assume the weight of the substance is m_0 (g) at the beginning of the experiment.

11. The velocity of a liquid flowing out from a container is $v = c\sqrt{2gh}$, where g is the acceleration of gravity, h is the height of surface of the liquid from the opening of the container, and $c = 0.6$ is an experimental coefficient. Now we have a container in the shape of cylinder. The diameter of the base is 1 m and the height is 2 m. Fill the entire container with water and then let the water flow out from a round hold with diameter 1 cm on the base. How long will it take to let all water flow out of the container?

12. A rod has uniform mass, and its linear density is a constant ρ. The length of the rod is $2l$. A particle M is placed on the perpendicular bisector of the rod at distance a from the rod. Find the gravitational force of the rod to the particle.

Chapter 4

Ordinary Differential Equations

4.1 First Order Differential Equations

4.1.1 *Concepts*

Some rules of motion are formulated in terms of differential equations—the relation between derivatives or differentials, even higher order derivatives or differentials of a function. Solving a differential equation is to find the actual function from these relations. Let's look at some examples first.

Example 1. An object with temperature $100°C$ is placed into a medium with temperature $0°C$ to cool off. By the law of cooling, the rate of cooling is directly proportional to the temperature T. Determine the relation between the temperature T and the time t.

Solution: By the law of cooling, the equation is

$$\frac{dT}{dt} = -kT, \tag{1}$$

where k is a positive constant.

Example 2. A submarine is diving into the water. The resistance is directly proportional to the velocity of diving. If the velocity is zero at the beginning, determine the equation of the movement.

Solution: The submarine sinks into the water by its weight W overcoming the resistance. Suppose the velocity of diving is $v(t)$ at time t. Then the resistance is kv (where k ia an constant). By the second law of motion,

$$\frac{W}{g}\frac{dv}{dt} = W - kv. \tag{2}$$

Example 3. A particle with mass m is at one end of a string having a pendular movement. The length of the string is l (Figure 4.1). Suppose the movement is on the same plane at any time. Let the natural position of the pendulum be the origin of the coordinates, and the length of the arc from the position of the moving particle M to O be s. Consider that $s > 0$ when the point M is at the right side of O, and $s < 0$ when the point M is at the left side of O. Suppose the pendulum is at the natural position at the beginning, and there is no resistance. Also, denote the maximum arc from M to O as s_0. Determine the equation of the movement.

Solution: Decomposing the gravity mg along the tangential direction and normal direction. The force in the tangential direction is $-mg\sin\theta$. By the second law of motion, we have the equation

$$m\frac{d^2s}{dt^2} = -mg\sin\theta.$$

Therefore,

$$l\frac{d^2\theta}{dt^2} = -g\sin\theta \qquad (3)$$

since $s = l\theta$.

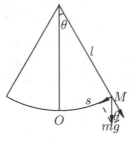

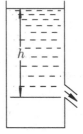

Fig. 4.1   Fig. 4.2

Example 4. A bucket contains water. Its cross-sectional area is $1\,\text{m}^2$ (Figure 4.2). Suppose water flows out from a $2\,\text{cm}$ diameter hole on the bucket $1\,\text{m}$ below the surface of the water. How much time is needed for all the water above the hole to flow out of the bucket if the resistance is ignored.

Solution: The velocity of water flowing out of the hole is the same as the free-fall velocity from the height h. It is approximately equal to

$$v = 0.6\sqrt{2gh} = 26.57\sqrt{h}\,\text{cm/s}.$$

Thus, the amount of water flowing out of the hold during the time dt is $v\,dt$ multiplied with the area of the hole:

$$\pi v\,dt.$$

If we assume that the water surface height decreased dh during the time dt, then we have

$$-10000\,dh = \pi v\,dt = \pi 26.57\sqrt{h}\,dt. \tag{4}$$

The problem can be solved if the function relation between the height h and the time t can be obtained.

Example 5. Determine a curve such that the tangent line at any point (x, y) on the curve is perpendicular to the radius vector of the same point.

Solution: The slope of the tangent line at any point of the curve is $\dfrac{dy}{dx}$, and the slope of the radius vector is $\dfrac{y}{x}$. Thus,

$$\frac{dy}{dx} = -\frac{x}{y}. \tag{5}$$

From these examples we can see that, to find the answer of these problems is to find the solution of these differential equations. A *differential equation* is an equation that involves the first or higher order derivatives or differentials of a function. For example:

$$\frac{dx}{dt} = x^2 + t^2; \tag{6}$$

$$\frac{dy}{dx} = xy; \tag{7}$$

$$y'' + ay' + by = \sin x; \tag{8}$$

$$(x + y)dx + (x - y)dy = 0. \tag{9}$$

These equations and the equations in examples 1–5 are all differential equations.

The highest order of the derivative of the unknown function of the equation is called the order of the equation. Equations (1), (2), (4), (5), (6), (7) and (9) are all first-order differential equations. Equations (3) and (8) are second-order differential equations.

If in a differential equation, the function y and its derivative all appear in the first degree, and the equation does not contain any product terms like yy', $y'y''$, $y'y'''$ or some functions of y, y', $\cdots$ such as $\sin y$, $e^{y'}$, $\cdots$, then the equation is called a linear differential equation, otherwise it is called nonlinear. Equations (1) through (9) are all linear equations. The general form of an nth order linear differential equation is

$$a_0(x)\frac{d^n y}{dx^n} + a_1(x)\frac{d^{n-1}y}{dx^{n-1}} + \cdots + a_n(x)y = f(x),$$

where $a_0(x)$, $a_1(x)$, $\cdots$, $a_n(x)$ and $f(x)$ are all functions of x.

The equations

$$\frac{d^2 y}{dx^2} + y\frac{dy}{dx} + y^2 = 0,$$

$$\frac{d^2 y}{dx^2} + \left(\frac{dy}{dx}\right)^{\frac{1}{2}} + y = 0$$

are second-order nonlinear differential equations.

In this chapter, we mainly introduce the first- and the second-order linear differential equations, and then discuss some higher-order differential equations and systems of equations.

If a function satisfies a differential equation, then it is a solution of the equation. For instance, $y = \cos 2x$ and $y = 1$ are solutions of the differential equation

$$y'' - 2(\cot 2x)y' = 0$$

in the interval $\left(0, \frac{\pi}{2}\right)$.

If the domain intervals of a solution are not mentioned, then the solution is on the entire real number line.

To solve a differential equation is to find all solution functions of the equation or to find solution functions that satisfy some given conditions.

4.1.2 *Separation of Variables*

By the Fundamental Theorem of Calculus, to find the integral of a function $f(x)$, is to find a function y such that

$$\frac{dy}{dx} = f(x).$$

Therefore, to find the integral of a function is to solve a first-order differential equation.

We start our study from here.

The above equation can be rewritten as

$$dy = f(x)dx. \tag{10}$$

Integrating both sides of the equation, we have

$$\int dy = \int f(x)\,dx,$$

that is

$$y = \int f(x)\,dx + C.$$

If we consider an equation such as

$$g(y)dy = f(x)dx, \tag{11}$$

then we only need to integrate on both side of the equation,

$$\int g(y)\,dy = \int f(x)\,dx + C.$$

Solving y as a function of x from this relation, we have the solution of (11). Thus, we know how to solve equations of the form (11) or of the form

$$\frac{dy}{dx} = \psi(y)f(x) \left(\psi(y) = \frac{1}{g(y)} \right).$$

The different variables of this kind of equations can be separated to two sides of the equal sign, and the method we use to solve this kind of equations is called *separation of variables*. This is one of the basic methods of solving differential equations.

Example 1. Find the solution of the problem in Example 1 of Section 4.1.1.

Solution: The equation that needs to be solved is

$$\frac{dT}{dt} = -kT,$$

that is

$$\frac{1}{T}\,dT = -k\,dt.$$

The temperature of the object is $100°C$ at $t = 0$. Thus

$$\int_{100}^{T} \frac{1}{T}\,dT = \int_{0}^{t} (-k)\,dt.$$

It follows that

$$\ln \frac{T}{100} = -kt.$$

Therefore,

$$T = 100e^{-kt},$$

and this is what we want to find.

Example 2. Find the solution of the problem in Example 2 of Section 4.1.1.

Solution: The equation that needs to be solved is

$$\frac{W}{g}\frac{dv}{dt} = W - kv.$$

Separating the variables, we get

$$\frac{W}{g(W - kv)}\, dv = dt.$$

The velocity $v = 0$ when $t = 0$. Thus

$$\int_0^v \frac{W}{g(W - kv)}\, dv = \int_0^t dt,$$

that is

$$\frac{-W}{kg}\ln\frac{W - kv}{W} = t.$$

It follows that

$$W - kv = We^{-kgt/W}.$$

Therefore,

$$v = \frac{W}{k}(1 - e^{-kgt/W}).$$

Since $s(0) = 0$, we have

$$\int_0^s ds = \int_0^t \frac{W}{k}(1 - e^{-kgt/W})\, dt.$$

Therefore, The distance with respect to time t is

$$s(t) = \frac{W}{k}\left(t + \frac{W}{kg}e^{-kgt/W} - \frac{W}{kg}\right).$$

From the above examples we can see that, the solutions not only need to satisfy the equation but also need to have certain values at some given points. This is called the *initial condition* of the equation. This kind of problems are called *initial value problems*.

Example 3. Find the solution of the problem in Example 4 of Section 4.1.1.

Solution: The problem becomes to solve the differential equation (4). Separating the variables, we get

$$dt = -119.8 \frac{1}{\sqrt{h}} \, dh.$$

This is an initial value problem, and the initial condition is: $h = 100$ cm when $t = 0$. Thus,

$$\int_0^t dt = -\int_{100}^h 119.8 \frac{1}{\sqrt{h}} \, dh.$$

It follows that

$$t = -239.6\sqrt{h} + 239.6\sqrt{100}.$$

Therefore,

$$t = 2396\text{s} = 39.9 \text{ min.}$$

when $h = 0$.

Example 4. Find the solution of the equation in Example 5 of Section 4.1.1.

Solution: Separating the variables in equation (5) we have

$$y\,dy = -x\,dx.$$

Integrating both sides, we have

$$y^2 + x^2 = C,$$

where C is a constant. This implies that any circle centered at the origin satisfies the equation. If we add a condition that the solution curve should pass the point $(3, 4)$, then

$$x^2 + y^2 = 25$$

is the solution we want.

Example 5. Find the solution of equation (7).

Solution: Separating the variables in equation (7), we have

$$\frac{1}{y}dy = x\,dx.$$

Integrating both sides, we have

$$\ln|y| = \frac{1}{2}x^2 + C'.$$

That is

$$|y| = e^{\frac{1}{2}x^2 + C'} = Ce^{\frac{1}{2}x^2}.$$

where $C = e^{C'}$.

We can also write the solution as

$$y = C_1 e^{\frac{1}{2}x^2},$$

where $C_1 = \pm C$ is an arbitrary constant.

Some equations are not of the form of (11), but can be transformed to the form of (11). Consider the equation

$$y' = f\left(\frac{y}{x}\right). \tag{12}$$

Let $y = xu$ in the equation where u is a new variable. Then

$$y' = u + xu'.$$

Substituting into the equation (12), we have

$$u + xu' = f(u).$$

Thus

$$\frac{dx}{x} = \frac{du}{-u + f(u)}.$$

Solving the equation we have

$$\ln|x| = \int \frac{du}{f(u) - u} + C_1.$$

Therefore,

$$x = Ce^{\int \frac{du}{f(u) - u}} = C\Phi(u),$$

where $C = \pm e^{C_1}$ is an arbitrary constant. Substituting back to y, we get $x = C\Phi\left(\frac{y}{x}\right).$

Example 6. Find the solution of equation (9).

Solution: The equation (9) can be written as

$$\frac{dy}{dx} = -\frac{x + y}{x - y} = -\frac{1 + \dfrac{y}{x}}{1 - \dfrac{y}{x}}.$$

Let $y = xu$, then we have

$$u + x\frac{du}{dx} = -\frac{1 + u}{1 - u}.$$

Thus,

$$\frac{dx}{x} = \frac{1-u}{-1-2u+u^2} du.$$

Integrating both sides, we have

$$-\frac{1}{2}\ln|-1-2u+u^2| = \ln|x| + C'.$$

That is

$$-1-2u+u^2 = Cx^{-2},$$

where $C = \pm e^{-2C'}$ is an arbitrary constant. Substitute $u = \frac{y}{x}$ into the above equation and the solution

$$y^2 - 2xy - x^2 = C$$

follows.

Example 7. Solve the equation

$$\frac{dy}{dx} = \frac{y + \sqrt{x^2 + y^2}}{x}.$$

Solution: Let $y = ux$, then we have

$$u + x\frac{du}{dx} = u + \sqrt{1 + u^2},$$

or

$$\frac{du}{\sqrt{1 + u^2}} = \frac{dx}{x}.$$

Integrating both sides, we have

$$\ln(u + \sqrt{1 + u^2}) = \ln Cx.$$

That is

$$u + \sqrt{1 + u^2} = Cx.$$

Substituting $u = \frac{y}{x}$ into the equation, we have

$$y + \sqrt{x^2 + y^2} = Cx^2.$$

Example 8. A small boat sails in a river. Suppose the instantaneous speed of the boat at every moment is v_1, and the boat always sails towards the point O on the shore that is directly opposite to the starting point. If the speed of the water is v_2, find the path of the boat (Figure 4.3).

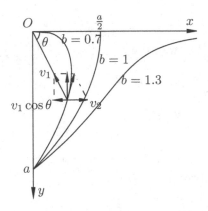

Fig. 4.3

Solution: Let the width of the river be a and set the coordinate system as in Figure 4.3, then the components of the speed of the boat along the two axis are

$$\frac{dx}{dt} = v_2 - v_1 \cos\theta,$$

$$\frac{dy}{dt} = -v_1 \sin\theta$$

respectively. Thus,

$$\frac{dy}{dx} = \frac{-v_1 \sin\theta}{v_2 - v_1 \cos\theta} = \frac{\sin\theta}{\cos\theta - b},$$

where $b = \dfrac{v_2}{v_1}$. Since

$$\sin\theta = \frac{y}{\sqrt{x^2 + y^2}}$$

and

$$\cos\theta = \frac{x}{\sqrt{x^2 + y^2}},$$

the equation can be written as

$$\frac{dy}{dx} = \frac{y}{x - b\sqrt{x^2 + y^2}},$$

and it follows that

$$\frac{dx}{dy} = \frac{x}{y} - b\sqrt{1 + \left(\frac{x}{y}\right)^2}.$$

Let $x = uy$. Then we have

$$\frac{du}{\sqrt{1+u^2}} = -b\frac{dy}{y}.$$

Integrating both sides, we get

$$\ln(u + \sqrt{1+u^2}) = -b\ln y + C.$$

Obviously, $x = 0$ (the boat at the start point) as $y = a$, so we have $u = 0$. Thus,

$$C = b\ln a$$

and therefore

$$\ln(u + \sqrt{1+u^2}) = b\ln\frac{a}{y}.$$

Solving for u, we get

$$u = \frac{1}{2}\left[\left(\frac{a}{y}\right)^b - \left(\frac{y}{a}\right)^b\right].$$

Substituting $u = \frac{x}{y}$ into the equation, we have the solution

$$x = \frac{a}{2}\left[\left(\frac{a}{y}\right)^{b-1} - \left(\frac{y}{a}\right)^{b+1}\right].$$

Now we discuss about this solution.

(i) If $v_1 > v_2$, then $b < 1$ and $x \to 0$ as $y \to 0$. The boat can reach the point O on the opposite shore.

(ii) If $v_1 < v_2$, then $b > 1$ and $x \to \infty$ as $y \to 0$. The boat will be washed off by the water and reach the opposite shore at infinity.

(iii) If $v_1 = v_2$, then $b = 1$. The trace of the boat is the parabola

$$x = \frac{a}{2}\left(1 - \frac{y^2}{a^2}\right).$$

We can see that $x \to \frac{a}{2}$ as $y \to 0$. In other words, the boat will reach the opposite shore at a point $\frac{a}{2}$ down stream from the point O. In fact, the boat will never reach the opposite shore since the time it spend is infinity.

To have a clear view of this conclusion, let's go back to the original equation

$$\frac{dy}{dx} = -v_1 \sin \theta.$$

Since

$$\sin \theta = \frac{y}{\sqrt{x^2 + y^2}},$$

we have

$$dt = -\frac{1}{v_1} \frac{\sqrt{x^2 + y^2}}{y} \, dy.$$

Substituting $x = \frac{a}{2} \left(1 - \frac{y^2}{a^2} \right)$ into the above equation we have

$$dt = -\frac{1}{v_1} \frac{\sqrt{\frac{a^2}{4} \left(1 - \frac{y^2}{a^2} \right)^2 + y^2}}{y} \, dy.$$

Let T be the time when then boat reach the opposite shore, then

$$T = \int_a^0 -\frac{1}{v_1} \frac{\sqrt{\frac{a^2}{4} \left(1 - \frac{y^2}{a^2} \right)^2 + y^2}}{y} \, dy = \frac{1}{v_1} \int_0^a \frac{\sqrt{\frac{a^2}{4} \left(1 - \frac{y^2}{a^2} \right)^2 + y^2}}{y} \, dy.$$

This is a *divergent improper integral* (a detailed discussion about divergent improper integrals can be found in Chapter 9).

We can at least do some estimation now. Since

$$\sqrt{\frac{a^2}{4} \left(1 - \frac{y^2}{a^2} \right)^2 + y^2} > \frac{a}{2},$$

we have

$$\int_0^a \frac{\sqrt{\frac{a^2}{4} \left(1 - \frac{y^2}{a^2} \right)^2 + y^2}}{y} \, dy > \frac{a}{2} \int_0^a \frac{1}{y} \, dy.$$

From the definition, we have

$$\int_0^a \frac{1}{y} \, dy = \lim_{\varepsilon \to 0^+} \int_\varepsilon^a \frac{1}{y} \, dy = \lim_{\varepsilon \to 0^+} \left[\ln |a| - \ln |\varepsilon| \right] = +\infty.$$

Therefore, $T = +\infty$.

4.1.3 Linear Differential Equations

In this section, we find the general solution of first order linear differential equations. The general form of the first order linear differential equations is the equation (15) below. Again, we start from the Fundamental Theorem of Calculus.

Consider the equation

$$d(\Psi(x)y) = f(x)dx. \tag{13}$$

Integrating this equation, we get

$$\Psi(x)y = \int f(x)\,dx + C''.$$

So the solution of equation (13) is

$$y = \frac{1}{\Psi(x)} \int f(x)\,dx + \frac{C''}{\Psi(x)}. \tag{14}$$

Equation (13) is

$$\Psi(x)\frac{dy}{dx} + \Psi'(x)y = f(x).$$

This equation can also be written as

$$\frac{dy}{dx} + P(x)y = Q(x), \tag{15}$$

where

$$P(x) = \frac{\Psi'(x)}{\Psi(x)}, \text{ and } Q(x) = \frac{f(x)}{\Psi(x)}. \tag{16}$$

Equation (15) is the general form of a first order linear equation. For any $P(x)$ and $Q(x)$, there always exist $\Psi(x)$ and $f(x)$ that satisfy equations of (16). In fact, from the two equations of (16), we have

$$\Psi(x) = C'e^{\int P(x)dx}, \text{ and } f(x) = C'Q(x)e^{\int P(x)dx}.$$

Thus, the solution of (15) is

$$y = e^{-\int P(x)dx}\left[\int Q(x)e^{\int P(x)dx}dx + C\right], \tag{17}$$

where $C = \dfrac{C''}{C'}$.

The integral $\displaystyle\int P(x)dx$ represents only one antiderivative of $P(x)$.

When $Q(x) = 0$, the equation becomes

$$\frac{dy}{dx} + P(x)y = 0,$$

and is called a *homogeneous equation*. The solution of this equation is

$$y = Ce^{-\int P(x)dx}.$$

This is the general solution of the homogeneous equation. Obviously,

$$e^{-\int P(x)dx} \int Q(x)e^{\int P(x)dx} dx$$

is a solution of (15), called a special solution of the equation. The general solution of (15) is the function (17). And it includes two parts, one is a special solution of the equation and the other is the general solution of the corresponding homogeneous equation.

Example 1. Solve the equation

$$\frac{dy}{dx} + y\cot x = x^2 \csc x.$$

Solution: Compare with (15), we have $P(x) = \cot x$ and $Q(x) = x^2 \csc x$. By (17), the solution is

$$y = e^{-\int \cot x dx}\left[\int x^2(\csc x)e^{\int \cot x dx}\, dx + C\right]$$

$$= e^{-\ln \sin x}\left[\int x^2(\csc x)e^{\ln \sin x}\, dx + C\right]$$

$$= \frac{1}{\sin x}\left(\int x^2\, dx + C\right) = \frac{1}{\sin x}\left(\frac{x^3}{3} + C\right)$$

$$= \frac{1}{3}x^3 \csc x + C \csc x.$$

Example 2. Solve the equation

$$x\frac{dy}{dx} = y + x^3 + 3x^2 - 2x.$$

Solution: Compare with (15), we have

$$P(x) = -x^{-1}, \text{ and } Q(x) = x^2 + 3x - 2.$$

By (17), the solution is

$$y = e^{\int \frac{1}{x}dx}\left[\int (x^2 + 3x - 2)e^{-\int \frac{1}{x}dx} dx + C\right]$$

$$= x\int \left(x + 3 - \frac{2}{x}\right) dx + Cx = \frac{x^3}{2} + 3x^2 - 2x\ln|x| + Cx.$$

Example 3. Solve the equation

$$y \ln y \, dx + (x - \ln y) dy = 0.$$

Solution: Consider y as the independent variable and x as the dependent variable. The equation can be written as

$$\frac{dx}{dy} + \frac{x}{y \ln y} = \frac{1}{y}.$$

Compare with (15), we have $P(y) = \dfrac{1}{y \ln y}$ and $Q(x) = \dfrac{1}{y}$. By (17), the solution is

$$x = e^{-\int \frac{1}{y \ln y} dy} \left[\int \frac{1}{y} e^{\int \frac{1}{y \ln y} dy} dy + C \right]$$

$$= e^{-\ln \ln y} \left[\int \frac{1}{y} e^{\ln \ln y} dy + C \right] = \frac{1}{\ln y} \left[\int \frac{\ln y}{y} dy + C \right]$$

$$= \frac{1}{2} \ln y + \frac{C}{\ln y}.$$

Some nonlinear equations can be converted to linear equations, for instance, the *Bernoulli equation*

$$\frac{dy}{dx} + P(x)y = Q(x)y^n, \quad (n \neq 0, 1).$$

Dividing both side by y^n, we have

$$y^{-n} \frac{dy}{dx} + P(x)y^{-n+1} = Q(x).$$

Let $z = y^{-n+1}$, then the equation becomes

$$\frac{dz}{dx} + (-n+1)P(x)z = (-n+1)Q(x).$$

This is a linear equation.

Example 4. Solve the equation

$$\frac{dy}{dx} - xy = -e^{-x^2} y^3.$$

Solution: Let $u = y^{-2}$. Then

$$\frac{1}{2} \frac{du}{dx} + xu = e^{-x^2}.$$

This is a linear equation with the solution

$$u = e^{-x^2}(2x + C).$$

Thus, the solution of the original equation is

$$y^2 = e^{x^2}(2x + C)^{-1}.$$

The general solution of the first order linear equation consists of two parts, a special solution of the equation (15) and the general solution of the corresponding homogeneous equation. This result can also be obtained by the *method of variation of constants*.

By the method of separation of variables, we get the general solution

$$y = Ce^{-\int P(x)dx}$$

of the homogeneous equation $\dfrac{dy}{dx} + P(x)y = 0$, where C is a constant. To find a special solution of (15), we replace C by an unknown function $C(x)$, and substitute

$$C(x)e^{-\int P(x)dx}$$

into (15), to get a differential equation for $C(x)$

$$C'(x) = Q(x)e^{\int P(x)dx}.$$

The solution of this equation is

$$C(x) = \int Q(x)e^{\int P(x)dx}\, dx.$$

Thus, we obtain a special solution of (15)

$$e^{-\int P(x)dx}\int Q(x)e^{\int P(x)dx}dx.$$

The sum of the general solution of the homogeneous equation and this special solution is the general solution of (15).

Exercises 4.1

1. Find the general solution of the following equations by separation of variables.

   (1) $(1 + y^2)\, dx = x\, dy$;   (3) $y - xy' = a(y + x^2 y)$;

   (2) $y' = e^{2x-y}$;   (4) $xyy' = 1 - x^2$;

   (5) $(\tan x)(\sin^2 y)\, dx + (\cos^2 x)(\cot y)\, dy = 0$;

   (6) $(xy^2 + x)\, dx + (y - x^2 y)\, dy = 0$;

   (7) $yy' = \dfrac{1 - 2x}{y}$;

   (8) $xy' + y = y^2$.

2. Find the special solution of the following equations satisfying the given initial conditions.

   (1) $y' \sin x = y \ln y,$ $\quad y\big|_{x=\frac{\pi}{2}} = 1;$

   (2) $y' = \dfrac{1+y^2}{1+x^2},$ $\quad y\big|_{x=0} = 1;$

   (3) $(\sin y)(\cos x)dy = (\cos y)(\sin x)\,dx,$ $\quad y\big|_{x=0} = \dfrac{\pi}{4};$

   (4) $y - xy' = 6(1 + x^2 y'),$ $\quad y\big|_{x=1} = 1.$

3. Find a curve that passes the point $(2, 3)$, and whose tangent line segments between the two axes are bisected by the tangent point.

4. Find all the curves that have tangent lines that intersect with both x-axis and y-axis, and the y-intersection point bisects the tangent line segment between the tangent point on the curve and the x-intersection point.

5. The cooling rate of an object is directly proportional to the difference between the temperature of the object and the temperature of the medium that the object is in. The proportionality coefficient is $k = k_0(1 + at)$. Let the temperature of the object be $\theta = \theta_0$, and the temperature of the medium be $\theta = \theta_1$ when $t = 0$. Determine the relation between the temperature θ and the time t.

6. The rate of chemical reaction of a kind of liquid is directly proportional to the remainder of the liquid. Assume the total amount of liquid is a. Determine the relation between the amount of liquid that has already participated in the chemical reaction at time t.

7. Under a given force, a particle with mass $1\,$g moves along a straight line. Starting from $t = 0$ s, the force is directly proportional to time, and is inversely proportional to the velocity of the particle. Assume that the velocity of the particle is 50 cm/s, and the force is 4 dyn at $t = 10$ s. Determine the differential equation of the movement of the particle. What is the velocity after one minute.

8. A motorboat is driven in still water with velocity $v = 10$ km/h. The engine is shut down after it reaches full power. The velocity of the motorboat reduce to $v_1 = 6$ km/h after 20 s. Assume that the resistance of water is directly proportional to the velocity of the motorboat. Find the velocity of the motorboat two minutes after the engine is shut down. Find the distance that the motorboat traveled one minute after the engine is shut down.

9. The following equations have the form $y' = f\left(\dfrac{y}{x}\right)$. Solve these equations.

(1) $y' = \dfrac{y^2}{x^2} - 2$;

(2) $y' = \dfrac{2xy}{3x^2 - y^2}$;

(3) $y' = \dfrac{x}{y} + \dfrac{y}{x}$;

(4) $y^2 + x^2 y' = xyy'$;

(5) $xy' = y \ln \dfrac{y}{x}$;

(6) $y' = e^{\frac{y}{x}} + \dfrac{y}{x}$;

(7) $y' = \dfrac{y}{x}\left(1 + \ln \dfrac{y}{x}\right)$.

10. Solve the equation

$$\frac{dy}{dx} = f\left(\frac{ax + by + c}{a_1 x + b_1 y + c_1}\right).$$

Hint: Let $x = \xi + \alpha$ and $y = \eta + \beta$ if the line $ax + by + c = 0$ intersects with the line $a_1 x + b_1 y + c_1 = 0$ at the point (α, β). Let $z = ax + by$ if these two lines are parallel.

11. Solve the equation

$$\frac{dy}{dx} = \frac{3y - 7x + 7}{3x - 7y - 3}.$$

12. Solve the equation

$$\frac{dy}{dx} = \frac{2x + 4y + 3}{x + 2y + 1}.$$

13. Determine the equation of a curve such that the length of the radius vector at an arbitrary point M on the curve is equal to the distance from the origin to the y-intersection of the tangent line at M.

14. A light source is placed by a reflector. The reflection rays from the reflector are parallel. Determine the shape of the reflector.

15. Find the solution of the following equations. Hint: Consider y as the dependent variable in (4), (5) and (6).

(1) $(1 + x^2)y' - 2xy = (1 + x^2)^2$; (2) $y' + ay = e^{mx}$;

(3) $y' + \dfrac{1 - 2x}{x^2} y = 1$;

(4) $(y^2 - 6x)y' + 2y = 0$;

(5) $y' = \dfrac{1}{2x - y^2}$;

(6) $y' = \dfrac{y}{2y \ln y + y - x}$;

(7) $dy + (xy - xy^3)dx = 0$;

(8) $y' + 2xy = 2x^3 y^3$;

(9) $(xy + x^2 y^3)y' = 1$;

(10) $3y^2 y' - ay^3 = x + 1$.

16. Suppose there is a curve on the xy-plane and l is the tangent line at an arbitrary point (x, y) on the curve. Let b be the y-intersection of l. Suppose the area of the rectangle with the line segment from the origin $(0, 0)$ to $(0, b)$ as one side, and the line segment from the origin $(0, 0)$ to $(x, 0)$ as the other side, is a constant a^2. Determine the expression of the curve.

17. A particle with mass m moves along a straight line. A force F is applied on the particle starting when the velocity of the particle is v_0. The force F is directly proportional to the cube of the time with proportionality coefficient k_1. The particle also received a resistance from the medium that is inversely proportional to the product of the velocity and the time with proportionality coefficient k. Determine the relation between the velocity and the time.

4.2 Second Order Differential Equations

4.2.1 *Reducible Differential Equations*

Some second order equations can be transformed into first order equations. Let's look at two simple situations:

(1) The second order equation does not involve the unknown function explicitly:

$$y'' = f(x, y').$$

Let $y' = p$. Then we have

$$p' = f(x, p),$$

and this substitution reduces the order of the equation to one. The solution of this first order equation is $p = \varphi(x, C_1)$. Since

$$\frac{dy}{dx} = \varphi(x, C_1),$$

the solution of the original equation is

$$y = \psi(x, C_1, C_2).$$

Example 1. Solve the equation $xy'' + y' = 4x$.

Solution: Let $y' = p$. Then we get

$$xp' + p = 4x,$$

or

$$\frac{d}{dx}(xp) = 4x.$$

Thus,

$$xp = 2x^2 + C_1,$$

and that is

$$\frac{dy}{dx} = 2x + \frac{C_1}{x}.$$

Integrating one more time, we have

$$y = x^2 + C_1 \ln |x| + C_2.$$

(2) The second order equation does not involve the independent variable explicitly:

$$y'' = f(y, y').$$

Again let $y' = p$. Since

$$\frac{d^2y}{dx^2} = \frac{dp}{dx} = \frac{dp}{dy} \cdot \frac{dy}{dx} = p\frac{dp}{dy},$$

the equation can be written as

$$\frac{dp}{dy} = \varphi(y, p).$$

Solve this equation and integrate the solution with respect to x, the solution of the original equation follows.

Example 2. Solve the equation

$$2y\frac{d^2y}{dx^2} = 1 + \left(\frac{dy}{dx}\right)^2.$$

Solution: Let $\dfrac{dy}{dx} = p$, then

$$2py\frac{dp}{dy} = 1 + p^2.$$

Separating the variables, we have

$$\frac{2p}{1+p^2}dp = \frac{dy}{y}.$$

Integrating this equation, we get

$$\ln(1+p^2) = \ln C_1 y,$$

or

$$p = \pm\sqrt{C_1 y - 1}.$$

That is

$$\frac{dy}{dx} = \sqrt{C_1 y - 1}, \text{ or } \frac{dy}{dx} = -\sqrt{C_1 y - 1}.$$

Integrating the above equations yields the solution of the original equation

$$2\sqrt{C_1 y - 1} = \pm C_1 x + C_2.$$

Example 3. Find the equation of the catenary. A catenary is the shape of an uniform non-elastic rope hanging from two points of equal height under its own weight.

Solution: Set the coordinate system as in Figure 4.4, and assume the equation we want to find is $y = f(x)$. Consider the part of the hanging rope between the points $M(x, y)$ and $M_1(x + \Delta x, y + \Delta y)$. There are three forces acting on this segment:

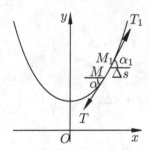

Fig. 4.4

(i) The tension T at M. This force is in the direction of the tangent line to the curve at M and has an angle α with the x-axis. We have $\tan\alpha = f'(x)$.

(ii) The tension T_1 at M_1. This force is in the direction of the tangent line to the curve at M_1 and has an angle α_1 with the x-axis. We have $\tan \alpha_1 = f'(x + \Delta x)$.

(iii) The weight of this segment $\rho g \Delta s$, where ρ is the mass per unit length, g is the acceleration of gravity, and Δs is the length of the segment. When the two points M and M_1 are very close, we have

$$\Delta s \approx \sqrt{(\Delta x)^2 + (\Delta y)^2} = \sqrt{1 + \left(\frac{\Delta y}{\Delta x}\right)^2} \, \Delta x.$$

The sum of these three forces is zero when the chain is in equilibrium. It follows that the sum of the horizontal components of the three forces as well as the sum of the vertical components of the three forces are also zero. Thus, we have that

$$T \cos \alpha = T_1 \cos \alpha_1 = T_0,$$
$$T \sin \alpha + \rho g \Delta s = T_1 \sin \alpha_1.$$

From the first equation above, we have

$$T = \frac{T_0}{\cos \alpha} \text{ and } T_1 = \frac{T_0}{\cos \alpha_1}.$$

Substituting into the second equation above, we get

$$T_0 \tan \alpha + \rho g \Delta s = T_0 \tan \alpha_1,$$

or

$$T_0(\tan \alpha_1 - \tan \alpha) \approx \rho g \sqrt{1 + \left(\frac{\Delta y}{\Delta x}\right)^2} \, \Delta x.$$

By the Mean-Value Theorem,

$$\tan \alpha_1 - \tan \alpha = f'(x + \Delta x) - f'(x) = f''(x + \theta \Delta x)\Delta x,$$

where $0 < \theta < 1$. Thus

$$T_0 f''(x + \theta \Delta x) \approx \rho g \sqrt{1 + \left(\frac{\Delta y}{\Delta x}\right)^2}.$$

Therefore,

$$\frac{d^2 y}{dx^2} = \frac{\rho g}{T_0} \sqrt{1 + y'^2}$$

as $\Delta x \to 0$.

This is a second-order equation that does not have the independent variable explicitly. It can be transformed to a first-order equation

$$\frac{dp}{dx} = k\sqrt{1 + p^2},$$

where $k = \dfrac{\rho g}{T_0}$, by letting $y' = p$. Separating variables, we obtain

$$\frac{dp}{\sqrt{1 + p^2}} = kdx.$$

Integrating the equation, we get

$$\ln(p + \sqrt{1 + p^2}) = kx + C_1,$$

or

$$p = \frac{1}{2}(e^{kx+C_1} - e^{-kx-C_1}).$$

That is

$$\frac{dy}{dx} = \frac{1}{2}(e^{kx+C_1} - e^{-kx-C_1}).$$

Integrating one more time, we have

$$y = \frac{1}{2k}(e^{kx+C_1} + e^{-kx-C_1}) + C_2.$$

The minimum value of the curve is reached at $x = 0$. It follows that $y' = 0$ at this point, which implies $C_1 = 0$. Thus, the solution of the equation is

$$y = \frac{1}{2k}(e^{kx} + e^{-kx}) + C_2.$$

If we move the x-axis in such a way that the distance between the lowest point of the rope and the origin of the coordinate system is equal to $\dfrac{1}{k}$, in other words, $y = \dfrac{1}{k}$ as $x = 0$, then we have $C_2 = 0$. The solution becomes

$$y = \frac{1}{k}\cosh kx = \frac{1}{2k}(e^{kx} + e^{-kx}).$$

This is the equation of a *catenary*.

4.2.2 *Second Order Linear Differential Equations*

All the second order equations discussed in Section 4.2.1 can be reduced to first order equations through some transformations. In this section, we study second order linear differential equations in general.

The general form of a second order linear differential equation is

$$y'' + p(x)y' + q(x)y = f(x), \tag{1}$$

where $p(x)$, $q(x)$ and $f(x)$ are continuous functions, and y, y' and y'' appear in the first degree. If $f(x) = 0$, then the equation is homogeneous, otherwise it is non-homogeneous.

The homogeneous equation corresponding to (1) is

$$y'' + p(x)y' + q(x)y = 0. \qquad (2)$$

The initial value problem of homogeneous second order linear differential equation is

$$\begin{cases} y'' + p(x)y' + q(x)y = 0. \\ y(x_0) = \alpha, \ \ y'(x_0) = \beta. \end{cases} \qquad (I)$$

In Chapter 10, we will prove that the solution of problem (I) exists and is unique.

Based on this existence and uniqueness theorem, we have the following structure theorem for solutions:

If $y_1(x)$, $y_2(x)$ are linearly independent solutions of the equation

$$y'' + p(x)y' + q(x)y = 0,$$

then

$$C_1 y_1(x) + C_2 y_2(x)$$

is also a solution of the equation for arbitrary constants C_1, C_2. On the other hand, any solution of the equation can be written in the form of

$$y = C_1 y_1(x) + C_2 y_2(x).$$

In other words, $C_1 y_1(x) + C_2 y_2(x)$ is the general solution of the equation.

Before proving this structure theorem, we need to know the definition of linearly independent solutions.

Two functions $\varphi(x)$ and $\psi(x)$ are linearly dependent on $[a, b]$ if there exist two constants C_1 and C_2 which are not zero simultaneously, such that

$$C_1 \varphi(x) + C_2 \psi(x) = 0$$

on $[a, b]$, otherwise they are linearly independent.

If two functions $\varphi(x)$ and $\psi(x)$ satisfy

$$C_1 \varphi(x) + C_2 \psi(x) = 0,$$

then it is also true that

$$C_1 \varphi'(x) + C_2 \psi'(x) = 0.$$

Thus, there exist two constants C_1 and C_2 (not zero simultaneously), if and only if

$$\begin{vmatrix} \varphi(x) & \psi(x) \\ \varphi'(x) & \psi'(x) \end{vmatrix} = 0.$$

This is called the *Wronskian determinant* of $\varphi(x)$ and $\psi(x)$, denoted by $W(\varphi, \psi)$:

$$W(\varphi, \psi) = \begin{vmatrix} \varphi(x) & \psi(x) \\ \varphi'(x) & \psi'(x) \end{vmatrix}.$$

Now we can conclude that if the functions $\varphi(x)$ and $\psi(x)$ are linearly dependent, then $W(\varphi, \psi) = 0$.

The converse is not necessarily true. For instance, let

$$\varphi(x) = \begin{cases} (x-1)^2, & 0 \leqslant x \leqslant 1; \\ 0, & 1 < x \leqslant 2. \end{cases}$$

$$\psi(x) = \begin{cases} 0, & 0 \leqslant x \leqslant 1; \\ (x-1)^2 & 1 < x \leqslant 2. \end{cases}$$

These two functions are linearly independent on $[0,2]$ since $C_1 = 0$ when the equation

$$C_1 \varphi(x) + C_2 \psi(x) = 0$$

holds on $[0,1]$, and $C_2 = 0$ when the same equation holds on $[1,2]$. Thus, we must have $C_1 = C_2 = 0$ when this equation holds on $[0,2]$. By definition, $\varphi(x)$ and $\psi(x)$ are linearly independent on $[0,2]$, but $W(\varphi, \psi)$ is equal to zero on $[0,2]$.

The above discussion is for arbitrary functions. The conclusion is different for functions that are solutions of homogeneous linear equations.

Suppose $y_1(x)$ and $y_2(x)$ are solutions of the homogeneous linear equation

$$y'' + p(x)y' + q(x)y = 0.$$

If $W(y_1, y_2)$ is equal to zero at a point in $[a,b]$, then $y_1(x)$ and $y_2(x)$ are linearly dependent on $[a,b]$. In other words, if $y_1(x)$ and $y_2(x)$ are linearly independent, then $W(y_1, y_2)$ is not zero for every point in $[a,b]$.

In fact, if there exists a point x_0 such that

$$\begin{vmatrix} y_1(x_0) & y_2(x_0) \\ y_1'(x_0) & y_2'(x_0) \end{vmatrix} = 0,$$

then we can find constants C_1 and C_2 that are not zero simultaneously, such that

$$\begin{cases} C_1 y_1(x_0) + C_2 y_2(x_0) = 0, \\ C_1 y_1'(x_0) + C_2 y_2'(x_0) = 0. \end{cases}$$

Consider the function

$$y(x) = C_1 y_1(x) + C_2 y_2(x),$$

which is also a solution of the homogeneous equation, and satisfies the initial conditions

$$y(x_0) = 0, \quad y'(x_0) = 0.$$

On the other hand, $\tilde{y}(x) = 0$ is a solution of the homogeneous equation and satisfies the initial conditions $\tilde{y}(x_0) = 0$, $\tilde{y}'(x_0) = 0$. Thus, $\tilde{y}(x)$ and $y(x)$ satisfy the same initial conditions. By the existence and uniqueness theorem of solutions,

$$y(x) = \tilde{y}(x) = 0.$$

This implies that $y_1(x)$ and $y_2(x)$ are linearly dependent.

Suppose $y_1(x)$ and $y_2(x)$ are solutions of a homogeneous equation, and let

$$W(x) = W(y_1(x), y_2(x)).$$

The derivative of $W(x)$ is

$$\frac{dW}{dx} = \frac{d}{dx}(y_1 y_2' - y_1' y_2) = y_1 y_2'' - y_1'' y_2.$$

Since

$$y_1'' = -p(x)y_1' - q(x)y_1,$$

and

$$y_2'' = -p(x)y_2' - q(x)y_2,$$

we have

$$\frac{dW}{dx} = y_1(-py_2' - qy_2) - y_2(-py_1' - qy_1) = -pW.$$

Solving the equation we have

$$W(x) = W(x_0)e^{-\int_{x_0}^{x} p(t)dt} \tag{3}$$

or

$$W(x) = Ce^{-\int p(x)dx}.$$

This is known as the *Liouville formula*. It gives the relation between the coefficient $p(x)$ of a homogeneous equation and the Wronskian determinant of its solutions y_1 and y_2. Also, $W(x)$ is equal to zero at a point is equivalent to $W(x)$ is equal to zero everywhere.

Hence, the Wronskian determinant of $y_1(x)$ and $y_2(x)$ is not equal to .zero or is zero everywhere, corresponding to the linear independence or linear dependence of the two solutions.

With the above preparations, now we are ready to prove the structure theorem of the solutions of homogeneous equations:

The equation

$$y'' + p(x)y' + q(x)y = 0$$

is homogeneous, so if $y_1(x)$, $y_2(x)$ are solutions of the equation, then for any constants C_1, C_2, $C_1 y_1(x) + C_2 y_2(x)$ is also a solution of the equation. On the other hand, if $y_1(x)$, $y_2(x)$ are linearly independent, then there exists x_0, such that

$$\begin{vmatrix} y_1(x_0) & y_2(x_0) \\ y_1'(x_0) & y_2'(x_0) \end{vmatrix} \neq 0.$$

Thus, for any solution $y(x)$, we can always find constants C_1 and C_2 such that

$$C_1 y_1(x_0) + C_2 y_2(x_0) = y(x_0),$$

$$C_1 y_1'(x_0) + C_2 y_2'(x_0) = y'(x_0).$$

Since both $y(x)$ and $C_1 y_1(x) + C_2 y_2(x)$ are solutions, and from the above equations, they satisfy the same initial conditions. By the uniqueness theorem, we have

$$y(x) = C_1 y_1(x) + C_2 y_2(x).$$

This proves the structure theorem for homogeneous second order linear differential equations. Thus, the problem of finding the general solution of the homogeneous differential equation is equivalent to the problem of finding two linearly independent solutions.

Now we prove the existence of such solutions. In fact, for any two groups of constants α_1, β_1; α_2, β_2 that satisfy

$$\begin{vmatrix} \alpha_1 & \alpha_1 \\ \beta_1 & \beta_1 \end{vmatrix} \neq 0,$$

solve the following two initial value problems

$$\begin{cases} y'' + p(x)y' + q(x)y = 0, \\ y(x_0) = \alpha_1, \ y'(x_0) = \beta_1; \end{cases}$$

and

$$\begin{cases} y'' + p(x)y' + q(x)y = 0, \\ y(x_0) = \alpha_2, \ y'(x_0) = \beta_2. \end{cases}$$

By the existence and uniqueness theorem for homogeneous second order linear differential equations, the above initial value problem have solution y_1, y_2. And they are linearly independent because when $x = x_0$

$$W(y_1(x_0), y_2(x_0)) = \begin{vmatrix} \alpha_1 & \alpha_2 \\ \beta_1 & \beta_2 \end{vmatrix} \neq 0.$$

This proves the existence of the two linearly independent solutions of equation (2).

When it comes to finding actual solutions, however, there is no general method to find such linearly independent solutions. However, if we can find a solution in some special ways like observation or a method that only fits the specific equation we want to solve, then we can obtain the other solution by the Liouville formula such that the two solutions are linearly independent.

Suppose $y_1(x)(\not\equiv 0)$ is the solution we have found and we need to find another solution y_2. By the Liouville formula, we have

$$W(y_1, y_2) = Ce^{-\int p\,dx} \quad (C \neq 0).$$

Multiplying $\dfrac{1}{y_1^2}$ on both sides of this equation, we have

$$\frac{y_1 y_2' - y_1' y_2}{y_1^2} = \frac{C}{y_1^2} e^{-\int p\,dx},$$

or

$$\frac{d}{dx}\left(\frac{y_2}{y_1}\right) = \frac{C}{y_1^2} e^{-\int p\,dx}.$$

Integrating both sides of this equation, we obtain

$$\frac{y_2}{y_1} = C\int \frac{1}{y_1^2} e^{-\int p\,dx}\,dx.$$

Therefore,

$$y_2 = Cy_1 \int \frac{1}{y_1^2} e^{-\int p\,dx}\,dx. \tag{4}$$

Obviously, y_1 and y_2 are linearly independent.

If a special solution of (2), y_1, can be found, then by formula (4), we can also find the other special solution of (2), y_2, and y_1 and y_2 are linearly independent. Thus, $C_1 y_1(x) + C_2 y_2(x)$ is the general solution of (2), where C_1, C_2 are arbitrary constants.

The problem of finding the general solution of (2) becomes the problem of finding a special solution of (2).

Example 1. Find the general solution of the equation $xy'' - y' = 0$.

Solution: It is easy to see that $y_1(x) = 1$ is a solution of this equation. By equation (4), the other solution y_2 is

$$y_2(x) = \int e^{-\int p \, dx} \, dx = \int e^{-\int \frac{1}{x} dx} \, dx = \frac{x^2}{2}$$

if we take $C = 1$. Thus, the general solution is $y(x) = C_1 + C_2 \dfrac{x^2}{2}$.

Now we study the solutions of non-homogeneous equation (1)

$$y'' + p(x)y' + q(x)y = f(x).$$

If we have a special solution y of the non-homogeneous equation and the general solution u of the corresponding homogeneous equation

$$y'' + p(x)y' + q(x)y = 0,$$

then $y + u$ is the general solution of the non-homogeneous equation.

In fact, substituting $y + u$ into the non-homogeneous equation shows that it satisfies the equation. On the other hand, if z is a solution of the non-homogeneous equation, then $z - y$ is a solution of the corresponding homogeneous equation.

It follows that, any solution of the non-homogeneous equation have the form of $y + u$. Since the general solution of

$$y'' + p(x)y' + q(x)y = 0$$

is $C_1 y_1(x) + C_2 y_2(x)$, where $y_1(x)$ and $y_2(x)$ are linearly independent, the general solution of

$$y'' + p(x)y' + q(x)y = f(x)$$

is $C_1 y_1(x) + C_2 y_2(x) + y^*$ where y^* is a special solution of this equation.

How do we find a special solution of the non-homogeneous second-order linear differential equation (1)?

In the end of Section 4.1.3, we introduced the method of variation of constants. That is, in order to find a special solution of a non-homogeneous first order linear differential equation

$$y' + P(x)y = Q(x),\tag{5}$$

we replace the constant C in the general solution

$$Ce^{-\int p(x)dx}$$

of the corresponding homogeneous linear equation

$$y' + p(x)y = 0$$

with a function $C(x)$. Substituting the function $C(x)e^{-\int p(x)dx}$ into the non-homogeneous equation

$$y' + p(x)y = q(x)$$

and solve for $C(x)$ to get a special solution of (5).

The same method can be used to find a special solution of the non-homogeneous second-order linear differential equation (1) from the general solution of homogeneous second-order linear differential equation (2).

Suppose $y(x) = C_1 y_1(x) + C_2 y_2(x)$ is the general solution of the homogeneous equation (2). If the non-homogeneous second-order linear differential equation (1) has a special solution of the form

$$y^*(x) = C_1(x)y_1(x) + C_2(x)y_2(x),$$

then we have

$$\frac{dy^*}{dx} = C_1(x)y_1'(x) + C_2(x)y_2'(x) + [C_1'(x)y_1(x) + C_2'y_2(x)].$$

Let

$$C_1'(x)y_1(x) + C_2'y_2(x) = 0.$$

Then

$$\frac{dy^*}{dx} = C_1(x)y_1'(x) + C_2(x)y_2'(x).$$

Differentiate again,

$$\frac{d^2y^*}{dx^2} = C_1(x)y_1''(x) + C_2(x)y_2''(x) + [C_1'(x)y_1'(x) + C_2'y_2'(x)].$$

Substituting into

$$y'' + p(x)y' + q(x)y = f(x)$$

we have

$$C_1(x)[y_1'' + p(x)y_1' + q(x)y_1] + C_2(x)[y_2'' + p(x)y_2' + q(x)y_2]$$
$$+ [C_1'(x)y_1'(x) + C_2'(x)y_2'(x)] = f(x).$$

Since y_1 and y_2 are solutions of the homogeneous equation, we have

$$C_1'(x)y_1'(x) + C_2'(x)y_2'(x) = f(x).$$

Consider this equation and the equation

$$C_1'(x)y_1(x) + C_2'(x)y_2(x) = 0$$

as simultaneous equations. Since $y_1(x)$ and $y_2(x)$ are linearly independent, The determinant

$$\begin{vmatrix} y_1(x) & y_2(x) \\ y_1'(x) & y_2'(x) \end{vmatrix} \neq 0.$$

Solving the equations, we obtain

$$\begin{cases} C_1'(x) = \dfrac{-y_2(x)f(x)}{y_1(x)y_2'(x) - y_1'(x)y_2(x)}, \\ C_2'(x) = \dfrac{y_1(x)f(x)}{y_1(x)y_2'(x) - y_1'(x)y_2(x)}. \end{cases}$$

Integrating these equations, we have

$$\begin{cases} C_1(x) = -\displaystyle\int \dfrac{-y_2(x)f(x)}{y_1(x)y_2'(x) - y_1'(x)y_2(x)}\, dx, \\ C_2(x) = \displaystyle\int \dfrac{y_1(x)f(x)}{y_1(x)y_2'(x) - y_1'(x)y_2(x)}\, dx. \end{cases}$$

Therefore, we obtain a special solution

$$y^*(x) = C_1(x)y_1(x) + C_2(x)y_2(x)$$
$$= -y_1(x) \int \frac{y_2(x)f(x)}{y_1(x)y_2'(x) - y_1'(x)y_2(x)}\, dx$$
$$+ y_2(x) \int \frac{y_1(x)f(x)}{y_1(x)y_2'(x) - y_1'(x)y_2(x)}\, dx. \tag{6}$$

In summary, the problem of finding the general solution of the non-homogeneous second-order linear differential equation (1) is equivalent to the problem of finding a special solution of the homogeneous second-order linear differential equation (2). If a special solution $y_1(x)$ of (2) is obtained, then another special solution of (2), $y_2(x)$, that is linearly independent of $y_1(x)$ can be obtained by formula (4). From y_1 and y_2, a special solution

$y^*(x)$ of (1) can be obtained by formula (6). And the general solution of (1) is then

$$C_1 y_1(x) + C_2 y_2(x) + y^*(x).$$

Example 2. Find the solution of the equation

$$\begin{cases} xy'' - y' = x^2, \\ y(1) = 1, \quad y'(1) = 1. \end{cases}$$

Solution: From Example 1, we know that $y_1 = 1$ and $y_2 = \dfrac{x^2}{2}$ are linearly independent solutions of the corresponding homogeneous equation.

By the method of variation of constants, we have

$$\begin{cases} \dfrac{dC_1(x)}{dx} + \dfrac{x^2}{2}\dfrac{dC_2(x)}{dx} = 0, \\ x\dfrac{dC_2(x)}{dx} = x. \end{cases}$$

Thus, $C_2(x) = x$, $C_1(x) = -\dfrac{x^3}{6}$. And the equation has a special solution

$$C_1(x)y_1(x) + C_2(x)y_2(x) = -\frac{x^3}{6} + \frac{x^3}{2} = \frac{x^3}{3}.$$

The general solution of the original equation is

$$y = k_1 + \frac{k_2}{2}x^2 + \frac{1}{3}x^3.$$

From the initial condition, we have

$$\begin{cases} k_1 + \dfrac{1}{2}k_2 + \dfrac{1}{3} = 1, \\ k_2 + 1 = 1. \end{cases}$$

Therefore, $k_2 = 0$ and $k_1 = \dfrac{2}{3}$. And the special solution that satisfies the initial condition is

$$y = \frac{2}{3} + \frac{1}{3}x^3.$$

4.2.3 *Linear Differential Equations with Constant Coefficients*

The problem of finding the general solution of a non-homogeneous second-order linear differential equation

$$y'' + p(x)y' + q(x)y = f(x),$$

is equivalent to the problem of finding a special solution of a corresponding homogeneous second-order linear differential equation

$$y'' + p(x)y' + q(x)y = 0.$$

If $p(x)$ and $q(x)$ are real constants a and b respectively, then the equation is called a linear equation with constant coefficients

$$y'' + ay' + by = f(x). \tag{7}$$

This is an important category of differential equations. We study how to find the general solutions of this kind of equations in this section.

To find the general solution of (7), as we studied before, we only need to find a special solution of the corresponding homogeneous second order linear differential equation

$$y'' + ay' + by = 0. \tag{8}$$

This is not very hard. Observation tells us it may have solutions of the form

$$y = e^{\lambda x}.$$

Substituting $y - e^{\lambda x}$ into equation (8), we get

$$(\lambda^2 + a\lambda + b)e^{\lambda x} = 0.$$

Since $e^{\lambda x} \neq 0$, we have

$$\lambda^2 + a\lambda + b = 0. \tag{9}$$

Equation (9) is called the *characteristic equation* of equation (8). The roots of (9) are called the *characteristic roots* of (8), and $\lambda^2 + a\lambda + b$ is called the *characteristic polynomial* of (8).

There are three possibilities for characteristic roots: two distinct real roots; one double real root; and two distinct complex conjugate roots.

1. If equation (9) has two distinct real roots λ_1 and λ_2, then $a^2 - 4b > 0$ and

$$\lambda_1 = \frac{1}{2}(-a + \sqrt{a^2 - 4b}),$$

$$\lambda_2 = \frac{1}{2}(-a - \sqrt{a^2 - 4b}).$$

Thus, the equation (8) has solutions $e^{\lambda_1 x}$ and $e^{\lambda_2 x}$.

Since the Wronskian determinant of these solutions is not zero,

$$W(e^{\lambda_1 x}, e^{\lambda_2 x}) = \begin{vmatrix} e^{\lambda_1 x} & e^{\lambda_2 x} \\ \lambda_1 e^{\lambda_1 x} & \lambda_2 e^{\lambda_2 x} \end{vmatrix} = (\lambda_2 - \lambda_1)e^{(\lambda_1 + \lambda_2)x} \neq 0$$

they are linearly independent. Therefore, the general solution of (8) is

$$y = C_1 e^{\lambda_1 x} + C_2 e^{\lambda_2 x}$$

where C_1 and C_2 are constants.

2. If equation (9) has one double real root, then $a^2 - 4b = 0$ and

$$\lambda = -\frac{a}{2}.$$

Thus,

$$y_1(x) = e^{-\frac{a}{2}x}$$

is a solution of (8). By formula (4) from the last section, we obtain the other solution

$$y_2(x) = e^{-\frac{a}{2}x} \int e^{ax} e^{-\int a\,dx}\,dx = xe^{-\frac{a}{2}x}.$$

Obviously, $y_1(x)$ and $y_2(x)$ are linearly independent and therefore, the general solution of (8) is

$$y(x) = e^{-\frac{a}{2}x}(C_1 + C_2 x),$$

where C_1 and C_2 are constants.

3. If equation (9) has a pair of complex conjugate roots $\lambda_1 = \alpha + i\beta$ and $\lambda_2 = \alpha - i\beta$, then $a^2 - 4b < 0$ and

$$\alpha = -\frac{a}{2}, \quad \beta = \frac{1}{2}\sqrt{4b - a^2}.$$

Using *Euler's formula* $\cos\beta + i\sin\beta = e^{i\beta}$, the solutions of equation (8) can be written as

$$\tilde{y}_1(x) = e^{(\alpha + i\beta)x} = e^{\alpha x}(\cos\beta x + i\sin\beta x),$$
$$\tilde{y}_2(x) = e^{(\alpha - i\beta)x} = e^{\alpha x}(\cos\beta x - i\sin\beta x).$$

Therefore, we have real valued solutions of equation (8)

$$y_1(x) = \frac{1}{2}(\tilde{y}_1(x) + \tilde{y}_2(x)) = e^{\alpha x}\cos\beta x,$$
$$y_2(x) = \frac{1}{2i}(\tilde{y}_1(x) - \tilde{y}_2(x)) = e^{\alpha x}\sin\beta x.$$

Since the Wronskian determinant is not equal to zero,

$$W(y_1(x), y_2(x))$$

$$= \begin{vmatrix} e^{\alpha x} \cos \beta x & e^{\alpha x} \sin \beta x \\ \alpha e^{\alpha x} \cos \beta x - \beta e^{\alpha x} \sin \beta x & \alpha e^{\alpha x} \sin \beta x + \beta e^{\alpha x} \cos \beta x \end{vmatrix}$$

$$= \beta e^{2\alpha x} \neq 0$$

the functions $y_1(x)$ and $y_2(x)$ are linearly independent. Therefore, the general solution of (8) is

$$y = e^{\alpha x}(C_1 \cos \beta x + C_2 \sin \beta x).$$

Example 1. Find the general solution of the equation $y'' + 2y' + 2y = 0$.

Solution: Since the corresponding characteristic equation

$$\lambda^2 + 2\lambda + 2 = 0$$

has complex roots $-1 + i$ and $-1 - i$, the general solution of the equation is

$$y(x) = e^{-x}(C_1 \cos x + C_2 \sin x).$$

Example 2. Find the general solution of the equation $y'' + 4y' + 4y = 0$.

Solution: Since the corresponding characteristic equation

$$\lambda^2 + 4\lambda + 4 = 0$$

has a double real root -2, there are two linearly independent special solu tions e^{-2x} and xe^{-2x}. Thus, the general solution of the equation is

$$y(x) = e^{-2x}(C_1 + C_2 x).$$

We have obtained all possible general solutions of the homogeneous second-order linear differential equation with constant coefficients (8). We can find a special solution of the non-homogeneous second-order linear differential equation with constant coefficients (7) using formula (6) from the last section. This then will give rise to the general solution of (7).

1. If the equation (9) has two different real roots λ_1 and λ_2, then by formula (6) of the last section,

$$\frac{1}{\lambda_1 - \lambda_2} \left[e^{\lambda_1 x} \int^x e^{-\lambda_1 t} f(t) \, dt - e^{\lambda_2 x} \int^x e^{-\lambda_2 t} f(t) \, dt \right]$$

is a special solution of equation (7). (The notation $\int^{x}$ means the antiderivative of the integrand is a function of x.) Therefore, the general solution of (7) is

$$y(x) = \frac{1}{\lambda_1 - \lambda_2} \left[e^{\lambda_1 x} \int^{x} e^{-\lambda_1 t} f(t) \, dt - e^{\lambda_2 x} \int^{x} e^{-\lambda_2 t} f(t) \, dt \right]$$
$$+ C_1 e^{\lambda_1 x} + C_2 e^{\lambda_2 x}, \tag{10}$$

where C_1 and C_2 are constants.

2. If equation (9) has a double real root $\lambda = -\dfrac{a}{2}$, then by formula (6) of the last section,

$$e^{\lambda x} \int^{x} (x - t) e^{-\lambda t} f(t) \, dt$$

is a special solution of equation (7). Therefore, the general solution of (7) is

$$y(x) = e^{\lambda x} \int^{x} (x - t) e^{-\lambda t} f(t) \, dt + C_1 x e^{\lambda x} + C_2 e^{\lambda x}, \tag{11}$$

where C_1 and C_2 are constants.

3. If equation (9) has two complex conjugate roots $\lambda_1 = \alpha + i\beta$ and $\lambda_2 = \alpha - i\beta$, then by formula (6) of the last section,

$$\frac{1}{\beta} \int^{x} e^{-\frac{a}{2}(x-t)} \sin \beta(x - t) f(t) \, dt$$

is a special solution of equation (7), so the general solution is

$$y(x) = \frac{1}{\beta} \int^{x} e^{-\frac{a}{2}(x-t)} \sin \beta(x - t) f(t) \, dt$$
$$+ C_1 e^{-\frac{ax}{2}} \cos \beta x + C_2 e^{-\frac{ax}{2}} \cos \beta x, \tag{12}$$

where C_1 and C_2 are constants.

Example 3. Solve the equation $y'' + ay' + by = \sin x$. (This is equation (8) in Section 4.1.1.)

Solution: The corresponding characteristic equation is $\lambda^2 + a\lambda + b = 0$.

(i) The characteristic equation has two distinct real roots λ_1 and λ_2.

The solution is

$$y(x) = \frac{1}{\lambda_1 - \lambda_2}\left[e^{\lambda_1 x}\int^x e^{-\lambda_1 t}\sin t\, dt - e^{\lambda_2 x}\int^x e^{-\lambda_2 t}\sin t\, dt\right] + C_1 e^{\lambda_1 x} + C_2 e^{\lambda_2 x}$$

$$= \frac{1}{\lambda_1 - \lambda_2}\left[\frac{-\lambda_1 \sin x - \cos x}{\lambda_1^2 + 1} + \frac{\lambda_2 \sin x + \cos x}{\lambda_2^2 + 1}\right] + C_1 e^{\lambda_1 x} + C_2 e^{\lambda_2 x}$$

$$= \frac{(\lambda_1 + \lambda_2)\cos x + (\lambda_1 \lambda_2 - 1)\sin x}{(\lambda_1^2 + 1)(\lambda_2^2 + 1)} + C_1 e^{\lambda_1 x} + C_2 e^{\lambda_2 x}$$

$$= \frac{-a\cos x + (b-1)\sin x}{(b-1)^2 + a^2} + C_1 e^{\lambda_1 x} + C_2 e^{\lambda_2 x}.$$

(ii) The characteristic equation has one double real root λ. The solution is

$$y(x) = e^{\lambda x}\int^x (x-t)e^{-\lambda t}\sin t\, dt + C_1 x e^{\lambda x} + C_2 e^{\lambda x}$$

$$= x\frac{-\lambda \sin x - \cos x}{\lambda^2 + 1} + \frac{x\cos x + \lambda x \sin x}{\lambda^2 + 1}$$

$$\quad - \frac{1}{(\lambda^2 + 1)^2}[-\lambda \cos x + \sin x + \lambda(-\lambda \sin x - \cos x)] + C_1 x e^{\lambda x} + C_2 e^{\lambda x}$$

$$= \frac{(\lambda^2 - 1)\sin x + 2\lambda \cos x}{(\lambda^2 + 1)^2} + C_1 x e^{\lambda x} + C_2 e^{\lambda x}$$

$$= \frac{(b-1)\sin x - a\cos x}{(b-1)^2 + a^2} + C_1 x e^{\lambda x} + C_2 e^{\lambda x}.$$

(iii) The characteristic equation has two complex conjugate roots $-\frac{a}{2} + i\beta$ and $-\frac{a}{2} - i\beta$. The solution is

$$y(x) = \frac{1}{\beta}\int^x e^{-\frac{a}{2}(x-t)}\sin \beta(x-t)\sin t\, dt + C_1 e^{-\frac{a}{2}x}\cos \beta x + C_2 e^{-\frac{a}{2}x}\sin \beta x$$

$$= \frac{1}{2\beta}\int^x e^{-\frac{a}{2}(x-t)}(\cos(\beta x - (\beta+1)t) - \cos(\beta x - (\beta-1)t))\, dt$$

$$\quad + C_1 e^{-\frac{a}{2}x}\cos \beta x + C_2 e^{-\frac{a}{2}x}\sin \beta x$$

$$= \frac{(b-1)\sin x - a\cos x}{(b-1)^2 + a^2} + C_1 e^{-\frac{a}{2}x}\cos \beta x + C_2 e^{-\frac{a}{2}x}\sin \beta x.$$

Clearly, the above solutions also require $a \neq 0$ or $b \neq 1$. If $a = 0$ and $b = 1$, the equation to be solved becomes

$$y'' + y = \sin x.$$

By formula (6), the solution is

$$y = \int^x \sin(x - t) \sin t \, dt + C_1 \cos x + C_2 \sin x$$

$$= \frac{-x}{2} \cos x + C_1 \cos x + C_2' \sin x.$$

Example 4. Solve the equation $y'' - 5y' + 6y = xe^{2x}$.

Solution: The two roots of the corresponding characteristic equation $\lambda^2 - 5\lambda + 6 = 0$ are 2 and 3. By formula (10), the general solution of the equation is

$$y = e^{3x} \int^x e^{-3t} t e^{2t} \, dt - e^{2x} \int^x e^{-2t} t e^{2t} \, dt + C_1 e^{3x} + C_2 e^{2x}$$

$$= e^{3x} \int^x t e^{-t} \, dt - e^{2x} \int^x t \, dt + C_1 e^{3x} + C_2 e^{2x}$$

$$= -xe^{2x} - e^{2x} - \frac{x^2}{2} e^{2x} + C_1 e^{3x} + C_2 e^{2x}$$

$$= \left(-\frac{x^2}{2} - x \right) e^{2x} + C_1 e^{3x} + C_3 e^{2x}$$

where $C_3 = C_2 - 1$.

Example 5. Solve the equation $y'' - 4y' + 4y = e^{2x} \sin x$.

Solution: The corresponding characteristic equation $\lambda^2 - 4\lambda + 4 = 0$ have a double root 2. By formula (11), the general solution of the equation is

$$y = e^{2x} \int^x (x - t) e^{-2t} e^{2t} \sin t \, dt + C_1 x e^{2x} + C_2 e^{2x}$$

$$= e^{2x} \int^x (x - t) \sin t \, dt + C_1 x e^{2x} + C_2 e^{2x}$$

$$= e^{2x} (-\sin x) + C_1 x e^{2x} + C_2 e^{2x}.$$

Example 6. Solve the equation $y'' + y = x \cos x$.

Solution: The two complex roots of the corresponding characteristic equation $\lambda^2 + 1 = 0$ are i and $-i$. By formula (12), the general solution of the equation is

$$y = \int^x t \sin(x - t) \cos t \, dt + C_1 \cos x + C_2 \sin x$$

$$= \frac{1}{2} \int^x t[\sin x - \sin(2t - x)] \, dt + C_1 \cos x + C_2 \sin x$$

$$= \frac{x^2}{4} \sin x + \frac{x}{4} \cos x + C_1 \cos x + C_3 \sin x.$$

We have found the general solutions of second order linear differential equations with constant coefficients by integration. For some $f(x)$, the special solutions of the non-homogeneous second-order linear differential equation with constant coefficients (7) can be found without using the method of integration.

Consider the following forms of functions:

(i) $f(x) = \varphi_m(x)$ (a polynomial of degree m);

(ii) $f(x) = \varphi_m(x)e^{\alpha x}$ (where α is a real number);

(iii) $f(x) = \varphi_m(x)e^{\alpha x}\cos\beta x$ or $f(x) = \varphi_m(x)e^{\alpha x}\sin\beta x$,   (where β is a real number).

These three forms of functions can be considered as one form of function with different constants:

$$f(x) = \varphi_m(x)e^{(\alpha+i\beta)x}.$$

When $\alpha = \beta = 0$, it is form (i), when $\beta = 0$, it is form (ii), and the third form is the real part and the imaginary part of this function. Thus, we only need to study $f(x)$ of the form

$$f(x) = \varphi_m(x)e^{\rho x} \tag{13}$$

where ρ is a complex number.

Suppose there is a special solution of equation (7) of the form

$$y^* = z(x)e^{\rho x}, \tag{14}$$

where $z(x)$ is unknown.

Since

$$y^{*\prime} = e^{\rho x}(\rho z(x) + z'(x)),$$
$$y^{*\prime\prime} = e^{\rho x}(\rho^2 z(x) + 2\rho z'(x) + z''(x)),$$

substituting into (7), we have

$$[z''(x) + (2\rho + a)z'(x) + (\rho^2 + a\rho + b)z(x)]e^{\rho x} = \varphi_m(x)e^{\rho x}.$$

Let $l(\rho) = \rho^2 + a\rho + b$, and divide both sides of the equation by $e^{\rho x}$. Then the above equation becomes

$$z''(x) + l'(\rho)z'(x) + l(\rho)z(x) = \varphi_m(x). \tag{15}$$

(i) If ρ is not a root of the corresponding characteristic equation, that is, $l(\rho) \neq 0$, then we can assume that $z(x)$ is a polynomial of the same degree as $\varphi_m(x)$. Thus, we can write

$$y^* = Q_m(x)e^{\rho x},$$

where $Q_m(x)$ is an mth degree polynomial:

$$Q_m(x) = b_0 x^m + b_1 x^{m-1} + \cdots + b_{m-1} x + b_m$$

where $b_0, b_1, \cdots, b_m$ are indeterminate coefficients which can be easily determined by (15).

(ii) If ρ is a single root of the characteristic equation, that is, $l(\rho) = 0$, but $l'(\rho) \neq 0$, then we should assume that $z(x)$ is a polynomial one degree higher than $\varphi_m(x)$. Thus, we can write

$$y^* = x Q_m(x) e^{\rho x}.$$

(iii) If ρ is a double root of the characteristic equation, that is, $l(\rho) = 0$, and $l'(\rho) = 0$, then we should assume that $z(x)$ is a polynomial two degrees higher than $\varphi_m(x)$. Thus, we can write

$$y^* = x^2 Q_m(x) e^{\rho x}.$$

In summary, if $f(x) = \varphi_m(x) e^{\rho x}$, then we can assume that there is a special solution of the equation (7) of the form

$$y(x) = x^k Q_m(x) e^{\rho x},$$

where $Q_m(x)$ is a polynomial having the same degree as $\varphi_m(x)$, and the value of k is 2 when ρ is a double root of the corresponding characteristic equation, is 1 when ρ is a single root, and is 0 when ρ is not a root.

Example 7. Solve the equation $y'' - 5y' + 6y = x e^{2x}$. (Same equation in Example 4.)

Solution: The two roots of the corresponding characteristic equation

$$\lambda^2 - 5\lambda + 6 = 0$$

are $\lambda_1 = 2$ and $\lambda_2 = 3$.

The general solution of the corresponding homogeneous equation is

$$\bar{y} = C_1 e^{2x} + C_2 e^{3x}.$$

Since $\varphi_m(x) = x$ and $\rho = 2$, we can assume that

$$z(x) = x(b_0 x + b_1).$$

By equation (15), we have

$$z''(x) + l'(2)z'(x) + l(2)z(x) = 2b_0 - (2b_0 x + b_1) = x.$$

Thus,

$$b_0 = -\frac{1}{2}, \quad b_1 = -1.$$

Therefore, the special solution is

$$y^* = \left(-\frac{1}{2}x^2 - x\right)e^{2x},$$

and the general solution is

$$y = \bar{y} + y^* = C_1 e^{2x} + C_2 e^{3x} + \left(-\frac{1}{2}x^2 - x\right)e^{2x}.$$

Example 8. Find the general solution of the equation $y'' - y = 4x\sin x$.

Solution: The two roots of the corresponding characteristic equation

$$\lambda^2 - 1 = 0$$

are $\lambda_1 = 1$ and $\lambda_2 = -1$.

The general solution of the corresponding homogeneous equation is

$$\bar{y} = C_1 e^x + C_2 e^{-x}.$$

Consider the equation

$$y'' - y = 4xe^{ix}. \tag{16}$$

Since i is not a root of the corresponding characteristic equation, we can assume its special solution is of the form

$$y_1^* = (b_0 x + b_1)e^{ix}.$$

Substituting

$$z(x) = b_0 x + b_1$$

into (15) we get

$$b_0 = -2, b_1 = -2i.$$

Thus, the special solution of equation (16) is

$$y_1^* = (-2x - 2i)e^{ix} = -2(x+i)(\cos x + i\sin x),$$

and its imaginary part

$$y^* = -2(\cos x + x\sin x)$$

is the special solution we want. Therefore, the general solution of the equation is

$$y = \bar{y} + y^* = C_1 e^x + C_2 e^{-x} - 2(\cos x + x\sin x).$$

Example 9. Find the general solution of the equation

$$y'' - 2y' - 3y = 3x + 1 + e^{-x}.$$

Solution: The two roots of the corresponding characteristic equation

$$\lambda^2 - 2\lambda - 3 = 0$$

are $\lambda_1 = -1$ and $\lambda_2 = 3$.

The general solution of the corresponding homogeneous equation is

$$\bar{y} = C_1 e^{-x} + C_2 e^{3x}.$$

A special solution of the equation

$$y'' - 2y' - 3y = 3x + 1$$

is

$$y_1^* = -x + \frac{1}{3}$$

and a special solution of the equation

$$y'' - 2y' - 3y = e^{-x}$$

is

$$y_2^* = -\frac{1}{4} x e^{-x}.$$

Thus, a special solution of the original equation is

$$y^* = y_1^* + y_2^* = -x + \frac{1}{3} - \frac{1}{4} x e^{-x}.$$

Therefore, the general solution of the original equation is

$$y = \bar{y} + y^* = C_1 e^{-x} + C_2 e^{3x} - x + \frac{1}{3} - \frac{1}{4} x e^{-x}.$$

The method used in the above solutions of the equations is called the method of *undetermined coefficients*.

4.2.4 *Mechanical Vibration*

We are now ready to study an example in physics.

Consider a mechanical system consists of a spring with an oscillating body M attached. The mass of the oscillating body is m. The origin ($x = 0$) is set to be the equilibrium position of M. The position of M is indicated by its displacement x from the equilibrium position.

By Hooke's law, the restoring force f and the displacement x have the relation

$$f = -bx,$$

where b is the spring constant.

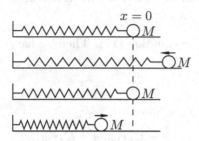

Fig. 4.5

By Newton's second law, we have

$$f = m\frac{d^2x}{dt^2}.$$

Thus, the differential equation of the motion is

$$m\frac{d^2x}{dt^2} + bx = 0. \tag{17}$$

This is the equation of simple harmonic motion, and it is a second-order homogeneous equation with constant coefficients. If the initial conditions are given:

$$x(0) = 0, \quad \left.\frac{dx}{dt}\right|_{t=0} = v_0 > 0,$$

then the motion is uniquely determined.

Suppose the oscillatory system is in an environment with damping. The damping force is linearly related to the velocity of the oscillating body,

$$f = -a\frac{dx}{dt},$$

where a is the damping coefficient. The equation of the motion becomes

$$m\frac{d^2x}{dt^2} + bx = -a\frac{dx}{dt}. \tag{18}$$

This is also a second-order homogeneous equation with constant coefficients.

Suppose a periodic force is applied to the oscillator

$$f_1 = F\sin\tilde{\omega}t,$$

where F is a constant. Then the displacement x satisfies the equation

$$m\frac{d^2x}{dt^2} + a\frac{dx}{dt} + bx = -F\sin\tilde{\omega}t. \tag{19}$$

It is a second-order linear equation with constant coefficients.

We now solve these equations.

1. Let $\omega = \sqrt{\dfrac{b}{m}}$ in equation (17). Then we have

$$\frac{d^2x}{dt^2} + \omega^2 x = 0.$$

By (12), we obtain the solution

$$x = C_1 \cos \omega t + C_2 \sin \omega t.$$

Rewrite the solution as

$$x = \sqrt{C_1^2 + C_2^2} \left(\frac{C_1}{\sqrt{C_1^2 + C_2^2}} \cos \omega t + \frac{C_2}{\sqrt{C_1^2 + C_2^2}} \sin \omega t \right).$$

Let $\sin r = \dfrac{C_1}{\sqrt{C_1^2 + C_2^2}}$ and $A = \sqrt{C_1^2 + C_2^2}$. Then we have

$$x = A \sin(\omega t + r)$$

where A is the amplitude and r is the initial phase.

This solution implies that the change of the distance between the oscillator and its equilibrium position is sinusoidal in time with period

$$T = \frac{2\pi}{\omega} = 2\pi \sqrt{\frac{m}{b}}.$$

The amplitude A and the initial phase r are determined by the initial conditions of the equation.

2. Next, in equation (18), let $\beta = \dfrac{a}{2m}$ and $\omega = \sqrt{\dfrac{b}{m}}$, then we have

$$\frac{d^2x}{dt^2} + 2\beta \frac{dx}{dt} + \omega^2 x = 0.$$

The corresponding characteristic equation is

$$\lambda^2 + 2\beta\lambda + \omega^2 = 0.$$

Its solutions are different depending on whether $\beta > \omega$, $\beta = \omega$, or $\beta < \omega$.

(i) If $\beta > \omega$, the equation has two real roots

$$\lambda_1 = -\beta + \sqrt{\beta^2 - \omega^2}, \ \lambda_2 = -\beta - \sqrt{\beta^2 - \omega^2}.$$

By (10), the general solution of (18) is

$$x = C_1 e^{(-\beta + \sqrt{\beta^2 - \omega^2})t} + C_2 e^{(-\beta - \sqrt{\beta^2 - \omega^2})t},$$

where C_1 and C_2 are determined by the initial conditions. For instance, if the initial conditions are $x(0) = 0$ and $x'(0) = 1$, then we have

$$x = \frac{1}{2\sqrt{\beta^2 - \omega^2}}[e^{(-\beta + \sqrt{\beta^2 - \omega^2})t} - e^{(-\beta - \sqrt{\beta^2 - \omega^2})t}].$$

Since $\sqrt{\beta^2 - \omega^2} < \beta$, we have $-\beta + \sqrt{\beta^2 - \omega^2} < 0$ and $-\beta - \sqrt{\beta^2 - \omega^2} < 0$. Therefore,

$$\lim_{t \to \infty} x = 0.$$

This implies that the oscillator tends to the equilibrium position monotonically as time goes by, and no vibration happens.

(ii) When $\beta = \omega$, the equation has a double root $\lambda = -\beta$. By (11), the general solution of equation (18) is

$$x = (C_1 + C_2 t)e^{-\beta t}.$$

Obviously, we also have

$$\lim_{t \to \infty} x = 0.$$

This implies that the oscillator tends to the equilibrium position monotonically as time goes by, and no vibration happens.

(iii) If $\beta < \omega$, the equation has two complex conjugate roots

$$\lambda_1 = -\beta + i\sqrt{\omega^2 - \beta^2}, \ \lambda_2 = -\beta - i\sqrt{\omega^2 - \beta^2}.$$

By (12), the general solution of equation (18) is

$$x = e^{-\beta t}(C_1 \cos \sqrt{\omega^2 - \beta^2}t + C_2 \sin \sqrt{\omega^2 - \beta^2}t).$$

Let $A = \sqrt{C_1^2 + C_2^2}$ and $\sin r = \dfrac{C_1}{\sqrt{C_1^2 + C_2^2}}$. Then

$$x = Ae^{-\beta t} \sin(\sqrt{\omega^2 - \beta^2}t + r).$$

Again, we also have

$$\lim_{t \to \infty} x = 0.$$

Since the derivative of x is

$$x' = Ae^{-\beta t}[-\beta \sin(\sqrt{\omega^2 - \beta^2}t + r) + \sqrt{\omega^2 - \beta^2} \cos(\sqrt{\omega^2 - \beta^2}t + r)],$$

it is equal to 0 when

$$t = \frac{1}{\sqrt{\omega^2 - \beta^2}} \left(\arctan \frac{\sqrt{\omega^2 - \beta^2}}{\beta} - r + n\pi \right).$$

This implies x reaches its extreme values as the oscillator reaches the equilibrium position. That is, the oscillator vibrates "infinitely" many times, each time with a smaller amplitude, before it stops (Figure 4.6.) This is different from cases (i) and (ii).

In this solution

$$Ae^{\beta t} = Ae^{-\frac{a}{2m}t}$$

is the amplitude, r is the initial phase. The values of A and r can be determined by the initial conditions. Its frequency

$$\omega' = \sqrt{\frac{b}{m} - \frac{a^2}{4m^2}}$$

is called the natural frequency of the system. Its period is

$$T = \frac{2\pi}{\omega'} = \frac{2\pi}{\sqrt{\dfrac{b}{m} - \dfrac{a^2}{4m^2}}}.$$

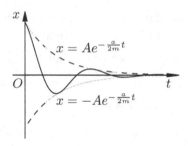

Fig. 4.6

(3) Finally, we look at forced vibrations.

(i) Suppose the damping coefficient $a \neq 0$ in equation (19). Then the equation has a special solution

$$x_1 = F\frac{a\tilde{\omega}\cos\tilde{\omega}t - (b - \tilde{\omega}^2 m)\sin\tilde{\omega}t}{(b - m\tilde{\omega}^2)^2 + a^2\tilde{\omega}^2}.$$

Since $b = m\omega^2$, $\beta = \dfrac{a}{2m}$, the solution becomes

$$x_1 = Fm^{-1}\frac{2\beta\tilde{\omega}\cos\tilde{\omega}t - (\omega^2 - \tilde{\omega}^2)\sin\tilde{\omega}t}{(\omega^2 - \tilde{\omega}^2)^2 + 4\beta^2\tilde{\omega}^2} = \tilde{F}\sin(\tilde{\omega}t + \varphi),$$

where

$$\tilde{F} = \frac{Fm^{-1}}{\sqrt{(\omega^2 - \tilde{\omega}^2)^2 + (2\beta\tilde{\omega})^2}},$$

$$\cos\varphi = \frac{\tilde{\omega}^2 - \omega^2}{\sqrt{(\omega^2 - \tilde{\omega}^2)^2 + (2\beta\tilde{\omega})^2}}.$$

If we denote the general solution of (18) as $x_0(t)$, then the general solution of (19) is

$$x(t) = x_0(t) + \tilde{F}\sin(\tilde{\omega}t + \varphi).$$

Since $\lim\limits_{t\to\infty} x_0(t) = 0$, we see that the effect of $x_0(t)$ is prominent only for small t. As time goes by, the vibration looks more and more like a vibration with a frequency that matches the frequency of the external force.

(ii) Suppose the damping coefficient $a = 0$ in equation (19).

(a) If $\omega \neq \tilde{\omega}$, then the general solution of (19) is

$$x(t) = A\sin(\omega t + r) - \frac{F}{m(\omega^2 - \tilde{\omega}^2)}\sin\tilde{\omega}t.$$

This is the superposition of two simple vibration. The first is the natural vibration of the oscillator itself. The second is caused by the external force. The composite motion has variable amplitude and period, but the amplitude remains bounded.

(b) If $\omega = \tilde{\omega}$, then the general solution of (19) is

$$x(t) = A\sin(\omega t + r) + \frac{F}{2m\omega}t\sin\left(\omega t + \frac{\pi}{2}\right).$$

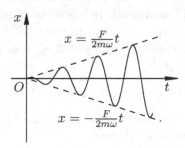

Fig. 4.7

Obviously, the amplitude of the first term is a constant, the amplitude of the second term increases infinitely as time increases (Figure 4.7). This is called the resonance phenomenon. Resonance happens quite often in nature. Engineers seek to reduce resonance in architecture, and to take advantage of it in electronics.

4.2.5   General Linear Differential Equations and Systems of Linear Equations

The general form of an nth-order non-homogeneous linear differential equation is

$$y^{(n)} + p_1(x)y^{(n-1)} + \cdots + p_n(x)y = f(x), \tag{20}$$

where $p_1(x), \cdots, p_n(x), f(x)$ are continuous functions.

The corresponding homogeneous equation is

$$y^{(n)} + p_1(x)y^{(n-1)} + \cdots + p_n(x)y = 0. \tag{21}$$

We studied the structure of the solution of second-order linear differential equations in Section 4.2.2. Now we study the structure of the solutions of nth-order linear differential equations.

If $y_1(x), y_2(x), \cdots, y_n(x)$ are n linearly independent solutions of (21), and $\tilde{y}(x)$ is a special solution of (20), then the general solution of (21) is

$$y(x) = C_1 y_1(x) + C_2 y_2(x) + \cdots + C_n y_n(x)$$

and the general solution of (20) is

$$y(x) = C_1 y_1(x) + C_2 y_2(x) + \cdots + C_n y_n(x) + \tilde{y}(x).$$

The concept of "n linearly independent solutions" can be defined similarly to the concept of "two linearly independent solutions" in section 4.2.2.

Suppose $\varphi_1(x)$, $\varphi_2(x)$, $\cdots$, $\varphi_n(x)$ are functions defined on the interval $[a, b]$. If there exist n constants c_1, c_2, $\cdots$, c_n that are not equal to zero simultaneously such that

$$c_1 \varphi_1(x) + c_2 \varphi_2(x) + \cdots + c_1 \varphi_1(x) = 0$$

on $[a, b]$, then we say that $\varphi_1(x)$, $\varphi_2(x)$, $\cdots$, $\varphi_n(x)$ are linearly dependent. Otherwise, they are linearly independent.

We study only the nth-order linear differential equation with constant coefficients.

The general form of the nth-order linear differential equation with constant coefficients is

$$y^{(n)} + a_1 y^{(n-1)} + \cdots + a_{n-1} y' + a_n y = f(x), \tag{22}$$

where $a_1, \cdots, a_{n-1}, a_n$ are real constants.

Now we solve equation (22) using the results from section 4.2.3.

The corresponding characteristic equation of (22) is

$$\lambda^n + a_1\lambda^{n-1} + \cdots + a_{n-1}\lambda + a_n = 0.$$

The left side of the above equation can be factored as the product of some first-degree and second-degree polynomials. That is, the equation can be written as

$$(\lambda - \alpha_1) \cdots (\lambda - \alpha_r)(\lambda^2 - p_1\lambda + q_1)(\lambda^2 - p_2\lambda + q_2) \cdots = 0,$$

where $\alpha_1, \cdots, \alpha_r, p_1, q_1, p_2, q_2, \cdots$ are real numbers. Thus, equation (22) can be rewritten as

$$\left(\frac{d}{dx} - \alpha_1\right) \cdots \left(\frac{d}{dx} - \alpha_r\right)\left(\frac{d^2}{dx^2} - p_1\frac{d}{dx} - q_1\right) \cdots y = f(x).$$

This is equivalent to

$$\frac{dy_1}{dx} - \alpha_1 y_1 = f(x), \quad \frac{dy_2}{dx} - \alpha_2 y_2 = y_1,$$

$$\cdots$$

$$\frac{dy_r}{dx} - \alpha_r y_r = y_{r-1}, \quad \frac{d^2 y_{r+1}}{dx^2} - p_1\frac{dy_{r+1}}{dx} - q_1 y_{r+1} = y_r,$$

$$\cdots$$

This is a system of first order and second order linear equations with constant coefficients. We know how to solve these equations from sections 4.2.2 and 4.2.3. Solve the first equation, and substitute the solution into the second equation. By continuing this process, we can get the final solution y.

Example 1. Solve the equation $ay''' - by'' + ay' - by = \sin x$.

Solution: The corresponding characteristic equation is

$$a\lambda^3 - b\lambda^2 + a\lambda - b = 0.$$

Factoring the left side, we have

$$(a\lambda - b)(\lambda^2 + 1) = 0.$$

Thus, the equation can be written as

$$\left(a\frac{d}{dx} - b\right)\left(\frac{d^2}{dx^2} + 1\right)y = \sin x.$$

Let $z = \dfrac{d^2y}{dx^2} + y$. Then the equation becomes

$$a\frac{dz}{dx} - bz = \sin x.$$

Therefore, we have

$$z = \frac{1}{a}e^{\frac{b}{a}x}\left(\int \sin x e^{-\frac{b}{a}x}\,dx + C_1\right)$$

$$= \frac{1}{a}e^{\frac{b}{a}x}\left[\frac{-\dfrac{b}{a}\sin x - \cos x}{\dfrac{b^2}{a^2}+1}e^{-\frac{b}{a}x} + C_1\right]$$

$$= \frac{-b\sin x - a\cos x}{b^2 + a^2} + C_1 e^{\frac{b}{a}x}.$$

Next we solve the equation

$$\frac{d^2y}{dx^2} + y = \frac{-b\sin x - a\cos x}{b^2 + a^2} + C_1 e^{\frac{b}{a}x} = \frac{-\sin(x+\theta)}{\sqrt{a^2+b^2}} + C_1 e^{\frac{b}{a}x},$$

where

$$\theta = \arccos\frac{b}{\sqrt{a^2+b^2}}.$$

By (12), the solution is

$$y = \int^x \sin(x-t)\left[\frac{-\sin(t+\theta)}{\sqrt{a^2+b^2}} + C_1 e^{\frac{b}{a}t}\right]dt + C_2 \cos x + C_3 \sin x$$

$$= \frac{x\cos(x+\theta)}{2\sqrt{a^2+b^2}} + \frac{a^2 C_1 e^{\frac{b}{a}x}}{a^2+b^2} + C_2' \cos x + C_3' \sin x.$$

The method of undetermined coefficients can be used to solve the equation (22) with $f(x) = \varphi(x)e^{(\alpha+i\beta)x}$, where $\varphi(x)$ is a polynomial, and α, β are real numbers.

Suppose an nth order linear differential equation with constant coefficients is

$$y^{(n)} + a_1 y^{(n-1)} + a_2 y^{(n-2)} + \cdots + a_{n-1}y + a_n = \varphi_m(x)e^{\rho x}, \qquad (23)$$

where $a_1, a_2, \cdots, a_n$ are real numbers, ρ is a complex number and $\varphi_m(x)$ is a polynomial of degree m.

The corresponding homogeneous equation is

$$y^{(n)} + a_1 y^{(n-1)} + a_2 y^{(n-2)} + \cdots + a_{n-1}y + a_n = 0. \qquad (24)$$

The corresponding characteristic equation is

$$l(\lambda) = \lambda^n + a_1\lambda^{n-1} + a_2\lambda^{n-2} + \cdots + a_n = 0.$$

(i) If the characteristic equation has n distinct real roots $\lambda_1, \lambda_2, \cdots, \lambda_n$, then the general solution of the homogeneous equation (24) is

$$\bar{y} = C_1 e^{\lambda_1 x} + C_2 e^{\lambda_2 x} + \cdots + C_n e^{\lambda_n x}.$$

(ii) If the characteristic equation has a real root λ with multiplicity r, then the homogeneous equation (24) has r linearly independent solutions

$$e^{\lambda x}, xe^{\lambda x}, \cdots, x^{r-1} e^{\lambda x}.$$

(iii) If the characteristic equation has complex conjugate roots $\alpha + i\beta$ and $\alpha - i\beta$ with multiplicity r, then equation (24) has $2r$ linearly independent solutions

$$e^{\alpha x} \cos \beta x, xe^{\alpha x} \cos \beta x, \cdots, x^{r-1} e^{\alpha x} \cos \beta x,$$
$$e^{\alpha x} \sin \beta x, xe^{\alpha x} \sin \beta x, \cdots, x^{r-1} e^{\alpha x} \sin \beta x.$$

To find the special solution y^* of the non-homogeneous equation (23), we again assume that

$$y^* = z(x)e^{\rho x}.$$

Substituting into equation (23), we have

$$z^{(n)} + \frac{1}{(n-1)!} l^{(n-1)}(\rho) z^{(n-1)} + \frac{1}{(n-2)!} l^{(n-2)}(\rho) z^{(n-2)} + \cdots$$
$$+ l'(\rho) z' + l(\rho) z = \varphi_m(x).$$

Assume that

$$z(x) - x^r Q_m(x) c^{\rho x},$$

where $Q_m(x)$ is a polynomial of degree m, if ρ is a root of multiplicity r of the characteristic equation. Substituting $y^* = z(x)e^{\rho x}$ into (23) and compare coefficients of the expressions on both side of the equation, then we can find y^*.

Example 2. Solve the equation $y^{(4)} + 2y'' + y = 9 \sin 2x$.

Solution: The corresponding characteristic equation is

$$\lambda^4 + 2\lambda^2 + 1 = 0.$$

It has roots $\lambda_1 = i$ and $\lambda_2 = -i$, each of multiplicity two. Thus, the general solution of the corresponding homogeneous equation is

$$\bar{y} = (C_1 + C_2 x) \cos x + (C_3 + C_4 x) \sin x.$$

To find the special solution of the non-homogeneous equation, consider the equation

$$y^{(4)} + 2y'' + y = 9e^{2ix}.$$

Since $\rho = 2i$ is not a root of the corresponding characteristic equation, we can assume that the special solution has the form

$$y_1^* = Ae^{2ix}.$$

Substituting y_1^* into the equation, we obtain $A = 1$. It follows that $y_1^* = e^{2ix}$ and its imaginary part is the special solution of the non-homogeneous equation. Thus, $y^* = \sin 2x$, and the general solution of the equation is

$$y = \bar{y} + y^* = (C_1 + C_2 x)\cos x + (C_3 + C_4 x)\sin x + \sin 2x.$$

We now discuss briefly about systems of differential equations.

A system of simultaneous differential equations that involve one independent variable and two or more dependent variables is called a system of differential equations. For instance, the following

$$\begin{cases} \dfrac{dx}{dt} = y, \\[2mm] \dfrac{dy}{dt} = x; \end{cases} \tag{25}$$

$$\begin{cases} \dfrac{dx}{dt} = x(x^2 + y^2)^{\frac{1}{2}}(x^2 + y^2 - 1)^2 + y, \\[2mm] \dfrac{dy}{dt} = y(x^2 + y^2)^{\frac{1}{2}}(x^2 + y^2 - 1)^2 - x; \end{cases}$$

are systems of differential equations.

We introduce *elimination method* of solving systems of differential equations through a few examples.

Example 3. Solve the system of equations (25).

Solution: Differentiating the first equation, we have

$$\frac{d^2 x}{dt^2} = \frac{dy}{dt}.$$

Combining with the second equation, we obtain an equation of higher order

$$\frac{d^2 x}{dt^2} - x = 0.$$

Its solution is

$$x = C_1 e^t + C_2 e^{-t}.$$

By the first equation, we get

$$y = \frac{dx}{dt} = C_1 e^t - C_2 e^{-t}.$$

Therefore, the solution of the system is

$$\begin{cases} x = C_1 e^t + C_2 e^{-t}, \\ y = C_1 e^t - C_2 e^{-t}. \end{cases}$$

In general, consider the system

$$\begin{cases} \dfrac{dx}{dt} = f(t, x, y), & \text{(26)} \\[2mm] \dfrac{dy}{dt} = g(t, x, y). & \text{(27)} \end{cases}$$

Differentiating (26), we have

$$\frac{d^2 x}{dt^2} = \frac{df}{dt}.$$

Since x and y are functions of t, the expression of $\dfrac{df}{dt}$ includes the terms $\dfrac{dx}{dt}$ and $\dfrac{dy}{dt}$, and therefore is also a function of t, x, y after substituting (26) and (27) into it. Denote

$$\frac{d^2 x}{dt^2} = F(t, x, y), \tag{28}$$

then we get a new system

$$\begin{cases} \dfrac{dx}{dt} = f(t, x, y), & \text{(26)} \\[2mm] \dfrac{d^2 x}{dt^2} = F(t, x, y). & \text{(28)} \end{cases}$$

Eliminating y from the system, we get a second order equation in x. We can find its general solution

$$x = \varphi(t, C_1, C_2).$$

Substituting into (26) and solve for y, we get the solution of the original system (26), (27).

The same method can be applied to solve systems of equations with three or more dependent variables.

Example 4. Solve the system of equations

$$\begin{cases} \dfrac{dx}{dt} = y + z, \\[2mm] \dfrac{dy}{dt} = z + x, \\[2mm] \dfrac{dz}{dt} = x + y. \end{cases}$$

Solution: Differentiating the first equation, we have

$$\frac{d^2x}{dt^2} = \frac{dy}{dt} + \frac{dz}{dt} = 2x + y + z.$$

Eliminate y, z using the first equation, we have

$$\frac{d^2x}{dt^2} - \frac{dx}{dt} - 2x = 0.$$

Its solution is

$$x = C_1 e^{-t} + C_2 e^{2t}.$$

Thus,

$$\frac{dx}{dt} = -C_1 e^{-t} + 2C_2 e^{2t}.$$

Substituting into the first and the third equations and eliminating y, we obtain

$$\frac{dz}{dt} + z = 3C_2 e^{2t}.$$

Solving this equation, we get

$$z = C_3 e^{-t} + C_2 e^{2t}.$$

From the first equation we have

$$y = -(C_1 + C_3)e^{-t} + C_2 e^{2t}.$$

Therefore, the solution of the system is

$$\begin{cases} x = C_1 e^{-t} + C_2 e^{2t}, \\ y = -(C_1 + C_3)e^{-t} + C_2 e^{2t}, \\ z = C_3 e^{-t} + C_2 e^{2t}. \end{cases}$$

Example 5. Find the solution of the system

$$\begin{cases} \dfrac{dy_1}{dx} = y_1 + y_2 + x, \\[2mm] \dfrac{dy_2}{dx} = -4y_1 - 3y_2 + 2x, \end{cases}$$

that satisfies the initial condition $y_1(0) = 1$ and $y_2(0) = 0$.

Solution: Differentiating the first equation we have

$$\frac{d^2 y_1}{dx^2} = \frac{dy_1}{dx} + \frac{dy_2}{dx} + 1.$$

Substituting $\dfrac{dy_1}{dx}$ and $\dfrac{dy_2}{dx}$ into this equation we have

$$\frac{d^2 y_1}{dx^2} = -3y_1 - 2y_2 + 3x + 1.$$

From the first equation of the system,

$$y_2 = \frac{dy_1}{dx} - y_1 - x.$$

Substituting into the above, we have

$$\frac{d^2 y_1}{dx^2} + 2\frac{dy_1}{dx} + y_1 = 5x + 1.$$

Solving for y_1 by (11), we get

$$y_1 = (C_1 + C_2 x)e^{-x} + 5x - 9.$$

Thus,

$$y_2 = \frac{dy_1}{dx} - y_1 - x = (C_2 - 2C_1 - 2C_2 x)e^{-x} - 6x + 14.$$

The initial conditions $y_1(0) = 1$ and $y_2(0) = 0$ implies

$$\begin{cases} 1 = C_1 - 9, \\ 0 = C_2 - 2C_1 + 14 \end{cases}$$

Therefore $C_1 = 10$ and $C_2 = 6$, and the solution we want is

$$\begin{cases} y_1 = (10 + 6x)e^{-x} + 5x - 9, \\ y_2 = (-14 - 12x)e^{-x} - 6x + 14. \end{cases}$$

Exercises 4.2

1. Find the general solution of the following second-order equations.

   (1) $y'' = x + \sin x$;
           (2) $xy'' = y'$;

   (3) $y'' = y' + x$;
           (5) $(1 + x^2)y'' + (y')^2 + 1 = 0$;

   (4) $y'' = \dfrac{y'}{x} + x$;
           (6) $xy'' = y' \ln \dfrac{y'}{x}$;

(7) $2xy'y'' = (y')^2 + 1$;

(8) $1 + (y')^2 = 2yy''$;

(9) $(y')^2 + 2yy'' = 0$;

(10) $a^2y'' - y = 0$;

(11) $y'' + \dfrac{2}{1-y}(y')^2 = 0$;

(12) $yy'' + (y')^2 = 1$;

(13) $yy'' = (y')^2$.

2. An object of mass m is thrown upwards with initial velocity v_0. Assume the air resistance is kv^2. Determine the equation of the movement of the object. Also, find the velocity at the time when the object reaches the ground.

3. A bullet is lodged into a board with velocity $v_0 = 200$ m/s. The thickness of the board is $h = 10$ cm. The velocity of the bullet is $v_1 = 80$ m/s after it penetrates the board. Assume the resistance is directly proportional to the square of velocity. Find the time that is needed for the bullet to penetrate the board.

4. Solve the following equations.

(1) $y'' - 2y' - y = 0$;

(2) $4y'' - 8y' - 5y = 0$;

(3) $y'' + y' + y = 0$;

(4) $y'' - 2y + y = 0$;

(5) $\dfrac{d^2r}{dt^2} + 2\dfrac{dr}{dt} + 10r = 0$.

5. Solve the following initial value problems.

(1) $\dfrac{d^2s}{dt^2} + 2\dfrac{ds}{dt} + 5s = 0$, $s(0) = 1$, $s'(0) = 1$;

(2) $\dfrac{d^2s}{dt^2} - 4\dfrac{ds}{dt} + 3s = 0$, $s(0) = 6$, $s'(0) = 10$;

(3) $y'' + 4y' + 29y = 0$, $y(0) = 0$, $y'(0) = 15$;

(4) $4y'' + 4y' + y = 0$, $y(0) = 2$, $y'(0) = 0$.

6. Find the general solution of the following equations.

(1) $2y'' + 5y' = \cos^2 x$;

(2) $2y'' + 5y' = 29x \sin x$;

(3) $y'' - 3y' + 2y = e^x(3 - 4x)$;

(4) $y'' - 3y' + 2y = 2e^x \cos \dfrac{x}{2}$;

(5) $y'' - 2y' + y = 2e^{-x} + 3e^x$;

(6) $y''' + y = e^{-x}$;

(7) $y''' - y = \sin x$;

(8) $y''' - y = e^{-x}$;

(9) $y'' + y' - 2y = e^{5x}(x - 2) + e^{-x}(x^3 - 2x + 3)$;

(10) $y'' - y = 4x \sin x$;

(11) $3y'' + 2y' - 8y = 5 \cos x$;

(12) $y'' + 4y = e^x$;

(13) $y''' - 3y'' + 3y' - y = e^x$;

(14) $y'' + y = x^2 - x + 2$;

(15) $y'' + y' - 12y = x^2 e^x$.

7. Solve the following initial value problems.

   (1) $y'' - 7y' + 10y = x^2 e^x$, $y(0) = 0$, $y'(0) = 0$;

   (2) $y'' - 6y' + 9y = e^x \sin x$, $y(0) = 0$, $y'(0) = 1$;

   (3) $y'' + y' + y = x \cos x$, $y(0) = 1$, $y'(0) = 0$;

   (4) $y''' - y = \cos x$, $y(0) = 0$, $y'(0) = 0$, $y''(0) = 0$;

   (5) $y'' + 9y = 5 \sin x$, $y(0) = 1$, $y'(0) = 1$.

8. Solve the equation (Euler equation)

$$x^2 y'' + axy' + by = f(x),$$

where a and b are constants.

Hint: Let $x = e^t$, then the equation becomes

$$\frac{d^2 y}{dt^2} + (a - 1)\frac{dy}{dt} + by = f(e^t).$$

9. Find the general solution of the following equations.

   (1) $x^2 y'' + 3xy' + y = 0$;

   (2) $x^2 y'' - 4xy' + 6y = x$;

   (3) $x^2 y'' - xy' + y = 0$;

   (4) $\dfrac{d^2 R}{dr^2} + \dfrac{2}{r}\dfrac{dR}{dr} - \dfrac{n(n+1)}{r^2} R = 0$;

   (5) $x^2 y'' - xy' + 2y = x \ln x$, $(x > 0)$;

   (6) $x^3 y''' - x^2 y'' + 2xy' - 2y = x^3 + 3x$.

10. Solve the following initial value problems:

   (1) $\begin{cases} \dfrac{dx}{dt} - 2\dfrac{dy}{dt} = \sin t, \\[2mm] \dfrac{dx}{dt} + \dfrac{dy}{dt} = \cos t. \end{cases}$ $x(0) = 0$, $y(0) = 1$;

   (2) $\begin{cases} \dfrac{dx}{dt} = y, \\[2mm] \dfrac{dy}{dt} = -x. \end{cases}$ $x(0) = 0$, $y(0) = 1$;

   (3) $\begin{cases} \dfrac{d^2 x}{dt^2} = y, \\[2mm] \dfrac{d^2 y}{dt^2} = x. \end{cases}$ $x(0) = 2$, $x'(0) = 2$, $y(0) = 2$, $y'(0) = 2$;

$$(4) \begin{cases} \dfrac{d^2x}{dt^2} + 2\dfrac{dy}{dt} - x = 0, \\ \dfrac{dx}{dt} + y = 0. \end{cases} \qquad x(0) = 1,\; y(0) = 0.$$

11. Find the general solution of the following systems of equations.

(1)   (2)

$$\begin{cases} \dfrac{dx}{dt} + y = \cos t, \\ \dfrac{dy}{dt} + x = \sin t. \end{cases} \qquad\qquad \begin{cases} \dfrac{dx}{dt} + \dfrac{dy}{dt} = -x + y + 3, \\ \dfrac{dx}{dt} - \dfrac{dy}{dt} = x + y - 3. \end{cases}$$

(3)

$$\begin{cases} \dfrac{dx}{dt} = -x + y + z, \\ \dfrac{dy}{dt} = x - y + z, \\ \dfrac{dz}{dt} = x + y - z. \end{cases}$$

12. From the prospect of the xy-coordinate system, a shell is fired from the origin with initial velocity v_0, and initial angle α between its direction and the x-axis. (Assume the movement is in the xy-plane.) Suppose the resistance is directly proportional to the velocity of shell. (The proportional coefficient is k.) Determine the system of differential equations about the movement of the shell. Find the trajectory of the shell.

13. An object with mass m is thrown out from a flying airplane. The airplane flies horizontally with speed v_0. Assume the resistance of air is directly proportional to the speed. (The proportional coefficient is k.) Determine the equation of the movement of the object. Find the trajectory of the object.

14. A particle with mass $m = 1$ is attached to a spring. The elastic coefficient of the spring is k. Assume the resistance is directly proportional to the velocity of the particle, and the proportional coefficient is μ. Suppose a periodic force $f(t) = A_0 \sin 3t$, $(A_0 > 0)$, is applied on the particle. Let x be the distance between the particle and its equilibrium position. Find the following:
    (1) the differential equation of the movement;
    (2) the equation of the movement if $\mu = 10$ and $k = 16$;
    (3) the equation of the movement if $\mu = 10$ and $k = 25$;

(4) the equation of the movement if $\mu = 8$ and $k = 25$;

(5) the equation of the movement if $\mu = 0$ and $k = 9$.

(6) Interpret the above results if no force is applied on the particle ($A_0 = 0$.)

15. Which groups of functions are linearly independent?

(1) x, e^x;

(2) $\cos x$, $\sin x$;

(3) $\cos x$, $\cos^3 x$, $\cos 3x$;

(4) x, x^3, e^x.

16. The equation $y'' \sin^2 x = 2y$ has a special solution $y_1 = \cot x$. Find the general solution of this equation.

17. Find the general solution of the following equations.

(1) $x^2 y'' - 2xy' + 2y = 0$, $(x \neq 0)$;

(2) $xy'' - (1 + x)y' + y = 0$, $(x \neq 0)$;

(3) $(1 + x^2)y'' - 2xy' + 2y = 0$;

(4) $x^2 y'' - xy' = 3x^3$, $(x \neq 0)$.

18. The equation

$$(1 + x^2)y'' + 2xy' - 6x^2 - 2 = 0$$

has a special solution $y = x^2$. Find the special solution that satisfies the initial condition $y|_{x=-1} = y'|_{x=-1} = 0$.

19. Solve the initial value problem

$$\begin{cases} (2x - x^2)y'' + (x^2 - 2)y' + 2(1 - x)y = 0, & (x \neq 0, 2), \\ y|_{x=1} = 0, \ y'|_{x=1} = 1. \end{cases}$$

Chapter 5

Vector Algebra and Analytic Geometry in Three-Dimensional Space

5.1 Coordinate System of Three-Dimensional Space and Concept of Vectors

5.1.1 *Rectangular Coordinate System*

Consider three lines that are mutually perpendicular at a point O. We label the three lines x-axis, y-axis, and z-axis with the order following the right-hand rule (Figure 5.1): Hold your right hand up, with the fingers curled. If the curl of your fingers point from the x-axis to the y-axis, then your thumb points to the z-axis.

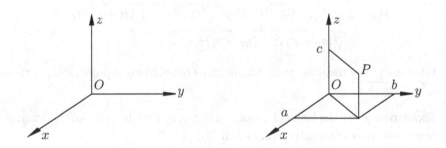

Fig. 5.1   Fig. 5.2

Suppose P is an arbitrary point in space. Construct three planes passing through P and perpendicular to the x-axis, y-axis, and z-axis respectively. The three planes intersect their axis at $x = a$, $y = b$, $z = c$. This gives rise to a correspondence from P to a three tuple of real numbers (a, b, c).

On the other hand, for any given three tuple of real numbers (a, b, c), we can construct three planes that are perpendicular to x-axis, y-axis, and z-axis and pass through the points $x = a$, $y = b$, $z = c$ respectively. The three planes intersect at a point P. This gives rise to the inverse of the above correspondence. Therefore, there is an one-to-one correspondence between points P and three tuples (a, b, c). The tuple (a, b, c) is called the coordinates of P (Figure 5.2.) This is called the *rectangular coordinate system*.

The plane that contains the x-axis and the y-axis is defined as the xy-plane. The yz-plane and zx-plane are defined in a similar way. These three mutually perpendicular planes are called the coordinate planes (Figure 5.3) and they partition the space into eight octants.

Suppose $P_1(x_1, y_1, z_1)$ and $P_2(x_2, y_2, z_2)$ are two arbitrary points in space. Construct three planes passing through P_1 and perpendicular to x-axis, y-axis, z-axis respectively, and construct another three planes passing through P_2 the same way (Figure 5.4). We can see a rectangular parallelepiped with the line segment $P_1 P_2$ as its diagonal, and all the sides of this rectangular parallelepiped are parallel to the coordinate axes. Since

$$P_1 A = x_2 - x_1, \quad AB = y_2 - y_1, \quad BP_2 = z_2 - z_1,$$

we have the following distance formula of two points in space:

$$P_1 P_2 = \sqrt{(P_1 B)^2 + (BP_2)^2} = \sqrt{(P_1 A)^2 + (AB)^2 + (BP_2)^2}$$
$$= \sqrt{(x_2 - x_1)^2 + (y_2 - y_1)^2 + (z_2 - z_1)^2}.$$

Obviously, the distance from the origin $O(0, 0, 0)$ to a point $P(x, y, z)$ is $\sqrt{x^2 + y^2 + z^2}$.

Example: Find the locus of a point $M(x, y, z)$ that has the same distance from two fixed points $A(1, -2, 1)$ and $B(2, 1, -2)$.

Solution: Since $AM = BM$, we have

$$\sqrt{(x - 1)^2 + (y + 2)^2 + (z - 1)^2} = \sqrt{(x - 2)^2 + (y - 1)^2 + (z + 2)^2}.$$

The simplified version of this equation is

$$2x + 6y - 6z - 3 = 0.$$

This is an equation of a plane.

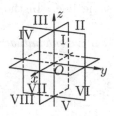

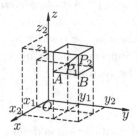

Fig. 5.3 Fig. 5.4

5.1.2 *Addition and Scalar Multiplication of Vectors*

A quantity that has both magnitude and direction is called a vector quantity. Velocity, acceleration, force, displacement, ..., are examples of vectors. A *vector* can be expressed as a directed line segment whose length is equal to the magnitude of the vector, and whose direction is the direction of the vector. If a line segment has the initial point $P(x_0, y_0, z_0)$, and the end point $Q(x_0 + a_1, y_0 + a_2, z_0 + a_3)$, then it determines a vector $\overrightarrow{PQ}$ (Figure 5.5).

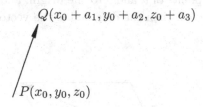

$$Q(x_0 + a_1, y_0 + a_2, z_0 + a_3)$$

$$P(x_0, y_0, z_0)$$

Fig. 5.5

There are two kinds of vectors in general: A *free vector* represents all vectors with the same direction and the same magnitude, regardless of initial points. If we move the initial point P to the origin, then the coordinates of Q becomes (a_1, a_2, a_3) and $\overrightarrow{PQ}$ can be represented by $\vec{a} =$

(a_1, a_2, a_3). A *sliding vector* represents all vectors with the same direction and the same magnitude, and lying on the same line. In the following discussion, we restrict to free vectors.

Suppose $\vec{a} = (a_1, a_2, a_3)$ is a free vector. Then the *magnitude* or length of the vector is

$$|\vec{a}| = \sqrt{a_1^2 + a_2^2 + a_3^2},$$

and its direction is determined by

$$\cos \alpha = \frac{a_1}{|\vec{a}|}, \quad \cos \beta = \frac{a_2}{|\vec{a}|}, \quad \cos \gamma = \frac{a_3}{|\vec{a}|},$$

where a_1, a_2 and a_3 are projections of vector $\vec{a}$ on the x-axis, y-axis and z-axis respectively, and α, β and γ are the angles between $\vec{a}$ and the positive x-axis, y-axis, and z-axis respectively.

The numbers, $\cos \alpha$, $\cos \beta$, $\cos \gamma$ are called the *direction cosines* of the vector $\vec{a}$. They satisfy

$$\cos^2 \alpha + \cos^2 \beta + \cos^2 \gamma = 1.$$

Some examples of vectors in the real world are: displacement, velocity, force.

The sum of two vectors $\vec{a} = (a_1, a_2, a_3)$ and $\vec{b} = (b_1, b_2, b_3)$ is defined to be

$$\vec{a} + \vec{b} = (a_1 + b_1, a_2 + b_2, a_3 + b_3).$$

Moving the initial points of $\vec{a}$ and $\vec{b}$ to the origin, and construct a parallelogram as in Figure 5.6, then the diagonal is the vector $\vec{a} + \vec{b}$.

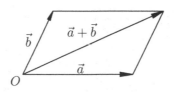

Fig. 5.6

The *zero vector* is $\vec{0} = (0,0,0)$, and this is the only vector without a well-defined direction. If a vector has the same magnitude but opposite direction of the vector $\vec{a}$, then it is the *negative vector* of $\vec{a}$, denoted as

$$-\vec{a} = (-a_1, -a_2, -a_3).$$

For any real number λ, define the *scalar product* of λ and a vector $\vec{a}$ as

$$\lambda\vec{a} = (\lambda a_1, \lambda a_2, \lambda a_3).$$

The scalar product of a vector changes the magnitude, but not the direction of the vector if $\lambda > 0$. A negative λ will change the direction of a vector to its opposite direction. Especially, If

$$\lambda = \frac{1}{\sqrt{a_1^2 + a_2^2 + a_3^2}},$$

then

$$\lambda\vec{a} = \left(\frac{a_1}{\sqrt{a_1^2 + a_2^2 + a_3^2}}, \frac{a_2}{\sqrt{a_1^2 + a_2^2 + a_3^2}}, \frac{a_3}{\sqrt{a_1^2 + a_2^2 + a_3^2}}\right).$$

Obviously, the magnitude of this vector is 1 and it is called the *unit vector* $\vec{a}^0$. With this notation, any vector $\vec{a}$ can be written as $\vec{a} = |\vec{a}|\vec{a}^0$.

Let $\vec{i} = (1,0,0)$, $\vec{j} = (0,1,0)$, $\vec{k} = (0,0,1)$. Then $\vec{i}, \vec{j}, \vec{k}$ are unit vectors and for any vector $\vec{a} = (a_1, a_2, a_3)$, we have

$$\vec{a} = a_1\vec{i} + a_2\vec{j} + a_3\vec{k}.$$

That is, $\vec{a}$ can be expressed as a linear combination of these three unit vectors. Furthermore, for any vectors $\vec{a}$, $\vec{b}$, $\vec{c}$ and any real numbers λ, μ, the following equations hold.

(1) $\vec{a} + \vec{b} = \vec{b} + \vec{a}$;

(2) $\vec{a} + (\vec{b} + \vec{c}) = (\vec{a} + \vec{b}) + \vec{c}$;

(3) $\lambda(\vec{a} + \vec{b}) = \lambda\vec{a} + \lambda\vec{b}$;

(4) $\lambda(\mu\vec{a}) = (\lambda\mu)\vec{a}$;

(5) Vector $\vec{a}$ and vector $\vec{b}$ are collinear if and only if there exists a real number λ such that $\vec{a} = \lambda\vec{b}$;

(6) $|\vec{a} + \vec{b}| \leqslant |\vec{a}| + |\vec{b}|$.

Example 1. Show that if the two diagonals of a quadrilateral bisect each other, then this quadrilateral is a parallelogram.

Proof: Since $\overrightarrow{AO} = \overrightarrow{OC}$, $\overrightarrow{BO} = \overrightarrow{OD}$, we have

$$\overrightarrow{AO} + \overrightarrow{OD} = \overrightarrow{BO} + \overrightarrow{OC}.$$

That is $\overrightarrow{AD} = \overrightarrow{BC}$ (Figure 5.7). This implies that $\overrightarrow{AD}$ is parallel to $\overrightarrow{BC}$ and that they have the same magnitude. Therefore, the quadrilateral $ABCD$ is a parallelogram.

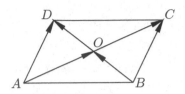

Fig. 5.7

Example 2. Given two points $M_1(x_1, y_1, z_1)$, $M_2(x_2, y_2, z_2)$ and a point $M(x, y, z)$ on the line segment $\overline{M_1M_2}$, such that the ratio of $|M_1M|$ over $|MM_2|$ is λ. Find the coordinates of M.

Solution: Let the coordinates of M be (x, y, z), and

$$\overrightarrow{OM_1} = \vec{r}_1, \quad \overrightarrow{OM_2} = \vec{r}_2, \quad \overrightarrow{OM} = \vec{r}.$$

Then (Figure 5.8)

$$\overrightarrow{M_1M} = \vec{r} - \vec{r}_1, \quad \overrightarrow{MM_2} = \vec{r}_2 - \vec{r}.$$

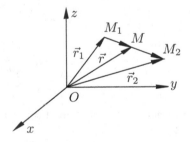

Fig. 5.8

Since $\overrightarrow{M_1M} = \lambda \overrightarrow{MM_2}$, we have

$$\vec{r} - \vec{r}_1 = \lambda(\vec{r}_2 - \vec{r}).$$

That is

$$\vec{r} = \frac{\vec{r}_1 + \lambda \vec{r}_2}{1 + \lambda}.$$

Hence, the coordinates of the point M are:

$$x = \frac{x_1 + \lambda x_2}{1 + \lambda}, \quad y = \frac{y_1 + \lambda y_2}{1 + \lambda}, \quad z = \frac{z_1 + \lambda z_2}{1 + \lambda}.$$

If $\lambda = 1$, we get the coordinates of the mid-point between M_1 and M_2:

$$x = \frac{x_1 + x_2}{2}, \quad y = \frac{y_1 + y_2}{2}, \quad z = \frac{z_1 + z_2}{2}.$$

Exercises 5.1

1. (1) Find the distance between the point $M(4, -3, 5)$ and the origin. Also, find the distance between the point and each coordinate axis.

   (2) Given the point $A(4, -7, 1)$, find a point $B(6, 2, z)$ such that the distance between A and B is 11.

   (3) Given the point $A(2, 3, 4)$, find a point $B(x, -2, 4)$ such that the distance between A and B is 5.

2. Find a point on the yz-plane such that it has equal distance from the points $A(3, 1, 2)$, $B(4, -2, -2)$ and $C(0, 5, 1)$.

3. Given two vectors $\vec{a}$ and $\vec{b}$, graph the following:

   (1) $\vec{a} - \vec{b}$;   (2) $\vec{b} - \vec{a}$;   (3) $-\vec{a} - \vec{b}$;   (4) $3\vec{a}$;

   (5) $-\dfrac{1}{2}\vec{b}$;   (6) $2\vec{a} + \dfrac{1}{3}\vec{b}$;   (7) $\dfrac{1}{2}\vec{a} - 3\vec{b}$.

4. Suppose $\vec{a}$ is perpendicular to $\vec{b}$, and $|a| = 5$, $|b| = 12$. Find $|\vec{a} + \vec{b}|$ and $|\vec{a} - \vec{b}|$.

5. Let the angle between $\vec{a}$ and $\vec{b}$ be $\alpha = 60°$, and $|\vec{a}| = 5$, $|\vec{b}| = 8$. Find $|\vec{a} + \vec{b}|$ and $|\vec{a} - \vec{b}|$.

6. Let the angle between $\vec{P}$ and $\vec{Q}$ be $\varphi = 120°$, and $|\vec{P}| = 4$, $|\vec{Q}| = 5$. Find the magnitude and the direction of $\vec{R} = \vec{P} + \vec{Q}$.

7. Suppose $|\vec{a}| = 13$, $|\vec{b}| = 19$, and $|\vec{a} + \vec{b}| = 24$. Find $|\vec{a} - \vec{b}|$.

8. Interpret the geometric meaning of each of the following equations:

   (1) $\lambda(\vec{a} + \vec{b}) = \lambda\vec{a} + \lambda\vec{b}$;   (3) $|\vec{a} - \vec{b}| \geqslant ||\vec{a}| - |\vec{b}||$.
   (2) $(\vec{a} + \vec{b}) + \vec{c} = \vec{a} + (\vec{b} + \vec{c})$;

9. The four vertices of a parallelogram are denoted as A, B, C and D counterclockwise. Assume that $\overrightarrow{AB} = \vec{a}$, $\overrightarrow{AD} = \vec{b}$, and M is the intersecting point of the two diagonals. Express the vectors $\overrightarrow{MA}$, $\overrightarrow{MB}$, $\overrightarrow{MC}$ and $\overrightarrow{MD}$ in terms of $\vec{a}$ and $\vec{b}$.

10. Suppose M is the middle point of a triangle ABC. Show that for any point O,
$$\overrightarrow{OM} = \frac{1}{3}(\overrightarrow{OA} + \overrightarrow{OB} + \overrightarrow{OC}).$$

11. The four vertices of a tetrahedron are denoted as A, B, C and D. Suppose M, N are mid-points of AB and CD respectively. Show that
$$\overrightarrow{MN} = \frac{1}{2}(\overrightarrow{AD} + \overrightarrow{BC}).$$

12. Given two points $P_1(a_1, b_1, c_1)$ and $P_2(a_2, b_2, c_2)$. Find the coordinates, magnitude, and direction cosines of the vector $\overrightarrow{P_1 P_2}$.

13. How does a vector relate with the three coordinate axes and the three coordinate planes if its direction cosine satisfies
(1) $\cos\alpha = 0$, or (2) $\cos\beta = 1$, or (3) $\cos\alpha = \cos\beta = 0$?

14. Suppose the angle between a vector $\vec{a}$ and the x-axis is the same as the angle between $\vec{a}$ and the y-axis, and the angle between $\vec{a}$ and the z-axis is twice the former angle. Find the direction cosines of $\vec{a}$.

15. Draw a line segment AB from the point $A(2, -1, 7)$ and along the direction of the vector
$$\vec{a} = 8\vec{i} + 9\vec{j} - 12\vec{k}.$$
The length of the line segment is 34. Find the coordinates of the point B.

16. Suppose $\vec{a} = 3\vec{i} + 4\vec{j} - 12\vec{k}$. Determine the coordinates of the unit vector
$$\vec{a}^0 = \frac{\vec{a}}{|\vec{a}|}.$$

17. Suppose the angles between $\vec{a}$ and the x-axis and the y-axis are $\alpha = 60°$ and $\beta = 120°$ respectively. The magnitude of $\vec{a}$ is $|\vec{a}| = 2$. Find the coordinates of $\vec{a}$.

18. Given two points $M_1(2, 5, -3)$, $M_2(3, -2, 5)$. Let M be a point on the line segment $M_1 M_2$ such that $\overrightarrow{M_1 M} = 3\overrightarrow{MM_2}$. Determine the coordinates of M.

19. Given three vectors $\vec{p} = (3, -2, 1)$, $\vec{q} = (-1, 1, -2)$, and $\vec{r} = (2, 1, -3)$. Determine the decomposition expression of the vector $\vec{a} = (11, 6, 5)$ with respect to $\vec{p}$, $\vec{q}$, $\vec{r}$. That is, determine the numbers α, β, γ such that $\vec{a} = \alpha\vec{p} + \beta\vec{q} + \gamma\vec{r}$.

20. The three vertices of a triangle are $A(2, 5, 0)$, $B(11, 3, 8)$, and $C(5, 11, 12)$. Find the lengths of the medians of the triangle.

5.2   Products of Vectors

5.2.1   *Inner Products of Vectors*

The *inner product* (or *dot product*) of two vectors $\vec{a} = (a_1, a_2, a_3)$ and $\vec{b} = (b_1, b_2, b_3)$ is defined as

$$\vec{a} \cdot \vec{b} = a_1 b_1 + a_2 b_2 + a_3 b_3.$$

For any vectors $\vec{a}$, $\vec{b}$, $\vec{c}$ and any real numbers λ, μ, we have:

$\vec{a} \cdot \vec{b} = \vec{b} \cdot \vec{a}$ (commutative law);

$\vec{a} \cdot (\vec{b} + \vec{c}) = \vec{a} \cdot \vec{b} + \vec{a} \cdot \vec{c}$ (distributive law);

$(\lambda \vec{a}) \cdot (\mu \vec{b}) = \lambda \mu \vec{a} \cdot \vec{b}$ (associative law of number multiplication);

$\vec{a} \cdot \vec{a} = |\vec{a}|^2.$

If θ is the angle between vectors $\vec{a}$ and $\vec{b}$, by the cosine theorem, we have (Figure 5.9)

$$|\vec{a} - \vec{b}|^2 = |\vec{a}|^2 + |\vec{b}|^2 - 2|\vec{a}||\vec{b}| \cos \theta.$$

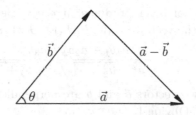

Fig. 5.9

On the other hand,

$$|\vec{a} - \vec{b}|^2 = (\vec{a} - \vec{b}) \cdot (\vec{a} - \vec{b}) = \vec{a} \cdot \vec{a} + \vec{b} \cdot \vec{b} - 2\vec{a} \cdot \vec{b}$$
$$= |\vec{a}|^2 + |\vec{b}|^2 - 2\vec{a} \cdot \vec{b}.$$

Thus

$$\vec{a} \cdot \vec{b} = |\vec{a}||\vec{b}| \cos \theta.$$

Therefore, the inner product of vectors $\vec{a}$ and $\vec{b}$ can be considered as the product of the magnitude of $\vec{a}$ and the projection of $\vec{b}$ on $\vec{a}$ (Figure 5.10).

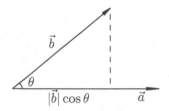

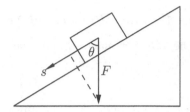

Fig. 5.10   Fig. 5.11

Inner product is an important concept in mathematics and science. For instance, an object slides down a slope due to gravity $\vec{F}$ (Figure 5.11). The direction of $\vec{F}$ is vertically downward. The direction of displacement $\vec{s}$ is parallel to the slope. If the angle between $\vec{F}$ and the positive direction of of $\vec{s}$ is θ, then the work done by $\vec{F}$ is

$$w = |\vec{F}||\vec{s}|\cos\theta = \vec{F} \cdot \vec{s}.$$

By the definition of inner products, if the angle between the vector $\vec{a} = (a_1, a_2, a_3)$ and the vector $\vec{b} = (b_1, b_2, b_3)$ is θ, then

$$\cos\theta = \frac{a_1 b_1 + a_2 b_2 + a_3 b_3}{\sqrt{a_1^2 + a_2^2 + a_3^2}\sqrt{b_1^2 + b_2^2 + b_3^2}}$$

It follows that two vectors $\vec{a}$ and $\vec{b}$ are perpendicular to each other if and only if $\vec{a} \cdot \vec{b} = 0$. Obviously,

$$\vec{i} \cdot \vec{i} = \vec{j} \cdot \vec{j} = \vec{k} \cdot \vec{k} = 1, \quad \vec{i} \cdot \vec{j} = \vec{j} \cdot \vec{k} = \vec{k} \cdot \vec{i} = 0.$$

Example 1. Let $|\vec{a}| = 2$, and $|\vec{b}| = 1$, and the angle θ between $\vec{a}$ and $\vec{b}$ be $\dfrac{\pi}{3}$. Find the angle φ between vectors $\vec{c} = 2\vec{a} + 3\vec{b}$ and $\vec{d} = 3\vec{a} - \vec{b}$.

Solution: Since $\cos\varphi = \dfrac{\vec{c} \cdot \vec{d}}{|\vec{c}||\vec{d}|}$, and

$$\vec{c} \cdot \vec{d} = (2\vec{a} + 3\vec{b}) \cdot (3\vec{a} - \vec{b}) = 6\vec{a} \cdot \vec{a} - 2\vec{a} \cdot \vec{b} + 9\vec{b} \cdot \vec{a} - 3\vec{b} \cdot \vec{b}$$

$$= 6|\vec{a}|^2 + 7\vec{a} \cdot \vec{b} - 3|\vec{d}|^2 = 6(2^2) + 7(2)\cos\frac{\pi}{3} - 3(1^2) = 28,$$

by similar calculations

$$\vec{c} \cdot \vec{c} = (2\vec{a} + 3\vec{b}) \cdot (2\vec{a} + 3\vec{b}) = 37,$$

$$\vec{d} \cdot \vec{d} = (3\vec{a} - \vec{b}) \cdot (3\vec{a} - \vec{b}) = 31,$$

we have,

$$\cos \theta = \frac{28}{\sqrt{37}\sqrt{31}} = 0.8268\cdots, \quad \varphi = 34°14'.$$

Example 2. Find a unit vector that is perpendicular to both $\vec{a} = (1, -3, 1)$ and $\vec{b} = (2, -1, 3)$.

Solution: Let $\vec{x} = (x_1, x_2, x_3)$ be the unit vector we want. Then

$$\vec{x} \cdot \vec{a} = 0, \ \vec{x} \cdot \vec{b} = 0, \ \vec{x} \cdot \vec{x} = 1.$$

It follows that

$$\begin{cases} x_1 - 3x_2 + x_3 = 0, \\ 2x_1 - x_2 + 3x_3 = 0, \\ x_1^2 + x_2^2 + x_3^2 = 1. \end{cases}$$

Solving the system, we get

$$x_1 = \pm\frac{8}{3\sqrt{10}}, \ x_2 = \pm\frac{1}{3\sqrt{10}}, \ x_3 = \pm\frac{5}{3\sqrt{10}}.$$

Therefore,

$$\vec{x} = \pm\frac{1}{3\sqrt{10}}(8, 1, -5).$$

Example 3. Prove the inequality

$$(a_1b_1 + a_2b_2 + a_3b_3)^2 \leqslant (a_1^2 + a_2^2 + a_3^2)(b_1^2 + b_2^2 + b_3^2).$$

Proof: Let $\vec{a} = (a_1, a_2, a_3)$, $\vec{b} = (b_1, b_2, b_3)$. By the definition of inner products we have

$$\vec{a} \cdot \vec{b} = |\vec{a}||\vec{b}| \cos \theta.$$

Since $|\cos \theta| \leqslant 1$, we have

$$(\vec{a} \cdot \vec{b})^2 \leqslant |\vec{a}|^2 |\vec{b}|^2.$$

5.2.2 *Cross Products of Vectors*

The *cross product* of two vectors $\vec{a} = (a_1, a_2, a_3)$ and $\vec{b} = (b_1, b_2, b_3)$ is defined as

$$\vec{a} \times \vec{b} = (a_2 b_3 - a_3 b_2, a_3 b_1 - a_1 b_3, a_1 b_2 - a_2 b_1).$$

It can also be written as

$$\vec{a} \times \vec{b} = \begin{vmatrix} \vec{i} & \vec{j} & \vec{k} \\ a_1 & a_2 & a_3 \\ b_1 & b_2 & b_3 \end{vmatrix}.$$

For any vectors $\vec{a}$, $\vec{b}$, $\vec{c}$ and any real numbers λ, μ, the following equations hold.

$$(\vec{a} + \vec{b}) \times \vec{c} = \vec{a} \times \vec{c} + \vec{b} \times \vec{c} \quad \text{(distributive law)};$$
$$(\lambda \vec{a}) \times \mu \vec{b} = \lambda \mu (\vec{a} \times \vec{b}) \quad \text{(associative law)};$$
$$\vec{a} \times \vec{b} = -\vec{b} \times \vec{a} \quad \text{(commutative law is not satisfied)}.$$

Especially,

$$\vec{a} \times \vec{a} = -\vec{a} \times \vec{a},$$

thus

$$\vec{a} \times \vec{a} = \vec{0}.$$

If $\vec{a} \times \vec{b} = \vec{0}$, then $\vec{a}$ and $\vec{b}$ are collinear.

The geometric meaning of cross products can be interpreted as follows:

(1) The vector $\vec{a} \times \vec{b}$ is perpendicular to both $\vec{a}$ and $\vec{b}$, that is, it is perpendicular to the plane spanned by $\vec{a}$ and $\vec{b}$. This is because

$$\vec{a} \cdot (\vec{a} \times \vec{b}) = a_1(a_2 b_3 - a_3 b_2) + a_2(a_3 b_1 - a_1 b_3) + a_3(a_1 b_2 - a_2 b_1)$$
$$= \begin{vmatrix} a_1 & a_2 & a_3 \\ a_1 & a_2 & a_3 \\ b_1 & b_2 & b_3 \end{vmatrix} = 0.$$

Similarly, $\vec{b} \cdot (\vec{a} \times \vec{b}) = 0$.

(2) The length of the vector $\vec{a} \times \vec{b}$ is equal to the area of the parallelogram with sides $\vec{a}$ and $\vec{b}$, and

$$|\vec{a} \times \vec{b}|^2 = |\vec{a}|^2 |\vec{b}|^2 - (\vec{a} \cdot \vec{b})^2.$$

Indeed,

$$|\vec{a} \times \vec{b}|^2 = (\vec{a} \times \vec{b}) \cdot (\vec{a} \times \vec{b})$$
$$= (a_2b_3 - a_3b_2)^2 + (a_3b_1 - a_1b_3)^2 + (a_1b_2 - a_2b_1)^2$$
$$= (a_1^2 + a_2^2 + a_3^2)(b_1^2 + b_2^2 + b_3^2) - (a_1b_1 + a_2b_2 + a_3b_3)^2.$$

That is

$$|\vec{a} \times \vec{b}|^2 = |\vec{a}|^2|\vec{b}|^2 - (\vec{a} \cdot \vec{b})^2 = |\vec{a}|^2|\vec{b}|^2(1 - \cos^2\theta) = |\vec{a}|^2|\vec{b}|^2 \sin^2\theta.$$

(3) Vectors $\vec{a}$, $\vec{b}$ and $\vec{a} \times \vec{b}$ follow the right-hand rule (Figure 5.12).

In fact, let $\vec{a}$ be a vector on the x-axis, with the same direction as the positive x-axis, and let xy-plane contain $\vec{a}$ and $\vec{b}$. Suppose θ is the angle from $\vec{a}$ to $\vec{b}$ ($|\theta| \leqslant \pi$). Then

$$\vec{a} = |\vec{a}|(1, 0, 0), \quad \vec{b} = |\vec{b}|(\cos\theta, \sin\theta, 0).$$

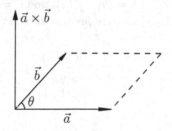

Fig. 5.12

Thus

$$\vec{c} = \vec{a} \times \vec{b} = |\vec{a}||\vec{b}|(0, 0, \sin\theta).$$

This implies that the vector $\vec{c}$ is on the z-axis (of course we need to assume that the initial points of all three vectors are at the origin). The direction of $\vec{c}$ depends on the sign of θ. It points to the positive direction of the z-axis if θ is positive, and it points to the negative direction of the z-axis if θ is negative. So they follow the right-hand rule.

Especially, we have

$$\vec{i} \times \vec{i} = \vec{0}, \quad \vec{j} \times \vec{j} = \vec{0}, \quad \vec{k} \times \vec{k} = \vec{0};$$
$$\vec{i} \times \vec{j} = \vec{k}, \quad \vec{j} \times \vec{k} = \vec{i}, \quad \vec{k} \times \vec{i} = \vec{j}.$$

Example: Suppose $A(1, 2, 3)$, $B(3, 4, 5)$, and $C(-1, -2, 7)$ are vertices of a triangle $\triangle ABC$. Find the area of the triangle.

Solution: The area of the triangle S is $\frac{1}{2}|\overrightarrow{AB} \times \overrightarrow{AC}|$. Since

$$\overrightarrow{AB} = (2, 2, 2), \quad \overrightarrow{AC} = (-2, -4, 4)$$

we have

$$\overrightarrow{AB} \times \overrightarrow{AC} = \begin{vmatrix} \vec{i} & \vec{j} & \vec{k} \\ 2 & 2 & 2 \\ -2 & -4 & 4 \end{vmatrix} = 16\vec{i} - 12\vec{j} - 4\vec{k}.$$

Thus,

$$S = \frac{1}{2}\sqrt{16^2 + (-12)^2 + (-4)^2} = 2\sqrt{26}.$$

5.2.3   Scalar Triple Products of Vectors

For vectors $\vec{a} = (a_x, a_y, a_z)$, $\vec{b} = (b_x, b_y, b_z)$, and $\vec{c} = (c_x, c_y, c_z)$, the product $(\vec{a} \times \vec{b}) \cdot \vec{c}$ is called the *scalar triple product* of the three vectors.

Since

$$\vec{a} \times \vec{b} = \begin{vmatrix} \vec{i} & \vec{j} & \vec{k} \\ a_x & a_y & a_z \\ b_x & b_y & b_z \end{vmatrix}$$

we have

$$(\vec{a} \times \vec{b}) \cdot \vec{c} = \begin{vmatrix} a_x & a_y & a_z \\ b_x & b_y & b_z \\ c_x & c_y & c_z \end{vmatrix}.$$

On the other hand, from Figure 5.13

$$(\vec{a} \times \vec{b}) \cdot \vec{c} = |\vec{a} \times \vec{b}||\vec{c}|\cos\varphi = \pm|\vec{a} \times \vec{b}|h = \pm V,$$

where V is the volume of the parallelepiped with sides $\vec{a}$, $\vec{b}$ and $\vec{c}$.

The angle φ is a positive acute angle when $\vec{a}$, $\vec{b}$ and $\vec{c}$ follow the right-hand rule, otherwise, it is negative. Hence,

$$(\vec{a} \times \vec{b}) \cdot \vec{c} = (\vec{b} \times \vec{c}) \cdot \vec{a} = (\vec{c} \times \vec{a}) \cdot \vec{b}$$

$$= -(\vec{b} \times \vec{a}) \cdot \vec{c} = -(\vec{c} \times \vec{b}) \cdot \vec{a} = -(\vec{a} \times \vec{c}) \cdot \vec{b}.$$

The absolute value of the scalar triple product is the volume of the parallelepiped with sides $\vec{a}$, $\vec{b}$ and $\vec{c}$. This volume becomes zero when the three vectors are coplanar. It is equivalent to

$$(\vec{a} \times \vec{b}) \cdot \vec{c} = 0,$$

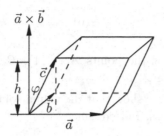

Fig. 5.13

or

$$\begin{vmatrix} a_x & a_y & a_z \\ b_x & b_y & b_z \\ c_x & c_y & c_z \end{vmatrix} = 0.$$

Furthermore,

$$(\vec{a} \times \vec{b}) \times \vec{c} = (\vec{a} \cdot \vec{c})\vec{b} - (\vec{b} \cdot \vec{c})\vec{a},$$

$$\vec{a} \times (\vec{b} \times \vec{c}) = (\vec{a} \cdot \vec{c})\vec{b} - (\vec{a} \cdot \vec{b})\vec{c}.$$

Hence, the equation

$$(\vec{a} \times \vec{b}) \times \vec{c} = \vec{a} \times (\vec{b} \times \vec{c})$$

is not true in general.

Example 1. Find the volume of an irregular tetrahedron with vertices $A(1,1,1,)$, $B(3,4,4)$, $C(3,5,5)$ and $D(2,4,7)$.

Solution: The volume of the parallelepiped with $\overrightarrow{AB}$, $\overrightarrow{AC}$ and $\overrightarrow{AD}$ as its sides is six times the volume of the irregular tetrahedron we need to find. So

$$V = \frac{1}{6}|\overrightarrow{AB} \cdot (\overrightarrow{AC} \times \overrightarrow{AD})|.$$

Since

$$\overrightarrow{AB} = (2,3,3), \quad \overrightarrow{AC} = (2,4,4), \quad \overrightarrow{AD} = (1,3,6),$$

we have

$$\overrightarrow{AB} \cdot (\overrightarrow{AC} \times \overrightarrow{AD}) = \begin{vmatrix} 2 & 3 & 3 \\ 2 & 4 & 4 \\ 1 & 3 & 6 \end{vmatrix} = 6.$$

Thus

$$V = \frac{1}{6}(6) = 1.$$

Example 2. Show that the points $A(1,1,1)$, $B(4,5,6)$, $C(2,3,3)$, and $D(10,15,17)$ are coplanar.

Solution: We only need to show that the volume of the parallelepiped with sides $\overrightarrow{AB}$, $\overrightarrow{AC}$, $\overrightarrow{AD}$ is zero. Indeed, since

$$\overrightarrow{AB} = (3,4,5), \quad \overrightarrow{AC} = (1,2,2), \quad \overrightarrow{AD} = (9,14,16),$$

we have

$$\overrightarrow{AB} \cdot (\overrightarrow{AC} \times \overrightarrow{AD}) = \begin{vmatrix} 3 & 4 & 5 \\ 1 & 2 & 2 \\ 9 & 14 & 16 \end{vmatrix} = 0.$$

Hence, these four points are on the same plane.

Exercises 5.2

1. Suppose the angle between $\vec{a}$ and $\vec{b}$ is $\varphi = \frac{2\pi}{3}$, and $|\vec{a}| = 3$, $|\vec{b}| = 4$.
   Find (1) $\vec{a} \cdot \vec{b}$; (2) $(3\vec{a} - 2\vec{b}) \cdot (\vec{a} + 2\vec{b})$.

2. Suppose $\vec{a} = (4,-2,4)$, $\vec{b} = (6,-3,2)$. Find
   (1) $\vec{a} \cdot \vec{b}$; (2) $(2\vec{a} - 3\vec{b}) \cdot (\vec{a} + 2\vec{b})$; (3) $|\vec{a} - \vec{b}|^2$.

3. Find the projection of the vector $\vec{a} = (4,-3,4)$ on the vector $\vec{b} = (2,3,1)$.

4. Two sides of a triangle are $\overrightarrow{AB} = (2,1,-2)$ and $\overrightarrow{BC} = (3,2,6)$. Find the measurement of the three interior angles of this triangle.

5. Suppose the angle between $\vec{a}$ and $\vec{b}$ is $\frac{\pi}{3}$, and $|\vec{a}| = 5$, $|\vec{b}| = 2$. Find the magnitude of the vector $2\vec{a} - 3\vec{b}$.

6. If $\vec{a} + 3\vec{b}$ is perpendicular to $7\vec{a} - 5\vec{b}$, and $\vec{a} - 4\vec{b}$ is perpendicular to $7\vec{a} - 2\vec{b}$, find the angle between $\vec{a}$ and $\vec{b}$.

7. Suppose $\vec{a} = (2,-3,1)$, $\vec{b} = (1,-1,3)$, $\vec{c} = (1,-2,0)$. Find

   (1) $(\vec{a} \cdot \vec{b})\vec{c} - (\vec{a} \cdot \vec{c})\vec{b}$;   (3) $(\vec{a} \times \vec{b}) \cdot \vec{c}$;
   (2) $(\vec{a} + \vec{b}) \times (\vec{b} + \vec{c})$;   (4) $(\vec{a} \times \vec{b}) \times \vec{c}$.

8. Suppose $\vec{a}$ and $\vec{b}$ are perpendicular to each other, and $|\vec{a}| = 3$, $|\vec{b}| = 4$.
   Compute (1) $|(\vec{a} + \vec{b}) \times (\vec{a} - \vec{b})|$;   (2) $|(3\vec{a} - \vec{b}) \times (\vec{a} - 2\vec{b})|$.

9. Suppose the angle between $\vec{a}$ and $\vec{b}$ is $\dfrac{2\pi}{3}$, and $|\vec{a}| = 1$, $\vec{b} = 2$. Find

   (1) $|\vec{a} \times \vec{b}|^2$; (2) $|(\vec{a} + 3\vec{b}) \times (3\vec{a} - \vec{b})|^2$.

10. Find the area of the triangle with vertices $A(3, 4, -1)$, $B(2, 0, 3)$ and $C(-3, 5, 4)$.

11. Two sides of a parallelogram are vectors $\vec{a} = (2, 1, -1)$ and $\vec{b} = (1, -2, 1)$. Find the sine of the angle between its two diagonals.

12. Find the volume of the irregular tetrahedron with vertices $A(2, -1, 1)$, $B(5, 5, 4)$, $C(3, 2, -1)$, $D(4, 1, 3)$.

13. Decide whether $\vec{a}$, $\vec{b}$ and $\vec{c}$ are coplanar if

    (1) $\vec{a} = (2, 3, -1)$, $\vec{b} = (1, -1, 3)$, $\vec{c} = (1, 9, -11)$;

    (2) $\vec{a} = (3, -2, 1)$, $\vec{b} = (2, 1, 2)$, $\vec{c} = (3, -1, -2)$.

14. Suppose $\vec{a} + \vec{b} + \vec{c} = 0$ and $|\vec{a}| = 3$, $|\vec{b}| = 5$, $|\vec{c}| = 7$. Find the angle between $\vec{a}$ and $\vec{b}$.

15. Suppose $\vec{a} + \vec{b} + \vec{c} = 0$. Show that $\vec{a} \times \vec{b} = \vec{b} \times \vec{c} = \vec{c} \times \vec{a}$.

16. Prove the following equations.

    (1) $(\vec{a} \times \vec{b}) \times \vec{c} = (\vec{a} \cdot \vec{c})\vec{b} - (\vec{b} \cdot \vec{c})\vec{a}$;

    (2) $\vec{a} \times (\vec{b} \times \vec{c}) = (\vec{a} \cdot \vec{c})\vec{b} - (\vec{a} \cdot \vec{b})\vec{c}$;

    (3) $(\vec{a} \times \vec{b}) \times \vec{c} + (\vec{b} \times \vec{c}) \times \vec{a} + (\vec{c} \times \vec{a}) \times \vec{b} = \vec{0}$.

17. Show that $|\vec{a} \times \vec{b}|^2 = |\vec{a}|^2 |\vec{b}|^2 - (\vec{a} \cdot \vec{b})^2$. More generally,

    $(\vec{a} \times \vec{b}) \cdot (\vec{c} \times \vec{d}) = (\vec{a} \cdot \vec{c})(\vec{b} \cdot \vec{d}) - (\vec{a} \cdot \vec{d})(\vec{b} \cdot \vec{c})$.

18. Show that $|(\vec{a} + \vec{b}) \times (\vec{a} - \vec{b})| = 2|\vec{a} \times \vec{b}|$.

19. Show that if $\vec{a} \times \vec{b} + \vec{b} \times \vec{c} + \vec{c} \times \vec{a} = 0$, then $\vec{a}$, $\vec{b}$ and $\vec{c}$ are coplanar.

20. Let a, b, c be the length of the medians of a triangle, and a_1, a_2, a_3 be the length of the three sides. Show that

$$m_1^2 + m_2^2 + m_3^2 = \frac{3}{4}(a_1^2 + a_2^2 + a_3^2).$$

21. Prove by using vectors.

    (1) The law of cosines is true.

    (2) The inscribed angle of a half circle is a right angle.

    (3) If the diagonals of a parallelogram are perpendicular to each other, then the parallelogram is a rhombus.

22. Suppose that vectors $\vec{a}$, $\vec{b}$ and $\vec{c}$ are given, $\vec{a} \cdot \vec{b} \neq 0$ and r is a real number. If a vector $\vec{x}$ satisfies the following equations $\vec{x} \cdot \vec{a} = r$, $\vec{x} \times \vec{b} = \vec{c}$, find $\vec{x}$.

23. If there exists a vector $\vec{x}$ that satisfies the equations $\vec{a}_1 \times \vec{x} = \vec{b}_1$ and $\vec{a}_2 \times \vec{x} = \vec{b}_2$, show that $\vec{a}_1 \cdot \vec{b}_2 + \vec{a}_2 \cdot \vec{b}_1 = 0$.

5.3   Planes and Lines

5.3.1   *Equations of Planes*

Equations of planes and lines can be expressed succinctly using vectors.

(1) Suppose we have a point $M_0(x_0, y_0, z_0)$ and a vector $\vec{n} = (A, B, C)$. Find the equation of the plane that passes M_0 and is perpendicular to $\vec{n}$.

Let $M(x, y, z)$ be an arbitrary point on the plane. Then $\overrightarrow{M_0M}$ is perpendicular to $\vec{n}$ (Figure 5.14). Denote $\overrightarrow{OM} = \vec{r}$, $\overrightarrow{OM_0} = \vec{r}_0$. Then

$$(\vec{r} - \vec{r}_0) \cdot \vec{n} = 0.$$

This is the *equation of the plane*, where $\vec{n}$ is called the *normal vector of the plane*. This equation can also be written as

$$A(x - x_0) + B(y - y_0) + C(z - z_0) = 0,$$

or

$$Ax + By + Cz + D = 0,$$

where $D = -(Ax_0 + By_0 + Cz_0)$.

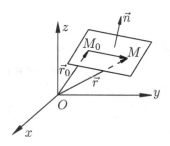

Fig. 5.14

Conversely, all the points that satisfy the equation

$$Ax + By + Cz + D = 0$$

consist a plane if A, B and C are not all zeros.

In fact, since not all A, B and C are zero, we can always find x_0, y_0 and z_0 such that

$$Ax_0 + By_0 + Cz_0 + D = 0.$$

Thus

$$D = -(Ax_0 + By_0 + Cz_0).$$

If we denote (x_0, y_0, z_0) as M_0, $M(x, y, z)$ as any point that satisfies $Ax + By + Cz + D = 0$, $\overrightarrow{OM} = \vec{r}$ and $\overrightarrow{OM_0} = \vec{r}_0$, then the equation

$$Ax + By + Cz + D = 0$$

is the equation

$$A(x - x_0) + B(y - y_0) + C(z - z_0) = 0,$$

which is

$$(\vec{r} - \vec{r}_0) \cdot \vec{n} = 0.$$

This implies that $\vec{r} - \vec{r}_0$ is always perpendicular to $\vec{n}$, so M is on the plane that passes M_0 and is perpendicular to $\vec{n}$. Hence, the equation

$$Ax + By + Cz + D = 0$$

represent a plane.

(2) Given three points $M_1(x_1, y_1, z_1)$, $M_2(x_2, y_2, z_2)$, $M_3(x_3, y_3, z_3)$. Find the equation of the plane that passes the three points.

Let $M(x, y, z)$ be an arbitrary point on the plane. Let $\overrightarrow{OM_i} = \vec{r}_i$ $(i = 1, 2, 3)$, and $\overrightarrow{OM} = r$. Since $\vec{r} - \vec{r}_1$, $\vec{r}_2 - \vec{r}_1$, $\vec{r}_3 - \vec{r}_1$ are coplanar, we have

$$(\vec{r} - \vec{r}_1) \cdot [(\vec{r}_2 - \vec{r}_1) \times (\vec{r}_3 - \vec{r}_1)] = 0.$$

That is

$$\begin{vmatrix} x - x_1 & y - y_1 & z - z_1 \\ x_2 - x_1 & y_2 - y_1 & z_2 - z_1 \\ x_3 - x_1 & y_3 - y_1 & z_3 - z_1 \end{vmatrix} = 0.$$

Especially, when the coordinates of M_1, M_2, M_3 are $(a, 0, 0)$, $(0, b, 0)$, $(0, 0, c)$ respectively, the above equation becomes

$$\begin{vmatrix} x - a & y & z \\ -a & b & 0 \\ -a & 0 & c \end{vmatrix} = 0.$$

This equation can be simplified as

$$\frac{x}{a} + \frac{y}{b} + \frac{z}{c} = 1.$$

Conversely, any equation of the form $\frac{x}{a} + \frac{y}{b} + \frac{z}{c} = 1$ represents a plane which intersects the x-axis, y-axis, and z-axis at $(a, 0, 0)$, $(0, b, 0)$, $(0, 0, c)$ respectively. This is called the intercept form of the equation of a plane.

Example 1. Find the equation of a plane that passes two points $M_1(8, -3, 1)$, $M_2(4, 7, 2)$, and is perpendicular to the plane $3x + 5y - 7z + 21 = 0$.

Solution: Suppose $\vec{n}$ is the normal vector of the plane we want to find, then it is perpendicular to the vector $\overrightarrow{M_1 M_2}$ and to the normal vector of the given plane $\vec{n}_1 = (3, 5, -7)$ as well. Thus, we can let

$$\vec{n} = \vec{n}_1 \times \overrightarrow{M_1 M_2} = \begin{vmatrix} \vec{i} & \vec{i} & \vec{i} \\ 3 & 5 & -7 \\ -4 & 10 & 1 \end{vmatrix} = 25(3\vec{i} + \vec{j} + 2\vec{k}).$$

By (1), the equation of the plane is

$$3(x - 8) + (y + 3) + 2(z - 1) = 0.$$

That is

$$3x + y + 2z - 23 = 0.$$

Example 2. Find the equation of the plane that passes the point $P(4, -3, -1)$ and the x-axis.

Solution: Any two points on the x-axis can be chosen to do our calculation. We can chose $(0, 0, 0)$ and $(1, 0, 0)$. By (2), the equation of the plane is

$$\begin{vmatrix} x & y & z \\ 1 & 0 & 0 \\ 4 & -3 & -1 \end{vmatrix} = 0.$$

That is

$$y - 3z = 0.$$

Example 3. Find the distance from the point $P_1(x_1, y_1, z_1)$ to the plane $Ax + By + Cz + D = 0$.

Solution: Draw a line from P_1 perpendicular to the given plane and denote the intersection point as $P_0(x_0, y_0, z_0)$, then the distance to be found is $|P_0 P_1|$. Since P_0 is on the plane, we have

$$Ax_0 + By_0 + Cz_0 + D = 0.$$

Since the vector $\overrightarrow{P_0 P_1}$ and the normal vector of the given plane $\vec{n} = (A, B, C)$ are collinear, we have

$$\vec{n} \cdot \overrightarrow{P_0 P_1} = \pm |\vec{n}| |\overrightarrow{P_0 P_1}|.$$

On the other hand,

$$\vec{n} \cdot \overrightarrow{P_0 P_1} = A(x - x_0) + B(y - y_0) + C(z - z_0)$$
$$= Ax_1 + By_1 + Cz_1 + D.$$

Hence

$$|P_0 P_1| = \frac{|\vec{n} \cdot \overrightarrow{P_0 P_1}|}{|\vec{n}|} = \frac{|Ax_1 + By_1 + Cz_1 + D|}{\sqrt{A^2 + B^2 + C^2}}.$$

5.3.2 *Equations of Lines*

(1) Find the equation of a line that passes the point $M_0(x_0, y_0, z_0)$ and is parallel to the vector $\vec{s} = (l, m, n)$ (Figure 5.15).

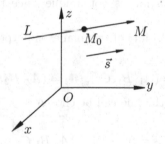

Fig. 5.15

Suppose a point $M(x, y, z)$ is on the line, then the vector $\overrightarrow{M_0 M}$ is parallel to $\vec{s}$. Thus, there is a scalar λ, such that

$$\overrightarrow{M_0 M} = \lambda \vec{s}.$$

That is

$$(x - x_0, y - y_0, z - z_0) = \lambda(l, m, n).$$

Thus

$$\begin{cases} x = x_0 + \lambda l, \\ y = y_0 + \lambda m, \\ z = z_0 + \lambda n. \end{cases}$$

This is the *parametric equation of the line* with parameter λ. It also can be written as

$$\frac{x - x_0}{l} = \frac{y - y_0}{m} = \frac{z - z_0}{n}$$

This is called the *symmetric equations of the line*. The vector $\vec{s} = (l, m, n)$ is the *direction vector* of the line. The numbers l, m, n are called the *direction numbers* of the line.

Notice that if any one of the constants l, m or n is zero in these equations, the corresponding numerator must be zero.

(2) A line in the space can be considered as the intersection of two planes

$$\begin{cases} A_1 x + B_1 y + C_1 z + D_1 = 0, \\ A_1 x + B_1 y + C_1 z + D_2 = 0. \end{cases}$$

This system of equations is the general form of equations of a line.

This expression of a line is not unique since there are infinite many planes passing through one line.

Since the intersection line of two planes is perpendicular to the normal vectors of both planes

$$\vec{n}_1 = (A_1, B_1, C_1), \quad \vec{n}_2 = (A_2, B_2, C_2),$$

the direction vector of the line can be chosen as

$$\vec{s} = \vec{n}_1 \times \vec{n}_2 = \begin{vmatrix} \vec{i} & \vec{j} & \vec{k} \\ A_1 & B_1 & C_1 \\ A_2 & B_2 & C_2 \end{vmatrix}.$$

Example 1. Convert the general form of equations of the line

$$\begin{cases} 2x - 3y + z - 5 = 0, \\ 3x + y - 2z - 2 = 0; \end{cases}$$

to symmetric equations.

Solution: We need to find a point on the line, and we need to find the direction vector of the line.

Let $z = 0$ in the system, then by solving

$$\begin{cases} 2x - 3y = 5, \\ 3x + y = 2; \end{cases}$$

we have $x = 1$, $y = -1$. This implies that the point $(1, -1, 0)$ is on the line. The direction vector of the line is

$$\vec{s} = \begin{vmatrix} \vec{i} & \vec{j} & \vec{k} \\ 2 & -3 & 1 \\ 3 & 1 & -2 \end{vmatrix} = 5\vec{i} + 7\vec{j} + 11\vec{k}.$$

Thus, the symmetric equations of the line are

$$\frac{x - 1}{5} = \frac{y + 1}{7} = \frac{z}{11}.$$

Example 2. Find the equation of a line that passes two points $M_1(x_1, y_1, z_1)$ and $M_2(x_2, y_2, z_2)$.

Solution: The vector $\overrightarrow{M_1 M_2}$ can be considered as the direction vector of the line. So the symmetric equations of the line can be written as

$$\frac{x - x_1}{x_2 - x_1} = \frac{y - y_1}{y_2 - y_1} = \frac{z - z_1}{z_2 - z_1}.$$

Exercises 5.3

1. Graph the following planes.

   (1) $2x + 2y + 2z = -1$; (3) $4y - 7z - 0$;
   (2) $2x - 3y + 20 = 0$; (4) $3x - 2 = 0$.

2. Find the equations of the following planes:

   (1) passes the point $(2, 0, -3)$ and is perpendicular to two planes
   $$x - 2y + 4z - 7 = 0, \quad 3x + 5y - 2z + 1 = 0;$$

   (2) passes two points $M_1(1, 2, 1)$, $M_2(2, -1, 2)$ and is parallel to the vector $\vec{n} = (3, 2, 1)$;

   (3) passes the point $M(1, 3, 2)$ and is parallel to two vectors
   $$\vec{n}_1 = (1, -1, 1), \quad \vec{n}_2 = (3, 1, 2);$$

   (4) contains the x-axis and is perpendicular to the plane
   $$5x - 4y - 2z + 3 = 0;$$

   (5) passes three points $A(1, 1, -1)$, $B(-2, -2, 2)$, $C(1, -1, 2)$;

   (6) passes two points $M_1(2, -1, 1)$, $M_2(3, 1, 2)$ and is parallel to the y-axis.

3. Find the cosine of angles between the plane $2x - 2y + z + 5 = 0$ and each coordinate planes.

4. Find the equation of a plane that passes the points $(0, -1, 0)$ and $(0, 0, 1)$ and makes a $60°$ angle with the xy-plane.

5. Find the distance between the given point and the given plane.
   (1) $M(2, -1, -1)$, $16x - 12y + 15z - 4 = 0$;
   (2) $M(1, 2, -3)$, $5x - 3y + z + 4 = 0$;
   (3) $M(9, 2, -2)$, $12y - 5z + 5 = 0$.

6. Find the distance between the two parallel planes $2x - y + 2z + 9 = 0$ and $4x - 2y + 4z - 21 = 0$.

7. Find a point inside the tetrahedron bounded by the plane

$$x + y + z - 1 = 0$$

and the three coordinate planes, such that the point has the same distance from each plane.

8. Find a point on the y-axis, such that the distance from the point to the plane $x + 2y - 2z - 2 = 0$ is 4.

9. Find the equation of a plane that is parallel to the plane

$$2x + y + 2z + 5 = 0,$$

and bounds an irregular tetrahedron of volume 1 with the three coordinate planes.

10. Find the equations of a line that passes the two given points.
    (1) $(1, -2, 1)$, $(3, 1, -1)$;   (2) $(3, -1, 0)$, $(1, 0, -3)$.

11. Find a set of parametric equations of the line:

$$\begin{cases} 2x + 3y - z - 4 = 0, \\ 3x - 5y + 2z + 1 = 0. \end{cases}$$

12. Find the intersection point of the given line and plane.
    (1) $\dfrac{x-1}{1} = \dfrac{y+1}{-2} = \dfrac{z}{6}$, $2x + 3y + z - 1 = 0$;
    (2) $\dfrac{x+2}{-2} = \dfrac{y-1}{3} = \dfrac{z-3}{2}$, $x + 2y - 2z + 6 = 0$.

13. Find the angle between the two given lines:
    (1)

$$\begin{cases} 2x - 2y - z + 3 = 0, \\ x + 2y - 2z + 1 = 0. \end{cases}$$

and

$$\begin{cases} 4x + y + 3z - 21 = 0, \\ 2x + 2y - 3z + 15 = 0. \end{cases}$$

(2) $\dfrac{x-2}{4} = \dfrac{y-3}{-12} = \dfrac{z-1}{3}$ and $\dfrac{x}{2} = \dfrac{y-3}{-1} = \dfrac{z-8}{-2}$.

14. Find the angle between the line

$$\begin{cases} 3x - 2y = 24, \\ 3x - z = -4. \end{cases}$$

and the plane $6x + 15y - 10z + 31 = 0$.

15. Find the equation of a line that passes the point $(0, 2, 4)$, and is parallel to the two planes $x + 2z - 1 = 0$, $y - 3z - 2 = 0$.

16. Find the equation of the plane that passes the point $(1, -2, 1)$ and is perpendicular to the line

$$\begin{cases} x - 2y + z - 3 = 0, \\ x + y - z + 2 = 0. \end{cases}$$

17. Determine the equation of the plane that passes the line of intersection of the planes $x + 5y + z = 0$ and $x - z + 4 = 0$, and makes a $45°$ angle with the plane $x - 4y - 8z + 12 = 0$.

18. Find the equation of the plane that passes the point $M(1, -2, 3)$ and the intersection line of two planes

$$2x - 3y + z - 3 = 0, \qquad x + 3y + 2z + 1 = 0.$$

19. Find the equations of the line that passes the point $M(-1, 2, -3)$, is perpendicular to the vector $\vec{s} = (6, -2, -3)$, and intersects with the line

$$\dfrac{x-1}{3} = \dfrac{y+1}{2} = \dfrac{z-3}{-5}.$$

20. Find the equation of the plane that contains two parallel lines

$$\dfrac{x+3}{3} = \dfrac{y+2}{-2} = z, \qquad \dfrac{x+3}{3} = \dfrac{y+4}{-2} = z+1.$$

21. Find the equation of the plane that passes the point $(1, 2, -3)$ and is parallel to two lines

$$\dfrac{x-7}{2} = \dfrac{y+7}{-3} = \dfrac{z-7}{3}, \qquad \dfrac{x+5}{3} = \dfrac{y-2}{-2} = \dfrac{z+3}{-1}.$$

22. Find the equation of the plane that passes the line

$$\begin{cases} x = 1 + 3t, \\ y = 3 + 2t, \\ z = -2 - t; \end{cases}$$

and is parallel to the line

$$\begin{cases} 2x - y + z - 3 = 0, \\ x + 2y - z - 5 = 0. \end{cases}$$

23. Find the distance between the two parallel lines

$$\begin{cases} 2x + 2y - z - 10 = 0, \\ x - y - z - 22 = 0; \end{cases}$$

and $\dfrac{x + 7}{3} = \dfrac{y - 5}{-1} = \dfrac{z - 9}{4}$.

24. Find the distance between the two lines $\dfrac{x - 1}{4} = \dfrac{y + 2}{-3} = z - 3$ and

$$\begin{cases} x + 1 = \dfrac{z - 1}{-3}, \\ y - 2 = 0. \end{cases}$$

25. Find the equation of a plane such that the distance from the origin to the plane is 6, and the ratio of its x-, y-, and z-intercepts is $1 : 3 : 2$.

5.4 Quadric Surfaces

The general form of an equation of a surface in the three dimensional space is $F(x, y, z) = 0$

We know that a second-degree equation in two variables represents a conic section. In three-dimensional space, a second-degree equation represents a *quadric surface*. We now study some of the simplest types.

5.4.1 Cylindrical Surfaces

A *cylindrical surface* is generated by a line moving parallel to a fixed line L as to pass through the points on a fixed plane curve C.

The moving line is called a *generating line* or *generatrix*; the fixed plane curve is called the *directrix* (Figure 5.16).

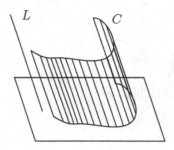

Fig. 5.16

Suppose the generating line of a cylindrical surfacer is parallel to the z-axis, and the directrix C is a curve on the xy-plane with the equation $f(x, y) = 0$ (Figure 5.17). For any point $M(x, y, z)$ on the cylindrical surface, the point $M_1(x, y, 0)$ is on the curve C. That is, M satisfies the equation $f(x, y) = 0$. On the other hand, any point $M(x, y, z)$ which satisfies the equation $f(x, y) = 0$ for the first two variables, is on the generating line that passes through the point $M_1(x, y, 0)$, that is, such $M(x, y, z)$ is on the cylindrical surface. Therefore, $f(x, y) = 0$ is the equation of a cylindrical surface with directrix C, and generating lines parallel to the z-axis.

Similarly, equations of the form $g(y, z) = 0$ represent cylindrical surfaces with generating lines parallel to the x-axis; and equations of the form $h(z, x) = 0$ represent cylindrical surfaces with generating lines parallel to the y-axis.

Remark: To distinguish from the cylindrical surface equation $f(x, y) = 0$, we should use

$$\begin{cases} f(x, y) = 0, \\ z = 0; \end{cases}$$

to express the equations of its directrix C.

We give some equations of cylindrical surfaces with generating lines parallel to the z-axis:

Elliptic cylindrical surfaces: $\dfrac{x^2}{a^2} + \dfrac{y^2}{b^2} = 1$ (Figure 5.18);

Hyperbolic cylindrical surfaces: $\dfrac{x^2}{a^2} - \dfrac{y^2}{b^2} = 1$ (Figure 5.19);

Parabolic cylindrical surfaces: $y^2 = 2px$ (Figure 5.20).

The equations of above types of cylindrical surfaces with generating

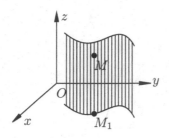

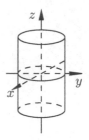

Fig. 5.17   Fig. 5.18

lines parallel to the x-axis or y-axis can be written in similar ways.

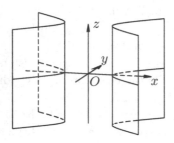

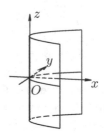

Fig. 5.19   Fig. 5.20

5.4.2   *Surfaces of Revolution*

A surface generated by revolving a plane curve about a fixed line in its plane is called a *surface of revolution*. The plane curve is called a *generating curve* or *generatrix* and the fixed line is called the *axis of revolution*.

Suppose a surface of revolution is generated by revolving a curve C: $f(x, y) = 0$ about the x-axis. Let $P_0(x_0, y_0, z_0)$ be an arbitrary point on the surface. Then the intersection of the surface and the plane $x = x_0$ (which passes P_0) is a circle with center on the x-axis and radius

$$r = \sqrt{y_0^2 + z_0^2}.$$

This circle is generated by revolving the point (x_0, r) on the plane curve C about x-axis, so we have $f(x_0, r) = 0$. Thus, P_0 satisfies the equation

$$f(x_0, \sqrt{y_0^2 + z_0^2}) = 0.$$

Since P_0 is an arbitrary point on the surface, the equation of the surface of revolution is

$$f(x, \sqrt{y^2 + z^2}) = 0.$$

Similarly, the equation of the surface of revolution about the y-axis is

$$f(\sqrt{x^2 + z^2}, y) = 0.$$

Some quadric surfaces of revolution are:

(1) *Paraboloid of revolution* (Figure 5.21)

$$x^2 + y^2 = 2pz$$

was generated by revolving the parabola

$$\begin{cases} y^2 = 2pz, \\ x = 0; \end{cases}$$

about z-axis.

(2) *Ellipsoid of revolution* (Figure 5.22)

$$\frac{x^2 + y^2}{a^2} + \frac{z^2}{b^2} = 1$$

was generated by revolving the ellipse $\dfrac{y^2}{a^2} + \dfrac{z^2}{b^2} = 1$ about z-axis.

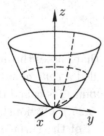

Fig. 5.21

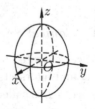

Fig. 5.22

(3) *Hyperboloid of revolution of one sheet* (Figure 5.23)

$$\frac{x^2 + y^2}{a^2} - \frac{z^2}{b^2} = 1$$

was generated by revolving the hyperbola

$$\begin{cases} \dfrac{x^2}{a^2} - \dfrac{z^2}{b^2} = 1, \\ y = 0; \end{cases}$$

about the z-axis.

The same hyperbola revolving about the x-axis generates the surface of revolution

$$\frac{x^2}{a^2} - \frac{y^2 + z^2}{b^2} = 1.$$

This is a *hyperboloid of revolution of two sheets* (Figure 5.24).

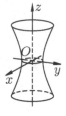

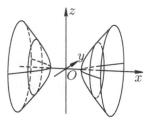

Fig. 5.23   Fig. 5.24

5.4.3 *Conical Surfaces*

A surface generated by a moving straight line L that passes through a fixed point Q and intersects with a fixed curve C is called a *conical surface*. The straight line L is called the *generating line*, the point Q is called the *vertex*, and the curve C is called the *directrix* (Figure 5.25.) Obviously, the line which connects the vertex and any point (x, y, z) of the surface is on the surface. If the vertex is at the origin, then the coordinates of points on the line is (tx, ty, tz), where t is any real number. Thus, for a conical surface with vertex at the origin, if a point (x, y, z) is on the surface, then the point

(tx, ty, tz) also satisfies the equation of the surface for any real number t. If $F(tx, ty, tz) = t^l F(x, y, z)$, where l is an integer, then $F(x, y, z)$ is called a *homogeneous function*.

Suppose $F(x, y, z)$ is a homogeneous polynomial. Then $F(x, y, z) = 0$ is a conical surface with vertex at the origin. For instance, the equation

$$\frac{x^2}{a^2} + \frac{y^2}{b^2} - \frac{z^2}{c^2} = 0, \quad (a \neq 0, b \neq 0, c \neq 0)$$

represents a conical surface with vertex at the origin. The intersection of the conical surface with the plane $z = c$ is a directrix.

$$\begin{cases} \dfrac{x^2}{a^2} + \dfrac{y^2}{b^2} - 1 = 0 \\ z = c. \end{cases}$$

Obviously, it is an ellipse, but the cross sections of the conical surface on the planes $x = a$ or $y = b$ are hyperbolas (Figure 5.26).

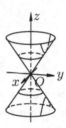

Fig. 5.25   Fig. 5.26

5.4.4   *Ellipsoid*

The standard form of the equation of *ellipsoid* is

$$\frac{x^2}{a^2} + \frac{y^2}{b^2} + \frac{z^2}{c^2} = 1 \quad (a > 0, b > 0, c > 0).$$

It is easy to see from the equation that

$$\frac{x^2}{a^2} \leqslant 1, \quad \frac{y^2}{b^2} \leqslant 1, \quad \frac{z^2}{c^2} \leqslant 1;$$

or, equivalently,

$$|x| \leqslant a, \quad |y| \leqslant b, \quad |z| \leqslant c.$$

This implies that all points on this surface are in a rectangular parallelepiped bounded by

$$x = \pm a, \quad y = \pm b, \quad z = \pm c.$$

The equation of the section of the surface by the plane $z = h$ is

$$\begin{cases} \dfrac{x^2}{a^2} + \dfrac{y^2}{b^2} = 1 - \dfrac{h^2}{c^2}, \\ z = h. \end{cases}$$

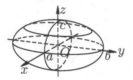

Fig. 5.27

The section is an ellipse for $0 \leqslant |h| < c$ (Figure 5.27), and the largest one is on the xy-plane

$$\begin{cases} \dfrac{x^2}{a^2} + \dfrac{y^2}{b^2} = 1, \\ z = 0. \end{cases}$$

The section gets smaller along the two sides of the z-axis as the value of $|h|$ gets greater. The section becomes a point $(0, 0, c)$ or $(0, 0, -c)$ when $|h| = c$.

Similarly, the sections of the surface by a plane parallel to the yz-plane or the zx-plane can be described in a similar way.

In an equation of an ellipsoid, if any two of the three numbers a, b, and c are the same, for instance, if $a = b$, then it represents an ellipsoid of revolution generated by revolving the ellipse

$$\frac{y^2}{a^2} + \frac{z^2}{c^2} = 1$$

on yz-plane about z-axis. If $a = b = c$, then we have the equation of a sphere

$$x^2 + y^2 + z^2 = a^2.$$

5.4.5 *Hyperbolic Paraboloid*

The standard form of equation of a *hyperbolic paraboloid* is

$$\frac{x^2}{a^2} - \frac{y^2}{b^2} = 2z.$$

The section of the surface by the plane $z = h$ is

$$\begin{cases} \dfrac{x^2}{a^2} - \dfrac{y^2}{b^2} = 2h, \\ z = h. \end{cases}$$

This is a hyperbola for $h > 0$, with its *transverse axis* parallel to the x-axis, and its *conjugate axis* parallel to the y-axis.

The section becomes two straight lines on the xy-plane that intersect at the origin for $h = 0$.

$$\begin{cases} \dfrac{x}{a} + \dfrac{y}{b} = 0, \\ z = 0; \end{cases} \qquad\qquad \begin{cases} \dfrac{x}{a} - \dfrac{y}{b} = 0, \\ z = 0. \end{cases}$$

The section is again a hyperbola for $h < 0$, but with its transverse axis parallel to the y-axis, and its conjugate axis parallel to the x-axis.

The section of the surface by the plane $x = h$ is

$$\begin{cases} \dfrac{y^2}{b^2} = \dfrac{h^2}{a^2} - 2z, \\ x = h. \end{cases}$$

It is a parabola on the yz-plane with vertex at the origin and opens downward for $h = 0$. It is a parabola opening downward for $|h| > 0$. The position of the vertex gets higher as $|h|$ gets greater.

The section of the surface by the plan $y = h$ is a parabola opening upward

$$\begin{cases} \dfrac{x^2}{a^2} = 2z + \dfrac{h^2}{b^2}, \\ y = h. \end{cases}$$

With all the descriptions above, we can see that the graph of the surface is shaped like a saddle (Figure 5.28). It is also called a *saddle surface*.

By the same analyzing method, we can find the shape of the following surfaces.

5.4.6 *Hyperboloid of One Sheet*

The standard form of the equation of the *hyperboloid of one sheet* is

$$\frac{x^2}{a^2} + \frac{y^2}{b^2} - \frac{z^2}{c^2} = 1.$$

The graph is Figure 5.29.

5.4.7 *Hyperboloid of Two Sheets*

The standard form of the equation of the *hyperboloid of two sheets* is

$$\frac{x^2}{a^2} + \frac{y^2}{b^2} - \frac{z^2}{c^2} = -1.$$

The graph is Figure 5.30.

5.4.8 *Elliptic Paraboloid*

The standard form of the equation of the *elliptic paraboloid* is

$$\frac{x^2}{a^2} + \frac{y^2}{b^2} = 2z.$$

The graph is Figure 5.31.

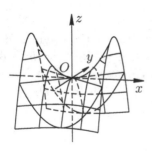

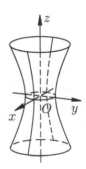

Fig. 5.28   Fig. 5.29

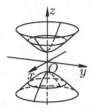

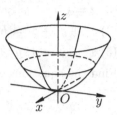

Fig. 5.30   Fig. 5.31

Exercises 5.4

1. Which of the following represents a surface of revolution? If so, how are these surfaces of revolution generated?

   (1) $\dfrac{x^2}{4} + \dfrac{y^2}{9} + \dfrac{z^2}{9} = 1$;

   (2) $x^2 + y^2 + z^2 = 1$;   (3) $x^2 + 2y^2 + 3z^2 = 1$;

   (4) $x^2 - \dfrac{y^2}{4} - z^2 = 1$;   (5) $\dfrac{x^2}{9} + \dfrac{y^2}{16} - \dfrac{z^2}{25} = 1$;

   (6) $x^2 - y^2 - z^2 = 1$.

2. Find the equation of the following surfaces of revolution.

   (1) Curve $4x^2 - 9y^2 = 36$, $z = 0$ revolving about the x-axis and the y-axis.

   (2) Curve $y = kx$, $z = 0$ revolving about the x-axis and the y-axis.

   (3) Curve $y = \sin x$, $(0 \leqslant x \leqslant \pi)$, $z = 0$ revolving about the x-axis.

3. Find the graph of the following equations in two-dimensional plane and in three-dimensional space.

   (1) $x = 2$;   (3) $x^2 + y^2 = 4$;

   (2) $y = x + 1$;   (4) $x^2 - y^2 = 1$;

   (5) $y - 5x - 1 = 0$, $y - 2x + 3 = 0$;

   (6) $\dfrac{x^2}{4} + \dfrac{y^2}{9} = 1$, $y = 2$.

4. Graph the following solids bounded by the given surfaces.

   (1) $\dfrac{x}{3} + \dfrac{y}{2} + z = 1$ and the coordinate planes;

(2) $z = x^2 + y^2$, $x = 0$, $y = 0$, $z = 0$, $x + y - 1 = 0$;

(3) $z = \sqrt{x^2 + y^2}$, $x^2 + y^2 + z^2 = R^2$;

(4) $x^2 + y^2 + z^2 = R^2$, $x^2 + y^2 + (z - R)^2 = R^2$;

(5) $x^2 + y^2 = 1$, $y^2 + z^2 = 1$.

5. Find the equations of the sections of the surface

$$x^2 + y^2 - \frac{z^2}{9} = 0$$

by the planes

$$z = 0, \ z = 3, \ x = 0, \ x = \frac{1}{3}, \ y = 0, \ y = \frac{1}{3},$$

and graph the surface.

6. Find the equations of the sections of the surface

$$\frac{x^2}{9} - \frac{y^2}{25} + \frac{z^2}{4} = 1$$

by the planes $x = 2$, $y = 0$, $y = 5$, $z = 1$, $z = 2$, and graph the surface.

7. The distance from a moving point to the origin is the same as the distance from the point to the plane $z = 4$. Find the equation for the locus of this moving point. Determine what kind of quadric surface this equation represents.

5.5 Transformations of Coordinates

In the last section, we talked about some special second degree equations in three variables. In this section, we convert the general form of the equations of quadric surfaces to the forms of equations in the last section by translation of coordinates and rotation of coordinates (except for the degenerate equations).

5.5.1 *Translation of Axes*

Suppose there are two rectangular coordinate systems in a three-dimensional space $Oxyz$, $O_1x_1y_1z_1$ (Figure 5.32).

Suppose also the corresponding axes of two systems are parallel and have the same directions. Let the coordinates of the origin of one system O_1 in the other system $Oxyz$ be (a, b, c). Then there are two sets of coordinates for a point M in the space: (x, y, z) in $Oxyz$ and (x_1, y_1, z_1) in $O_1x_1y_1z_1$.

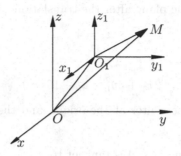

Fig. 5.32

Since

$$\overrightarrow{OM} = \overrightarrow{OO_1} + \overrightarrow{O_1M},$$

we have

$$x\vec{i} + y\vec{j} + z\vec{k} = (a\vec{i} + b\vec{j} + c\vec{k}) + (x_1\vec{i} + y_1\vec{j} + z_1\vec{k}).$$

Thus, the formula of the *translation of axes* follows

$$\begin{cases} x = x_1 + a, \\ y = y_1 + b, \\ z = z_1 + c. \end{cases}$$

Example: Show that the surface

$$2x^2 + 3y^2 + z^2 + 4x - 12y + 4z + 2 = 0$$

is tangent to the plane $z = 2$.

Solution: The left side of the equation can be written as

$$2x^2 + 3y^2 + z^2 + 4x - 12y + 4z + 2$$
$$= 2(x + 1)^2 + 3(y - 2)^2 + (z + 2)^2 - 16.$$

Let $a = -1$, $b = 2$, $c = -2$ in the translation of axes formula, we have

$$x = x_1 - 1, \quad y = y_1 + 2,$$
$$z = z_1 - 2.$$

Substituting into the equation, we get the equation of the surface after the translation:

$$2x_1^2 + 3y_1^2 + z_1^2 - 16 = 0$$

and the equation of the plane after the translation

$$z_1 = 4.$$

The equation

$$2x_1^2 + 3y_1^2 + z_1^2 - 16 = 0$$

is an ellipsoid with the center at the origin, and the lengths of the three half axes are $2\sqrt{2}$, $\dfrac{4}{\sqrt{3}}$, 4.

Obviously, the plane $z_1 = 4$ is tangent to

$$2x_1^2 + 3y_1^2 + z_1^2 - 16 = 0.$$

The plane remains tangent to the surface regardless the translation of coordinates. Therefore, the conclusion follows.

5.5.2   *Rotation of Axes*

Suppose there are two coordinate systems in the three-dimensional space $Oxyz$, $Ox_1y_1z_1$ with a common origin (Figure 5.33). Let M be a point with coordinates (x, y, z) in $Oxyz$ and coordinates (x_1, y_1, z_1) in $Ox_1y_1z_1$. Assume that the unit vectors of the system $Oxyz$ are $\vec{i}$, $\vec{j}$, $\vec{k}$, and the unit vectors of the system $Ox_1y_1z_1$ are $\vec{i}_1$, $\vec{j}_1$, $\vec{k}_1$. If the nine angles between each axis of the two systems are as shown in the following table

	$\vec{i}$	$\vec{j}$	$\vec{k}$
$\vec{i}_1$	α_1	β_1	γ_1
$\vec{j}_1$	α_2	β_2	γ_2
$\vec{k}_1$	α_3	β_3	γ_3

then we have

$$\vec{i}_1 = \cos\alpha_1\vec{i} + \cos\beta_1\vec{j} + \cos\gamma_1\vec{k},$$

$$\vec{j}_1 = \cos\alpha_2\vec{i} + \cos\beta_2\vec{j} + \cos\gamma_2\vec{k},$$

$$\vec{k}_1 = \cos\alpha_3\vec{i} + \cos\beta_3\vec{j} + \cos\gamma_3\vec{k}.$$

In the form of matrix, this is

$$(\vec{i}_1, \vec{j}_1, \vec{k}_1) = (\vec{i}, \vec{j}, \vec{k}) \begin{pmatrix} \cos\alpha_1 & \cos\alpha_2 & \cos\alpha_3 \\ \cos\beta_1 & \cos\beta_2 & \cos\beta_3 \\ \cos\gamma_1 & \cos\gamma_2 & \cos\gamma_3 \end{pmatrix}.$$

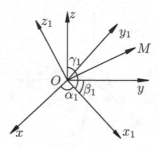

Fig. 5.33

It is obvious that

$$\cos^2 \alpha_1 + \cos^2 \beta_1 + \cos^2 \gamma_1 = 1,$$
$$\cos^2 \alpha_2 + \cos^2 \beta_2 + \cos^2 \gamma_2 = 1,$$
$$\cos^2 \alpha_3 + \cos^2 \beta_3 + \cos^2 \gamma_3 = 1,$$
$$\cos^2 \alpha_1 + \cos^2 \alpha_2 + \cos^2 \alpha_3 = 1,$$
$$\cos^2 \beta_1 + \cos^2 \beta_2 + \cos^2 \beta_3 = 1,$$
$$\cos^2 \gamma_1 + \cos^2 \gamma_2 + \cos^2 \gamma_3 = 1.$$

The coordinates of the vector $\overrightarrow{OM}$ in the two coordinate systems are

$$x\vec{i} + y\vec{j} + z\vec{k}$$

and

$$x_1\vec{i}_1 + y_1\vec{j}_1 + z_1\vec{k}_1.$$

Thus,

$$x\vec{i} + y\vec{j} + z\vec{k} = x_1\vec{i}_1 + y_1\vec{j}_1 + z_1\vec{k}_1.$$

It follows that

$$\begin{cases} x = x_1 \cos \alpha_1 + y_1 \cos \alpha_2 + z_1 \cos \alpha_3, \\ y = x_1 \cos \beta_1 + y_1 \cos \beta_2 + z_1 \cos \beta_3, \\ z = x_1 \cos \gamma_1 + y_1 \cos \gamma_2 + z_1 \cos \gamma_3, \end{cases}$$

or in the form of matrix

$$(x, y, z) = (x_1, y_1, z_1) \begin{pmatrix} \cos \alpha_1 & \cos \beta_1 & \cos \gamma_1 \\ \cos \alpha_2 & \cos \beta_2 & \cos \gamma_2 \\ \cos \alpha_3 & \cos \beta_3 & \cos \gamma_3 \end{pmatrix}.$$

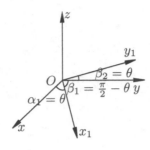

Fig. 5.34

This is the formula of *rotation of axes* in the three-dimensional rectangular coordinate system.

If the z-axis in system $Oxyz$ coincide with z_1-axis in system $Ox_1y_1z_1$, and the angle between the x-axis and the x_1-axis is θ, then

$$\alpha_1 = \theta, \quad \alpha_2 = \frac{\pi}{2} + \theta, \quad \beta_1 = \frac{\pi}{2} - \theta, \quad \beta_2 = \theta,$$

$$\gamma_1 = \gamma_2 = \alpha_3 = \beta_3 = \frac{\pi}{2}, \quad \gamma_3 = 0.$$

Thus, the formula of rotation of axes becomes

$$\begin{cases} x = x_1 \cos \theta - y_1 \sin \theta, \\ y = x_1 \sin \theta - y_1 \cos \theta, \\ z = z_1. \end{cases}$$

Example: Suppose the coordinate system $Ox_1y_1z_1$ is obtained by rotating the coordinate system $Oxyz$ by $\dfrac{\pi}{4}$ clockwise about the z-axis. Find the formula of rotation of axes between these two systems, and the new equation of $z = xy$ with respect to the coordinate system $Ox_1y_1z_1$.

Solution: The angles between each corresponding axis in these two systems are

	$\vec{i}$	$\vec{j}$	$\vec{k}$
$\vec{i_1}$	$\dfrac{\pi}{4}$	$\dfrac{3}{4}\pi$	$\dfrac{\pi}{2}$
$\vec{j_1}$	$\dfrac{\pi}{4}$	$\dfrac{\pi}{4}$	$\dfrac{\pi}{2}$
$\vec{k_1}$	$\dfrac{\pi}{2}$	$\dfrac{\pi}{2}$	0

Therefore the formula of rotation of axes is

$$\begin{cases} x = x_1 \cos \dfrac{\pi}{4} + y_1 \cos \dfrac{\pi}{4} + z_1 \cos \dfrac{\pi}{2} = \dfrac{1}{\sqrt{2}}(x_1 + y_1), \\[2mm] y = x_1 \cos \dfrac{3\pi}{4} + y_1 \cos \dfrac{\pi}{4} + z_1 \cos \dfrac{\pi}{2} = \dfrac{1}{\sqrt{2}}(-x_1 + y_1), \\[2mm] z = x_1 \cos \dfrac{\pi}{2} + y_1 \cos \dfrac{\pi}{2} + z_1 \cos 0 = z_1. \end{cases}$$

The equation $z = xy$ becomes

$$z_1 = \frac{1}{\sqrt{2}}(x_1 + y_1)\frac{1}{\sqrt{2}}(-x_1 + y_1) = -\frac{x_1^2}{2} + \frac{y_1^2}{2}.$$

This is a hyperbolic paraboloid. (Figure 5.35)

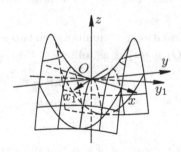

Fig. 5.35

Exercises 5.5

1. Find the coordinates of the following points with respect to the new coordinate system obtained by moving the origin to $(2, 4, -1)$.

   (1) $A(-3, 4, 1)$;   (3) $C(6, 7, -3)$.
   (2) $B(2, -3, -5)$;

2. Find the equations of the following surfaces with respect to the new coordinate system obtained by moving the origin to $(1, 2, 3)$.

   (1) $3x - 5y + 6z = 1$;   (3) $x^2 + y^2 + z^2 - 2x - 4y - 6z - 11 = 0$.
   (2) $\dfrac{x-1}{3} = \dfrac{y-2}{4} = \dfrac{z-3}{2}$;

3. The rotation of axes formula between the rectangular systems $Oxyz$ and $Ox_1y_1z_1$ is

$$\begin{cases} x = x_1 \cos \dfrac{\pi}{3} + y_1 \sin \dfrac{\pi}{3}, \\ y = -x_1 \sin \dfrac{\pi}{3} + y_1 \cos \dfrac{\pi}{3}, \\ z = z_1. \end{cases}$$

   Find the coordinates of the point M in $Oxyz$ if the coordinates of M is $(6, 1, -2)$ in $Ox_1y_1z_1$.

4. Suppose the coordinates of the origin O_1 of the rectangular coordinate system $O_1x_1y_1z_1$ is (x_0, y_0, z_0) in the coordinate system $Oxyz$. Suppose also the x-axis and the x_1-axis are parallel with the same positive direction and the y-axis after rotating $45°$ counterclockwise about the x-axis is parallel to the y_1-axis. Find the transformation formula between the coordinate systems.

5. Suppose the transformation relation between two rectangular coordinate systems $Oxyz$ and $O_1x_1y_1z_1$ is as follows

$$\begin{cases} x = x_1 \cos \theta + y_1 \sin \theta + x_0, \\ y = -x_1 \sin \theta + y_1 \cos \theta + y_0, \\ z = z_1 + z_0. \end{cases}$$

   Explain the geometric meaning of this transformation. Find the coordinates of the point M in the system $Oxyz$ if the coordinates of this point is $(0, 0, 1)$ or $(\cos \theta, \sin \theta, 0)$ in the system $O_1x_1y_1z_1$.

Chapter 6

Multiple Integrals and Partial Derivatives

6.1 Multiple Integrals

6.1.1 *Limits and Continuity of Functions of Several Variables*

With the preparation of Chapter 5, we now study the theory of Calculus for functions of several variables. We pay more attention to the differences between the results about single variable functions and the results about multi-variable functions rather than their similarities.

Let S be a set of points $(x_1, \cdots, x_n)$ in the n dimensional space. A function defined on S is a relation that to each point of S associates a number. We write

$$z = f(x_1, \cdots, x_n).$$

In this chapter, we focus on functions with two or three variables. The domain of a two variable function is usually the interior points of one or more closed curves, called open regions. If the points on the bounding curves are added, then we get a closed region. The domains of the functions we study in this chapter are usually connected, which means that for any two different points in the region, there exists a curve completely lying in the region that connects them.

For instance, the domain of the function

$$z = \frac{1}{(9 - x^2 - y^2)^{\frac{1}{2}}}$$

is the disc without edge

$$x^2 + y^2 < 9$$

which is open and connected.

For a function $z = f(x,y)$, if as a point $P(x,y)$ in the domain approaches a fixed point $A(a,b)$ regardless paths, the function $f(x,y)$ always tends to the same value l, then we say that the number l is a limit of $f(x,y)$ as $P(x,y)$ tends to $A(a,b)$, and denote it as

$$\lim_{\substack{x \to a \\ y \to b}} f(x,y) = l,$$

or

$$\lim_{P \to A} f(x,y) = l.$$

For instance, the function $xy + x^2 + y^2$ tends to 0 as $(x,y) \to (0,0)$ regardless of the paths.

Notice that the difference between the limit of functions of more than one variable and the limit of functions of a single variable is: there are infinity many paths between P and A, and the limit of $f(x,y)$ exists and remain the same regardless the paths that P takes as it approaches A. There exist functions where a limit of the function exists for every different path from P to A, but the limits are different for each path. For example, the function

$$z = f(x,y) = \frac{xy}{x^2 + y^2}$$

becomes

$$f(x,y) = \frac{\lambda x^2}{x^2 + \lambda^2 x^2} = \frac{\lambda}{1 + \lambda^2}$$

when $y = \lambda x$. Thus, the limit of the function exists when $P = (x,y)$ tends to $A = (0,0)$ along the straight line $y = \lambda x$, but they are different for different straight lines. Therefore, the limit of the function does not exists (refer to Section 9.2.1).

The properties of the limit of multi-variable functions are similar to the corresponding properties of single variable functions. We omit this part of the discussion.

When $P(x,y)$ tends to $A(a,b)$, if the limit l of the function $f(x,y)$ exits and is equal to $f(a,b)$, that is

$$\lim_{P \to A} f(x,y) = f(a,b),$$

then we say that $f(x,y)$ is continuous at $A(a,b)$.

If $f(x,y)$ is continuous at every point of its domain, then we say that $f(x,y)$ is a continuous function.

For example, $z = xy + x^2 + y^2$ is continuous at $(0,0)$. It is continuous at any point of the plane. For an example of a discontinuous function, consider the function

$$z = \begin{cases} \dfrac{xy}{x^2 + y^2}, & x^2 + y^2 \neq 0; \\ 0, & x = y = 0. \end{cases}$$

The function is continuous when $x^2 + y^2 \neq 0$. The function is not continuous at the point $(0,0)$ because the limit of the function as $P(x, y)$ tends to $(0,0)$ does not exist. For any fixed x, $z = \dfrac{xy}{x^2 + y^2}$ is a continuous function of y, and for any fixed y, $z = \dfrac{xy}{x^2 + y^2}$ is a continuous function of x. Although this function is continuous for each independent variable separately, it is not continuous as a multi-variable function.

Many of the properties of continuous functions of a single variable remain true for continuous functions of several variables. There exist the maximum and minimum values for a continuous function defined on a closed region. The sum, difference, product and quotient (with non-zero denominator) of two or more continuous functions are still continuous functions.

6.1.2 *Multiple Integration*

The idea of the integration of functions in one variable can be extended to functions in several variables in several ways. In this chapter, we study multiple integrals. We will study line integrals and surface integrals in the next chapter.

Consider the problem of finding the volume of a column with a bounded region D on the xy-plane as the base and a surface $z = f(x, y)$ at the top (Figure 6.1).

Suppose $D_1, \cdots, D_n$ is a partition T of the domain D, and $\Delta A_1, \cdots, \Delta A_n$ are the areas of $D_1, \cdots, D_n$ respectively.

The diameter of D_i is the maximum distance between any two points in D_i. The diameter of the partition T is the maximum of all diameters of D_i. We denote the diameter of T as $\lambda(T)$.

Let (x_i, y_i) be a point in D_i. Then $f(x_i, y_i)\Delta A_i$ is the volume of the small column with base D_i, and $\displaystyle\sum_{i=1}^{n} f(x_i, y_i)\Delta A_i$ is the approximate volume of the column. The actual volume of the column is the limit of the sum

$$\sum_{i=1}^{n} f(x_i, y_i)\Delta A_i$$

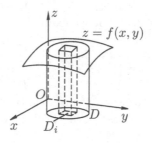

Fig. 6.1

when $\lambda(T)$ tends to zero. That is

$$\lim_{\lambda(T) \to 0} \sum_{i=1}^{n} f(x_i, y_i) \Delta A_i.$$

If for all the partitions T of D where $\lambda(T) \to 0$, and for all the points (x_i, y_i) in D_i, the above limit exists and is equal to the same value, then we say that $f(x, y)$ is integrable on D. The limit is called the *double integral* of $f(x, y)$ on D and is denoted as

$$I = \iint\limits_{D} f(x, y) \, dA = \lim_{\lambda(T) \to 0} \sum_{i=1}^{n} f(x_i, y_i) \Delta A_i.$$

The function $f(x, y)$ is called an *integrand*, D is called the *region of integration* and dA is called the *area element*.

If a function is continuous on a bounded closed region D, then the function is integrable.

Moreover, if a function is bounded on a bounded closed region D, and is continuous on D except for finite number of points and finite number of simple curves, then this function is integrable on D (refer to Section 9.3). (A simple curve is defined parametrically by $x = \varphi(t)$, $y = \psi(t)$, where $\alpha \leqslant t \leqslant \beta$, and $\varphi(t)$, $\psi(t)$ are continuously differentiable functions with finitely many maximum and minimum values.)

In contrast to double integrals, the integration of functions with one variable are called single integrals.

Now we define the definite integral in a higher dimensional space.

Suppose we need to find the total mass of an object V. Let $\rho = \rho(M)$ be the density function of V. Let T be a partition of V that divides V into $V_1, \cdots, V_n$, and denote the volume of V_i as ΔV_i (Figure 6.2). Suppose M_i

is a point in V_i. Then the mass of V_i is approximately equal to $\rho(M_i)\Delta V_i$. The total mass of the object is approximately equal to

$$\sum_{i=1}^{n} \rho(M_i)\Delta V_i.$$

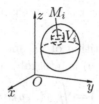

Fig. 6.2

Let $\lambda(T)$ be the diameter of T. Then

$$\lim_{\lambda(T)\to 0} \sum_{i=1}^{n} \rho(M_i)\Delta V_i$$

is the total mass of the object V.

Suppose V is a closed and bounded region in the three dimensional space. For a function $f(x, y, z)$ defined on V, if for any partition T of V, $\lambda(T) \to 0$, and for any point (x_i, y_i, z_i) in V_i, the limit

$$\lim_{\lambda(T)\to 0} \sum_{i=1}^{n} f(x_i, y_i, z_i)\Delta V_i$$

exists and has the same value, then we say that $f(x, y, z)$ is integrable on V, and the limit is called the *triple integral* of $f(x, y, z)$ on V, and is denoted as

$$I = \iiint_V f(x, y, z)\, dV$$

where $f(x, y, z)$ is called the integrand, V the region of integration and dV the *volume element*.

The following properties of double integrals can be derived from the definition.

(1) If $f(x, y)$ and $g(x, y)$ are integrable functions on a region D, then the function

$$\alpha f(x, y) + \beta g(x, y)$$

is also integrable on D for arbitrary constants α and β, and

$$\iint_D (\alpha f(x, y) + \beta g(x, y))\, dA = \alpha \iint_D f(x, y)\, dA + \beta \iint_D g(x, y)\, dA.$$

(2) $\displaystyle \iint_D 1\, dA = A$ where A is the area of the region D.

(3) If D_1, D_2 is a partition of D, and $f(x, y)$ is integrable on D_1 and D_2 respectively, then $f(x, y)$ is integrable on D and

$$\iint_D f(x, y)\, dA = \iint_{D_1} f(x, y)\, dA + \iint_{D_2} f(x, y)\, dA.$$

(4) Suppose $f(x, y)$ and $g(x, y)$ are integrable functions and $f(x, y) \leqslant g(x, y)$, then

$$\iint_D f(x, y)\, dA \leqslant \iint_D g(x, y)\, dA.$$

Especially, if there exist two constants m and M, such that

$$m \leqslant f(x, y) \leqslant M,$$

then

$$mA \leqslant \iint_D f(x, y)\, dA \leqslant MA.$$

From the above, we get the *Mean-Value Theorem of Double Integrals*:

(5) If $f(x, y)$ is continuous on a bounded closed region D, then there exists a point (ξ, η) in D such that

$$I = \iint_D f(x, y)\, dA = f(\xi, \eta)A.$$

Indeed, assume that m and M are minimum and maximum values of $f(x, y)$ on D respectively. Then by property (4), I is between mA and MA, that is, there exists an constant α, such that

$$I = \alpha A$$

where $m \leqslant \alpha \leqslant M$.

Since a continuous function attains all values between its minimum and maximum values on a bounded closed region, there exists a point (ξ, η) in D, such that $\alpha = f(\xi, \eta)$. That is

$$\iint\limits_{D} f(x, y)\, dA = f(\xi, \eta) A$$

We leave the interpretation of the geometric meaning of this Mean-Value Theorem to our readers.

Furthermore, if a function $\varphi(x, y)$ is integrable on D and does not change its sign for all (x, y) in D, then there exists a point (ξ, η) in D such that

$$\iint\limits_{D} f(x, y)\varphi(x, y)\, dA = f(\xi, \eta) \iint\limits_{D} \varphi(x, y)\, dA$$

where $f(x, y)$ is a continuous function on D.

(6) If $f(x, y)$ is integrable on D, then $|f(x, y)|$ is also integrable on D, and

$$\left| \iint\limits_{D} f(x, y)\, dA \right| \leqslant \iint\limits_{D} |f(x, y)|\, dA.$$

The above properties also hold true for triple integrals and higher dimensional multiple integrals.

6.1.3 *Calculation of Multiple Integrals*

Consider a continuous function $f(x, y)$ on a region D, then the double integral

$$\iint\limits_{D} f(x, y)\, dA$$

exists, and represents the volume of a column with base D and top surface $z = f(x, y)$.

Assume that D is between two straight lines $x = a$ and $x = b$ (where $a < b$), and every line $x = x_0$ ($a < x_0 < b$) parallel to the y-axis intersects the boundary of D at only two points: one is where the straight line enters D (denoted by M_1), and the other is where it exits D (denoted by M_2). (Figure 6.3) When x passes through all the values between a and b, the points M_1 and M_2 form two boundary curves $y = \varphi_1(x)$ and $y = \varphi_2(x)$.

Consider the intersection of the column and the plane $x = x_0$, which is a "parallelogram" $M_1 M_2 N_1 N_2$ with one curved side ($N_1 N_2$: $z = f(x_0, y)$). (Figure 6.4).

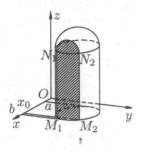

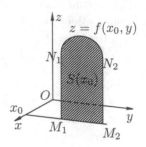

Fig. 6.3   Fig. 6.4

The area of this "parallelogram" is

$$S(x_0) = \int_{M_1}^{M_2} f(x_0, y)\, dy = \int_{\varphi_1(x_0)}^{\varphi_2(x_0)} f(x_0, y)\, dy.$$

Let $x_0, x_1, \cdots, x_n$ be a partition of the interval (a, b). Then the planes $x = x_0, \cdots, x = x_n$ divide the column into thin "slices" between them with the approximate volume $S(x_i)\Delta x_i$, where $\Delta x_i = x_i - x_{i-1}$ for $i = 1, \cdots, n$. Thus, the volume of the column is approximately equal to

$$\sum_{i=1}^{n} S(x_i)\Delta x_i.$$

The actual volume of the column is the limit

$$\lim_{\lambda \to 0} \sum_{i=1}^{n} S(x_i)\Delta x_i = \int_a^b S(x)\, dx$$

Therefore,

$$V = \iint_D f(x, y)\, dA = \int_a^b S(x)\, dx = \int_a^b \left(\int_{\varphi_1(x)}^{\varphi_2(x)} f(x, y)\, dy \right) dx.$$

In this way, the calculation of a double integral is converted to the iteration of two single integrals.

We can also write the right side of the above equation as

$$\int_a^b dx \int_{\varphi_1(x)}^{\varphi_2(x)} f(x, y)\, dy.$$

Similarly, if the region D is between two straight lines $y = c$ and $y = d$ ($c < d$), and every line parallel to the x-axis, $y = y_0$, ($c < y_0 < d$) intersects the boundary of D in at most two points, then when y moves through every value between c and d, the two intersection points from two curves $x = \psi_1(y)$ and $x = \psi_2(y)$ of the boundary of D. Then, we have

$$\iint\limits_{D} f(x,y)\, dA = \int_{c}^{d} \left(\int_{\psi_1(y)}^{\psi_2(y)} f(x,y)\, dx \right) dy = \int_{c}^{d} dy \int_{\psi_1(y)}^{\psi_2(y)} f(x,y)\, dx.$$

In light of the above iterative integral formulas, the double integral

$$\iint\limits_{D} f(x,y)\, dA$$

can also be written as $\iint\limits_{D} f(x,y)\, dx\, dy$, where $dx\, dy$ is an area element.

For more complicated shapes of region D, we can divide it to smaller parts (Figure 6.5), such that the iterative integral method of calculating double integrals can be applied on each part, and the integral is the sum of the integral values on each part.

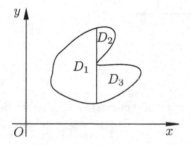

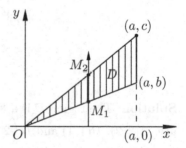

Fig. 6.5   Fig. 6.6

Example 1. Find $\iint\limits_{D} (x^2 + y^2)\, dx\, dy$, where D is the triangle bounded by $ay = bx$, $ay = cx$ and $x = a$ (Figure 6.6).

Solution: The vertices of the triangle are $(0,0)$, (a,b), (a,c) and the three sides are

$$y = \frac{b}{a}x, \quad y = \frac{c}{a}x, \quad x = a.$$

We have that

$$\iint_D (x^2 + y^2)\, dx\, dy = \int_0^a dx \int_{\frac{b}{a}x}^{\frac{c}{a}x} (x^2 + y^2)\, dy$$

$$= \frac{c-b}{a}\left(1 + \frac{b^2 + bc + c^2}{3a^2}\right) \int_0^a x^3\, dx$$

$$= \frac{1}{12}a(c-b)(3a^2 + b^2 + c^2 + bc).$$

Example 2. Find $\displaystyle\iint_D \frac{x^2}{y^2}\, dx\, dy$, where the plane region D is bounded by $x = 2$, $y = x$, and $xy = 1$ (Figure 6.7).

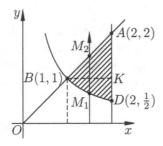

Fig. 6.7

Solution: The region D is a "triangle" with one curved side. The vertices are: $A(2,2)$, $B(1,1)$ and $D\left(2, \dfrac{1}{2}\right)$.

If we want to integrate first with respect to y, holding x constant, then we can view the region D as

$$\frac{1}{x} \leqslant y \leqslant x, \quad 1 \leqslant x \leqslant 2.$$

Thus

$$\iint_D \frac{x^2}{y^2}\, dx\, dy = \int_1^2 dx \int_{\frac{1}{x}}^x \frac{x^2}{y^2}\, dy$$

$$= \int_1^2 \left[-\frac{x^2}{y}\right]\Big|_{y=\frac{1}{x}}^{y=x}\, dx = \int_1^2 (-x + x^3)\, dx$$

$$= \left[-\frac{1}{2}x^2 + \frac{1}{4}x^4\right]\Big|_1^2 = \frac{9}{4}.$$

Now we calculate this integral in a different order of x and y. That is, we integrate first with respect to x.

It is necessary to add a horizontal line BK to separate the region to two parts: triangle ABK which consists of all points (x, y) such that

$$y \leqslant x \leqslant 2, \quad 1 \leqslant y \leqslant 2,$$

and triangle (with one curved side) BDK which consists of all points (x, y) such that

$$\frac{1}{y} \leqslant x \leqslant 2, \quad \frac{1}{2} \leqslant y \leqslant 1.$$

Thus

$$\iint_D \frac{x^2}{y^2} \, dx \, dy = \iint_{\triangle ABK} \frac{x^2}{y^2} \, dx \, dy + \iint_{\triangle BDK} \frac{x^2}{y^2} \, dx \, dy$$

$$= \int_1^2 dy \int_y^2 \frac{x^2}{y^2} \, dx + \int_{\frac{1}{2}}^1 dy \int_{\frac{1}{y}}^2 \frac{x^2}{y^2} \, dx$$

$$= \int_1^2 \left(\frac{8}{3y^2} - \frac{1}{3}y \right) dy + \int_{\frac{1}{2}}^1 \left(\frac{8}{3y^2} - \frac{1}{3y^5} \right) dy$$

$$= \frac{5}{6} + \frac{17}{12} = \frac{9}{4}.$$

From this example we can see that, when the iteration method is used to calculate a double integral, the order of integration sometimes matters. An careless choice will lead to some tedious calculation.

Example 3. Find

$$\iint_D y^2 \sqrt{1 - x^2} \, dx \, dy$$

where the region D is bounded by $y = x$, $y = -x$ and $x^2 + y^2 = 1$, and $x > 0$ for all x in D, that is, D is the area in quadrants I and IV (Figure 6.8).

Solution: A vertical line BC is needed to separate D to two parts D_1 and D_2. Since the equation of BC is $x = \dfrac{1}{\sqrt{2}}$, the integral can be calculated as

$$\iint_{D_1} y^2 \sqrt{1 - x^2} \, dx \, dy = \int_0^{\frac{1}{\sqrt{2}}} dx \int_{-x}^x y^2 \sqrt{1 - x^2} \, dy = \frac{4}{45} - \frac{7\sqrt{2}}{180},$$

$$\iint_{D_2} y^2 \sqrt{1 - x^2} \, dx \, dy = \int_{\frac{1}{\sqrt{2}}}^1 dx \int_{-\sqrt{1-x^2}}^{\sqrt{1-x^2}} y^2 \sqrt{1 - x^2} \, dy = \frac{16}{45} - \frac{43\sqrt{2}}{180}.$$

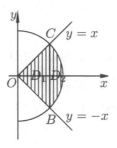

Fig. 6.8

Thus

$$\iint_D y^2 \sqrt{1 - x^2}\, dx\, dy = \frac{4}{9} - \frac{5}{18}\sqrt{2}.$$

Triple integrals can also be converted to a single integral and a double integral, and then to three single integrals by the iteration method.

Consider a triple integral

$$\iiint_V f(x, y, z)\, dV$$

where $f(x, y, z)$ is an integrable function. The region V is a three-dimensional region. Let D be the projection of V on the xy-plane. Suppose the region V has the property that every straight line parallel to the z-axis intersects the boundary surface of V in at most two points, $M_1(x, y, z)$ and $M_2(x, y, z)$, where M_1 is the point where the straight line enters V and M_2 is the point it exits V. When (x, y) moves over all the points in D, M_1 and M_2 form the lower and the upper boundary surfaces of V (Figure 6.9)

$$z = z_1(x, y), \quad z = z_2(x, y),$$

respectively.

For every point (x, y, z) in V, the first two coordinates indicate a point (x, y) in D, the coordinate z is between $z_1(x, y)$ and $z_2(x, y)$. Thus, the triple integral is

$$\iiint_V f(x, y, z)\, dV = \iint_D \left(\int_{z_1(x,y)}^{z_2(x,y)} f(x, y, z)\, dz \right) dx\, dy.$$

That is a double integral following a single integral and it can also be written as

$$\iint_D dx\, dy \int_{z_1(x,y)}^{z_2(x,y)} f(x, y, z)\, dz.$$

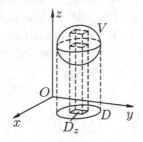

Fig. 6.9

In light of this, the triple integral can also be written as

$$\iiint_V f(x,y,z)\,dV = \iiint_V F(x,y,z)\,dx\,dy\,dz,$$

where $dx\,dy\,dz$ is the volume element dV in rectangular coordinate system.

We can also consider a cross-section D_z of V which is the intersection of V and a plane perpendicular to the z axis. If V is located between the planes $z = g$ and $z = h$, then we have

$$\iiint_V f(x,y,z)\,dV = \int_g^h \left(\iint_{D_z} f(x,y,z)\,dx\,dy \right) dz$$

$$= \int_g^h dz \iint_{D_z} f(x,y,z)\,dx\,dy.$$

If the projection of V on the yz-plane is E, and the projection of V on the zx-plane is G, then we have

$$\iiint_V f(x,y,z)\,dV = \iint_E \left(\int_{x_1(y,z)}^{x_2(y,z)} f(x,y,z)\,dx \right) dy\,dz$$

$$= \iint_E dy\,dz \int_{x_1(y,z)}^{x_2(y,z)} f(x,y,z)\,dx,$$

where $x_1(y,z)$ and $x_2(y,z)$ are boundary surfaces of V.

$$\iiint_V f(x,y,z)\,dV = \iint_G \left(\int_{y_1(x,z)}^{y_2(x,z)} f(x,y,z)\,dy \right) dx\,dz$$

$$= \iint_G dx\,dz \int_{y_1(x,z)}^{y_2(x,z)} f(x,y,z)\,dy,$$

where $y_1(x, z)$ and $y_2(x, z)$ are boundary surfaces of V.

If D_x is a cross-section of V and a plane perpendicular to the x axis, and D_y is a cross-section of V and a plane that is perpendicular to the y axis, and V is between $x = a$ and $x = b$, and $y = c$ and $y = d$, then the triple integral can also be written as

$$\iiint_V f(x, y, z)\, dV = \int_a^b \left(\iint_{D_x} f(x, y, z)\, dy\, dz \right) dx$$

$$= \int_a^b dx \iint_{D_x} f(x, y, z)\, dy\, dz,$$

$$\iiint_V f(x, y, z)\, dV = \int_c^d \left(\iint_{D_y} f(x, y, z) dz\, dx \right) dy$$

$$= \int_c^d dy \iint_{D_y} f(x, y, z)\, dz\, dx.$$

Example 4. Evaluate $\displaystyle\iiint_V z\, dx\, dy\, dz$, where V is a three-dimensional region bounded by the planes $x = 0$, $y = 0$, $z = 0$ and $2x + y + z = 1$ (Figure 6.10).

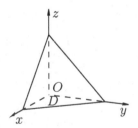

Fig. 6.10

Solution: The projection of V on the xy-plane is a triangle D with three sides $x = 0$, $y = 0$ and $2x + y = 1$. The boundary surfaces of V traced by

M_1, M_2 are $z = 0$ and $z = 1 - 2x - y$. Thus

$$\iiint\limits_V z\,dx\,dy\,dz = \iint\limits_D dx\,dy \int_0^{1-2x-y} z\,dz$$

$$= \frac{1}{2} \iint\limits_D (4x^2 + y^2 + 4xy - 4x - 2y + 1)\,dx\,dy$$

$$= \frac{1}{2} \int_0^{\frac{1}{2}} dx \int_0^{1-2x} (4x^2 + y^2 + 4xy - 4x - 2y + 1)\,dy$$

$$= \frac{1}{6} \int_0^{\frac{1}{2}} (1 - 2x)^3\,dx$$

$$= -\frac{1}{12} \left.\frac{(1 - 2x)^4}{4}\right|_0^{\frac{1}{2}} = \frac{1}{48}.$$

Example 5. Evaluate $\iiint\limits_V z\,dx\,dy\,dz$, where V is bounded by the cone $R^2 z^2 = h^2(x^2 + y^2)$ and the plane $z = h$ (Figure 6.11).

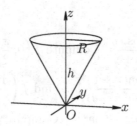

Fig. 6.11

Solution: The projection of V on the xy-plane is a disc bounded by the circle $x^2 + y^2 = R^2$. The upper and lower limits of the definite integral with respect z are $z = h$ and $z = \frac{h}{R}\sqrt{x^2 + y^2}$. Hence we have

$$\iiint_V z\,dx\,dy\,dz = \iint_{x^2+y^2\leqslant R^2} dx\,dy \int_{\frac{h}{R}\sqrt{x^2+y^2}}^h z\,dz$$

$$= \frac{1}{2}\iint_{x^2+y^2\leqslant R^2}\left(h^2 - \frac{h^2}{R^2}(x^2+y^2)\right)dx\,dy$$

$$= \frac{1}{2}h^2\pi R^2 - \frac{h^2}{2R^2}\iint_{x^2+y^2\leqslant R^2}(x^2+y^2)\,dx\,dy$$

$$= \frac{1}{2}h^2\pi R^2 - \frac{2h^2}{R^2}\int_0^R x^2\sqrt{R^2-x^2}\,dx - \frac{2h^2}{3R^2}\int_0^R (R^2-x^2)^{\frac{3}{2}}\,dx$$

$$= \frac{1}{4}\pi R^2 h^2.$$

There is an easier way to calculate this integral.

Since V is between planes $z = 0$ and $z = h$, and for any z, $(0 \leqslant z \leqslant h)$, the cross-section of V and the plane at z is the disc D_z on the plane:

$$x^2 + y^2 \leqslant \frac{R^2}{h^2}z^2.$$

Therefore,

$$\iiint_V z\,dV = \int_0^h z\,dz \iint_{x^2+y^2\leqslant \frac{R^2}{h^2}z^2} dx\,dy = \frac{\pi R^2}{h^2}\int_0^h z^3\,dz = \frac{\pi R^2 h^2}{4}.$$

Exercises 6.1

1. Suppose $f(x,y) = \dfrac{\cos xy}{x+y}$. Compute $f\left(0,\dfrac{\pi}{2}\right)$, $f(1,0)$, $f\left(2,\dfrac{\pi}{4}\right)$.

2. Let $f(x,y) = \left(\dfrac{\arctan(x+y)}{\arctan(x-y)}\right)^2$. Find $f\left(\dfrac{1+\sqrt{3}}{2}, \dfrac{1-\sqrt{3}}{2}\right)$.

3. Suppose $f(u,v,w) = u^w + w^{u+v}$. Compute $f(x+y, x-y, xy)$.

4. Find $f(x)$ if $f\left(\dfrac{y}{x}\right) = \dfrac{\sqrt{x^2+y^2}}{x}$ where $x > 0$.

5. Show that the function $F(x,y) = \ln x \ln y$ satisfies the equation

$$F(xy, uv) = F(x,u) + F(x,v) + F(y,u) + F(y,v).$$

6. Determine the domain of the following functions.

(1) $z = \sqrt{x-1} + \sqrt{y}$;

(2) $z = \dfrac{1}{\sqrt{1-x^2}} + \sqrt{y^2-1}$;

(3) $z = \ln xy$;

(4) $z = \dfrac{1}{\sqrt{x+y}} + \dfrac{1}{\sqrt{x-y}}$;

(5) $u = \sqrt{1 - \dfrac{x^2}{a^2} - \dfrac{y^2}{b^2}}$;

(6) $u = \arcsin \dfrac{y}{x}$;

(7) $z = \sqrt{x - \sqrt{y}}$;

(8) $z = \dfrac{\sqrt{4x-y^2}}{\ln(1-x^2-y^2)}$;

(9) $u = \sqrt{\sin(x^2+y^2)}$;

(10) $u = \sqrt{\dfrac{x^2+y^2-x}{2x-x^2-y^2}}$.

7. Suppose $f(x,y) = \dfrac{2xy}{x^2-y^2}$. Compute $f(\cos t, \sin t)$.

8. Assume $f(x,y) = x^y$, $\varphi(x,y) = x+y$, $\psi(x,y) = x-y$. Find

$$f(\varphi(x,y), \psi(x,y)); \quad \varphi(f(x,y), \ \psi(x,y)); \quad \psi(\varphi(x,y), f(x,y)).$$

9. Suppose $f(x,y) = \dfrac{x-y}{x+y}$. Show that

(1) $\lim\limits_{\substack{x\to 0 \\ y\to 0}} f(x,y)$ does not exist;

(2) $\lim\limits_{x\to 0}(\lim\limits_{y\to 0} f(x,y)) = 1$;

(3) $\lim\limits_{y\to 0}(\lim\limits_{x\to 0} f(x,y)) = -1$.

10. Find the points in the domain of the following functions where the function is not continuous.

(1) $z = \dfrac{2}{x^2+y^2}$;

(3) $z = \dfrac{y^2+2x}{y^2-2x}$.

(2) $z = \dfrac{1}{x-y}$;

11. Convert the double integral $\displaystyle\iint\limits_{D} f(x,y)\,dx\,dy$ to iterated integrals, where

D is given as the following regions:

(1) The triangle with vertices $O(0,0)$, $A(2,1)$ and $B(-2,1)$;

(2) The region bounded by $x+y=1$, $x-y=1$, $x=0$;

(3) The region bounded by $y = x^2$, $y = 4 - x^2$;

(4) $x^2 + y^2 \leqslant y$;

(5) $(x-1)^2 + (y-1)^2 \leqslant 1$;

(6) The region bounded by $y - 2x = 0$, $2y - x = 0$, $xy = 2$ in the first quadrant.

12. Change the order of the following integrals.

(1) $\displaystyle\int_1^e dx \int_0^{\ln x} f(x,y)\,dy,$ $\qquad$ (3) $\displaystyle\int_{-6}^2 dx \int_{\frac14 x^2-1}^{2-x} f(x,y)\,dy,$

(2) $\displaystyle\int_0^a dx \int_x^{\sqrt{2ax-x^2}} f(x,y)\,dy,$ $\qquad$ (4) $\displaystyle\int_{-1}^1 dx \int_{-\sqrt{1-x^2}}^{1-x^2} f(x,y)\,dy,$

(5) $\displaystyle\int_0^{2a} dx \int_{\sqrt{2ax-x^2}}^{\sqrt{2ax}} f(x,y)\,dy, \quad a>0,$

(6) $\displaystyle\int_0^1 dx \int_0^{x^2} f(x,y)\,dy + \int_1^3 dx \int_0^{\frac12(3-x)} f(x,y)\,dy.$

13. If $f(x,y)$ is continuous in its domain, prove that

$$\int_a^b dx \int_a^x f(x,y)dy = \int_a^b dy \int_y^b f(x,y)dx.$$

Evaluate the following double integrals:

14. $\displaystyle\iint_D e^{x+y}\,dx\,dy, \quad D:0\leqslant x \leqslant 1, 0 \leqslant y \leqslant 1.$

15. $\displaystyle\iint_D \frac{y}{x}\,dx\,dy,$ where D is bounded by $y=2x$, $y=x$, $x=4$, $x=2$.

16. $\displaystyle\iint_D xy\,dx\,dy,$ where D is bounded by $y=x^2$, $x=y^2$.

17. $\displaystyle\iint_D (x+y)\,dx\,dy,$ where D is bounded by $y^2=4x$, $x=1$.

18. $\displaystyle\iint_D xy\,dx\,dy,$ where D is bounded by $y=x$, $y=x+a$, $y=a$, $y=3a$
    $(a>0).$

19. $\displaystyle\iint_D \frac{x}{y+1}\,dx\,dy,$ where D is bounded by $y=x^2+1$, $y=2x$, $x=0$.

20. $\displaystyle\iint_D y\,dx\,dy,$ where D is the part of the disc $x^2+y^2 \leqslant a^2$ in the first
    quadrant.

21. $\displaystyle\iint_D xy\,dx\,dy,$ where D is bounded by $x^2-y^2=1$, $x^2+y^2=9$, and
    contains the origin $(0,0).$

22. $\displaystyle\iint_D |xy|\,dx\,dy, \; D: x^2+y^2 \leqslant a^2.$

23. $\displaystyle\iint\limits_{D} |xy|\, dx\, dy$, where D is bounded by $|x| + |y| = 1$.

24. If $f(x)$ is a continuous function on the closed interval $[a, b]$, then the inequality

$$\left[\int_a^b f(x)\, dx\right]^2 \leqslant (b - a) \int_a^b f^2(x)\, dx$$

holds, and the equality holds if and only if $f(x)$ is a constant.

25. Convert the triple integral

$$\iiint\limits_{V} f(x, y, z)\, dx\, dy\, dz$$

into iterated integrals for each of the following regions V:

(1) $\dfrac{x^2}{a^2} + \dfrac{y^2}{b^2} + \dfrac{z^2}{c^2} \leqslant 1$;

(2) Bounded by $x^2 + y^2 = z^2$ and $z = 1$;

(3) Bounded by $x^2 + y^2 + z^2 = a^2$ and $x^2 + y^2 + (z - a)^2 = a^2$;

(4) Bounded by $x^2 + y^2 = R^2$, $z = 0$ and $z = H$, $\quad (H > 0)$.

Evaluate the following triple integrals:

26. $\displaystyle\iiint\limits_{V} \frac{1}{(x + y + z)^2}\, dx\, dy\, dz$, V: $1 \leqslant x \leqslant 2, 1 \leqslant y \leqslant 2, 1 \leqslant z \leqslant 2$.

27. $\displaystyle\iiint\limits_{V} xyz\, dx\, dy\, dz$, where V is bounded by

$$x = 0, \ y = 0, \ x + y + z = 1.$$

28. $\displaystyle\iiint\limits_{V} xyz\, dx\, dy\, dz$, where V is bounded by

$$z = xy, \ z = 0, \ x = -1, \ x = 1, \ y = 2, \ y = 3.$$

29. $\displaystyle\iiint\limits_{V} xyz\, dx\, dy\, dz$, where V is the part of the unit ball

$$x^2 + y^2 + z^2 \leqslant 1$$

in the first octant.

30. $\displaystyle\iiint\limits_{V} xy^2 z^3\, dx\, dy\, dz$, where V is bounded by

$$z = xy, \ y = x, \ x = 1, \ z = 0.$$

6.2   Partial Derivatives

6.2.1   *Partial Derivatives and Total Differentials*

For a function with two independent variables, the input is a point on a plane. The movement of a point can change not only in distance but also in direction. Thus, the derivative of a function is also related to directions.

We define

$$\lim_{\Delta x \to 0} \frac{f(x + \Delta x, y) - f(x, y)}{\Delta x} = \frac{\partial f}{\partial x} = f'_x(x, y)$$

as the *partial derivative* of $f(x, y)$ with respect to x, if the limit exists. This is the rate of change of the function $f(x, y)$ along the direction that is parallel to the x-axis.

We also define

$$\lim_{\Delta y \to 0} \frac{f(x, y + \Delta y) - f(x, y)}{\Delta y} = \frac{\partial f}{\partial y} = f'_y(x, y)$$

as the partial derivative of $f(x, y)$ with respect to y, if the limit exists. This is the rate of change of the function $f(x, y)$ along the direction that is parallel to the y-axis.

The *partial differential* of $f(x, y)$ with respect to x is $\dfrac{\partial f}{\partial x} dx$, and the partial differential of $f(x, y)$ with respect to y is $\dfrac{\partial f}{\partial y} dy$. Their sum

$$df = \frac{\partial f}{\partial x} dx + \frac{\partial f}{\partial y} dy$$

is called the *total differential* of $f(x, y)$ at a point (x, y).

The notation Δ is a difference operator. We define a new function

$$\Delta f = f(x + \Delta x, y + \Delta y) - f(x, y).$$

This function is called the difference of the function f with respect to the increments Δx and Δy.

If the partial derivatives of a function $f(x, y)$ are continuous at a point (x, y), then by the Mean-Value Theorem of Derivatives, we have

$$\Delta f = f(x + \Delta x, y + \Delta y) - f(x, y + \Delta y) + f(x, y + \Delta y) - f(x, y)$$
$$= f'_x(x + \theta_1 \Delta x, y + \Delta y)\Delta x + f'_y(x, y + \theta_2 \Delta y)\Delta y,$$

where θ_1, θ_2 are constants with values between 0 and 1.

Let

$$f'_x(x + \theta_1 \Delta x, y + \Delta y) - f'_x(x, y) = \varepsilon_1,$$
$$f'_y(x, y + \theta_2 \Delta y) - f'_y(x, y) = \varepsilon_2.$$

Then $\varepsilon_1 \to 0$ and $\varepsilon_2 \to 0$ as Δx and Δy tend to 0. Thus,

$$\Delta f = \frac{\partial f}{\partial x}dx + \frac{\partial f}{\partial y}dy + o(\rho) = df + o(\rho),$$

where $\rho = \sqrt{(\Delta x)^2 + (\Delta y)^2}$, and $o(\rho)$ is an higher order infinitesimal with respect to ρ. That is, $\dfrac{o(\rho)}{\rho}$ tends to zero as $\rho \to 0$. Thus, we can omit the higher order infinitesimal and approximate the difference by the total differential.

There are more properties for the total differential:

$$d(u \pm v) = du \pm dv, \quad d(uv) = u\,dv + v\,du,$$

$$d\left(\frac{u}{v}\right) = \frac{v\,du - u\,dv}{v^2}.$$

These formulas have the same form as the rules for differentials of a function of one variable.

Recall the formula of derivative of composition of functions of a single variable, the derivative of $z = F(\Phi(t))$ is $\dfrac{dF}{dt} = \dfrac{dF}{ds}\dfrac{d\Phi}{dt}$, where $s = \Phi(t)$.

For a function with two variables, the formula becomes: If x, y are functions of u, v,

$$x = \varphi(u, v), \quad y = \psi(u, v),$$

then the partial derivatives of the function

$$z = f(\varphi(u, v), \psi(u, v))$$

with respect to u, v are

$$\frac{\partial f}{\partial u} = \frac{\partial f}{\partial x}\frac{\partial x}{\partial u} + \frac{\partial f}{\partial y}\frac{\partial y}{\partial u},$$

$$\frac{\partial f}{\partial v} = \frac{\partial f}{\partial x}\frac{\partial x}{\partial v} + \frac{\partial f}{\partial y}\frac{\partial y}{\partial v}.$$

That is

$$\left(\frac{\partial f}{\partial u} \quad \frac{\partial f}{\partial v}\right) = \left(\frac{\partial f}{\partial x} \quad \frac{\partial f}{\partial y}\right) \begin{pmatrix} \dfrac{\partial x}{\partial u} & \dfrac{\partial x}{\partial v} \\ \dfrac{\partial y}{\partial u} & \dfrac{\partial y}{\partial v} \end{pmatrix}.$$

Here we assume that all the partial derivatives are continuous.

The matrix

$$\begin{pmatrix} \dfrac{\partial x}{\partial u} & \dfrac{\partial x}{\partial v} \\ \dfrac{\partial y}{\partial u} & \dfrac{\partial y}{\partial v} \end{pmatrix}$$

is called the *Jacobian matrix* of the transform

$$x = \varphi(u, v), \quad y = \psi(u, v)$$

of function $f(x, y)$, and its determinant is called the *Jacobian determinant* of f, and it is denoted as

$$\frac{\partial(x, y)}{\partial(u, v)} = \begin{vmatrix} \dfrac{\partial x}{\partial u} & \dfrac{\partial x}{\partial v} \\ \dfrac{\partial y}{\partial u} & \dfrac{\partial y}{\partial v} \end{vmatrix}.$$

Since

$$dx = \frac{\partial x}{\partial u}\, du + \frac{\partial x}{\partial v}\, dv, \quad dy = \frac{\partial y}{\partial u}\, du + \frac{\partial y}{\partial v}\, dv,$$

it follows that

$$df = \frac{\partial f}{\partial u}\, du + \frac{\partial f}{\partial v}\, dv = \frac{\partial f}{\partial x}\, dx + \frac{\partial f}{\partial y}\, dy.$$

This is called the *invariance of the differential*.

Now we give a simple proof of the equation of the derivative of composite functions. Let Δu be an increment of the variable u, and let the variable v remain unchanged. Then the increments of x and y are

$$\Delta x = \varphi(u + \Delta u, v) - \varphi(u, v),$$
$$\Delta y = \psi(u + \Delta u, v) - \psi(u, v).$$

The increment of f is

$$\begin{aligned}
\Delta f &= f(x + \Delta x, y + \Delta y) - f(x, y) \\
&= f(x + \Delta x, y + \Delta y) - f(x, y + \Delta y) + f(x, y + \Delta y) - f(x, y) \\
&= f'_x(x + \theta_1 \Delta x, y + \Delta y)\Delta x + f'_y(x, y + \theta_2 \Delta y)\Delta y,
\end{aligned}$$

where θ_1, θ_2 are constants with values between 0 and 1. Divide each term of the above equation by Δu, and take the limit as Δu tends 0, we get that

$$\frac{\partial f}{\partial u} = \frac{\partial f}{\partial x}\frac{\partial x}{\partial u} + \frac{\partial f}{\partial y}\frac{\partial y}{\partial u}$$

by the continuity of f'_x and f'_y.

The other equation

$$\frac{\partial f}{\partial v} = \frac{\partial f}{\partial x}\frac{\partial x}{\partial v} + \frac{\partial f}{\partial y}\frac{\partial y}{\partial v}$$

can be proved in a similar way.

For a function $z = f(x, y)$, the second order partial derivatives are

$$\frac{\partial}{\partial x}\left(\frac{\partial f}{\partial x}\right), \quad \frac{\partial}{\partial x}\left(\frac{\partial f}{\partial y}\right), \quad \frac{\partial}{\partial y}\left(\frac{\partial f}{\partial x}\right), \quad \frac{\partial}{\partial y}\left(\frac{\partial f}{\partial y}\right).$$

They can also be written as

$$\frac{\partial^2 f}{\partial x^2}, \quad \frac{\partial^2 f}{\partial x \partial y}, \quad \frac{\partial^2 f}{\partial y \partial x}, \quad \frac{\partial^2 f}{\partial y^2}$$

or

$$f''_{xx}, \quad f''_{yx}, \quad f''_{xy}, \quad f''_{yy}.$$

If $\dfrac{\partial^2 f}{\partial x \partial y}$ and $\dfrac{\partial^2 f}{\partial y \partial x}$ not only exist at a point (x, y) but are also continuous at this point, then

$$\frac{\partial^2 f}{\partial x \partial y} = \frac{\partial^2 f}{\partial y \partial x}.$$

That is, the order of differentiation with respect to different variables can be exchanged.

In fact, let

$$\Delta_x f = f(x + \Delta x, y) - f(x, y),$$
$$\Delta_y f = f(x, y + \Delta y) - f(x, y),$$

then

$$\Delta_y \Delta_x f = \Delta_y[f(x + \Delta x, y) - f(x, y)]$$
$$= f(x + \Delta x, y + \Delta y) - f(x + \Delta x, y) - f(x, y + \Delta y) + f(x, y),$$
$$\Delta_x \Delta_y f = \Delta_x[f(x, y + \Delta y) - f(x, y)]$$
$$= f(x + \Delta x, y + \Delta y) - f(x, y + \Delta y) - f(x + \Delta x, y) + f(x, y).$$

It follows that

$$\Delta_x \Delta_y f = \Delta_y \Delta_x f.$$

Let

$$g(y) = f(x + \Delta x, y) - f(x, y),$$

(variables x and Δx are considered to be fixed). Then by the Mean-Value Theorem of Derivatives, we have

$$\Delta_y \Delta_x f = \Delta_y[f(x + \Delta x, y) - f(x, y)] = \Delta_y g(y)$$
$$= g(y + \Delta y) - g(y) = g'(y + \theta_1 \Delta y)\Delta y$$
$$= [f'_y(x + \Delta x, y + \theta_1 \Delta y) - f'_y(x, y + \theta_1 \Delta y)]\Delta y$$
$$= f''_{yx}(x + \theta_2 \Delta x, y + \theta_1 \Delta y)\Delta x \Delta y,$$

where θ_1, θ_2 are constants in $(0,1)$. Similarly, we have

$$\Delta_x \Delta_y f = f''_{xy}(x + \theta_3 \Delta x, y + \theta_4 \Delta y) \Delta x \Delta y.$$

where θ_3, θ_4 are constants in $(0,1)$. Hence

$$f''_{yx}(x + \theta_2 \Delta x, y + \theta_1 \Delta y) = f''_{xy}(x + \theta_3 \Delta x, y + \theta_4 \Delta y).$$

By the continuity of f''_{yx} and f''_{xy}, we have $f''_{yx} = f''_{xy}$ when Δx and Δy tend to 0.

Now we study the partial derivatives of functions with three variables. A function $u = f(x, y, z)$ has three partial derivatives

$$\frac{\partial u}{\partial x}, \ \frac{\partial u}{\partial y}, \ \frac{\partial u}{\partial z},$$

and total differential

$$du = \frac{\partial u}{\partial x} dx + \frac{\partial u}{\partial y} dy + \frac{\partial u}{\partial z} dz,$$

and second order partial derivatives.

Example 1. The partial derivatives of $z = x^y$, $(x > 0)$, are

$$\frac{\partial z}{\partial x} = yx^{y-1}, \quad \frac{\partial z}{\partial y} = x^y \ln x.$$

The total differential is

$$dz = yx^{y-1} \, dx + x^y \ln x \, dy.$$

Example 2. Compute the approximate value of

$$\ln \left(\sqrt[3]{1.03} + \sqrt[4]{0.98} - 1 \right).$$

Solution: Consider the function with two variables

$$z = f(x, y) = \ln \left(\sqrt[3]{x} + \sqrt[4]{y} - 1 \right).$$

Let $x_0 = 1$, $\Delta x = 0.3$, $y_0 = 1$, and $\Delta y = -0.02$. Then

$$\ln \left(\sqrt[3]{1.03} + \sqrt[4]{0.98} - 1 \right) = f(x_0 + \Delta x, y_0 + \Delta y).$$

Since Δz is approximately equal to dz, we have

$$f(x_0 + \Delta x, y_0 + \Delta y) \approx f(x_0, y_0) + f'_x(x_0, y_0)\Delta x + f'_y(x_0, y_0)\Delta y.$$

Since

$$f(x_0, y_0) = f(1, 1) = 0,$$

$$f'_x(x_0, y_0) = f'_x(1, 1) = \frac{1}{3}, \ f'_y(x_0, y_0) = f'_y(1, 1) = \frac{1}{4},$$

we have

$$\ln(\sqrt[3]{1.03} + \sqrt[4]{0.98} - 1) \approx \frac{1}{3} \times 0.03 - \frac{1}{4} \times 0.02 = 0.005.$$

Example 3. Compute the partial derivatives and the total differential of the function

$$z = e^{xy} \sin(x + y).$$

Solution: Let $u = xy$, $v = x + y$. Then

$$z = e^u \sin v,$$

and

$$\frac{\partial u}{\partial x} = y, \quad \frac{\partial u}{\partial y} = x, \quad \frac{\partial v}{\partial x} = 1, \quad \frac{\partial v}{\partial y} = 1.$$

Since

$$\frac{\partial z}{\partial u} = e^u \sin v, \quad \frac{\partial z}{\partial v} = e^u \cos v,$$

we have

$$\frac{\partial z}{\partial x} = \frac{\partial z}{\partial u}\frac{\partial u}{\partial x} + \frac{\partial z}{\partial v}\frac{\partial v}{\partial x} = e^u \sin v \cdot y + e^u \cos v \cdot 1$$
$$= e^{xy}[y \sin(x + y) + \cos(x + y)],$$

$$\frac{\partial z}{\partial y} = \frac{\partial z}{\partial u}\frac{\partial u}{\partial y} + \frac{\partial z}{\partial v}\frac{\partial v}{\partial y} = e^u \sin v \cdot y + e^u \cos v \cdot 1$$
$$= e^{xy}[x \sin(x + y) + \cos(x + y)],$$

$$dz = \frac{\partial z}{\partial x}dx + \frac{\partial z}{\partial y}dy = e^{xy}[y \sin(x + y) + \cos(x + y)]dx$$
$$+ e^{xy}[x \sin(x + y) + \cos(x + y)]dy.$$

Example 4. Prove that the function $u = \dfrac{1}{r}$ satisfies the equation

$$\frac{\partial^2 u}{\partial x^2} + \frac{\partial^2 u}{\partial y^2} + \frac{\partial^2 u}{\partial z^2} = 0,$$

where $r = \sqrt{x^2 + y^2 + z^2}$.

Proof: Since

$$\frac{\partial u}{\partial x} = -\frac{1}{r^2}\frac{\partial r}{\partial x} = -\frac{1}{r^2}\frac{x}{\sqrt{x^2 + y^2 + z^2}} = -\frac{x}{r^3},$$

we have

$$\frac{\partial^2 u}{\partial x^2} = \frac{\partial}{\partial x}\left(-\frac{x}{r^3}\right) = \frac{r^3 - 3r^2\frac{\partial r}{\partial x}x}{r^6} = -\frac{1}{r^3} + \frac{3x^2}{r^5}.$$

By the symmetry of u in x, y, z, we also have

$$\frac{\partial^2 u}{\partial y^2} = -\frac{1}{r^3} + \frac{3y^2}{r^5}, \qquad \frac{\partial^2 u}{\partial z^2} = -\frac{1}{r^3} + \frac{3z^2}{r^5}.$$

Hence

$$\frac{\partial^2 u}{\partial x^2} + \frac{\partial^2 u}{\partial y^2} + \frac{\partial^2 u}{\partial z^2} = -\frac{3}{r^3} + \frac{3(x^2 + y^2 + z^2)}{r^5} = 0.$$

This equation is called the *Laplace's equation*. Functions that satisfy this equation are called *harmonic functions*.

Example 5. Prove that the function $u = \varphi(x - at) + \psi(x + at)$ satisfies the equation

$$\frac{\partial^2 u}{\partial t^2} = a^2 \frac{\partial^2 u}{\partial x^2},$$

where φ and ψ are functions that have second order derivatives.

Proof: Let $\xi = x - at$, $\eta = x + at$. Then

$$\frac{\partial u}{\partial x} = \frac{\partial \varphi}{\partial \xi}\frac{\partial \xi}{\partial x} + \frac{\partial \psi}{\partial \eta}\frac{\partial \eta}{\partial x} = \frac{\partial \varphi}{\partial \xi} + \frac{\partial \psi}{\partial \eta},$$

$$\frac{\partial u}{\partial t} = \frac{\partial \varphi}{\partial \xi}\frac{\partial \xi}{\partial t} + \frac{\partial \psi}{\partial \eta}\frac{\partial \eta}{\partial t} = -a\frac{\partial \varphi}{\partial \xi} + a\frac{\partial \psi}{\partial \eta},$$

$$\frac{\partial^2 u}{\partial x^2} = \frac{\partial^2 \varphi}{\partial \xi^2}\frac{\partial \xi}{\partial x} + \frac{\partial^2 \varphi}{\partial \eta^2}\frac{\partial \eta}{\partial x} = \frac{\partial^2 \varphi}{\partial \xi^2} + \frac{\partial^2 \psi}{\partial \eta^2},$$

$$\frac{\partial^2 u}{\partial t^2} = -a\frac{\partial^2 \varphi}{\partial \xi^2}\frac{\partial \xi}{\partial t} + a\frac{\partial^2 \psi}{\partial \eta^2}\frac{\partial \eta}{\partial t} = a^2\frac{\partial^2 \varphi}{\partial \xi^2} + a^2\frac{\partial^2 \psi}{\partial \eta^2}.$$

Compare the last two equations, we see that

$$\frac{\partial^2 u}{\partial t^2} = a^2 \frac{\partial^2 u}{\partial x^2}.$$

This equation is called the *wave equation*.

The function

$$u = \varphi(x - at) + \psi(x + at)$$

is the solution of this equation.

Example 6. Prove that the function

$$u = \frac{1}{\sqrt{t}}e^{-\frac{x^2}{4t}}, \quad (t > 0)$$

satisfies the equation

$$\frac{\partial u}{\partial t} = \frac{\partial^2 u}{\partial x^2}.$$

Proof: Since

$$\frac{\partial u}{\partial t} = \frac{x^2}{4t^2} e^{-\frac{x^2}{4t}} \cdot \frac{1}{\sqrt{t}} - \frac{1}{2} t^{-\frac{3}{2}} e^{-\frac{x^2}{4t}} = \left(\frac{x^2}{4t^{\frac{5}{2}}} - \frac{1}{2t^{\frac{3}{2}}} \right) e^{-\frac{x^2}{4t}},$$

and

$$\frac{\partial^2 u}{\partial x^2} = \left(\frac{x^2}{4t^{\frac{5}{2}}} - \frac{1}{2t^{\frac{3}{2}}} \right) e^{-\frac{x^3}{4t}},$$

we have

$$\frac{\partial u}{\partial t} = \frac{\partial^2 u}{\partial x^2}.$$

This equation is called *heat equation*. And the function in this example is a solution of this equation.

The above three differential equations are the major equations in mathematical physics.

Example 7. The equation $pv = RT$ is called the ideal gas equation in physical chemistry, where v is the volume of ideal gas, p is the pressure, T is the absolute temperature, and R is the ideal gas constant.

If T is considered as a function of p and v, then

$$\frac{\partial T}{\partial p} = \frac{v}{R}, \quad \frac{\partial T}{\partial v} = \frac{p}{R}.$$

If v is considered as a function of p and T, $v = \dfrac{RT}{p}$, then

$$\frac{\partial v}{\partial p} = -\frac{RT}{p^2}, \quad \frac{\partial v}{\partial T} = \frac{R}{p}.$$

If p is considered as a function of v and T, then

$$\frac{\partial p}{\partial v} = -\frac{RT}{v^2}, \quad \frac{\partial p}{\partial T} = \frac{R}{v}.$$

Therefore, we get an important equation in thermodynamics

$$\frac{\partial p}{\partial v} \cdot \frac{\partial v}{\partial T} \cdot \frac{\partial T}{\partial p} = -\frac{RT}{v^2} \cdot \frac{R}{p} \cdot \frac{v}{R} = -\frac{RT}{pv} = -1.$$

Now we study calculating partial derivatives for composite functions of several variables in some more complicated situations.

Suppose we have a function

$$u = f(x, y, t), \quad x = x(r, s, t), \quad y = y(r, s, t),$$

where x, y are intermediate variables, and r, s, t are independent variables.
The partial derivatives are

$$\frac{\partial u}{\partial r} = \frac{\partial f}{\partial x}\frac{\partial x}{\partial r} + \frac{\partial f}{\partial y}\frac{\partial y}{\partial r}, \quad \frac{\partial u}{\partial s} = \frac{\partial f}{\partial x}\frac{\partial x}{\partial s} + \frac{\partial f}{\partial y}\frac{\partial y}{\partial s},$$

$$\frac{\partial u}{\partial t} = \frac{\partial f}{\partial x}\frac{\partial x}{\partial t} + \frac{\partial f}{\partial y}\frac{\partial y}{\partial t} + \frac{\partial f}{\partial t}.$$

In the third equation, $\dfrac{\partial u}{\partial t}$ is the partial derivative of u with respect to
t, considering u as a function of r, s, t. The last term on the right hand
side of the equation, $\dfrac{\partial f}{\partial t}$, is the partial derivative of f with respect to t,
considering f as a function of x, y, t.

Example 8. Suppose

$$u = f(x, y, z), \quad y = \varphi(x, t), \quad t = \psi(x, z).$$

Find $\dfrac{\partial u}{\partial x}$ and $\dfrac{\partial u}{\partial z}$.

Solution: Notice that u is a function of x and z, while y and t are intermediate variables. The partial derivatives are

$$\frac{\partial u}{\partial x} = \frac{\partial f}{\partial x} + \frac{\partial f}{\partial y}\frac{\partial y}{\partial x} = \frac{\partial f}{\partial x} + \frac{\partial f}{\partial y}\left(\frac{\partial \varphi}{\partial x} + \frac{\partial \varphi}{\partial t}\frac{\partial \psi}{\partial x}\right),$$

$$\frac{\partial u}{\partial z} = \frac{\partial f}{\partial z} + \frac{\partial f}{\partial y}\frac{\partial y}{\partial z} = \frac{\partial f}{\partial z} + \frac{\partial f}{\partial y}\frac{\partial \varphi}{\partial t}\frac{\partial \psi}{\partial z}.$$

6.2.2 *Derivatives of Implicit Functions*

The functions we discussed so far are expressed explicitly in the form $z = f(x, y)$. They are called *explicit functions*. More generally, the relation
between independent variables x, y and dependent variable z may be given
by an equation

$$F(x, y, z) = 0.$$

Then we say that z is an *implicit function* of x and y.
For instance,

$$x^2 + y^2 + z^2 - 1 = 0,$$

$$e^z - xyz = 0,$$

define implicit functions.

In these functions, one variable can be defined as an implicit function of the other two variables. For an arbitrary implicit function $F(x, y, z) = 0$, the question is: Is it possible that one variable can always be solved from the equation as the explicit function of the other two variables? That is,

$$z = f(x, y),$$

or

$$x = g(y, z), \quad y = h(x, z).$$

For example, from $x^2 + y^2 + z^2 = 1$, we can solve for z or x or y

$$z = \pm\sqrt{1 - x^2 - y^2}, \quad x = \pm\sqrt{1 - y^2 - z^2}, \quad y = \pm\sqrt{1 - z^2 - x^2}.$$

If it can be solved, some follow on questions are: Is the function differentiable? If the function is differentiable, how to calculate the partial derivatives? We only study the last question here and leave the first two to Chapter 9.

Consider z as a function of x and y in the equation $F(x, y, z) = 0$. Substitute $z = z(x, y)$ into the equation, we have $F(x, y, z(x, y)) = 0$. Taking partial derivatives with respect to x and y, we get

$$\frac{\partial F}{\partial x} + \frac{\partial F}{\partial z}\frac{\partial z}{\partial x} = 0, \quad \frac{\partial F}{\partial y} + \frac{\partial F}{\partial z}\frac{\partial z}{\partial y} = 0.$$

If $F'_z \neq 0$, then we have

$$\frac{\partial z}{\partial x} = -\frac{F'_x}{F'_z}, \quad \frac{\partial z}{\partial y} = -\frac{F'_y}{F'_z}.$$

For $F(x, y, z) = 0$, the partial derivatives of $x = x(y, z)$ or $y = y(x, z)$ can be calculated in a similar way.

Example 1. Compute $\dfrac{\partial z}{\partial x}$ and $\dfrac{\partial z}{\partial y}$ for the equation $x^2 + y^2 + z^2 - 1 = 0$.

Solution: Let $F(x, y, z) = x^2 + y^2 + z^2 - 1 = 0$, and consider z as a function of x and y. Then

$$F'_x + F'_z\frac{\partial z}{\partial x} = 2x + 2z\frac{\partial z}{\partial x} = 0,$$

$$F'_y + F'_z\frac{\partial z}{\partial y} = 2y + 2z\frac{\partial z}{\partial y} = 0.$$

Hence

$$\frac{\partial z}{\partial x} = -\frac{x}{z}, \quad \frac{\partial z}{\partial y} = -\frac{y}{z}.$$

Example 2. Compute $\dfrac{\partial z}{\partial x}$ and $\dfrac{\partial z}{\partial y}$ for the equation $e^z - xyz = 0$.

Solution: It is not quite obvious how to solve z from this equation. Keeping in mind that z is a function of x and y, differentiating term by term with respect to x and then to y, we have

$$F'_x + F'_z \frac{\partial z}{\partial x} = -yz + (e^z - xy)\frac{\partial z}{\partial x} = 0,$$

$$F'_y + F'_z \frac{\partial z}{\partial y} = -xz + (e^z - xy)\frac{\partial z}{\partial y} = 0.$$

Thus

$$\frac{\partial z}{\partial x} = \frac{yz}{e^z - xy}, \qquad \frac{\partial z}{\partial y} = \frac{xz}{e^z - xy}.$$

Since $e^z = xyz$, the above results can be simplified to

$$\frac{\partial z}{\partial x} = \frac{yz}{xyz - xy} = \frac{z}{x(z - 1)},$$

$$\frac{\partial z}{\partial y} = \frac{xz}{xyz - xy} = \frac{z}{y(z - 1)}.$$

The same method can be applied to computing the derivatives or partial derivatives of the equations with two variables or more than three variables:

$$F(x, y) = 0, \qquad F(x, y, z, \cdots) = 0.$$

More generally, the same method of computing partial derivatives can also be applied to systems of equations. Consider, for instance, the system of two equations

$$\begin{cases} F(x, y, \cdots, u, v) = 0, \\ G(x, y, \cdots, u, v) = 0. \end{cases}$$

Consider u, v as functions of independent variables $x, y, \cdots$, then apply the same method. The partial derivative with respect to x is

$$\begin{cases} \dfrac{\partial F}{\partial x} + \dfrac{\partial F}{\partial u}\dfrac{\partial u}{\partial x} + \dfrac{\partial F}{\partial v}\dfrac{\partial v}{\partial x} = 0, \\[2mm] \dfrac{\partial G}{\partial x} + \dfrac{\partial G}{\partial u}\dfrac{\partial u}{\partial x} + \dfrac{\partial G}{\partial v}\dfrac{\partial v}{\partial x} = 0. \end{cases}$$

If

$$\frac{\partial(F, G)}{\partial(u, v)} = \begin{vmatrix} \dfrac{\partial F}{\partial u} & \dfrac{\partial F}{\partial v} \\[2mm] \dfrac{\partial G}{\partial u} & \dfrac{\partial G}{\partial v} \end{vmatrix} \neq 0,$$

then

$$\begin{cases} \dfrac{\partial u}{\partial x} = -\dfrac{\dfrac{\partial(F,G)}{\partial(x,v)}}{\dfrac{\partial(F,G)}{\partial(u,v)}}, \\[6mm] \dfrac{\partial v}{\partial x} = -\dfrac{\dfrac{\partial(F,G)}{\partial(u,x)}}{\dfrac{\partial(F,G)}{\partial(u,v)}}. \end{cases}$$

The partial derivatives $\dfrac{\partial u}{\partial y}$ and $\dfrac{\partial v}{\partial y}$, $\cdots$ can be calculated in a similar way.

Example 3. Compute $\dfrac{\partial u}{\partial x}$, $\dfrac{\partial v}{\partial x}$ for

$$\begin{cases} u^3 + xv - y = 0, \\ v^3 + yu - x = 0. \end{cases}$$

Solution: Let

$$\begin{cases} F = u^3 + xv - y = 0, \\ G = v^3 + yu - x = 0. \end{cases}$$

Then

$$\frac{\partial u}{\partial x} = -\frac{\dfrac{\partial(F,G)}{\partial(x,v)}}{\dfrac{\partial(F,G)}{\partial(u,v)}} = \frac{\begin{vmatrix} -v & x \\ 1 & 3v^2 \end{vmatrix}}{\begin{vmatrix} 3u^2 & x \\ y & 3v^2 \end{vmatrix}} = \frac{x + 3v^2}{xy - 9u^2v^2}.$$

$$\frac{\partial v}{\partial x} = -\frac{\dfrac{\partial(F,G)}{\partial(u,x)}}{\dfrac{\partial(F,G)}{\partial(u,v)}} = \frac{\begin{vmatrix} 3u^2 & -v \\ y & 1 \end{vmatrix}}{\begin{vmatrix} 3u^2 & x \\ y & 3v^2 \end{vmatrix}} = \frac{3u^2 + yv}{9u^2v^2 - xy}.$$

We now study the tangent lines of a curve, the tangent planes and normal vectors of a surface, where the curve and the surface are expressed by implicit functions.

Suppose

$$F(x, y) = 0$$

is a curve L on the xy-plane. If (x, y) is a point on L such that $F_y'(x, y) \neq 0$, then the slope of the tangent line of L at this point is

$$\frac{dy}{dx} = -\frac{F_x'(x, y)}{F_y'(x, y)}.$$

Similarly, if $F_x'(x, y) \neq 0$, then

$$\frac{dx}{dy} = -\frac{F_y'(x, y)}{F_x'(x, y)}.$$

Assume that (ξ, η) is a point on the tangent line of L at (x, y). Then the equation of this tangent line is

$$(\xi - x)F_x'(x, y) + (\eta - y)F_y'(x, y) = 0.$$

The equation of the normal line of L at (x, y) is

$$(\xi - x)F_y'(x, y) - (\eta - y)F_x'(x, y) = 0.$$

Example 4. Find the equation of the tangent line of the ellipse

$$\frac{x^2}{a^2} + \frac{y^2}{b^2} = 1$$

at a point (x, y).

Solution: Let $F(x, y) = \dfrac{x^2}{a^2} + \dfrac{y^2}{b^2} - 1 = 0$, then

$$F_x' = \frac{2x}{a^2}, \quad F_y' = \frac{2y}{b^2}.$$

The equation of the tangent line is

$$(\xi - x)\frac{x}{a^2} + (\eta - y)\frac{y}{b^2} = 0.$$

It can be simplified to

$$\frac{\xi x}{a^2} + \frac{\eta y}{b^2} = 1.$$

A point on the curve where F_x' and F_y' are zero simultaneously is called a *singular point* of L.

The above technique cannot be used to find the tangent lines at a singular point.

The origin $(0, 0)$ is a singular point of the lemniscate (Figure 6.12)

$$(x^2 + y^2)^2 - a^2(x^2 - y^2) = 0.$$

If a curve has no singular points, and F_x', F_y' are continuous, then it is called a smooth curve. A smooth curve has tangent lines that rotate smoothly as the tangent point moves along the curve.

Now we study tangent lines of a curve in three-dimensional space. Suppose the parametric equation of a curve L is

$$x = x(t), \quad y = y(t), \quad z = z(t).$$

Let

$$\vec{r}(t) = (x(t), y(t), z(t)).$$

Then

$$\frac{\vec{r}(t + \Delta t) - \vec{r}(t)}{\Delta t}$$
$$= \left(\frac{x(t + \Delta t) - x(t)}{\Delta t}, \frac{y(t + \Delta t) - y(t)}{\Delta t}, \frac{z(t + \Delta t) - z(t)}{\Delta t} \right).$$

If the limit of $\dfrac{\vec{r}(t + \Delta t) - \vec{r}(t)}{\Delta t}$ exists as Δt tends to zero, then it is the derivative of $\vec{r}(t)$ at the point (x, y, z), denoted $\dfrac{d\vec{r}(t)}{dt}$. Obviously, this is the tangent vector of the curve at point (x, y, z) (Figure 6.13).

Since both

$$\left(\frac{dx}{dt}, \frac{dy}{dt}, \frac{dz}{dt} \right)$$

and (dx, dy, dz) are tangent vectors, the line element of the curve is

$$ds = \sqrt{\left(\frac{dx}{dt} \right)^2 + \left(\frac{dy}{dt} \right)^2 + \left(\frac{dz}{dt} \right)^2}\, dt = \sqrt{(dx)^2 + (dy)^2 + (dz)^2}.$$

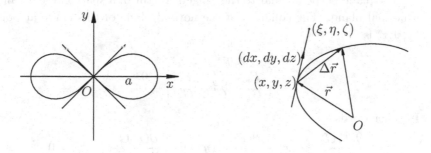

Fig. 6.12   Fig. 6.13

If a space curve L is expressed by a system of implicit functions

$$\begin{cases} F(x, y, z) = 0, \\ G(x, y, z) = 0, \end{cases}$$

then we have

$$\begin{cases} F'_x dx + F'_y dy + F'_z dz = 0, \\ G'_x dx + G'_y dy + G'_z dz = 0. \end{cases}$$

Denote $\vec{f} = (F'_x, F'_y, F'_z)$, and $\vec{g} = (G'_x, G'_y, G'_z)$. Then the above equations mean that the tangent vector of L, (dx, dy, dz) is perpendicular to $\vec{f}$ and $\vec{g}$. That is, it is parallel to

$$\vec{f} \times \vec{g} = \begin{vmatrix} \vec{i} & \vec{j} & \vec{k} \\ F'_x & F'_y & F'_z \\ G'_x & G'_y & G'_z \end{vmatrix} = \frac{\partial(F,G)}{\partial(y,z)}\vec{i} + \frac{\partial(F,G)}{\partial(z,x)}\vec{j} + \frac{\partial(F,G)}{\partial(x,y)}\vec{k}.$$

If (ξ, η, ζ) is a point on the tangent line, then the equation of the tangent line of L at point (x, y, z) is

$$\frac{\xi - x}{\dfrac{\partial(F,G)}{\partial(y,z)}} = \frac{\eta - y}{\dfrac{\partial(F,G)}{\partial(z,x)}} = \frac{\zeta - z}{\dfrac{\partial(F,G)}{\partial(x,y)}}.$$

On the other hand, the vector $(\xi - x, \eta - y, \zeta - z)$ is also on this tangent line, so the system of equations

$$\begin{cases} (\xi - x)F'_x + (\eta - y)F'_y + (\zeta - z)F'_z = 0, \\ (\xi - x)G'_x + (\eta - y)G'_y + (\zeta - z)G'_z = 0, \end{cases}$$

also represents the tangent line of L at (x, y, z).

The plane perpendicular to the tangent vector of a space curve is called a normal plane. The equation of the normal plane of this curve at point (x, y, z) is

$$\begin{vmatrix} \xi - x & \eta - y & \zeta - z \\ F'_x & F'_y & F'_z \\ G'_x & G'_y & G'_z \end{vmatrix} = 0.$$

In other words,

$$\frac{\partial(F,G)}{\partial(y,z)}(\xi - x) + \frac{\partial(F,G)}{\partial(z,x)}(\eta - y) + \frac{\partial(F,G)}{\partial(x,y)}(\zeta - z) = 0.$$

If

$$\frac{\partial(F,G)}{\partial(y,z)}, \frac{\partial(F,G)}{\partial(z,x)}, \frac{\partial(F,G)}{\partial(x,y)}$$

are zero simultaneously at a point of the curve L, then the point is called a singular point of the curve. The above technique cannot be used to discuss tangent lines at a singular point.

If a curve has no singular points, and the partial derivatives of F and G exist and are continuous, then the curve is called a smooth curve. A smooth curve has tangent lines that rotate smoothly as the tangent point moves along the curve.

Example 5. Find the equations of the tangent line and the normal plane of the curve

$$\begin{cases} x^2 + y^2 + z^2 = R^2, \\ x^2 + y^2 = Rx; \end{cases}$$

at a point (x, y, z).

Solution: Let

$$\begin{cases} F(x,y,z) = x^2 + y^2 + z^2 - R^2, \\ G(x,y,z) = x^2 + y^2 - Rx. \end{cases}$$

Then $F'_x = 2x$, $F'_y = 2y$, $F'_z = 2z$, $G'_x = 2x - R$, $G'_y = 2y$, $G'_z = 0$. Thus, the equation of the tangent line at (x, y, z) is

$$\begin{cases} x(\xi - x) + y(\eta - y) + z(\zeta - z) = 0, \\ (2x - R)(\xi - x) + 2y(\eta - z) = 0. \end{cases}$$

Simplify this equation and it becomes

$$\begin{cases} x\xi + y\eta + z\zeta = R^2, \\ (2x - R)\xi + 2\eta y = Rx. \end{cases}$$

Since

$$\frac{\partial(F,G)}{\partial(y,z)} = \begin{vmatrix} 2y & 2z \\ 2y & 0 \end{vmatrix} = -4yz,$$

$$\frac{\partial(F,G)}{\partial(z,x)} = \begin{vmatrix} 2z & 2x \\ 0 & 2x - R \end{vmatrix} = 4xz - 2zR,$$

$$\frac{\partial(F,G)}{\partial(x,y)} = \begin{vmatrix} 2x & 2y \\ 2x - R & 2y \end{vmatrix} = 2yR,$$

the equation of the tangent line can also be written as

$$\frac{\xi - x}{-2yz} = \frac{\eta - y}{2xz - Rz} = \frac{\zeta - z}{Ry}.$$

And the equation of the normal plane is

$$-2yz(\xi - x) + (2xz - Rz)(\eta - y) + Ry(\zeta - z) = 0.$$

Now we study tangent planes and normal vectors of surfaces.

Suppose M is a point on the surface Σ, L is an arbitrary curve on Σ, and the tangent line of L at the point M is l. As L goes through all curves on Σ that passes M, the tangent line l varies as well. If all the ls are in the same plane π, then π is called the tangent plane to Σ at M (Figure 6.14.) The normal vector of π is called the normal vector of the surface Σ at the point M. Just like a curve might not have a tangent line at every point, a surface might not have a tangent plane at every point.

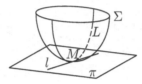

Fig. 6.14

Suppose

$$F(x, y, z) = 0$$

is the equation of the surface Σ, and a curve L:

$$x = x(t), \quad y = y(t), \quad z = (t),$$

is on Σ. Then

$$F(x(t), y(t), z(t)) = 0.$$

Thus

$$F'_x dx + F'_y dy + F'_z dz = 0.$$

That is, the tangent vector of L, (dx, dy, dz), is perpendicular to the vector (F'_x, F'_y, F'_z).

Since L is an arbitrary curve on Σ, by definition, (F'_x, F'_y, F'_z) is the normal vector of Σ. Hence, the equation of the tangent plane π is

$$(\xi - x)F'_x(x, y, z) + (\eta - y)F'_y(x, y, z) + (\zeta - z)F'_z(x, y, z) = 0.$$

Example 6. Find the equation of the tangent plane and the normal vector of the sphere

$$x^2 + y^2 + z^2 = R^2.$$

Solution: Let

$$F(x, y, z) = x^2 + y^2 + z^2 - R^2 = 0.$$

Then $F'_x = 2x$, $F'_y = 2y$, $F'_z = 2z$. Thus, the normal vector at the point (x, y, z) is $(2x, 2y, 2z)$, and the tangent plane is

$$2x(\xi - x) + 2y(\eta - y) + 2z(\zeta - z) = 0.$$

That is

$$x\xi + y\eta + z\zeta = R^2.$$

The equation of the tangent plane at the point $\left(\dfrac{R}{\sqrt{3}}, \dfrac{R}{\sqrt{3}}, \dfrac{R}{\sqrt{3}} \right)$ of the sphere is

$$\xi + \eta + \zeta = R\sqrt{3}.$$

Especially, if the equation of a surface is given explicitly

$$z = f(x, y),$$

then we can rewrite the equation as

$$F(x, y, z) = z - f(x, y) = 0,$$

and we have

$$F'_x = -f'_x, \quad F'_y = -f'_y, \quad F'_z = 1.$$

Hence, the normal vector is $(-f'_x, -f'_y, 1)$, and the tangent plane is

$$-(\xi - x)f'_x - (\eta - y)f'_y + (\zeta - z) = 0.$$

If F'_x, F'_y and F'_z are zero simultaneously at a point (x, y, z) of a surface Σ, then this point is called a singular point of Σ. If F'_x, F'_y and F'_z are continuous and are not zero simultaneously, then we say that Σ is a *smooth surface*. The normal vector and the tangent plane exist at every point of a smooth surface. A smooth surface has normal vectors that rotate smoothly as the point moves along the surface.

Suppose the parametric equations of a surface Σ are:

$$x = x(u, v), \quad y = y(u, v), \quad z = z(u, v),$$

or

$$\vec{r} = \vec{r}(u, v).$$

For a fixed value of the variable v, we get a curve on Σ, and it is called a u-curve. For a fixed value of the variable u, we get a curve on Σ, and it is called a v-curve. Then $\vec{r}'_u$ is the tangent vector of u-curve, and $\vec{r}'_v$ is the tangent vector of v-curve (Figure 6.15).

(We will use the notation $\vec{r}_u$ instead of $\vec{r}'_u$ when there is no risk of confusion.)

If we rewrite the parametric equations of this surface as

$$F(x, y, z) = 0,$$

then we have

$$\begin{cases} F'_x x'_u + F'_y y'_u + F'_z z'_u = 0, \\ F'_x x'_v + F'_y y'_v + F'_z z'_v = 0. \end{cases}$$

Hence, the normal vector of Σ, (F'_x, F'_y, F'_z), is perpendicular to $\vec{r}'_u$ and $\vec{r}'_v$. In other words, it is parallel to $\vec{r}'_u \times \vec{r}'_v$. Thus, $\vec{r}'_u \times \vec{r}'_v$ is a normal vector of Σ (Figure 6.16).

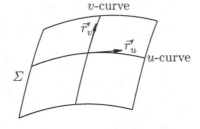 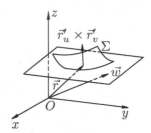

Fig. 6.15   Fig. 6.16

Moreover,

$$\vec{r}'_u \times \vec{r}'_v = \begin{vmatrix} \vec{i} & \vec{j} & \vec{k} \\ x'_u & y'_u & z'_u \\ x'_v & y'_v & z'_v \end{vmatrix} = \frac{\partial(y, z)}{\partial(u, v)}\vec{i} + \frac{\partial(z, x)}{\partial(u, v)}\vec{j} + \frac{\partial(x, y)}{\partial(u, v)}\vec{k}.$$

Let $\vec{\omega} = (\xi, \eta, \zeta)$ be an arbitrary point on the tangent plane. Then the equation of the tangent plane of Σ is

$$(\vec{\omega} - \vec{r}) \cdot (\vec{r}'_u \times \vec{r}'_v) = 0.$$

That is

$$\begin{vmatrix} \xi - x & \eta - y & \zeta - z \\ x'_u & y'_u & z'_u \\ x'_v & y'_v & z'_v \end{vmatrix} = 0.$$

If there exists a point on the surface Σ, such that

$$\vec{r}'_u \times \vec{r}'_v = 0,$$

in other words,

$$\frac{\partial(y, z)}{\partial(u, v)}, \quad \frac{\partial(z, x)}{\partial(u, v)}, \quad \frac{\partial(x, y)}{\partial(u, v)}$$

are zero simultaneously, then this point is called a singular point of Σ. If there are no singular points on Σ, and the partial derivatives of x, y, z are continuous, then we say Σ is a smooth surface. Such a surface has smoothly rotating normal vectors.

For instance, the parametric expression of the sphere $x^2 + y^2 + z^2 = R^2$ in Example 5 is

$$x = R \cos u \sin v, \quad y = R \sin u \sin v, \quad z = R \cos v,$$

and its tangent plane is

$$\begin{vmatrix} \xi - R \cos u \sin v & \eta - R \sin u \sin v & \zeta - R \cos v \\ -R \sin u \sin v & R \cos u \sin v & 0 \\ R \cos u \cos v & R \sin u \cos v & -R \sin v \end{vmatrix} = 0,$$

or

$$\xi x + \eta y + \zeta z = R^2.$$

Exercises 6.2

1. Find $f'_x(1, 1)$ and $f'_y(1, 1)$ if $f(x, y) = (1 + xy)^y$.

2. Find $f'_x\left(0, \dfrac{\pi}{4}\right)$ and $f'_y\left(0, \dfrac{\pi}{4}\right)$ if $f(x, y) = e^{-x} \sin(x + 2y)$.

3. Find the value of $\dfrac{\partial u}{\partial z}$ and $\dfrac{\partial u}{\partial t}$ at $z = b$, $t = a$, where $u = \sqrt{az^3 - bt^3}$.

In problems 4–16, find the partial derivatives with respect to each independent variable for the given function.

4. $z = \ln(x + y^2)$.

5. $z = (1 + xy)^y$.

6. $z = \left(\dfrac{1}{3}\right)^{-\frac{y}{x}}$.

9. $z = \dfrac{\cos x^2}{y}$.

7. $z = \ln \tan \dfrac{x}{y}$.

10. $z = \arctan \dfrac{y}{x}$.

8. $z = \dfrac{x}{\sqrt{x^2 + y^2}}$.

11. $z = e^{-xy} \sin \sqrt{\dfrac{y}{x}}$.

12. $u = \left(\dfrac{x}{y}\right)^z$.

13. $u = x^{y^z}$.

14. $u = \dfrac{1}{\sqrt{x^2 + y^2 + z^2}}$.

15. $z = 2\sqrt{\dfrac{1 - \sqrt{xy}}{1 + \sqrt{xy}}}$.

16. $z = \ln(xy^2 + yx^2 + \sqrt{1 + (xy^2 + yx^2)^2})$.

17. Suppose $z = \ln(\sqrt{x} + \sqrt{y})$. Show that $x\dfrac{\partial z}{\partial x} + y\dfrac{\partial z}{\partial y} = \dfrac{1}{2}$.

18. Suppose $z = e^{\frac{x}{y^2}}$. Show that $2x\dfrac{\partial z}{\partial x} + y\dfrac{\partial z}{\partial y} = 0$.

19. Suppose $z = x^y$. Show that $\dfrac{x}{y}\dfrac{\partial z}{\partial x} + \dfrac{1}{\ln x}\dfrac{\partial z}{\partial y} = 2z$.

In problem 20–24, find the second order partial derivatives with respect to the independent variables for the given function.

20. $z = \arcsin(xy)$.

22. $z = \ln(x + \sqrt{x^2 + y^2})$.

21. $z = \dfrac{x - y}{x + y}$.

23. $z = x^y$.

24. $z = x \ln(xy)$.

25. Suppose $z = \ln(e^x + e^y)$. Verify the following equations.

(1) $\dfrac{\partial z}{\partial x} + \dfrac{\partial z}{\partial y} = 1$;

(2) $\left(\dfrac{\partial^2 z}{\partial x^2}\right)\left(\dfrac{\partial^2 z}{\partial y^2}\right) - \left(\dfrac{\partial^2 z}{\partial x \partial y}\right)^2 = 0$.

26. Suppose $z = 2\cos^2\left(x - \dfrac{t}{2}\right)$. Show that $2\dfrac{\partial^2 z}{\partial t^2} + \dfrac{\partial^2 z}{\partial x \partial t} = 0$.

27. Suppose $r = \sqrt{x^2 + y^2 + z^2}$. Show that $\dfrac{\partial^2 r}{\partial x^2} + \dfrac{\partial^2 r}{\partial y^2} + \dfrac{\partial^2 r}{\partial z^2} = \dfrac{2}{r}$.

28. Find the total differential for the following functions.

(1) $z = \arcsin \dfrac{x}{y}$;

(2) $z = \ln(x^2 + y^2)$;

(5) $u = \dfrac{z}{x^2 + y^2}$.

(3) $u = \dfrac{s+t}{s-t}$;

(4) $u = xy + yz + zx$;

29. Find the total differential and total increment of the function $z = x^2 y^3$ when $x = 2$, $y = -1$, $\Delta x = 0.02$, and $\Delta y = -0.01$.

30. Find the total differential and total increment of the function $z = 2x^2 + 3y^2$ when $x = 10$, $y = 8$, $\Delta x = 0.2$, and $\Delta y = 0.3$.

31. Find the total differential and total increment of the function $z = \dfrac{y}{x}$ when $x = 2$, $y = 1$, $\Delta x = 0.1$, and $\Delta y = 0.2$.

32. Approximate the following values.

(1) $\sqrt{1.02^3 + 1.97^3}$;

(2) $\sin 29° \tan 46°$;

(3) $\dfrac{(1.03)^2}{\sqrt[3]{0.98}\sqrt[4]{1.05}}$;

(4) $(0.97)^{1.05}$.

33. Suppose $u = z^2 + y^2 + yz$ and $z = \sin t$, $y = t^3$. Find $\dfrac{du}{dt}$.

34. Suppose $u = \arctan(1 + xy)$ and $x = s + t$, $y = s - t$. Find $\dfrac{\partial u}{\partial s}$ and $\dfrac{\partial u}{\partial t}$.

35. Suppose $z = \arcsin(x - y)$ and $x = 3t$, $y = 4t^3$. Find $\dfrac{dz}{dt}$.

36. Suppose $u - \rho^2 + \varphi^2 + \theta^2$ and $\rho = \tan(\varphi\theta)$. Find $\dfrac{\partial u}{\partial \varphi}$ and $\dfrac{\partial u}{\partial \theta}$.

37. Suppose $u = \arcsin(x + y + z)$ and $z = \sin(xy)$. Find $\dfrac{\partial u}{\partial x}$ and $\dfrac{\partial u}{\partial y}$.

38. Suppose $u = \arcsin \dfrac{x}{z}$ and $z = \sqrt{x^2 + 1}$. Find $\dfrac{du}{dx}$.

39. Suppose $u = \dfrac{e^{ax}(y - z)}{a^2 + 1}$ and $y = a\sin x$, $z = \cos x$. Find $\dfrac{du}{dx}$.

40. Suppose $u = f(\xi, \eta)$ and $\xi = t^3$, $\eta = 2t^2$. Find $\dfrac{du}{dt}$.

41. Suppose $u = f(\xi, \eta, \zeta)$ and $\xi = x^2 + y^2$, $\eta = x^2 - y^2$, $\zeta = 2xy$. Find $\dfrac{\partial u}{\partial x}$ and $\dfrac{\partial u}{\partial y}$.

42. Suppose $u = f(x, y, z)$ and $x = t$, $y = t^2$, $z = t^3$. Find $\dfrac{du}{dt}$.

43. Suppose $u = f(x + y + z, x^2 + y^2 + z^2)$. Find $\Delta u = \dfrac{\partial^2 u}{\partial x^2} + \dfrac{\partial^2 u}{\partial y^2} + \dfrac{\partial^2 u}{\partial z^2}$.

44. Show that the equation

$$\frac{\partial^2 u}{\partial x^2} + 2\frac{\partial^2 u}{\partial x \partial y} - 3\frac{\partial^2 u}{\partial y^2} + 2\frac{\partial u}{\partial x} + 6\frac{\partial u}{\partial y} = 0$$

becomes

$$\frac{\partial^2 u}{\partial \xi \partial \eta} + \frac{1}{2}\frac{\partial u}{\partial \xi} = 0$$

after the substitution $\xi = x + y$ and $\eta = 3x - y$.

45. Show that the equation

$$\frac{\partial^2 u}{\partial x^2} + 2\cos x\frac{\partial^2 u}{\partial x \partial y} - \sin^2 x\frac{\partial^2 u}{\partial y^2} - \sin x\frac{\partial u}{\partial y} = 0$$

becomes

$$\frac{\partial^2 u}{\partial \xi \partial \eta} = 0$$

after the substitution $\xi = x - \sin x + y$ and $\eta = x + \sin x - y$.

46. Suppose $u = \sin x + F(\sin y - \sin x)$, where F is a differentiable function. Prove that

$$\frac{\partial u}{\partial y}\cos x + \frac{\partial u}{\partial x}\cos y = \cos x \cos y.$$

47. Suppose $z = xy + xF\left(\dfrac{y}{x}\right)$, where F is a differentiable function. Prove that

$$x\frac{\partial z}{\partial x} + y\frac{\partial z}{\partial y} = z + xy.$$

48. Suppose $z = x^n F\left(\dfrac{y}{x^2}\right)$, where F is a differentiable function. Prove that

$$x\frac{\partial z}{\partial x} + 2y\frac{\partial z}{\partial y} = nz.$$

49. Compute $\dfrac{dy}{dx}$ for the following equations.
   (1) $x^2 + 2xy - y^2 = a^2$;   (3) $xy - \ln y = a$;
   (2) $\ln(x^2 + y^2) = \arctan\dfrac{y}{x}$;   (4) $x^y = y^x$, $(x \neq y)$.

50. Find $\dfrac{\partial z}{\partial x}$ and $\dfrac{\partial z}{\partial y}$ from the following equations.

(1) $z^2 + 3xyz = a^3$;

(3) $x^3 + y^3 + z^3 - 3axyz = 0$;

(2) $\dfrac{x^2}{a^2} + \dfrac{y^2}{b^2} + \dfrac{z^2}{c^2} = 1$;

(4) $\dfrac{x}{z} = \ln \dfrac{z}{y}$.

51. Find $\dfrac{\partial^2 z}{\partial x^2}$ from the equation $x^2 + y^2 + z^2 = 4z$.

52. Find $\dfrac{\partial^2 z}{\partial x^2}$, $\dfrac{\partial^2 z}{\partial y^2}$, $\dfrac{\partial^2 z}{\partial x \partial y}$ from the equation $e^z = xyz$.

53. Suppose $2\sin(x + 2y - 3z) = x + 2y - 3z$. Show that $\dfrac{\partial z}{\partial x} + \dfrac{\partial z}{\partial y} = 1$.

54. Suppose that $\varphi(cx - az, cy - bz) = 0$, where φ is a differentiable function. Prove that

$$a\frac{\partial z}{\partial x} + b\frac{\partial z}{\partial y} = c.$$

55. Find the total differential dz of the function $z = z(x, y)$ defined by the equation $\cos^2 x + \cos^2 y + \cos^2 z = 1$.

56. Find the total differential dz of the function $z = z(x, y)$ defined by the equation $2xz - 2xyz + \ln xyz = 0$.

57. Show that if $F(x, y, z) = 0$ where F is differentiable, then

$$\frac{\partial x}{\partial y}\frac{\partial y}{\partial z}\frac{\partial z}{\partial x} = -1.$$

58. Show that if F is differentiable, and $F(x + zy^{-1}, y + zx^{-1}) = 0$, then

$$x\frac{\partial z}{\partial x} + y\frac{\partial z}{\partial y} = z - xy.$$

59. Show that if f is differentiable, and $x^2 + y^2 + z^2 = yf\left(\dfrac{z}{y}\right)$, then

$$(x^2 - y^2 - z^2)\frac{\partial z}{\partial x} + 2xy\frac{\partial z}{\partial y} = 2xz.$$

60. Suppose $z = \dfrac{y}{y^2 - a^2 x^2}$. Show that

$$\frac{\partial^2 z}{\partial x^2} = a^2 \frac{\partial^2 z}{\partial y^2}.$$

61. Suppose $z = f(x + \varphi(y))$, where f and φ are differentiable functions. Show that

$$\frac{\partial z}{\partial x}\frac{\partial^2 z}{\partial x \partial y} = \frac{\partial z}{\partial y}\frac{\partial^2 z}{\partial x^2}.$$

62. Suppose $u = u(x, y)$ and $v = v(x, y)$ are defined by the system of equations

$$\begin{cases} xu - yv = 0, \\ yu + xv = 1. \end{cases}$$

Find $\dfrac{\partial u}{\partial x}, \dfrac{\partial u}{\partial y}, \dfrac{\partial v}{\partial x}, \dfrac{\partial v}{\partial y}$.

63. Find the equation of the tangent line and the equation of the normal line of the curve $x^3 y + y^3 x = 3 - x^2 y^2$ at the point $(1, 1)$.

64. Find the equation of the tangent line and the equation of the normal line of the curve $\cos xy = x + 2y$ at the point $(1, 0)$.

65. Find the equation of the tangent line and the equation of the normal plane of the circle

$$\begin{cases} x^2 + y^2 + z^2 - 3x = 0, \\ 2x - 3y + 5z - 4 = 0; \end{cases}$$

at the point $(1, 1, 1)$.

66. Find a point on the curve $x = t$, $y = t^2$, $z = t^3$, such that the tangent line at the point is parallel to the plane $x + 2y + z = 4$.

67. Find the equation of the tangent plane and the normal vector of the surface $e^z - z + xy = 3$ at the point $(2, 1, 0)$.

68. Find the equation of the tangent plane and the normal vector of the surface

$$z = \sqrt{x^2 + y^2} - xy$$

at the point $(3, 4, -7)$.

69. Find the equation of the tangent plane of the ellipsoid $x^2 + 2y^2 + z^2 = 1$ that is parallel to the plane $x - y + 2z = 0$.

70. Find a point on the surface $z = xy$, such that the normal vector of the surface at the point is perpendicular to the plane $x + 3y + z + 9 = 0$.

71. Prove that the sum of the intercepts with the coordinate axes of the tangent plane at any point on the surface $\sqrt{x} + \sqrt{y} + \sqrt{z} = \sqrt{a}$ $(a > 0)$ is a.

72. Prove that the surface $x + 2y - \ln z + 4 = 0$ and the surface $x^2 - xy - 8x + z + 5 = 0$ are tangent to each other at the point $(2, -3, 1)$. In other words, they have a common tangent plane at the point.

6.3 Jacobian Determinants, Area Elements, Volume Elements

6.3.1 *Properties of Jacobian Determinant*

We introduced the concepts of Jacobian matrix and Jacobian determinant in the last section.

Suppose we have functions $u = u(x, y)$ and $v = v(x, y)$. The matrix that consists of the four first order partial derivatives

$$\begin{pmatrix} \dfrac{\partial u}{\partial x} & \dfrac{\partial u}{\partial y} \\ \dfrac{\partial v}{\partial x} & \dfrac{\partial v}{\partial y} \end{pmatrix}$$

is called the *Jacobian matrix* of functions $u = u(x, y)$ and $v = v(x, y)$; and its determinant is called the *Jacobian determinant*, and is denoted as

$$\frac{\partial(u, v)}{\partial(x, y)}.$$

The Jacobian matrix and Jacobian determinant for functions with more than two variables can be defined similarly. Suppose we have functions

$$u = u(x, y, z), \quad v = v(x, y, z), \quad w = w(x, y, z).$$

The matrix that consists of the nine first order partial derivatives

$$\begin{pmatrix} \dfrac{\partial u}{\partial x} & \dfrac{\partial u}{\partial y} & \dfrac{\partial u}{\partial z} \\ \dfrac{\partial v}{\partial x} & \dfrac{\partial v}{\partial y} & \dfrac{\partial v}{\partial z} \\ \dfrac{\partial w}{\partial x} & \dfrac{\partial w}{\partial y} & \dfrac{\partial w}{\partial z} \end{pmatrix}$$

is called the *Jacobian matrix* of functions $u = u(x, y, z)$, $v = v(x, y, z)$, $w = w(x, y, z)$; and its determinant is called the *Jacobian determinant*, and is denoted as

$$\frac{\partial(u, v, w)}{\partial(x, y, z)}.$$

Now we study the properties of Jacobian determinants.

(1) Suppose $u = u(x, y)$, $v = v(x, y)$ are two composite functions with $x = x(s, t)$ and $y = y(s, t)$. Then by the rule of derivatives for composite

functions, we have

$$\frac{\partial(u,v)}{\partial(s,t)} = \begin{vmatrix} \dfrac{\partial u}{\partial s} & \dfrac{\partial u}{\partial t} \\[2mm] \dfrac{\partial v}{\partial s} & \dfrac{\partial v}{\partial t} \end{vmatrix}$$

$$= \begin{vmatrix} \dfrac{\partial u}{\partial x}\dfrac{\partial x}{\partial s} + \dfrac{\partial u}{\partial y}\dfrac{\partial y}{\partial s} & \dfrac{\partial u}{\partial x}\dfrac{\partial x}{\partial t} + \dfrac{\partial u}{\partial y}\dfrac{\partial y}{\partial t} \\[2mm] \dfrac{\partial v}{\partial x}\dfrac{\partial x}{\partial s} + \dfrac{\partial v}{\partial y}\dfrac{\partial y}{\partial s} & \dfrac{\partial v}{\partial x}\dfrac{\partial x}{\partial t} + \dfrac{\partial v}{\partial y}\dfrac{\partial y}{\partial t} \end{vmatrix}$$

$$= \begin{vmatrix} \dfrac{\partial u}{\partial x} & \dfrac{\partial u}{\partial y} \\[2mm] \dfrac{\partial v}{\partial x} & \dfrac{\partial v}{\partial y} \end{vmatrix} \begin{vmatrix} \dfrac{\partial x}{\partial s} & \dfrac{\partial x}{\partial t} \\[2mm] \dfrac{\partial y}{\partial s} & \dfrac{\partial y}{\partial t} \end{vmatrix} = \frac{\partial(u,v)}{\partial(x,y)}\frac{\partial(x,y)}{\partial(s,t)}.$$

(2) From (1), we have

$$\frac{\partial(u,v)}{\partial(x,y)}\frac{\partial(x,y)}{\partial(u,v)} = \frac{\partial(u,v)}{\partial(u,v)} = \begin{vmatrix} 1 & 0 \\ 0 & 1 \end{vmatrix} = 1.$$

(3)

$$\frac{\partial(u,v)}{\partial(x,y)} = -\frac{\partial(v,u)}{\partial(x,y)}.$$

(4)

$$\frac{\partial(u,u)}{\partial(x,y)} = 0.$$

6.3.2 *Area Elements and Volume Elements*

Let D be a domain of double integral $\iint\limits_{D} f(x,y)\,dx\,dy$. Partition D into small rectangles by straight lines parallel to x-axis or y-axis. For each small rectangle in the partition, multiply its area $\Delta x \Delta y$ with the function value $f(x,y)$, for a point (x,y) in the rectangle. Then the double integral is the limit of the sum of these products. Here $dx\,dy$ is called the *area element under the rectangular coordinate system*.

In polar coordinates

$$x = \rho\cos\theta, \quad y = \rho\sin\theta,$$

we partition D by circles $\rho = \rho_1, \cdots, \rho = \rho_n$, and rays $\theta = \theta_1, \cdots, \theta = \theta_n$ (Figure 6.17). The area of a small quadrilateral with curved sides is

$$\frac{1}{2}[(\rho + \Delta\theta)^2 \Delta\theta - \rho^2 \Delta\theta] \approx \rho\,d\rho d\theta. \quad \text{(omit the higher order terms)}$$

The term on the right side, $\rho\,d\rho\,d\theta$, is an *area element in the polar coordinate system*. By the definition of double integrals, we have

$$\iint\limits_{D} f(x,y)\,dx\,dy = \iint\limits_{D'} F(\rho,\theta)\rho\,d\rho\,d\theta,$$

where $F(\rho,\theta) = f(\rho\cos\theta, \rho\sin\theta)$, and D' is the region under the coordinate transformation from D.

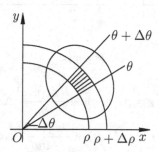

Fig. 6.17

More generally, suppose

$$u = \varphi(x,y), \quad v = \psi(x,y),$$

and x, y can be solved as

$$x = \varphi_1(u,v), \quad y = \psi_1(u,v).$$

Also, assume that for any point (x,y) on the plane, there is only one curve in each group of the curves $u = \varphi(x,y)$ and $v = \psi(x,y)$ that passes it. The coordinates (u,v) is called *curvilinear coordinates* of (x,y).

To determine the area element dA under the curvilinear coordinate (u,v), consider coordinate curves

$$\varphi(x,y) = u, \quad \varphi(x,y) = u + \Delta u;$$
$$\psi(x,y) = u, \quad \psi(x,y) = v + \Delta v.$$

We compute the area bounded by these curves, that is, compute the area of the curved quadrilateral $M_1M_2M_3M_4$ (Figure 6.18).

If all the partial derivatives of φ_1 and ψ_1 are continuous, then the coordinates of four points M_1, M_2, M_3, M_4 are

$$\begin{cases} x_1 = \varphi_1(u,v), \\ y_1 = \psi_1(u,v); \end{cases}$$

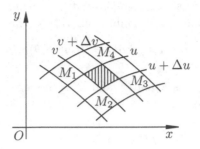

Fig. 6.18

$$\begin{cases} x_2 = \varphi_1(u + \Delta u, v) = \varphi_1(u, v) + \dfrac{\partial \varphi_1(u, v)}{\partial u} \Delta u + o(\rho), \\[2mm] y_2 = \psi_1(u + \Delta u, v) = \psi_1(u, v) + \dfrac{\partial \psi_1(u, v)}{\partial u} \Delta u + o(\rho); \end{cases}$$

$$\begin{cases} x_3 = \varphi_1(u + \Delta u, v + \Delta v) \;\; = \varphi_1(u, v) + \dfrac{\partial \varphi_1(u, v)}{\partial u} \Delta u \\[2mm] \qquad\qquad\qquad\qquad\qquad\quad + \dfrac{\partial \varphi_1(u, v)}{\partial v} \Delta v + o(\rho), \\[2mm] y_3 = \psi_1(u + \Delta u, v + \Delta v) \;\; = \psi_1(u, v) + \dfrac{\partial \psi_1(u, v)}{\partial u} \Delta u \\[2mm] \qquad\qquad\qquad\qquad\qquad\quad + \dfrac{\partial \psi_1(u, v)}{\partial v} \Delta v + o(\rho); \end{cases}$$

$$\begin{cases} x_4 = \varphi_1(u, v + \Delta v) = \varphi_1(u, v) + \dfrac{\partial \varphi_1(u, v)}{\partial v} \Delta v + o(\rho), \\[2mm] y_4 = \psi_1(u, v + \Delta v) = \psi_1(u, v) + \dfrac{\partial \varphi_1(u, v)}{\partial v} \Delta v + o(\rho), \end{cases}$$

where $\rho = \sqrt{\Delta u^2 + \Delta v^2}$.

Ignoring the higher order terms, we see that M_1, M_2, M_3, M_4 forms a parallelogram. We use the area of the parallelogram as the approximation of the area of the curved quadrilateral $M_1 M_2 M_3 M_4$. The area of the parallelogram is

$$|\overrightarrow{M_1 M_2} \times \overrightarrow{M_1 M_4}|.$$

Substituting the above coordinates expressions into this, we have

$$\left| \frac{\partial \varphi_1(u, v)}{\partial u} \frac{\partial \psi_1(u, v)}{\partial v} - \frac{\partial \varphi_1(u, v)}{\partial v} \frac{\partial \psi_1(u, v)}{\partial u} \right| du\, dv + o(\rho^2)$$

$$= \left| \frac{\partial(x, y)}{\partial(u, v)} \right| du\, dv + o(\rho^2).$$

Thus, the *area element in the curvilinear coordinates* is

$$dA = \left| \frac{\partial(x,y)}{\partial(u,v)} \right| du\, dv.$$

Therefore, by the definition of double integrals, we have

$$\iint\limits_{D} f(x,y)\, dA = \iint\limits_{\mathcal{D}} F(u,v) \left| \frac{\partial(x,y)}{\partial(u,v)} \right| du\, dv,$$

where $F(u,v) = f(\varphi_1(u,v), \psi_1(u,v))$ and $\mathcal{D}$ is the region D under the transformation of

$$\begin{cases} u = \varphi(x,y), \\ v = \psi(x,y). \end{cases}$$

We take the absolute value of the Jacobian determinant in the above formula because we have not introduced the concept of orientation, therefore we assume areas are always positive. We will introduce the concept of orientation in the next chapter and study this problem further.

The area element in polar coordinates is often used in evaluating double integrals.

Example 1. Evaluate the integral

$$\iint\limits_{D} \sqrt{x^2 + y^2}\, dx\, dy,$$

where D is the square in Figure 6.19.

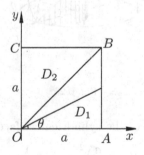

Fig. 6.19

Solution: Partition the square by the diagonal of square into two parts: D_1 and D_2. The part D_1 is between angles $\theta = 0$ and $\theta = \dfrac{\pi}{4}$. The equation of AB in polar coordinate is

$$r = \frac{a}{\cos\theta}.$$

Thus,

$$\iint\limits_{D_1} \sqrt{x^2 + y^2}\, dx\, dy = \int_0^{\frac{\pi}{4}} d\theta \int_0^{\frac{a}{\cos\theta}} r^2\, dr = \frac{a^3}{3} \int_0^{\frac{\pi}{4}} \frac{1}{\cos^3\theta}\, d\theta$$

$$= \frac{a^3}{3} \left(\frac{\sin\theta}{2\cos^2\theta} + \frac{1}{2} \ln\tan\left(\frac{\pi}{4} + \frac{\theta}{2} \right) \right) \Big|_0^{\frac{\pi}{4}} = \frac{a^3}{3} \left(\frac{1}{\sqrt{2}} + \frac{1}{2} \ln\tan\frac{3}{8}\pi \right).$$

It is easy to see that the integral value on D_2 is equal to the integral value on D_1. Therefore, we have

$$I = \frac{2}{3} a^3 \left(\frac{1}{\sqrt{2}} + \frac{1}{2} \ln\tan\frac{3}{8}\pi \right).$$

Example 2. The sphere $x^2 + y^2 + z^2 = a^2$ is sectioned by the cylindrical surface $x^2 + y^2 = ax$ as in Figure 6.20. Find the volume of the solid inside of the cylindrical surface.

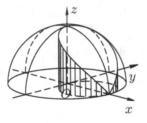

Fig. 6.20

Solution: The equation of the cylindrical surface can be written as

$$\left(x - \frac{a}{2} \right)^2 + y^2 = \frac{1}{4} a^2.$$

It is tangent to yz-plane at Oz-axis. Since the solid inside the cylindrical surface has four equal parts in the first, fourth, fifth and eighth octants, we only need to compute the volume of one part. The volume is

$$V = 4 \iint\limits_D \sqrt{a^2 - x^2 - y^2}\, dx\, dy,$$

where D is the half-disk as in Figure 6.20.

In the polar coordinate, D is between $\theta = 0$ and $\theta = \dfrac{\pi}{2}$. Since the equation of the circle is $r = a \cos \theta$, we have

$$
V = 4 \int_0^{\frac{\pi}{2}} d\theta \int_0^{a\cos\theta} \sqrt{x^2 - r^2}\, r\, dr
$$

$$
= \frac{4}{3} a^3 \int_0^{\frac{\pi}{2}} (1 - \sin^3 \theta)\, d\theta = \frac{4}{3} a^3 \left(\frac{\pi}{2} - \frac{2}{3} \right).
$$

For functions with more than two variables, the above method can also be applied. Suppose

$$
u = \varphi(x, y, z), \quad v = \psi(x, y, z), \quad w = \zeta(x, y, z),
$$

and x, y, z can be solved as

$$
x = \varphi_1(u, v, w), \quad y = \psi_1(u, v, w), \quad z = \zeta_1(u, v, w).
$$

Suppose these relations are one-to-one. If the function $f(x, y, z)$ is integrable in a domain V, then

$$
\iiint_V f(x, y, z)\, dx\, dy\, dz
$$

$$
= \iiint_{V'} f(\varphi_1(u, v, w), \psi_1(u, v, w), \zeta_1(u, v, w)) \left| \frac{\partial(x, y, z)}{\partial(u, v, w)} \right| du\, dv\, dw,
$$

where V' is V under the transformation

$$
u = \varphi(x, y, z), \quad v = \psi(x, y, z), \quad w = \zeta(x, y, z).
$$

Also,

$$
dV = \left| \frac{\partial(x, y, z)}{\partial(u, v, w)} \right| du\, dv\, dw
$$

is called the *volume element in the curvilinear coordinates* (u, v, w).

The spherical coordinates (or polar coordinates in space) is often used in coordinate transformations of space. Its relation with the rectangular coordinates is (Figure 6.21)

$$
\begin{cases}
x = \rho \sin \theta \cos \varphi, \\
y = \rho \sin \theta \sin \varphi, \\
z = \rho \cos \theta,
\end{cases}
$$

where $\rho \geqslant 0$, $0 \leqslant \theta \leqslant \pi$, $0 \leqslant \varphi \leqslant 2\pi$.

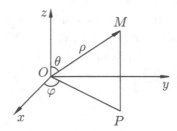

Fig. 6.21

The geometric meanings of ρ, φ and θ are: ρ is the length of the vector from the origin to a given point $M(x, y, z)$, θ is the angle between this vector and z-axis, φ is the angle between the projection OP of OM on the xy-plane and the positive x-axis.

This transformation is not an one-to-one correspondence. The plane $\rho = 0$ in (ρ, φ, θ) space corresponds to the origin $x = 0$, $y = 0$, $z = 0$. The straight line $\theta = 0$ (or π), $\rho =$ constant $\neq 0$ correspond to a point $x = y = 0$, $z = \rho$ (or $-\rho$). This transformation is an one-to-one correspondence for all other points.

There are three families of coordinate surfaces:

(1) Spheres centered at the origin, when ρ is equal to a constant;

(2) Conical surfaces with z-axis as its axis, when θ is equal to a constant;

(3) Half-planes that passes z-axis, when φ is equal to a constant.

The Jacobian determinant of this transformation is

$$\frac{\partial(x, y, z)}{\partial(\rho, \theta, \varphi)} = \begin{vmatrix} \sin\theta\cos\varphi & \sin\theta\sin\varphi & \cos\theta \\ \rho\cos\theta\cos\varphi & \rho\cos\theta\sin\varphi & -\rho\sin\theta \\ -\rho\sin\theta\sin\varphi & \rho\sin\theta\cos\varphi & 0 \end{vmatrix} = \rho^2\sin\theta.$$

Thus, the *volume element under the spherical coordinates* is $\rho^2 \sin\theta \, d\rho d\theta d\varphi$.

It can also be obtained by finding the volume of the solid bounded by the following surfaces (Figure 6.22):

$$\begin{cases} \rho = \rho_0, \\ \rho = \rho_0 + \Delta\rho; \end{cases} \qquad \begin{cases} \varphi = \varphi_0, \\ \varphi = \varphi_0 + \Delta\varphi; \end{cases} \qquad \begin{cases} \theta = \theta_0, \\ \theta = \theta_0 + \Delta\theta. \end{cases}$$

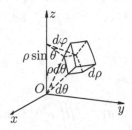

Fig. 6.22

Therefore, the integral formula under the spheroidal coordinates is

$$\iiint\limits_{V} f(x, y, z)\, dV = \iiint\limits_{V'} F(\rho, \theta, \varphi)\rho^2 \sin\theta\, d\rho d\varphi d\theta,$$

where $F(\rho, \theta, \varphi) = f(\rho\sin\theta\cos\varphi, \rho\sin\theta\sin\varphi, \rho\cos\theta)$.

Example 3. Evaluate the integral

$$\iiint\limits_{V} (x^2 + y^2 + z^2)\, dx\, dy\, dz,$$

where V is a region that is bounded by the part of conical surface $z^2 = x^2 + y^2$ above the xy-plane and the sphere $x^2 + y^2 + z^2 = R^2$.

Solution: The angle between each generatrix of the conical surface above the xy-plane and the z-axis is $\dfrac{\pi}{4}$. In spheroidal coordinates, the equation of the conical surface is $\theta = \dfrac{\pi}{4}$.

The variation ranges of ρ, θ and φ are:

$$0 \leqslant \rho \leqslant R, \quad 0 \leqslant \theta \leqslant \frac{\pi}{4}, \quad 0 \leqslant \varphi \leqslant 2\pi.$$

Thus,

$$\iiint\limits_{V} (x^2 + y^2 + z^2)\, dx\, dy\, dz = \iiint\limits_{V'} \rho^4 \sin\theta\, d\rho\, d\theta\, d\varphi$$

$$= \int_0^{2\pi} d\varphi \int_0^{\frac{\pi}{4}} d\theta \int_0^R \rho^4 \sin\theta\, d\rho = \frac{\pi R^5}{5}(2 - \sqrt{2}).$$

Example 4. Evaluate the integral

$$\iiint\limits_{V} (x^2 + y^2 + z^2)^\alpha\, dV,$$

where V is the solid ball $x^2 + y^2 + z^2 \leqslant R^2$, and α is a positive real number.

Solution: Since $(x^2 + y^2 + z^2)^\alpha = \rho^{2\alpha}$, and the variation ranges of ρ, θ and φ are

$$0 \leqslant \rho \leqslant R, \quad 0 \leqslant \theta \leqslant \pi, \quad 0 \leqslant \varphi \leqslant 2\pi,$$

we have

$$\iiint\limits_V (x^2 + y^2 + z^2)^\alpha \, dV = \iiint\limits_{V'} \rho^{2\alpha+2} \sin\theta \, d\rho d\theta d\varphi$$

$$= \int_0^{2\pi} d\varphi \int_0^\pi d\theta \int_0^R \rho^{2\alpha+2} \sin\theta \, d\rho$$

$$= \frac{4}{2\alpha+3} \pi R^{2\alpha+3}.$$

The cylindrical coordinate system is also often used in coordinate transformations (Figure 6.23)

$$x = r\cos\theta, \quad y = r\sin\theta, \quad z = z,$$

where $r \geqslant 0$, $0 \leqslant \theta < 2\pi$, $-\infty < z < \infty$.

These relations are one-to-one correspondences except that the straight line $r = 0$, $z = c$ (a constant) corresponds to the point $x = y = 0$, $z = c$.

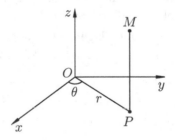

Fig. 6.23

There are three families of coordinate surfaces:

(1) The cylinders with generatrices parallel to the z-axis, and the directrix a circle on the xy-plane with center at the origin and radius r, when r is equal to a constant;

(2) Half-planes that passes the z-axis, when θ is equal to a constant;

(3) Planes parallel to xy-plane, when z is equal to a constant.

The Jacobian determinant of the transformation is

$$\frac{\partial(x, y, z)}{\partial(r, \theta, z)} = \begin{vmatrix} \cos\theta & \sin\theta & 0 \\ -r\sin\theta & r\cos\theta & 0 \\ 0 & 0 & 1 \end{vmatrix} = r.$$

Thus, the volume element under the cylindrical coordinates is

$$r \, dr \, d\theta \, dz.$$

It can also be calculated by finding the volume bounded by the three pairs of surfaces (Figure 6.24):

$$\begin{cases} r = r_0, \\ r = r_0 + \Delta r, \end{cases} \qquad \begin{cases} \theta = \theta_0, \\ \theta = \theta_0 + \Delta\theta, \end{cases} \qquad \begin{cases} z = z_0, \\ z = z_0 + \Delta z. \end{cases}$$

Therefore, the integral formula under the cylindrical coordinates is

$$\iiint_V f(x, y, z) \, dV = \iiint_{V'} F(r, \theta, z) r \, dr \, d\theta \, dz,$$

where $F(r, \theta, z) = f(r\cos\theta, r\sin\theta, z)$.

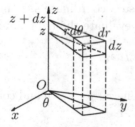

Fig. 6.24

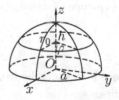

Fig. 6.25

Example 5. Find the mass of a spherical segment with uneven density (Figure 6.25).

Solution: Suppose the center of sphere is the origin, the radius of the sphere is a, the hight of the spherical segment is h and the radius of the base of the spherical segment is r_0.

The equation of the sphere is $r^2 + z^2 = a^2$ under the cylindrical coordinates. If the density function is $f(r, \varphi, z)$, then the mass of the spherical segment is

$$m = \iiint_V f(r, \varphi, z) r \, dr \, d\varphi \, dz$$

$$= \int_0^{2\pi} d\varphi \int_0^{r_0} r \, dr \int_{a-h}^{\sqrt{a^2-r^2}} f(r, \varphi, z) \, dz.$$

If $f(r, \varphi, z) = b + cz$, where b, c are given constants, then

$$m = \iiint_V (b + cz) r \, dr \, d\varphi \, dz$$

$$= \int_0^{2\pi} d\varphi \int_0^{r_0} r \, dr \int_{a-h}^{\sqrt{a^2-r^2}} (b + cz) \, dz$$

$$= 2\pi \int_0^{r_0} \left[bz + \frac{cz^2}{2} \right] \Big|_{a-h}^{z=\sqrt{a^2-r^2}} r \, dr = bV + c\pi \frac{r_0^4}{4},$$

where V is the volume of the spherical segment.

Of course, the calculation of double integrals is not restricted to using only polar coordinates, and the calculation of triple integrals is not restricted to using only spherical coordinates and cylindrical coordinates. For some particular integrals, we can choose a coordinate system for the convenience of calculation. For instance, to evaluate the triple integral

$$\iiint_V f(x, y, z) \, dx \, dy \, dz,$$

where V is a tetrahedron bounded by the plane $x + y + z = a$ and coordinate planes:

$$x > 0, \quad y > 0, \quad z > 0, \quad x + y + z < a,$$

we can choose new variables

$$x + y + z = q_1, \quad a(y + z) = q_1 q_2, \quad a^2 z = q_1 q_2 q_3.$$

That is

$$q_1 = x + y + z; \quad q_2 = \frac{a(y + z)}{x + y + z}; \quad q_3 = \frac{az}{y + z}.$$

Under this transformation, V becomes V':

$$0 < q_1 < a; \quad 0 < q_2 < a; \quad 0 < q_3 < a.$$

It is easy to see

$$x = \frac{q_1(a - q_2)}{a}; \quad y = \frac{q_1 q_2(a - q_3)}{a^2}; \quad z = \frac{q_1 q_2 q_3}{a^2}$$

and

$$\frac{\partial(x, y, z)}{\partial(q_1, q_2, q_3)} = \frac{1}{a^3} q_1^2 q_2.$$

Thus,

$$\iiint_V f(x, y, z) \, dx \, dy \, dz = \iiint_{V_1} F(q_1, q_2, q_3) \frac{1}{a^3} q_1^2 q_2 dq_1 \, dq_2 \, dq_3,$$

where $F(q_1, q_2, q_3)$ is the function $f(x, y, z)$ under the above transformation. Putting in the limits, we have

$$\int_0^a dx \int_0^{a-x} dy \int_0^{a-x-y} f(x, y, z) \, dz$$

$$= \frac{1}{a^3} \int_0^a q_1^2 \, dq_1 \int_0^a q_2 \, dq_2 \int_0^a F(q_1, q_2, q_3) \, dq_3.$$

Example 6. Evaluate

$$\iiint_V xyz(1 - x - y - z)^2 \, dx \, dy \, dz,$$

where V is defined by $x > 0$, $y > 0$, $z > 0$, $x + y + z < 1$.

Solution: By the transformation we discussed above, the integral is equal to

$$\int_0^1 d\xi \int_0^1 d\eta \int_0^1 \xi^5(1 - \xi)^2 \eta^3 (1 - \eta) \zeta(1 - \zeta) \, d\zeta$$

$$= \int_0^1 \xi^5(1 - \xi)^2 \, d\xi \int_0^1 \eta^3(1 - \eta) \, d\eta \int_0^1 \zeta(1 - \zeta) \, d\zeta = \frac{1}{20160}.$$

Now we calculate the area element of a spatial surface.

Suppose the parametric representation of a surface Σ is

$$x = x(u, v), \quad y = y(u, v), \quad z = z(u, v),$$

or in vector form

$$\vec{r} = \vec{r}(u, v),$$

where the range of variation of (u, v) is a closed region Δ on the uv-plane. Assume this surface is smooth. Partition Δ by two families of straight lines

$$u = u_i, \quad v = v_j, \quad i = 1, 2, \cdots, n; \quad j = 1, 2, \cdots, m,$$

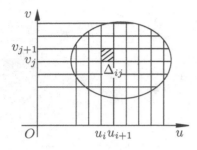

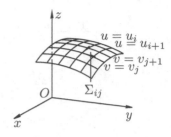

Fig. 6.26   Fig. 6.27

that are parallel to the coordinate axes. They correspond to two families of u-curves and v-curves on the surface Σ (Figure 6.26 and Figure 6.27).

The small region Δ_{ij} bounded by $u = u_i$, $u = u_{i+1}$, $v = v_j$, $v = v_{j+1}$ corresponds to the small region Σ_{ij} on the surface.

Consider the parallelogram that is bounded by the vectors

$$\vec{r}(u_{i+1}, v_j) - \vec{r}(u_i, v_j),$$
$$\vec{r}(u_i, v_{j+1}) - \vec{r}(u_i, v_j).$$

The area of this parallelogram is approximately equal to the area of Σ_{ij}. Also

$$\vec{r}(u_{i+1}, v_j) - \vec{r}(u_i, v_j)$$
$$= |x(u_{i+1}, v_j) - x(u_i, v_j)|\vec{i} + |y(u_{i+1}, v_j) - y(u_i, v_j)|\vec{j}$$
$$+ |z(u_{i+1}, v_j) - z(u_i, v_j)|\vec{k}$$
$$\approx x'_u(u_i, v_j)\Delta u_i \vec{i} + y'_u(u_i, v_j)\Delta u_i \vec{j} + z'_u(u_i, v_j)\Delta u_i \vec{k}$$
$$= \vec{r}'_u(u_i, v_j)\Delta u_i,$$

where $\vec{r}'_u = (x'_u, y'_u, z'_u)$ is the partial derivative of $\vec{r}$ with respect to u. Similarly,

$$\vec{r}(u_i, v_{j+1}) - \vec{r}(u_i, v_j) \approx \vec{r}'_v(u_i, v_j)\Delta v_j.$$

The high-order terms of Δu_i and Δv_j are omitted from the above two approximate equalities. The area of the parallelogram that is bounded by the vectors $\vec{r}'_u(u_i, v_j)\Delta u_i$ and $\vec{r}'_v(u_i, v_j)\Delta v_j$ is approximately equal to the area of Σ_{ij}, and the area of this parallelogram is

$$|\vec{r}'_u(u_i, v_j) \times \vec{r}'_v(u_i, v_j)|\Delta u_i \Delta v_j.$$

The area of the surface Σ is the limit (as Δu_i and Δv_j approach to zero) of the sum of all the areas of the parallelograms

$$S = \lim \sum_{i,j} |\vec{r}'_u(u_i, v_j) \times \vec{r}'_v(u_i, v_j)| \Delta u_i \Delta v_j$$

$$= \iint_\Delta |\vec{r}'_u \times \vec{r}'_v| \, du \, dv.$$

Thus, the element of area of the surface Σ is

$$dS = |\vec{r}'_u \times \vec{r}'_v| \, du \, dv.$$

Also, we know that

$$|\vec{r}'_u \times \vec{r}'_v|^2 = (\vec{r}'_u \cdot \vec{r}'_u)(\vec{r}'_v \cdot \vec{r}'_v) - (\vec{r}'_u \cdot \vec{r}'_v)^2.$$

Let

$$E = \vec{r}'_u \cdot \vec{r}'_u = (x'_u)^2 + (y'_u)^2 + (z'_u)^2,$$
$$G = \vec{r}'_v \cdot \vec{r}'_v = (x'_v)^2 + (y'_v)^2 + (z'_v)^2,$$
$$F = \vec{r}'_u \cdot \vec{r}'_v = x'_u x'_v + y'_u y'_v + z'_u z'_v.$$

Then the area of the surface can be written as

$$S = \iint_\Delta \sqrt{EG - F^2} \, du \, dv.$$

That is, the area element can be written as

$$dS = \sqrt{EG - F^2} \, du \, dv.$$

If the equation of the surface is an explicit expression

$$z = z(x, y),$$

and the domain is D, then

$$E = 1 + z'^2_x, \quad G = 1 + z'^2_y, \quad F = z'_x z'_y.$$

Thus, the area element is

$$dS = \sqrt{1 + z'^2_x + z'^2_y} \, dx \, dy.$$

The formula for computing the area is

$$S = \iint_D \sqrt{1 + z'^2_x + z'^2_y} \, dx \, dy.$$

Example 7. Find the area of the part of the hyperbolic paraboloid $z = xy$ that is sectioned by the cylindrical surface $x^2 + y^2 = R^2$.

Solution: The projection of this part of surface is a disk

$$x^2 + y^2 \leqslant R^2$$

on the xy-plane. Thus,

$$S = \iint\limits_{x^2+y^2 \leqslant R^2} \sqrt{1 + \left(\frac{\partial z}{\partial x}\right)^2 + \left(\frac{\partial z}{\partial y}\right)^2}\, dx\, dy$$

$$= \iint\limits_{x^2+y^2 \leqslant R^2} \sqrt{1 + x^2 + y^2}\, dx\, dy.$$

Evaluate this integral in polar coordinates, we have

$$S = \int_0^{2\pi} d\varphi \int_0^R r\sqrt{1 + r^2}\, dr = \frac{2\pi}{3}[(1 + R^2)^{\frac{3}{2}} - 1].$$

In spherical coordinates, the sphere of radius R is expressed as

$$x = R\sin\theta\cos\varphi, \quad y = R\sin\theta\sin\varphi, \quad z = R\cos\theta,$$

where $0 \leqslant \theta \leqslant \pi$, $0 \leqslant \varphi \leqslant 2\pi$.

Since

$$E = x_\varphi'^2 + y_\varphi'^2 + z_\varphi'^2 = R^2\sin^2\theta,$$
$$G = x_\theta'^2 + y_\theta'^2 + z_\theta'^2 = R^2,$$
$$F = x_\varphi'x_\theta' + y_\varphi'y_\theta' + z_\varphi'z_\theta' = 0,$$

the area element is

$$dS = R^2\sin\theta\, d\varphi\, d\theta.$$

From this we can also get the volume element

$$dV = R^2\sin\theta\, d\varphi\, d\theta\, dR.$$

Example 8. Find the area of the part of the sphere $x^2 + y^2 + z^2 = R^2$ that is sectioned by the cylinder $x^2 + y^2 = Rx$.

Solution: We only need to compute the area of the section Σ that is in the first octant. In spherical coordinates, the first octant is $0 \leqslant \varphi \leqslant \frac{\pi}{2}$, $0 \leqslant \theta \leqslant \frac{\pi}{2}$.

To find the intersection line of the sphere and the cylinder, substituting the parametric equation of the sphere into the equation of the cylinder $x^2 + y^2 = Rx$, we obtain

$$\sin\theta = \cos\varphi.$$

Thus,

$$\theta = \frac{\pi}{2} - \varphi.$$

This implies that the range of variation of the parameters of Σ is a triangle Δ (Figure 6.28, Figure 6.29). Hence the area of Σ is

$$S = \iint_{\Delta} R^2 \sin\theta \, d\varphi d\theta = R^2 \int_0^{\frac{\pi}{2}} d\varphi \int_0^{\frac{\pi}{2}-\varphi} \sin\theta \, d\theta$$

$$= R^2 \int_0^{\frac{\pi}{2}} (1 - \sin\varphi) \, d\varphi = R^2 \left(\frac{\pi}{2} - 1\right).$$

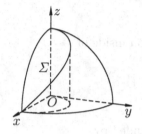

Fig. 6.28

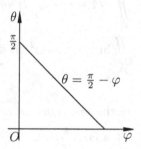

Fig. 6.29

Therefore, the area of the part of sphere that is sectioned by the cylinder is

$$4S = 2R^2(\pi - 2).$$

Exercises 6.3

1. Suppose the inverse functions of $x = x(u, v)$, $y = y(u, v)$ are $u = u(x, y)$, $v = v(x, y)$. Compute $\dfrac{\partial u}{\partial x}, \dfrac{\partial u}{\partial y}, \dfrac{\partial v}{\partial x}, \dfrac{\partial v}{\partial y}$.

2. Suppose $x = f(u, v)$, $y = g(u, v)$, $z = h(u, v)$. This defines z as the function of x and y. Find $\dfrac{\partial z}{\partial x}, \dfrac{\partial z}{\partial y}$.

3. Suppose $y = f(x, y)$, $F(x, y, t) = 0$. Prove that

$$\frac{dy}{dx} = \frac{\dfrac{\partial f}{\partial x}\dfrac{\partial F}{\partial t} - \dfrac{\partial f}{\partial t}\dfrac{\partial F}{\partial x}}{\dfrac{\partial f}{\partial t}\dfrac{\partial F}{\partial y} + \dfrac{\partial F}{\partial t}}.$$

4. Suppose $F(x, y) = 0$. Prove

$$\frac{d^2y}{dx^2} = -\frac{\dfrac{\partial^2 F}{\partial x^2}\left(\dfrac{\partial F}{\partial y}\right)^2 - 2\dfrac{\partial^2 F}{\partial x \partial y}\dfrac{\partial F}{\partial x}\dfrac{\partial F}{\partial y} + \dfrac{\partial^2 F}{\partial y^2}\left(\dfrac{\partial F}{\partial x}\right)^2}{\left(\dfrac{\partial F}{\partial y}\right)^2}.$$

Evaluate the following integrals.

5. $\displaystyle\iint_D \sqrt{1 + x^2 + y^2}\, dx\, dy, \quad D : x^2 + y^2 \leqslant R^2.$

6. $\displaystyle\iint_D \sqrt{1 - x^2 - y^2}\, dx\, dy, \quad D : x^2 + y^2 \leqslant x.$

7. $\displaystyle\iint_D \ln(1 + x^2 + y^2)\, dx\, dy, \quad D$: The region inside of

$$x^2 + y^2 = 1$$

in the first quadrant.

8. $\displaystyle\iint_D \arctan\frac{y}{x}\, dx\, dy, \quad D$: The region bounded by

$$x^2 + y^2 = 4, \ x^2 + y^2 = 1, \ y = x, \ y = 0$$

in the first quadrant.

9. $\displaystyle\iint_D \sqrt{x^2 + y^2}\, dx\, dy, \quad D$: Bounded by $x^2 + y^2 = x + y$.

10. $\displaystyle\iint_D \sin\sqrt{x^2 + y^2}\, dx\, dy, \quad D : \pi^2 \leqslant x^2 + y^2 \leqslant 4\pi^2.$

11. $\displaystyle\iint_D y\, dx\, dy, \quad D : \frac{x^2}{4} + \frac{(y-3)^2}{9} \leqslant 1.$

12. $\displaystyle\iint_D \sqrt{\frac{x^2}{a^2} + \frac{y^2}{b^2}}\, dx\, dy, \quad D$: The region bounded by

$$\frac{x^2}{a^2} + \frac{y^2}{b^2} = 4, \ y = 0, \ y = x$$

in the first quadrant.

13. $\displaystyle\iint_D (x^2 + y^2)\, dx\, dy, \quad D : 1 \leqslant \frac{x^2}{a^2} + \frac{y^2}{b^2} \leqslant 4.$

14. $\displaystyle\iint\limits_{D} \sqrt{1 - \frac{x^2}{a^2} - \frac{y^2}{b^2}}\, dx\, dy, \quad D: \frac{x^2}{a^2} + \frac{y^2}{b^2} \leqslant 1.$

15. $\displaystyle\iiint\limits_{V} (x^2 + y^2)\, dx\, dy\, dz, \quad V:$ The region bounded by

$$x^2 + y^2 = 2z \text{ and } z = 2.$$

16. $\displaystyle\iiint\limits_{V} xyz\, dx\, dy\, dz, \; V:$ The region bounded by

$$x^2 + y^2 + z^2 = 1, \; x = 0, \; y = 0, \; z = 0$$

in the first octant.

17. $\displaystyle\iiint\limits_{V} \sqrt{x^2 + y^2 + z^2}\, dx\, dy\, dz, \quad V: x^2 + y^2 + z^2 \leqslant z.$

18. $\displaystyle\iiint\limits_{V} \sqrt{1 - \frac{x^2}{a^2} - \frac{y^2}{b^2} - \frac{z^2}{c^2}}\, dx\, dy\, dz, \; V: \frac{x^2}{a^2} + \frac{y^2}{b^2} + \frac{z^2}{c^2} \leqslant 1.$

19. $\displaystyle\iiint\limits_{V} (x^2 + y^2)\, dx\, dy\, dz, \; V:$ The region bounded by

$$r^2 \leqslant x^2 + y^2 + z^2 \leqslant R^2, \; z \geqslant 0.$$

20. Find the areas of the regions bounded by the following plane curves.

   (1) $\dfrac{x^2}{a^2} + \dfrac{y^2}{b^2} = 1;$

   (2) $(x^2 + y^2)^2 = 2ax^3, \; (a > 0);$

   (3) $(x^2 + y^2)^2 = 2a^2(x^2 - y^2);$

   (4) $\left(\dfrac{x^2}{a^2} + \dfrac{y^2}{b^2}\right)^2 = \dfrac{xy}{c^2}.$

Find the volumes of the solids bounded by the following surfaces.

21. $x^2 + y^2 + z^2 = 2az, \; (a > 0)$ and $x^2 + y^2 = z^2$ (the part that includes the z-axis).

22. $z^2 = \dfrac{x^2}{4} + \dfrac{y^2}{9}$ and $2z = \dfrac{x^2}{4} + \dfrac{y^2}{9}.$

23. $x^2 + y^2 + z^2 = b^2, \; x^2 + y^2 + z^2 = a^2$ and $z = \sqrt{x^2 + y^2}, \; (b > a > 0).$

24. $\dfrac{x^2}{a^2} + \dfrac{y^2}{b^2} + \dfrac{z^2}{c^2} = 1, \; \dfrac{x^2}{a^2} + \dfrac{y^2}{b^2} = \dfrac{z^2}{c^2}$ (the part that includes the z-axis).

25. $(x^2 + y^2 + z^2)^2 = a^3 x$, $\quad (a > 0)$.

26. $\left(\dfrac{x^2}{a^2} + \dfrac{y^2}{b^2} + \dfrac{z^2}{c^2}\right)^2 = \dfrac{x}{h}$, $\quad (h > 0)$.

Evaluate the following integrals.

27. $\displaystyle\iint\limits_{D} xy \, dx \, dy$, $\quad$ D: The region bounded by

$$xy = 1, \; xy = 2, \; y = x, \; y = 2x$$

in the first quadrant.

28. $\displaystyle\iint\limits_{D} dx \, dy$, $\quad$ D: The region bounded by

$$y^2 = ax, \; y^2 = bx, \; x^2 = py, \; x^2 = qy, \; (0 < a < b, \, 0 < p < q)$$

in the first quadrant.

29. $\displaystyle\iint\limits_{D} dx \, dy$, $\quad$ D: The region bounded by

$$y^2 = 16 - 8x, \; y^2 = 1 - 2x, \; y^2 = 81 + 18x, \; y^2 = 1 + 2x$$

in the first and second quadrants.

Find the areas of the following surfaces.

30. The part of the surface $z = \sqrt{x^2 + y^2}$ that is inside the cylinder $x^2 + y^2 = 2x$.

31. The part of the surface $x^2 + y^2 = a^2$ sectioned by the plans

$$x + z = 0, \; x - z = 0, \; (x > 0, \, y > 0).$$

32. The part of the elliptic paraboloid $z = \dfrac{x^2}{2a} + \dfrac{y^2}{2b}$ sectioned by elliptic cylinder $\dfrac{x^2}{a^2} + \dfrac{y^2}{b^2} = c^2$.

33. The surface bounded by the sphere $x^2 + y^2 + z^2 = 3a^2$ and paraboloid $x^2 + y^2 = 2az$, $(z \geqslant 0)$.

34. The part of the helicoidal surface $x = r \cos\varphi, y = r \sin\varphi, z = h\varphi$ where $0 < r < a, \, 0 < \varphi < 2\pi$.

35. The part of the torus $x = (b + a \cos\psi)\cos\varphi$, $y = (b + a \cos\psi)\sin\varphi$, $z = a \sin\psi$, $(0 < a \leqslant b)$ between $\varphi = \varphi_1, \varphi = \varphi_2$ and $\psi = \psi_1, \psi = \psi_2$. What is the surface area of the torus.

36. The part of cylinder $x^2 + z^2 = a^2$ sectioned by $x^2 + y^2 = a^2$.

37. Suppose the equation of the plane π is $x \cos \alpha + y \cos \beta + z \cos \gamma + D = 0$, where $\cos^2 \alpha + \cos^2 \beta + \cos^2 \gamma = 1$. There is a bounded and closed region D on π with area S. Find the areas of the projections of D on the coordinate planes Oxy, Oyz and Ozx.

Chapter 7

Line Integrals, Surface Integrals and Exterior Differential Forms

7.1 Scalar Fields and Vector Fields

In this section, we study a physical concept *field*. It is a function of position and time.

A *scalar field* associates a scalar value with a point in space, and a *vector field* associates a vector with a point in space. A scalar field on a region define a function of each point of the region, a vector field define a vector function of each point on the region.

For Example, the temperature at each point in a room defines a scalar field, the velocity at each point of a region in the space where a fluid is flowing through defines a velocity vector field.

7.1.1 *Contour Surfaces and Gradient of a Scalar Field*

Suppose $u = u(M)$ is a scalar field on a spatial domain V. A surface $u(M) = C$ (a constant), is called a *contour surface* (or *level surface*) of the scalar field u.

For every point M_0 in V, there exists one and only one contour surface that passes this point, and there is no intersection between contour surfaces.

Under a rectangular coordinate system, a scalar field is a function of three variables

$$u = u(x, y, z), \quad (x, y, z) \in V.$$

For instance, the contour surface of the scalar field

$$u = u(x, y, z) = 4x^2 + 2y^2 + 3z^2$$

that passes the point $(-1, 0, 2)$ is

$$u(x, y, z) = u(-1, 2, 0) = 12.$$

That is

$$4x^2 + 2y^2 + 3z^2 = 12.$$

This is an ellipsoid.

Let M_0 be a point in V and l be a ray with the starting point M_0. Then the *directional derivative* of $u = u(M)$ in the direction of l is defined as

$$\left(\frac{\partial u}{\partial l}\right)_{M_0} = \lim_{\substack{M \to M_0 \\ M \in l}} \frac{u(M) - u(M_0)}{|MM_0|}.$$

Here $M \in l$ means M is a point on l.

At a point $M_0(x_0, y_0, z_0)$, the function $u(x, y, z)$ has directional derivatives in different directions. Suppose the direction cosines of l is $(\cos\alpha, \cos\beta, \cos\gamma)$. Then the coordinates of a point M on l can be written as

$$x = x_0 + t\cos\alpha, \quad y = y_0 + t\cos\beta, \quad z = z_0 + t\cos\gamma.$$

Thus, the directional derivative of $u(M)$ in the direction of l is (Figure 7.1)

$$\left(\frac{\partial u}{\partial l}\right)_{M_0} = \lim_{t \to 0} \frac{u(x_0 + t\cos\alpha, y_0 + t\cos\beta, z_0 + t\cos\gamma) - u(x_0, y_0, z_0)}{t}$$

$$= \frac{\partial u}{\partial x}\cos\alpha + \frac{\partial u}{\partial y}\cos\beta + \frac{\partial u}{\partial z}\cos\gamma.$$

This implies that

$$\left(\frac{\partial u}{\partial l}\right)_{M_0} = \left(\frac{\partial u}{\partial x}\vec{i} + \frac{\partial u}{\partial y}\vec{j} + \frac{\partial u}{\partial z}\vec{k}\right)_{M_0} \cdot (\cos\alpha\vec{i} + \cos\beta\vec{j} + \cos\gamma\vec{k}),$$

where $\vec{i}, \vec{j}$ and $\vec{k}$ are coordinate unit vectors.

Let

$$\vec{n} = \left(\frac{\partial u}{\partial x}\vec{i} + \frac{\partial u}{\partial y}\vec{j} + \frac{\partial u}{\partial z}\vec{k}\right)_{M_0},$$

$$\vec{l}^0 = \cos\alpha\vec{i} + \cos\beta\vec{j} + \cos\gamma\vec{k},$$

and θ be the angle between $\vec{l}^0$ and $\vec{n}$, then

$$\left(\frac{\partial u}{\partial l}\right)_{M_0} = \vec{n} \cdot \vec{l}^0 = |\vec{n}|\cos\theta.$$

The value of $\left(\dfrac{\partial u}{\partial l}\right)_{M_0}$ reaches the maximum $|\vec{n}|$ when $\theta = 0$. Also, the direction of $\vec{l}^0$ is the same as the direction of $\vec{n}$ when $\theta = 0$.

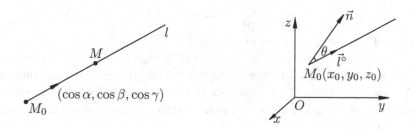

Fig. 7.1   Fig. 7.2

The vector $\vec{n}$ is called the *gradient* of u at the point M_0, and is denoted as $(\mathbf{grad}\, u)_{M_0}$,

$$(\mathbf{grad}\, u)_{M_0} = \vec{n} = \left(\frac{\partial u}{\partial x}\vec{i} + \frac{\partial u}{\partial y}\vec{j} + \frac{\partial u}{\partial z}\vec{k}\right)_{M_0}.$$

The direction of $\vec{n}$ is the direction of the normal vector of the tangent plane of the contour surface that passes the point M_0. In other words, the direction of the gradient at point M_0 is the direction of the normal vector of the contour surface at the point. Also, the value of the directional derivative in the direction of the gradient is the largest among all the values of directional derivatives at the point M_0. That is, if l is a ray in an arbitrary direction, and n is a ray with direction $\vec{n}$, then

$$\left(\frac{\partial u}{\partial n}\right)_{M_0} \geqslant \left(\frac{\partial u}{\partial l}\right)_{M_0}.$$

This property can be used as the definition of the gradient of a function at a point.

Now we look at an example. Suppose a scalar field is

$$u = u(x, y, z) = 4x^2 + 2y^2 + 3z^2.$$

Since

$$\frac{\partial u}{\partial x} = 8x, \quad \frac{\partial u}{\partial y} = 4y, \quad \frac{\partial u}{\partial z} = 6z,$$

the gradient is

$$\mathbf{grad}\, u = 8x\vec{i} + 4y\vec{j} + 6z\vec{k}.$$

From the above study, we know that

(1) The relation between the directional derivative and the gradient is:
$$\frac{\partial u}{\partial l} = (\mathbf{grad}\ u) \cdot \vec{l}^{\,0},$$
where $\vec{l}^{\,0}$ is the unit vector on the ray l.

This equation means that the rate of change of u in the direction l at the point M is equal to the projection of the gradient at the point on l (Figure 7.3). Thus, the largest rate of change of a scalar field is along the direction of the gradient. And, the magnitude of the gradient is the largest rate of change.

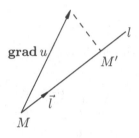

Fig. 7.3

(2) The gradient of $u = u(M)$ at a point M_0 is the normal vector of the contour surface passing through M_0, taken at M_0 (Figure 7.4). Thus, in the direction of the normal vector of the contour surface, the rate of change of a scalar field is the greatest, and the distance between contour surfaces are the smallest (Figure 7.5).

(3) Using the representation of gradients in the rectangular coordinate system, we derive the rules of operation of gradients:

(i) $\mathbf{grad}\ Cu = C\,\mathbf{grad}\ u$, where C is a constant;

(ii) $\mathbf{grad}\ (u_1 + u_2) = \mathbf{grad}\ u_1 + \mathbf{grad}\ u_2$;

(iii) $\mathbf{grad}\ u_1 u_2 = u_1\,\mathbf{grad}\ u_2 + u_2\,\mathbf{grad}\ u_1$;

(iv) $\mathbf{grad}\ f(u) = f'(u)\,\mathbf{grad}\ u$.

Now we introduce an operator ∇ (Nabla). In the rectangular coordinate system, it is defined as
$$\nabla = \vec{i}\,\frac{\partial}{\partial x} + \vec{j}\,\frac{\partial}{\partial y} + \vec{k}\,\frac{\partial}{\partial z}.$$
The result of applying this operator to a function $u = u(x, y, z)$ is
$$\nabla u = \vec{i}\,\frac{\partial u}{\partial x} + \vec{j}\,\frac{\partial u}{\partial y} + \vec{k}\,\frac{\partial u}{\partial z}.$$

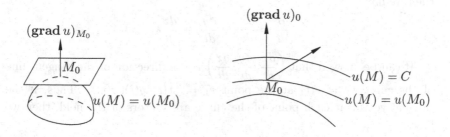

Fig. 7.4 Fig. 7.5

Therefore,

$$\nabla u = \mathbf{grad}\ u.$$

Example: Suppose O is a fixed point and $r = |OM|$. Find $\mathbf{grad}\ r$.

Solution: Let the point O be the origin of a rectangular coordinate system $Oxyz$, and the coordinate of the point M be (x, y, z). Then, we have

$$r = \sqrt{x^2 + y^2 + z^2}.$$

Thus,

$$\nabla r = \frac{1}{r}(x\vec{i} + y\vec{j} + z\vec{k}).$$

That is

$$\nabla r = \frac{\vec{r}}{r}.$$

7.1.2 *Streamlines of Vector Fields*

Suppose there is a vector field $\vec{v} = \vec{v}(M)$ in a region V. A curve in V, if its tangent vector at each point are vectors of $\vec{v}$, is called a *streamline* of $\vec{v}$.

There are infinity many streamlines for any vector field. Suppose a vector field in the rectangular coordinate system $Oxyz$ is

$$\vec{v} = P(x, y, z)\vec{i} + Q(x, y, z)\vec{j} + R(x, y, z)\vec{k}.$$

Let $\vec{r}_1 = \vec{r}(t)$ be the parametric equation of a streamline. If the position vector of a moving point on this curve is

$$\vec{r} = x\vec{i} + y\vec{j} + z\vec{k},$$

then we have

$$\frac{d\vec{r}}{dt} = \frac{dx}{dt}\vec{i} + \frac{dy}{dt}\vec{j} + \frac{dz}{dt}\vec{k}.$$

It can be showed that $\left(\dfrac{dx}{dt}, \dfrac{dy}{dt}, \dfrac{dz}{dt}\right)$ is the direction of the tangent line of the curve $\vec{r}_1 = \vec{r}(t)$ at the point $\vec{r}_1 = (x(t), y(t), z(t))$. Thus, if the tangent vector at each point of the curve are vectors of the field, then we have

$$\frac{d\vec{r}}{dt} = k\vec{v},$$

where k is a constant. That is,

$$\frac{dx}{dt} = kP(x, y, z), \quad \frac{dy}{dt} = kQ(x, y, z), \quad \frac{dz}{dt} = kR(x, y, z),$$

or

$$\frac{dx}{P} = \frac{dy}{Q} = \frac{dz}{R}.$$

Under certain restrictions on P, Q, R, the above system of differential equations has a unique solution for any given point. We omit the proof of this fact here. The solution of this system of differential equations is the family of streamlines. The existence and uniqueness theorem implies that, in a vector field, at every point, there is one and only one streamline that passes the point.

Example 1. Find the streamlines for the vector field

$$\vec{v} = \frac{-y}{\sqrt{x^2 + y^2}}\vec{i} + \frac{x}{\sqrt{x^2 + y^2}}\vec{j}$$

on the Oxy-plane.

Solution: The differential equation of the streamlines is

$$\frac{dx}{-y} = \frac{dy}{x},$$

that is

$$x\,dx = -y\,dy.$$

Thus, the family of streamlines is

$$x^2 + y^2 = a^2.$$

This is a family of circles centered at the origin (Figure 7.6).

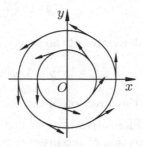

Fig. 7.6

Example 2. Find the streamlines for the vector field $\vec{v} = xy^2\vec{i} + x^2y\vec{j} + y^2z\vec{k}$.

Solution: The system of differential equations of the streamlines is

$$\frac{dx}{xy^2} = \frac{dy}{x^2y} = \frac{dz}{y^2z}.$$

The solution of this system is

$$\begin{cases} x^2 - y^2 = C_1, \\ z = Cx. \end{cases}$$

This is the family of streamlines.

Exercises 7.1

1. Find the contours for the following scalar fields.

   (1) $u = x + y + z$;   (2) $u = x^2 + y^2 + z^2$;

   (3) $u = x^2 + y^2 - z^2$;

   (4) $u = \dfrac{z}{\sqrt{x^2 + y^2}}$;

   (5) $u = \arcsin \dfrac{z}{\sqrt{x^2 + y^2}}$ and passes the point $(1, 1, 1)$;

   (6) $u = f(r)$, $r = \sqrt{x^2 + y^2 + z^2}$, and $f(r)$ is a monotone function of r.

2. Find the contour lines of the following scalar fields.

   (1) $z = \dfrac{y}{x}$;

   (2) $z = x^2 - y^2$;

(3) $z = \dfrac{1}{x^2 + 2y^2}$;

(4) $z = \dfrac{x}{x^2 + y^2}$.

3. Find the directional derivative of the function

$$z = x^3 - 3x^2y + 3xy^2 + 1$$

at the point $(3,1)$ in the direction of the vector from $(3,1)$ to $(6,5)$.

4. Find the directional derivative of the function $z = \arctan xy$ at the point $(1,1)$ in the direction of the bisector line of the first quadrant.

5. Find the directional derivative of the function $u = xy^2 + z^3 - xyz$ at the point $(1,1,2)$ in the direction that makes $60°$, $45°$, $60°$ angles with the three axes respectively.

6. Find the directional derivative of the function $u = xyz$ at the point $(5,1,2)$ in the direction of the vector from $(5,1,2)$ to $(9,4,14)$.

7. Find the directional derivative of the function $z = \ln(e^x + e^y)$ at the point $(0,0)$ and in the direction of the vector $\vec{a}$.

8. Find the directional derivative of the function $z = \ln(x+y)$ at the point $(1,2)$ in the direction of the tangent vector of the parabola $y^2 = 4x$ at the same point $(1,2)$.

9. Find the directional derivative of the function $z = \arctan \dfrac{y}{x}$ at the point $\left(\dfrac{1}{2}, \dfrac{\sqrt{3}}{2}\right)$ in the direction of the counterclockwise tangent vector of the circle $x^2 + y^2 - 2x = 0$ at the same point $\left(\dfrac{1}{2}, \dfrac{\sqrt{3}}{2}\right)$.

10. Find the directional derivative of the function $u = x+y+z$ at a point $M_0(x_0, y_0, z_0)$ on the sphere $x^2 + y^2 + z^2 = 1$ in the direction of the outward normal vector of the sphere.

11. Find the directional derivative of the function $u = x^2 + y^2 + z^2$ at a point $M_0(x_0, y_0, z_0)$ on the ellipsoid

$$\frac{x^2}{a^2} + \frac{y^2}{b^2} + \frac{z^2}{c^2} = 1$$

in the direction of the outward normal vector of the ellipsoid.

12. Find the rate of change of the two dimensional temperature field $T = x^2 + y^2$ at the point $(1,5)$ in directions that have $\dfrac{\pi}{6}$ and $\dfrac{7\pi}{6}$ angles to the Ox-axis.

13. Find the rate of change of the electric potential in the electric potential field $v = \ln \sqrt{x^2 + y^2 + z^2}$ at the point $(1, 2, 4)$ in the direction $(1, 1, 1)$. Also, find the electrical field's strength at this point $(\vec{E} = -\operatorname{\mathbf{grad}} v)$.

14. Suppose the electric potential of an electrostatic field generated by a point electric charge e is $v = \dfrac{e}{r}$ where $r = \sqrt{x^2 + y^2 + z^2}$. Find its equipotential surfaces and gradient. Also, find the electrical field strength.

15. Find the directional derivative of the field
$$u = \frac{x^2}{a^2} + \frac{y^2}{b^2} + \frac{z^2}{c^2}$$
at the point $M(x, y, z)$, and in the direction of the position vector $\vec{r}$ of the point. Under what conditions is the directional directive equal to the gradient?

16. Find a point in the scalar field $z = \ln\left(x + \dfrac{1}{y}\right)$ such that
$$\operatorname{\mathbf{grad}} z = \vec{i} - \frac{16}{9}\vec{j}$$
at this point.

17. Suppose $z = \sqrt{x^2 + y^2}$ and $z = x - 3y + \sqrt{3xy}$ are scalar fields. Find the angle between the gradients of these two fields at the point $(3, 4)$.

18. Find the angle between the gradients of the scalar field
$$u = \frac{x}{x^2 + y^2 + z^2}$$
at the points $A(1, 2, 2)$ and $B(-3, 1, 0)$.

19. Find the directional derivative of the scalar field $u = u(x, y, z)$ in the direction of the gradient of the scalar field $v = v(x, y, z)$. Under what conditions is this directional derivative equal to zero?

20. Find a point in the scalar field $u = \ln \dfrac{1}{r}$ such that the equation $|\operatorname{\mathbf{grad}} u| = 1$ holds (Here $r = \sqrt{(x-a)^2 + (y-b)^2 + (z-c)^2}$).

21. Suppose $\vec{r} = x\vec{i} + y\vec{j} + z\vec{k}$ and $r = |\vec{r}|$. Find the following gradients.

(1) $\operatorname{\mathbf{grad}} r$; (4) $\operatorname{\mathbf{grad}} \ln r$;
(2) $\operatorname{\mathbf{grad}} r^3$; (5) $\operatorname{\mathbf{grad}} f(r)$;
(3) $\operatorname{\mathbf{grad}} \dfrac{1}{r}$; (6) $\operatorname{\mathbf{grad}} f(r^2)$.

22. Prove
$$\operatorname{\mathbf{grad}} f(u, v) = \frac{\partial f}{\partial u} \operatorname{\mathbf{grad}} u + \frac{\partial f}{\partial v} \operatorname{\mathbf{grad}} v.$$

23. Suppose $u = \arctan \dfrac{z}{\sqrt{x^2 + y^2}}$ and $\vec{c} = \vec{i} + \vec{j} + \vec{k}$. Find $\vec{c} \cdot \nabla u$.

24. Find

(1) $(\vec{r} \cdot \nabla) r^n$,   (2) $(\vec{r} \cdot \nabla)\vec{r}$,   (3) $(\vec{a} \cdot \nabla)\vec{c}$,

where $\vec{r} = x\vec{i} + y\vec{j} + z\vec{k}$, $r = |\vec{r}|$, $\vec{a}$ and $\vec{c}$ are constant vectors, and
$\vec{r} \cdot \nabla = x\dfrac{\partial}{\partial x} + y\dfrac{\partial}{\partial y} + z\dfrac{\partial}{\partial z}$ is an operator.

25. Prove the formula $\nabla^2(uv) = u\nabla^2 v + v\nabla^2 u + 2\nabla u \cdot \nabla v$, where

$$\nabla = \vec{i}\frac{\partial}{\partial x} + \vec{j}\frac{\partial}{\partial y} + \vec{k}\frac{\partial}{\partial z},$$

$$\nabla^2 = \nabla \cdot \nabla = \frac{\partial^2}{\partial x^2} + \frac{\partial^2}{\partial y^2} + \frac{\partial^2}{\partial z^2} = \Delta.$$

26. Suppose $\vec{v} = \vec{\omega} \times \vec{r}$, where $\vec{\omega}$ is a constant vector and $\vec{r}$ is the position vector. Find $\mathbf{grad}\left(\dfrac{1}{2}\vec{v} \cdot \vec{v}\right)$.

27. Prove the following equations.
(1) $\mathbf{grad}\,(\alpha u + \beta v) = \alpha\,\mathbf{grad}\,u + \beta\,\mathbf{grad}\,v$, where α and β are constants.
(2) $\mathbf{grad}\,(uv) = u\,\mathbf{grad}\,v + v\,\mathbf{grad}\,u$.
(3) $\mathbf{grad}\,f(u) = f'(u)\,\mathbf{grad}\,u$.

28. Suppose $\vec{r}$ is the position vector, and $\vec{a}$, $\vec{b}$, $\vec{c}$ are constant vectors.
(1) Prove $\mathbf{grad}\,(\vec{a} \cdot \vec{r}) = \vec{a}$;
(2) Compute $\mathbf{grad}\,(\vec{r} \cdot \vec{a})(\vec{r} \cdot \vec{b})$;
(3) Compute $\mathbf{grad}\left\{\vec{c} \cdot \vec{r} + \dfrac{1}{2}\ln(\vec{c} \cdot \vec{r})\right\}$.

29. Find the streamlines of the two dimensional velocity field $\vec{v} = (x + t)\vec{i} + (t - y)\vec{j}$.

30. The electrostatic field $\vec{E} = \dfrac{e}{r^3}\vec{r}$ is generated by a point electric charge located at the origin. Find its electric field lines.

31. The magnetic field

$$\vec{H} = 2I\frac{-y\vec{i} + x\vec{j}}{x^2 + y^2}$$

is generated by an electric current I through a wire (Oz-axis) with infinite length. Find the magnetic field lines.

7.2 Line Integrals

7.2.1 *Line Integrals of the First Kind*

The line integral of the first kind is the line integral of a scalar field. It is a generalization of the definite integral to curves.

Suppose L is a curve with finite length. Partition L with $n + 1$ points $N_0, N_1, \cdots, N_n$ (Figure 7.7). Suppose that the length of the longest arc among all $\Delta s_i = \overset{\frown}{N_{i-1}N_i}$, $(i = 1, 2, \cdots, n)$, tends to zero when n tends to infinity.

Let M_i be a point on the arc $N_{i-1}N_i$, $(i = 1, 2, \cdots, n)$, and $f(M)$ be a function that is defined on L. If the limit

$$\lim_{n \to \infty} \sum_{i=1}^{n} f(M_i)\Delta s_i$$

exists and is independent from the choices of all N_i, and M_i, then the limit is called the *line integral of the first kind* of $f(M)$ on the curve L and is denoted as

$$\int_L f(M)\, ds = \lim_{n \to \infty} \sum_{i=1}^{n} f(M_i)\Delta s_i.$$

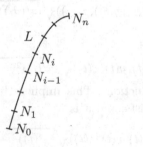

Fig. 7.7

Now we study the calculations of the line integral of the first kind.

In the rectangular coordinate system $Oxyz$, let the parametric equation of the curve L be

$$x = x(t), \quad y = y(t), \quad z = z(t), \quad \alpha \leqslant t \leqslant \beta,$$

and assume their the derivatives

$$\frac{dx}{dt} = x'(t), \quad \frac{dy}{dt} = y'(t), \quad \frac{dz}{dt} = z'(t)$$

are continuous. Also assume that L does not cross itself.

If a function $f(M) = f(x, y, z)$ is continuous on L, then the line integral of the first kind of $f(M)$ on L exists and

$$\int_L f(M)\,ds = \int_\alpha^\beta f(x(t), y(t), z(t))\sqrt{(x'(t))^2 + (y'(t))^2 + (z'(t))^2}\,dt.$$

In fact, partition L by $\alpha = t_0 < t_1 < \cdots < t_n = \beta$, and assume

$$\max_{1 \leqslant i \leqslant n} \Delta t_i \to 0$$

as $n \to \infty$.

Let $N_i = (x(t_i), y(t_i), z(t_i))$, $M_i = (x(\xi_i), y(\xi_i), z(\xi_i))$, where $t_{i-1} \leqslant \xi_i \leqslant t_i$, $i = 1, 2, \cdots, n$. Then by the formula of the length of arc and the Mean-Value Theorem of definite integrals, the length of the arc $\overset{\frown}{N_{i-1}N_i}$ is

$$\Delta s_i = \int_{t_{i-1}}^{t_i} \sqrt{(x'(t))^2 + (y'(t))^2 + (z'(t))^2}\,dt$$

$$= \sqrt{(x'(\theta_i))^2 + (y'(\theta_i))^2 + (z'(\theta_i))^2}\,\Delta t_i,$$

where $t_{i-1} \leqslant \theta_i \leqslant t_i$, $i = 1, 2, \cdots, n$. Thus,

$$\sum_{i=1}^n f(M_i)\Delta s_i$$

$$= \sum_{i=1}^n f(x(\xi_i), y(\xi_i), z(\xi_i))\sqrt{(x'(\theta_i))^2 + (y'(\theta_i))^2 + (z'(\theta_i))^2}\,\Delta t_i.$$

By the assumptions,

$$f(x(t), y(t), z(t))\sqrt{(x'(t))^2 + (y'(t))^2 + (z'(t))^2}$$

is a continuous function. This implies the limit of the right side of the above equation exists, and it is

$$\int_\alpha^\beta f(x(t), y(t), z(t))\sqrt{(x'(t))^2 + (y'(t))^2 + (z'(t))^2}\,dt.$$

Therefore, the limit of the left side of the equation also exists, and by the definition, it is the line integral of the first kind of $f(M)$ on L. Now we have

$$\int_L f(M)\,ds = \int_\alpha^\beta f(x(t), y(t), z(t))\sqrt{(x'(t))^2 + (y'(t))^2 + (z'(t))^2}\,dt.$$

It follows that, if the equation of the planar curve L is $y = y(x)$, where $a \leqslant x \leqslant b$, and $y'(x)$ is continuous, then

$$\int_L f(M)\,ds = \int_L f(x, y)\,ds = \int_a^b f(x, y(x))\sqrt{1 + (y'(x))^2}\,dx.$$

If the equation of a planar curve in polar coordinates is $r = r(\theta)$, where $\alpha \leqslant \theta \leqslant \beta$, and $r'(\theta)$ is continuous, then

$$\int_L f(M) \, ds = \int_L f(x, y) \, ds$$

$$= \int_\alpha^\beta f(r(\theta) \cos \theta, r(\theta) \sin \theta) \sqrt{(r(\theta))^2 + (r'(\theta))^2} \, d\theta.$$

Example: Evaluate

$$\int_L xy \, ds,$$

where L is the part of the ellipse $\dfrac{x^2}{a^2} + \dfrac{y^2}{b^2} = 1$ in the first quadrant.

Solution 1: Since the equation of L is

$$y = \frac{b}{a}\sqrt{a^2 - x^2}, \quad 0 \leqslant x \leqslant a,$$

we have

$$ds = \sqrt{1 + (y')^2} \, dx = \frac{1}{a}\sqrt{\frac{a^4 - (a^2 - b^2)}{a^2 - x^2}} \, dx.$$

Therefore,

$$\int_L xy \, ds = \int_0^a x \frac{b}{a}\sqrt{a^2 - x^2} \frac{1}{a}\sqrt{\frac{a^4 - (a^2 - b^2)}{a^2 - x^2}} \, dx$$

$$= \frac{b}{a^2}\int_0^a \sqrt{a^4 - (a^2 - b^2)x^2} \, x \, dx$$

$$= \frac{ab}{3} \frac{a^2 + ab + b^2}{a + b}.$$

Solution 2: The equation of L can also be written as

$$x = a \cos t, \quad y = b \sin t, \quad 0 \leqslant t \leqslant \frac{\pi}{2}.$$

So

$$ds = \sqrt{(x')^2 + (y')^2} \, dt = \sqrt{a^2 \sin^2 t + b^2 \cos^2 t} \, dt.$$

Thus,

$$\int_L xy \, ds = \int_0^{\frac{\pi}{2}} a \cos t \cdot b \sin t \sqrt{a^2 \sin^2 t + b^2 \cos^2 t} \, dt$$

$$= \frac{ab}{2}\int_\pi^2 \sin 2t \sqrt{a^2 \frac{1 - \cos 2t}{2} + b^2 \frac{1 + \cos 2t}{2}} \, dt.$$

Let $u = \cos 2t$, then

$$\sin 2t\, dt = -\frac{1}{2}\, du.$$

Therefore,

$$\int_L xy\, ds = \frac{ab}{4} \int_{-1}^{1} \sqrt{\frac{a^2 + b^2}{2} + \frac{b^2 - a^2}{2} u}\, du$$

$$= \frac{ab}{4} \cdot \frac{2}{b^2 - a^2} \cdot \frac{2}{3} \left[\frac{a^2 + b^2}{2} + \frac{b^2 - a^2}{2} u \right]^{\frac{3}{2}} \Bigg|_{-1}^{1}$$

$$= \frac{ab}{3} \cdot \frac{a^2 + ab + b^2}{a + b}.$$

7.2.2   Applications of Line Integrals of the First Kind (Areas of Surfaces of Revolution)

In order to simplify the definition of the surface of revolution, we suppose L is a curve with finite length on the xy-plane. The surface generated by revolving L about one coordinate axis is called a *surface of revolution*.

Suppose L is above the x-axis (Figure 7.8). Revolving L about the x-axis. We partition L by $n + 1$ points $N_0, N_1, \cdots, N_n$, and assume that the length of the longest arc among the lengths of all arcs $\Delta s_i = \overgroup{N_{i-1}N_i}$, $(i = 1, 2, \cdots, n)$, tends to zero as n tends to infinity.

Consider the arc $\overgroup{N_{i-1}N_i}$ as the line segment $\overline{N_{i-1}N_i}$. The area of the surface generated by revolving $\overline{N_{i-1}N_i}$ about x-axis is

$$\pi(y_{i-1} + y_i)\Delta s_i,$$

where y_{i-1} and y_i are the vertical coordinates of N_{i-1} and N_i. Thus, the total area of the surface of revolution generated by L is

$$\lim_{n \to \infty} \pi \sum_{i=1}^{n} (y_{i-1} + y_i)\Delta s_i.$$

By the definition of the line integral of the first kind, we have

$$\lim_{n \to \infty} \sum_{i=1}^{n} y_{i-1}\Delta s_i = \lim_{n \to \infty} \sum_{i=1}^{n} y_i \Delta s_i = \int_L y\, ds.$$

Therefore, the area of the surface of revolution generated by revolving L about x-axis is (Figure 7.9)

$$2\pi \int_L y\, ds.$$

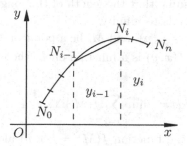

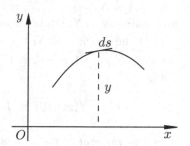

Fig. 7.8   Fig. 7.9

In general, for L in any position of xy-plane, if it is a single-valued function, then the area of the surface of revolution generated by revolving L about x-axis is

$$S = 2\pi \int_L |y|\, ds.$$

Example: Find the area of the surface of revolution generated by revolving the cycloid

$$x = a(t - \sin t), \quad y = a(1 - \cos t), \quad 0 \leqslant t \leqslant 2\pi$$

about x-axis.

Solution:

$$ds = \sqrt{(x')^2 + (y')^2}\, dt = a\sqrt{2 - 2\cos t}\, dt = 2a \sin \frac{t}{2}\, dt.$$

$$S = 2\pi \int_L y\, ds = 2\pi \int_0^{2\pi} a(1 - \cos t)2a \sin \frac{t}{2}\, dt$$

$$= 2\pi \left(4a^2 \int_0^{2\pi} \sin^3 \frac{t}{2}\, dt \right) = 16\pi a^2 \int_{-1}^1 (1 - u^2)\, du = \frac{64}{3}\pi a^2.$$

7.2.3   Line Integrals of the Second Kind

In the rectangular coordinate system, this is the integral of the projection of a curve on the coordinate axis.

Suppose L is a curve on xy-plane and it joins the points A and B (Figure 7.10). We divide up the arc into small pieces by means of the points $A = N_0, N_1, \cdots, N_n = B$ and assume that the length of the longest arc among $\overset{\frown}{N_{i-1}N_i}$ tends to zero as n tends to infinity.

Let the abscissa of N_i be x_i $(i = 1, 2, \cdots, n)$, and M_i be a point on the arc $\overset{\frown}{N_{i-1}N_i}$, $i = 1, 2, \cdots, n$. If $f(M) = f(x, y)$ is a function on L, then we define

$$\int_{L_{AB}} f(x, y) \, dx = \int_{L_{AB}} f(M) \, dx = \lim_{n \to \infty} \sum_{i=1}^{n} f(M_i) \Delta x_i$$

as a *line integral of the second kind* of the function $f(M) = f(x, y)$ along L with respect to x, under the assumption that the limit exists and is independent of the choices of the points of division N_i and the choice of each point M_i. The projection of the arc $\overset{\frown}{N_{i-1}N_i}$ on the x-axis is $\Delta x_i = x_i - x_{i-1}$.

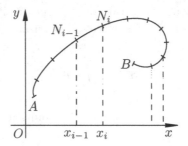

Fig. 7.10

Similarly, we can define the integral

$$\int_{L_{AB}} f(x, y) \, dy = \int_{L_{AB}} f(M) \, dy.$$

If L is a curve in the rectangular coordinate system $Oxyz$, then we can also define the integral

$$\int_{L_{AB}} f(x, y, z) \, dz = \int_{L_{AB}} f(M) \, dz.$$

Now we study some properties of the line integrals of the second kind:

(1)

$$\int_{L_{BA}} f(M) \, dx = -\int_{L_{AB}} f(M) \, dx.$$

In fact, when we choose the points of division from A to B, $A = N_0, N_1, \cdots, N_n = B$, the abscissae of each point are $x_0, x_1, \cdots, x_n$. Now we use the same set of points of division backwards, and we get a set of points of division from B to A, $B = \widetilde{N}_0, \widetilde{N}_1 = N_{n-1}, \cdots, \widetilde{N}_n = A$. The abscissae of $\widetilde{N}_0, \widetilde{N}_1, \cdots, \widetilde{N}_n$ are $x_n, x_{n-1}, \cdots, x_0$ (Figure 7.11). By the definition above, we have

$$\int_{L_{BA}} f(M)\, dx = \lim[f(M_n)(x_{n-1} - x_n) + f(M_{n-1})(x_{n-2} - x_{n-1})$$

$$+ \cdots + f(M_1)(x_0 - x_1)]$$

$$= -\lim[f(M_1)(x_1 - x_0) + \cdots + f(M_{n-1})(x_{n-1} - x_{n-2})$$

$$+ f(M_n)(x_n - x_{n-1})]$$

$$= -\int_{L_{AB}} f(M)\, dx.$$

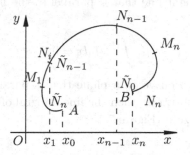

Fig. 7.11

This implies that the line integrals of the second kind has directivity. We have the following properties:

(2)

$$\int_{L_{AB}} dx = x_B - x_A. \quad \text{(Figure 7.12)}$$

(3)

$$\int_{L_{AB}} (af + bg)\, dx = a\int_{L_{AB}} f\, dx + \int_{L_{AB}} g\, dx,$$

where a and b are constants.

(4)

$$\int_{L_{AB}} f\, dx + \int_{L_{BC}} f\, dx = \int_{L_{AC}} f\, dx. \quad \text{(Figure 7.13)}$$

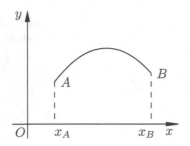

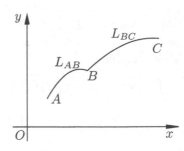

Fig. 7.12   Fig. 7.13

(5) If L is a straight line that is parallel to the y-axis on the xy-plane (Figure 7.14), then

$$\int_{L_{AB}} f\, dx = 0.$$

Similarly, if L is a curve on a plane that is parallel to the yz-plane in the coordinate system $Oxyz$, then the line integral of the second kind on L with respect to x is zero (Figure 7.15).

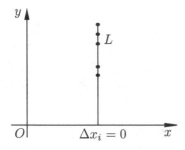

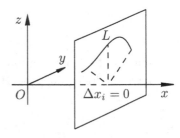

Fig. 7.14   Fig. 7.15

(6)

$$\left| \int_L f(M)\, dx \right| \le \int_L |f(M)|\, ds.$$

Especially, if $|f(M)| \le K$ (K is a constant), then the following inequality holds

$$\left| \int_L f(M)dx \right| \leqslant KL,$$

where the L on the right side of this inequality represent the total length of the curve L.

Indeed, as in Figure 7.16, since

$$\left| \int_L f(M)\, dx \right| = |\lim \sum f(M_i)\Delta x_i|$$

$$\leqslant \lim \sum |f(M_i)||\Delta x_i| \leqslant \lim \sum |f(M_i)|\Delta s_i$$

$$= \int_L |f(M)|\, ds \leqslant KL,$$

the conclusion follows.

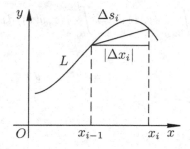

Fig. 7.16

7.2.4 *Calculation of Line Integrals of the Second Kind*

Suppose the parametric equations of a curve L in the rectangular coordinate system $Oxyz$ is

$$x = x(t), \quad y = y(t), \quad z = z(t)$$

where $\alpha \leqslant t \leqslant \beta$. Also, assume that $x'(t)$, $y'(t)$, $z'(t)$ are continuous. Let the coordinates of the two end points of L, A and B, be $A(x(\alpha), y(\alpha), z(\alpha))$ and $B(x(\beta), y(\beta), z(\beta))$ respectively. If $f(M) = f(x, y, z)$ is a continuous function defined on L, then the line integral of the second kind of $f(M)$ exists and

$$\int_{L_{AB}} f(M)\, dx = \int_\alpha^\beta f(x(t), y(t), z(t))x'(t)\, dt.$$

The line integral of the second kind of f with respect to y and z can be defined similarly.

When a particle moves along L, the coordinates of its position change. If the particle moves in a direction of parameter increases, then we define this direction as the positive direction of the curve.

By this definition, the above equation can be written as

$$\int_{L_+} f(M)\,dx = \int_\alpha^\beta f(x(t), y(t), z(t))x'(t)\,dt.$$

The proof of this formula is similar to the proof of the formula of line integrals of the first kind in Section 7.2.1, and we omit it here. We have the following corollaries:

If L is a curve $y = y(x)$ with $a \leqslant x \leqslant b$ on the xy-plane (Figure 7.17), then

$$\int_{L_{AB}} f(M)\,dx = \int_a^b f(x, y(x))\,dx.$$

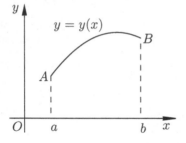

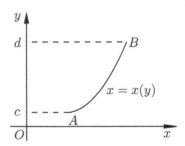

Fig. 7.17   Fig. 7.18

If L is a curve $x = x(y)$ with $c \leqslant y \leqslant d$ on xy-plane (Figure 7.18), then

$$\int_{L_{AB}} f(M)\,dy = \int_c^d f(x(y), y)\,dy.$$

Example: Evaluate

$$\int_L xy\,dx,$$

where L is the curve in Figure 7.19. Also evaluate the same integral for the curves Δ and C in Figure 7.19.

Solution:

$$\int_{LOA} xy\, dx = \int_0^1 x^2\, dx = \frac{1}{3};$$

$$\int_{\Delta OA} xy\, dx = \int_0^1 x\, x^2\, dx = \int_0^1 x^3\, dx = \frac{1}{4};$$

$$\int_{COA} xy\, dx = \int_{COB} xy\, dx + \int_{CBA} xy\, dx = \int_0^1 x0\, dx + 0 = 0.$$

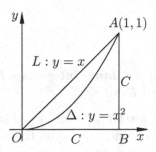

Fig. 7.19

If L is a closed curve, that is, its initial point A and its end point B coincide, then there are two directions that we can follow when we calculate line integrals of a function along the curve: Clockwise (Figure 7.20) or counter-clockwise (Figure 7.21). The counter-clockwise direction is called the positive direction of a closed curve.

The notations for line integrals of the second kind along these two directions are

$$\oint \quad \text{and} \quad \oint .$$

For example, if L is the circle

$$x = \cos t, \quad y = \sin t, \quad 0 \leqslant t \leqslant 2\pi,$$

on the xy-plane, then we have

$$\oint_L y^2\, dx = \int_0^{2\pi} \sin^2 t(-\sin t)\, dt = -\int_0^{2\pi} \sin^3 t\, dt = 0;$$

$$\oint_L x\, dy = \int_0^{2\pi} \cos t\, d\sin t = \int_0^{2\pi} \cos^2 t\, dt = \pi.$$

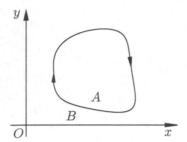
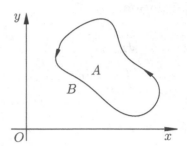

Fig. 7.20   Fig. 7.21

7.2.5   Relation Between Line Integrals of the First Kind and the Second Kind

Suppose L is a curve with starting point A and end point B in the three-dimensional coordinate system. Let s_0 be the length of L, and M be a moving point on L with coordinates (x, y, z) (Figure 7.22). If the length of the arc $\overparen{AM}$ is s, then x, y, z are functions of s:

$$x = x(s), \quad y = y(s), \quad z = z(s), \quad 0 \leqslant s \leqslant s_0.$$

Assume that $x'(s)$, $y'(s)$, $z'(s)$ are continuous in the interval $0 \leqslant s \leqslant s_0$, and $\cos\alpha_M$, $\cos\beta_M$, $\cos\gamma_M$ are direction cosines of unit tangent vector at point M that points towards the direction of point B (Figure 7.22). By the calculation formulas that we studied before, we have

$$\int_{L_{AB}} f(M)\,dx = \int_0^s f(x(s), y(s), z(s))x'(s)\,ds = \int_L f(M)\cos\alpha_M\,ds.$$

The integrals with respect to y and z can be obtained similarly. Therefore, the relations between the two types of line integrals are:

$$\int_{L_{AB}} f(M)dx = \int_L f(M)\cos\alpha_M\,ds;$$

$$\int_{L_{AB}} f(M)dy = \int_L f(M)\cos\beta_M\,ds;$$

$$\int_{L_{AB}} f(M)dz = \int_L f(M)\cos\gamma_M\,ds,$$

where $\alpha_M, \beta_M, \gamma_M$ are the direction angles of unit tangent vectors at point M that points towards the direction of point B.

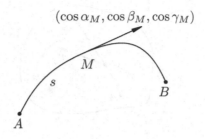

$(\cos \alpha_M, \cos \beta_M, \cos \gamma_M)$

M

s

B

A

Fig. 7.22

7.2.6 Circulations of Vector Fields and Line Integrals of Vectors

In this section, we are going to explain the practical meaning of line integrals of the two kinds in field theory.

Suppose $\vec{v} = \vec{v}(M)$ is a vector field in a spatial domain V, L is a curve in V that connects points A and B, and $\vec{t} = \vec{t}(M)$ is the unit tangent vector of L that points towards the direction of point B (Figure 7.23). The line integral

$$\int_L \vec{v} \cdot \vec{t}\, ds$$

is called the *circulation* of $\vec{v}$ along the curve L, and from point A to point B.

If the vector field is a force field, then the circulation along the curve L from A to B is the work done by the force field along L from A to B. If the vector field is an electric field, then the circulation along the curve L from A to B is the work done by the electric field in moving a unit charge along L from A to B.

To calculate the circulation of a vector field, consider a rectangular coordinate system $Oxyz$ in space, with base vectors $\vec{i}, \vec{j}, \vec{k}$ on each axes respectively. Vectors $\vec{v}$ and $\vec{t}$ can be written as:

$$\vec{v} = \vec{v}(M) = P(M)\vec{i} + Q(M)\vec{j} + R(M)\vec{k},$$
$$\vec{t} = \vec{t}(M) = \cos\alpha_M\vec{i} + \cos\beta_M\vec{j} + \cos\gamma_M\vec{k}.$$

By the formulas in Section 7.2.5, the circulation of the vector field $\vec{v}$ along

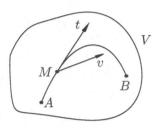

Fig. 7.23

L from A to B is

$$\int_L \vec{v} \cdot \vec{t}\,ds = \int_L (P\cos\alpha + Q\cos\beta + R\cos\gamma)\,ds$$

$$= \int_{L_{AB}} P\,dx + Q\,dy + R\,dz,$$

where

$$\int_{L_{AB}} P\,dx + Q\,dy + R\,dz = \int_{L_{AB}} P\,dx + \int_{L_{AB}} Q\,dy + \int_{L_{AB}} R\,dz.$$

Thus, the circulation can be calculated by finding values of three line integrals of the second kind.

Now we study the line integral of vectors. Suppose L is a curve with finite length that connects points A and B. The vector

$$\vec{v} = \vec{v}(M) = P(M)\vec{i} + Q(M)\vec{j} + R(M)\vec{k}$$

is defined on L. The definitions of N_i, M_i are the same as in Section 7.2.3 (Figure 7.24). The integral

$$\int_{L_{AB}} \vec{v} \cdot d\vec{r} = \lim \sum_{i=1}^{n} \vec{v}(M_i) \cdot \Delta \vec{r}_i$$

is called the line integral of the vector v along L from A to B, if the limit of the right side exits and is finite regardless of the choice of N_i and M_i.

For example, if $\vec{F}$ is a force field, then in mechanics, $\vec{F} \cdot d\vec{r}$ is called the work element.

The integral

$$\int_{L_{AB}} \vec{F} \cdot d\vec{r}$$

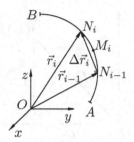

Fig. 7.24

is the work done by the force $\vec{F}$ along L from A to B.

Since

$$\Delta\vec{r}_i = \Delta x_i\vec{i} + \Delta y_i\vec{j} + \Delta z_i\vec{k},$$

by definition we have

$$\int_{L_{AB}} \vec{v} \cdot d\vec{r} = \int_{L_{AB}} P\,dx + Q\,dy + R\,dz.$$

Therefore, the circulation of a vector field can be expressed by a line integral of the vector.

For instance, if the vector field is

$$\vec{v} = (y^2 - x^2)\vec{i} + 2xy\vec{j} - x^2\vec{k},$$

and the parametric equation of the curve L is

$$x = t, \quad y = t^2, \quad z = t^3, \quad 0 \leqslant t \leqslant 1,$$

then

$$\int_{L_+} \vec{v} \cdot d\vec{r} = \int_{L_+} (y^2 - x^2)\,dx + 2xy\,dy - x^2\,dz$$

$$= \int_0^1 [(t^4 - t^2) + 2t^3\,2t - t^2\,3t^2]\,dt = \frac{1}{15}.$$

If we consider another vector field

$$\vec{v} = 2xy\vec{i} + x^2\vec{j},$$

and the curve L as in Figure 7.25, then

$$\oint_L \vec{v} \cdot d\vec{r} = \oint_L 2xy\,dx + x^2\,dy = \left(\int_{OB} + \int_{BA} + \int_{AO}\right) 2xy\,dx + x^2\,dy.$$

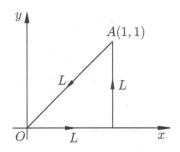

Fig. 7.25

Since

$$\int_{\overline{OB}} = 0, \quad \int_{\overline{BA}} = \int_0^1 dy = 1,$$

$$\int_{\overline{AO}} = \int_1^0 3x^2 \, dx = -1,$$

we have

$$\oint \vec{v} \cdot d\vec{r} = 0.$$

Exercises 7.2

1. Find $\int_L (x+y) \, ds$, where L is the boundary of the triangle with vertices $O(0,0)$, $A(1,0)$ and $B(0,1)$.

2. Find $\int_L y^2 \, ds$, where L is a cycloid defined by $x = a(t - \sin t)$, $y = a(1 - \cos t)$, $(0 \leqslant t \leqslant 2\pi)$.

3. Find $\int_L (x^2 + y^2) \, ds$, where L is the curve $x = a(\cos t + t \sin t)$, $y = a(\sin t - t \cos t)$, $(0 \leqslant t \leqslant 2\pi)$.

4. Find $\int_L xy \, ds$, where L is the circle $x^2 + y^2 = a^2$.

5. Find $\int_L \sqrt{x^2 + y^2} \, ds$, where L is the circle $x^2 + y^2 = ax$ $(a > 0)$.

6. Find $\int_L xy \, ds$, where L is an arc of the hyperbola $x = a \cosh t$, $y = a \sinh t$, $(0 \leqslant t \leqslant t_0)$.

7. Find $\int_L \dfrac{z^2}{x^2 + y^2}\, ds$, where L is a spiral defined by $x = a\cos t$, $y = a\sin t$, $z = at$, $(0 \leqslant t \leqslant 2\pi)$.

8. Find $\int_L z\, ds$, where L is a curve defined by the equations $x^2 + y^2 = z^2$, $y^2 = ax$ from $(0,0,0)$ to $(a, a, \sqrt{2}a)$, $(a > 0)$.

9. Find $\int_L (x^{\frac{4}{3}} + y^{\frac{4}{3}})\, ds$, where L is a hypocycloid in the first quadrant $x = a\cos^3 t$, $y = a\sin^3 t$, $(0 \leqslant t \leqslant \dfrac{\pi}{2})$.

10. Find $\int_L e^{\sqrt{x^2+y^2}}\, ds$, where L is the boundary of the region bounded by $x^2 + y^2 = a^2$, $y = x$ and the x-axis in the first quadrant.

11. Find $\int_L (x^2 + y^2)^n\, ds$, where L is the curve $x = a\cos t$, $y = a\sin t$, $(0 \leqslant t \leqslant 2\pi)$.

12. Find $\int_L y\, ds$, where L is the part of parabola $y^2 = 2px$ from $O(0,0)$ to $A(x_0, y_0)$.

13. Find $\int_L x^2\, ds$, where L is defined by equations
$$x^2 + y^2 + z^2 = a^2, \quad x + y + z = 0.$$

14. Find the length of the arc of the curve $x = 3t$, $y = 3t^2$, $z = 2t^3$ from $O(0,0,0)$ to $A(3,3,2)$.

15. Find the area of the surface generated by revolving the semicircle
$$x^2 + y^2 + R^2, \quad (x \geqslant 0),$$
about the y-axis.

16. Find $\int_L (x^2 - y^2)\, dx$, where L is an arc from $O(0,0)$ to $A(2,4)$ on the curve defined by the equation $y = x^2$.

17. Find $\int_L x\, dy$, where L is the boundary of the triangle bounded by the x-axis, the y-axis and
$$\frac{x}{2} + \frac{y}{3} = 1$$
in the counterclockwise direction.

18. Find $\int_L (x^2 + y^2)\, dx$, where L is the boundary of the rectangle bounded by $x = 1, y = 1, x = 3, y = 5$ in the counterclockwise direction.

19. Find $\int_L xy\,dx$, where L is an arc from $A(1,-1)$ to $B(1,1)$ on the curve defined by the equation $y^2 = x$.

20. Find $\int_L (x^2 - 2xy)\,dx + (y^2 - 2xy)\,dy$, where L is an arc defended by

$$y = x^2, \quad -1 \leqslant x \leqslant 1.$$

21. Find $\int_{(0,0)}^{(1,1)} 2xy\,dx + x^2\,dy$ along curves:

   (1) $y = x$,   (2) $y = x^2$,   (3) $y = x^3$,   (4) $y^2 = x$.

22. Find $\int_{(0,0)}^{(1,1)} x\,dy - y\,dx$ along curves:

   (1) $y = x$,   (2) $y = x^2$,

   (3) Line segments $y = 0, 0 \leqslant x \leqslant 1$ and $x = 1, 0 \leqslant y \leqslant 1$.

23. Find $\int_L y\,dx + x\,dy$, where L is an arc from $t_1 = 0$ to $t_2 = \dfrac{\pi}{2}$ on the curve defined by the parametric equations $x = R\cos t, y = R\sin t$.

24. Find $\int_L (x + y)\,dx + (x - y)\,dy$, where L is the ellipse

$$\frac{x^2}{a^2} + \frac{y^2}{b^2} = 1$$

in the counterclockwise direction.

25. Find $\int_L (x^2 - 2xy)\,dx + (y^2 - 2xy)\,dy$, where L is the boundary of the rectangle with vertices

$$M_1(0,-1), \quad M_2(2,-1), \quad M_3(2,2), \quad M_4(0,2)$$

in the counterclockwise direction.

26. Find $\int_L (y^2 - z^2)\,dx + 2yz\,dy - x^2\,dz$, where L is an arc from $t = 0$ to $t = 1$ on the curve defined by the parametric equations $x = t, y = t^2, z = t^3$.

27. Find $\int_L x\,dx + y\,dy + (x + y - 1)dz$, where L is a line segment from $A(1,1,1)$ to $B(2,3,4)$.

28. Find $\int_L \dfrac{y\,dx - x\,dy}{x^2 + y^2}$, where L is the circle $x^2 + y^2 = a^2, a > 0$, in the counterclockwise direction.

29. Find the work done by the force field $\vec{F} = 2xy\vec{i} - x^2\vec{j}$ along the curve that connects the two points $O(0,0)$ and $A(2,1)$ in the following different ways:

    (1) A straight line.

    (2) A parabola with x-axis as its axis of symmetry.

    (3) A parabola with y-axis as its axis of symmetry.

    (4) A broken line with a segment that connects $O(0,0)$ and $B(2,0)$, and another segment that connects $B(2,0)$ and $A(2,1)$.

    (5) A broken line with a segment that connects $O(0,0)$ and $C(0,1)$, and another segment that connects $C(0,1)$ and $A(2,1)$.

30. Suppose A is a rectangle with sides parallel to the coordinate axis, the side parallel to the y-axis has length a, and the other side has length b. Assume that a fluid flows in the positive direction of x-axis. The velocity at every point is directly proportional to the distance from this point to the x-axis with the proportionality coefficient m. Find the circulation K of this velocity vector field $\vec{v}$ on the rectangle A along the clockwise direction.

31. Find the circulation of the magnetic field $\vec{H}$ generated by a constant current passing through a straight wire (Oz-axis) of infinite length, along the curve $x^2 + y^2 = R^2$ on Oxy-plane in the positive direction.

32. Find the work done by the gravitational field $\vec{F} = -k\dfrac{\vec{r}}{r^3}$ that moves a particle with unit mass from $\vec{r_1}$ to $\vec{r_2}$. Also, find the circulation of this field along any closed curve.

7.3 Surface Integrals

7.3.1 *Surface Integrals of the First Kind*

The concept of the surface integral of the first kind is comparable to that of the line integral of the first kind.

Suppose Σ is a surface with finite area (Figure 7.26).

Let $\Sigma_1, \Sigma_2, \cdots, \Sigma_n$ be all the small sections of a partition of Σ. Assume that the largest area among all areas ΔS_i of Σ_i tends to zero when n tends to infinity. Assume also that the longest diameter among all diameters of Σ_i tends to zero when n tends to infinity. Let M_i be a point in Σ_i,

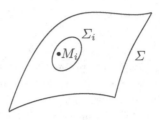

Fig. 7.26

$(i = 1, 2, \cdots, n)$. If $f(M)$ is a function defined on Σ, then we define

$$\iint_{\Sigma} f(M)\, dS = \lim_{n\to\infty} \sum_{i=1}^{n} f(M_i)\Delta S_i$$

as the *surface integral of the first kind* of $f(M)$ on the surface Σ. Of course, we assume that this limit is a finite number, and it is independent of the partitions and the choice of each M_i.

The calculation of the surface integrals of the first kind is also comparable to that of the line integrals of the first kind.

Suppose the parametric equations of a smooth surface Σ are

$$x = x(u,v), \quad y = y(u,v), \quad z = z(u,v), \quad (u,v) \in \Delta,$$

where Δ is a bounded closed region on the uv-plane. If a function $f(M) = f(x,y,z)$ is continuous on Σ, then the surface integral of the first kind of $f(M)$ on Σ exists, and the surface element is $dS = \sqrt{EG - F^2}\, du\, dv$. Thus

$$\iint_{\Sigma} f(M)\, dS = \iint_{\Delta} f(x(u,v), y(u,v), z(u,v))\sqrt{EG - F^2}\, du\, dv.$$

We omit the proof of this formula due to its similarity to the proof of the corresponding formula of the line integral of the first kind (refer to Chapter 6 Section 6.3.2). We also have the following conclusion: Suppose an explicit expression of a surface is $z = z(x,y)$ for $(x,y) \in D$, where D is a closed region on the xy-plane. If z'_x, z'_y are continuous, then

$$\iint_{\Sigma} f(M)\, dS = \iint_{D} f(x, y, z(x,y))\sqrt{1 + (z'_x)^2 + (z'_y)^2}\, dx\, dy.$$

Example 1. Let Σ be a spherical surface $x^2 + y^2 + z^2 = R^2$ ($x \geqslant 0, y \geqslant 0, z \geqslant 0$). Evaluate the surface integral

$$\iint_{\Sigma} (x^2 + y^2)\, dS.$$

Solution: The parametric equations of the surface are

$$x = R\cos\varphi\sin\theta, \ \ y = R\sin\varphi\sin\theta, \ \ z = R\cos\theta$$

where $0 \leqslant \varphi \leqslant \dfrac{\pi}{2}, 0 \leqslant \theta \leqslant \dfrac{\pi}{2}$.

The surface element of this spherical surface is

$$dS = R^2 \sin\theta \, d\varphi \, d\theta.$$

Thus,

$$\iint\limits_{\Sigma} (x^2 + y^2) \, dS = \iint\limits_{\substack{0\leqslant\varphi\leqslant\frac{\pi}{2}\\ 0\leqslant\theta\leqslant\frac{\pi}{2}}} R^2 \sin^2\theta R^2 \sin\theta d\varphi d\theta$$

$$= \frac{\pi}{2}R^4 \int_0^{\frac{\pi}{2}} \sin^3\theta \, d\theta = \frac{1}{3}\pi R^4.$$

Example 2. Let Σ be a part of the conical surface $z^2 = k^2(x^2+y^2) \ (z \geqslant 0)$ sectioned by the cylinder $x^2 + y^2 \leqslant 2ax \ (a > 0)$ (Figure 7.27). Evaluate the surface integral

$$\iint\limits_{\Sigma} (y^2z^2 + z^2x^2 + x^2y^2) \, dS.$$

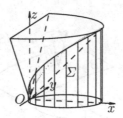

Fig. 7.27

Solution: Since

$$dS = \sqrt{1 + (z_x')^2 + (z_y')^2} \, dx \, dy = \sqrt{1 + k^2} \, dx \, dy,$$

we have

$$\iint_\Sigma (y^2 z^2 + z^2 x^2 + x^2 y^2)\, dS$$

$$= \sqrt{1+k^2} \iint_{x^2+y^2 \leqslant 2ax} [k^2(x^2+y^2)^2 + x^2 y^2]\, dx\, dy$$

$$= 2\sqrt{1+k^2} \iint_{\substack{x^2+y^2 \leqslant 2ax \\ y \geqslant 0}} [k^2(x^2+y^2)^2 + x^2 y^2]\, dx\, dy$$

$$= 2\sqrt{1+k^2} \int_0^{\frac{\pi}{2}} d\varphi \int_0^{2\cos\varphi} r^4(k^2 + \cos^2\varphi \sin^2\varphi) r\, dr$$

$$= \frac{\pi}{24} a^6 (80k^2 + 7)\sqrt{1+k^2}.$$

7.3.2   *Flux of Vector Fields, Surface Integrals of the Second Kind (Integral with respect to the projections of the area element)*

The flux of a vector field is a comparable concept to the circulation.

Suppose Σ is a surface in a spatial domain V. There are two sides of Σ, side A and side B (Figure 7.28).

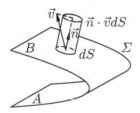

Fig. 7.28

For every point M on Σ, let $\vec{n} = \vec{n}_M$ be the normal unit vector of Σ, with the initial point M and points away from the side B.

Suppose $\vec{v} = \vec{v}(M)$ is a vector field in V. Then

$$\iint\limits_{\Sigma} \vec{v} \cdot \vec{n} \, dS$$

is the *flux* of the vector field $\vec{v}$ from the side A to the side B of Σ, or simply the flux on side B.

In physics, if $\vec{v}$ is a fluid (density $= 1$) velocity field, then the flux of the field is the amount of fluid that flows through the surface in a unit time interval. If $\vec{v}$ is an electric field, then the flux of this field is the amount of the electricity that flows through the surface in a unit time interval.

Suppose the base vectors of a rectangular coordinate system $Oxyz$ are $\vec{i}, \vec{j}, \vec{k}$. Then a vector field $\vec{v}$ can be decomposed as

$$\vec{v} = \vec{v}(M) = P(M)\vec{i} + Q(M)\vec{j} + R(M)\vec{k}.$$

Also, the normal unit vector $\vec{n}$ (that points away from side B) can be decomposed as

$$\vec{n} = \vec{n}(M) = \cos\alpha_M \vec{i} + \cos\beta_M \vec{j} + \cos\gamma_M \vec{k}.$$

Thus, the flux of $\vec{v}$ on side B of Σ can be expressed as

$$\iint\limits_{\Sigma} \vec{v} \cdot \vec{n} \, dS = \iint\limits_{\Sigma} [P(M)\cos\alpha_M + Q(M)\cos\beta_M + R(M)\cos\gamma_M] \, dS.$$

Therefore, to calculate the flux of a vector field is to calculate three surface integrals of the first kind on the right hand side of the above equation.

Since α, β and γ are direction angles of the normal unit vector $\vec{n}$, it is easy to see that $\cos\alpha \, dS$, $\cos\beta \, dS$ and $\cos\gamma \, dS$ are projections of the area element dS on the coordinate planes Oyz, Ozx and Oxy respectively (Figure 7.29). That is

$$\cos\alpha \, dS = \pm \, dy \, dz, \quad \cos\beta \, dS = \pm \, dz \, dx,$$
$$\cos\gamma \, dS = \pm \, dx \, dy.$$

Thus, we define

$$\iint\limits_{\Sigma_B} f(M) \, dy \, dz = \iint\limits_{\Sigma_B} f(x, y, z) \, dy \, dz = \iint\limits_{\Sigma} f(M) \cos\alpha_M \, dS,$$

$$\iint\limits_{\Sigma_B} f(M) \, dz \, dx = \iint\limits_{\Sigma_B} f(x, y, z) \, dz \, dx = \iint\limits_{\Sigma} f(M) \cos\beta_M \, dS,$$

$$\iint\limits_{\Sigma_B} f(M) \, dx \, dy = \iint\limits_{\Sigma_B} f(x, y, z) \, dx \, dy = \iint\limits_{\Sigma} f(M) \cos\gamma_M \, dS.$$

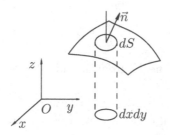

Fig. 7.29

as the *surface integral of the second kind* of the function $f(M)$ on the side B of Σ with respect to variables y and z, z and x, x and y. It follows that, the total flux of a vector field $\vec{v}$ across Σ from side A to side B is

$$\iint\limits_{\Sigma} \vec{v} \cdot \vec{n} \, dS = \iint\limits_{\Sigma_B} P \, dy \, dz + Q \, dz \, dx + R \, dx \, dy,$$

where $\vec{n}$ is the normal unit vector that points away from the side B of Σ, P, Q and R are the components of $\vec{v}$ on the x, y and z axis respectively. Hence, to find the flux of a vector field, is to calculate three surface integrals of the second kind.

There are two special properties of the surface integral of the second kind:

(1)

$$\iint\limits_{\Sigma_B} f(M) \, dx \, dy = - \iint\limits_{\Sigma_A} f(M) \, dx \, dy.$$

Similar equations also hold with respect to $dy \, dz$ and $dz \, dx$.

(2) If Σ a part of cylindrical surface parallel to the Oz axis (Figure 7.30), then $\cos\gamma = 0$. Thus, we have

$$\iint\limits_{\Sigma} f(M) \, dx \, dy = 0.$$

Similar equations also hold with respect to $dy \, dz$ and $dz \, dx$.

7.3.3   Calculation of Surface Integrals of the Second Kind

The calculation of surface integrals of the second kind follows from its geometric meaning explained in the last section. We consider the following cases:

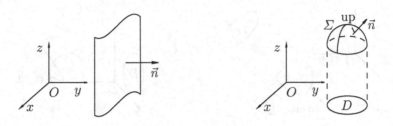

Fig. 7.30   Fig. 7.31

(1) The surface Σ has an explicit representation.

(1a) If the equation of the surface S has the form $z = z(x, y), (x, y) \in D$ (Figure 7.31), then

$$\iint_{\Sigma_{\text{up}}} f(x, y, z) \, dx \, dy = \iint_D f(x, y, z(x, y)) \, dx \, dy,$$

$$\iint_{\Sigma_{\text{down}}} f(x, y, z) \, dx \, dy = -\iint_D f(x, y, z(x, y)) \, dx \, dy,$$

(1b) If the equation of the surface S has the form $x = x(y, z), (y, z) \in D$ (Figure 7.32), then

$$\iint_{\Sigma_{\text{front}}} f(x, y, z) \, dy \, dz = \iint_D f(x(y, z), y, z) \, dy \, dz,$$

$$\iint_{\Sigma_{\text{back}}} f(x, y, z) \, dy \, dz = -\iint_D f(x(y, z), y, z) \, dy \, dz,$$

(1c) If the equation of the surface S has the form $y = y(z, x), (z, x) \in D$ (Figure 7.33), then

$$\iint_{\Sigma_{\text{right}}} f(x, y, z) \, dz \, dx = \iint_D f(x, y(z, x), z) \, dz \, dx,$$

$$\iint_{\Sigma_{\text{left}}} f(x, y, z) \, dz \, dx = -\iint_D f(x, y(z, x), z) \, dz \, dx.$$

We are not going to go over the detailed proofs of the above equations since they are corollaries of the proofs of the following calculation method.

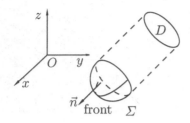
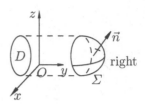

Fig. 7.32   Fig. 7.33

(2) The surface Σ has an parametric representation.

Suppose the parametric expressions of a smooth surface Σ are

$$x = x(u,v), \quad y = y(u,v), \quad z = z(u,v), \quad (u,v) \in \Delta.$$

The vector form of the above equations is

$$\vec{r} = \vec{r}(u,v), \quad (u,v) \in \Delta,$$

where Δ is a bounded closed region on the uv-plane.

If the function $f(M) = f(x,y,z)$ is continuous on Σ, then the three surface integrals of the second kind of $f(M)$ on Σ exist, and

$$\iint_{\Sigma_B} f(M)\, dx\, dy = \iint f(x(u,v), y(u,v), z(u,v)) \left(\pm \frac{\partial(x,y)}{\partial(u,v)} \right) du\, dv,$$

$$\iint_{\Sigma_B} f(M)\, dy\, dz = \iint f(x(u,v), y(u,v), z(u,v)) \left(\pm \frac{\partial(y,z)}{\partial(u,v)} \right) du\, dv,$$

$$\iint_{\Sigma_B} f(M)\, dz\, dx = \iint f(x(u,v), y(u,v), z(u,v)) \left(\pm \frac{\partial(z,x)}{\partial(u,v)} \right) du\, dv.$$

The sign of the three integrals on the right hand side of the above equations are determined by whether the direction of the vector $\vec{r}_u' \times \vec{r}_v'$ points away from the side B of Σ or not.

In fact, the vector $\vec{r}_u' \times \vec{r}_v'$ is also a normal vector of Σ. If $\vec{n}$ is a normal unit vector that points away from the B side, then

$$\vec{n} = \pm \frac{\vec{r}_u' \times \vec{r}_v'}{|\vec{r}_u' \times \vec{r}_v'|} = \pm \frac{1}{|\vec{r}_u' \times \vec{r}_v'|} \begin{vmatrix} \vec{i} & \vec{j} & \vec{k} \\ x_u' & y_u' & z_u' \\ x_v' & y_v' & z_v' \end{vmatrix}$$

$$= \frac{\pm 1}{|\vec{r}_u' \times \vec{r}_v'|} \cdot \left[\frac{\partial(y,z)}{\partial(u,v)} \vec{i} + \frac{\partial(z,x)}{\partial(u,v)} \vec{j} + \frac{\partial(x,y)}{\partial(u,v)} \vec{k} \right].$$

Thus, if the direction cosines of $\vec{n}$ are $\cos\alpha$, $\cos\beta$ and $\cos\gamma$, then

$$\cos\alpha = \pm\frac{\dfrac{\partial(y,z)}{\partial(u,v)}}{|\vec{r}'_u \times \vec{r}'_v|}, \quad \cos\beta = \pm\frac{\dfrac{\partial(z,x)}{\partial(u,v)}}{|\vec{r}'_u \times \vec{r}'_v|}, \quad \cos\gamma = \pm\frac{\dfrac{\partial(x,y)}{\partial(u,v)}}{|\vec{r}'_u \times \vec{r}'_v|}.$$

The sign of the terms on the right hand side of the above equations are determined by whether the directions of $\vec{r}'_u \times \vec{r}'_v$ and $\vec{n}$ are the same. That is, whether the vector $\vec{r}'_u \times \vec{r}'_v$ points away from the side B of Σ.

On the other hand,

$$dS = |\vec{r}'_u \times \vec{r}'_v|\, du\, dv.$$

By the definition of the surface integral of the second kind and the method of calculation of the surface integral of the first kind, we have

$$\iint_{\Sigma_B} f(M)\, dx\, dy = \iint_{\Sigma} f(M)\cos\gamma\, dS$$

$$= \iint_{\Sigma} f(x(u,v), y(u,v), z(u,v)) \left[\pm\frac{\dfrac{\partial(x,y)}{\partial(u,v)}}{|\vec{r}'_u \times \vec{r}'_v|}\right] |\vec{r}'_u \times \vec{r}'_v|\, du\, dv$$

$$= \iint_{\Sigma} f(x(u,v), y(u,v), z(u,v)) \left(\pm\frac{\partial(x,y)}{\partial(u,v)}\right) du\, dv.$$

The determination of the sign is the same as above, and the verifications of the other two equations can be processed in a similar way.

If P, Q, R are continuous functions on Σ, then we have

$$\iint_{\Sigma_B} P\, dy\, dz + Q\, dz\, dx + R\, dx\, dy = \iint \pm \begin{vmatrix} P & Q & R \\ x'_u & y'_u & z'_u \\ x'_v & y'_v & z'_v \end{vmatrix} du\, dv.$$

The determination of the sign is the same as above.

Example 1. Suppose Σ is the part of the cylinder $x^2 + y^2 = R^2$ between $0 \leqslant z \leqslant h$. Compute the surface integral

$$\iint_{\Sigma_{\text{outside}}} yz\, dy\, dz + zx\, dz\, dx + xy\, dx\, dy.$$

Solution: By the second property of the surface integral of the second kind in Section 7.3.2, the third term of this integral is zero. That is

$$\iint_{\Sigma_{\text{outside}}} xy\, dx\, dy = 0.$$

In order to calculate the second term, we divide Σ into two parts, Σ_1 and Σ_2 (Figure 7.34), where Σ_1 is the part on the right side of the xz-plane and Σ_2 is the part on the left side of the xz-plane. Thus, we have

$$\iint\limits_{\Sigma_{\text{outside}}} zx\,dz\,dx = \iint\limits_{\Sigma_{1\text{right}}} zx\,dz\,dx + \iint\limits_{\Sigma_{2\text{left}}} zx\,dz\,dx$$

$$= \iint\limits_{\substack{-R\leqslant x\leqslant R\\0\leqslant z\leqslant h}} zx\,dz\,dx - \iint\limits_{\substack{-R\leqslant x\leqslant R\\0\leqslant z\leqslant h}} zx\,dz\,dx = 0 - 0 = 0.$$

Similarly,

$$\iint\limits_{\Sigma_{\text{outside}}} yz\,dy\,dz = 0.$$

Therefore

$$\iint\limits_{\Sigma_{\text{outside}}} yz\,dy\,dz + zx\,dz\,dx + xy\,dx\,dy = 0.$$

Example 2. Compute the flux of the vector field $\vec{v} = x\vec{i} + y\vec{j} + z\vec{k}$ in the direction of the exterior normal of the sphere $x^2 + y^2 + z^2 = a^2$.

Solution: The parametric equations of the sphere Σ are:

$$\begin{cases} x = a\cos\varphi\sin\theta, \\ y = a\sin\varphi\sin\theta, \\ z = a\cos\theta. \end{cases}$$

where $0 \leqslant \theta \leqslant \pi, 0 \leqslant \varphi \leqslant 2\pi$, and the vector $\vec{r}'_\theta \times \vec{r}'_\varphi$ points towards outside of the sphere (Figure 7.35).
Thus, we have

$$\oiint\limits_{\Sigma} \vec{v}\cdot\vec{n}\,dS = \oiint\limits_{\Sigma_{\text{outside}}} x\,dy\,dz + y\,dz\,dx + z\,dx\,dy$$

$$= \iint\limits_{\substack{0\leqslant\varphi\leqslant 2\pi\\0\leqslant\theta\leqslant\pi}} \begin{vmatrix} x & y & z \\ x'_\theta & y'_\theta & z'_\theta \\ x'_\varphi & y'_\varphi & z'_\varphi \end{vmatrix} d\theta\,d\varphi$$

$$= \iint\limits_{\substack{0\leqslant\varphi\leqslant 2\pi\\0\leqslant\theta\leqslant\pi}} a^3\sin\theta\,d\theta\,d\varphi = 4\pi a^3.$$

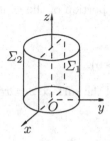

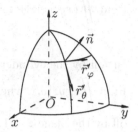

Fig. 7.34   Fig. 7.35

Exercises 7.3

1. Find $\iint\limits_{S} (x + y + z)\, dS$, where S is the surface of the cube

$$0 \leqslant x \leqslant 1,\ 0 \leqslant y \leqslant 1,\ 0 \leqslant z \leqslant 1.$$

2. Find $\iint\limits_{S} \left(z + 2x + \frac{4}{3}y \right) dS$, where S is the portion of the plane

$$\frac{x}{2} + \frac{y}{3} + \frac{z}{4} = 1$$

in the first octant.

3. Find $\iint\limits_{S} \dfrac{1}{(1 + x + y)^2}\, dS$, where S is the surface of the tetrahedron bounded by the three coordinate planes and the plane $x + y + z = 1$.

4. Find $\iint\limits_{S} (x + y + z)\, dS$, where S is the hemisphere

$$x^2 + y^2 + z^2 = a^2,\ (z \geqslant 0).$$

5. Find $\iint\limits_{S} \sqrt{R^2 - x^2 - y^2}\, dS$, where S is the hemisphere

$$x^2 + y^2 + z^2 = R^2,\ (z \geqslant 0).$$

6. Find $\iint\limits_{S} (x^2 + y^2)\, dS$, where S is the surface of the solid bounded by the surface $z = \sqrt{x^2 + y^2}$ and the plane $z = 1$.

7. Find $\iint\limits_{S} (xy + yz + zx)\, dS$, where S is the portion of the cone

$$z = \sqrt{x^2 + y^2}$$

cut off by the cylinder $x^2 + y^2 = 2ax$, $(a > 0)$.

8. Find $\iint\limits_{S} |xyz|\, dS$, where S is the portion of the surface $z = x^2 + y^2$ cut off by the plane $z = 1$.

9. Find the area of the top portion of the sphere $x^2 + y^2 + z^2 = R^2$ that is cut off by the cylinder $x^2 + y^2 = \rho^2$ $(\rho < R)$.

10. Find $\iint\limits_{S} (y - z)\, dy\, dz + (x + y + z)\, dx\, dy$, where S is the outside surface of the cube bounded by the three coordinate planes and the three planes $x = 1$, $y = 1$, $z = 1$.

11. Find $\iint\limits_{S} xy\, dy\, dz + yz\, dz\, dx + zx\, dx\, dy$, where S is the outside surface of the tetrahedron in Exercise 3.

12. Find $\iint\limits_{S} x^2\, dy\, dz + y^2\, dz\, dx + z^2\, dx\, dy$, where S is the outside surface of $(x - a)^2 + (y - b)^2 + (z - c)^2 = R^2$.

13. Find $\iint\limits_{S} x\, dy\, dz + y\, dz\, dx + z\, dx\, dy$, where S is the outside surface of the portion of the cylinder $x^2 + y^2 = 1$ that is cut off by the planes $z = 0$, $z = 3$.

14. Find $\iint\limits_{S} (y - z)\, dy\, dz + (z - x)\, dz\, dx + (x - y)\, dx\, dy$, where S is the outside surface of the solid bounded by the surface $z = \sqrt{x^2 + y^2}$ and the plane $z = k$ $(k > 0)$.

15. Find $\iint\limits_{S} (x + y)\, dy\, dz + (y + z)\, dz\, dx + (z + x)\, dx\, dy$, where S is the outside surface of the unit cube centered at the origin and with sides parallel to the coordinate planes.

16. Find $\iint\limits_{S} z^2\, dx\, dy$, where S is the outside of the ellipsoid

$$\frac{x^2}{a^2} + \frac{y^2}{b^2} + \frac{z^2}{c^2} = 1.$$

17. Find $\displaystyle\iint\limits_{S} \frac{1}{z}\,dx\,dy$, where S is the outside of the sphere $x^2 + y^2 + z^2 = a^2$.

18. Find $\displaystyle\iint\limits_{S} (x^2 \cos\alpha + y^2 \cos\beta + z^2 \cos\gamma)\,dS$, where S is the downside of the cone $x^2 + y^2 = z^2$ $(0 \leqslant z \leqslant h)$.

19. Find $\displaystyle\iint\limits_{S} (x \cos\alpha + y \cos\beta + z \cos\gamma)\,dS$, where S is the upside of the plane $x + y + z = 1$ in the first octant.

20. Find the flux of the vector $\vec{A} = yz\vec{i} + zx\vec{j} + xy\vec{k}$ across the outside of the surface S: $x^2 + y^2 = R^2$ $(0 \leqslant z \leqslant h)$.

21. Find the flux of the vector $\vec{A} = xy\vec{i} + yz\vec{j} + zx\vec{k}$ across the outside of the portion of the sphere S: $x^2 + y^2 + z^2 = 1$ in the first octant.

22. Find the flux of the vector $\vec{r}$ across the outside of the cylinder S:

$$x^2 + y^2 = R^2, \quad (0 \leqslant z \leqslant h).$$

7.4 Stokes Theorem

In this section, we study the connection between line integrals and surface integrals, as well as their physical significance.

7.4.1 *Green's Theorem*

The connection between line integrals and surface integrals on the two-dimensional plane can be well interpreted by Green's Theorem:

Green's Theorem Let D be a closed region bounded by a closed curve L on the xy plane. If the first order partial derivatives of $P(x,y)$ and $Q(x,y)$ are continuous on D, then

$$\oint\limits_{L} P\,dx + Q\,dy = \iint\limits_{D} \left(\frac{\partial Q}{\partial x} - \frac{\partial P}{\partial y} \right) dx\,dy.$$

Proof: First, we assume that the curve L satisfies the condition that if a line is parallel to the y-axis, then it intersects L at no more than two points (Figure 7.36.) Suppose $y = y_1(x)$, $a \leqslant x \leqslant b$ is the lower part L' of the curve L, and $y = y_2(x)$, $a \leqslant x \leqslant b$ is the upper part L'' of the curve L, in

Figure 7.36. Then we have

$$-\iint\limits_{D} \frac{\partial P}{\partial y}\, dx\, dy = -\int_{a}^{b} dx \int_{y_1(x)}^{y_2(x)} \frac{\partial P}{\partial y}\, dy$$

$$= -\int_{a}^{b} [P(x, y_2(x)) - P(x, y_1(x))]\, dx$$

$$= \int_{L''} P\, dx + \int_{L'} P\, dx = \oint_{L} P\, dx.$$

Similarly, we also have

$$\iint\limits_{D} \frac{\partial Q}{\partial x}\, dx\, dy = \oint_{L} Q\, dy.$$

Hence, the Green's Theorem is true for the kind of curve in Figure 7.36. In general, any region can be divided into small pieces as the one in Figure 7.36, and each small region can be treated in the same way above. As an example, we can divide D into two pieces D_1 and D_2 (Figure 7.37), then apply Green's Theorem on each piece. Thus,

$$-\iint\limits_{D_1} \frac{\partial P}{\partial y}\, dx\, dy = \int_{\overset{\frown}{ABCA}} P\, dx;$$

$$-\iint\limits_{D_2} \frac{\partial P}{\partial y}\, dx\, dy = \int_{\overset{\frown}{CDAC}} P\, dx.$$

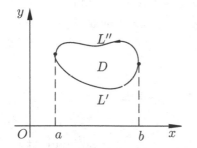

Fig. 7.36

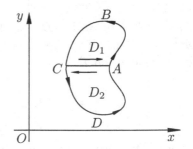

Fig. 7.37

Adding the above two equations, and by the fact that

$$\int_{\overline{CA}} P\,dx + \int_{\overline{AC}} P\,dx = 0,$$

we obtain

$$-\iint_D \frac{\partial P}{\partial y}\,dx\,dy = \oint_L P\,dx.$$

By a similar argument, we have

$$\iint_D \frac{\partial Q}{\partial x}\,dx\,dy = \oint_L Q\,dy.$$

Therefore, Green's Theorem is true for all closed regions bounded by an arbitrary closed curve L.

The general form of **Green's Theorem** is:

Suppose D is a region on the xy-plane bounded by $n+1$ closed curves $L, l_1, l_2, \cdots, l_n$, where $l_1, l_2, \cdots, l_n$ are enclosed by the curve L (Figure 7.38). If the functions $P(x, y)$, $Q(x, y)$ have continuous first order partial derivatives on D, then we have

$$\iint_D \left(\frac{\partial Q}{\partial x} - \frac{\partial P}{\partial y} \right) dx\,dy = \oint_L + \oint_{l_1} + \cdots + \oint_{l_n} P\,dx + Q\,dy.$$

Proof: We only need to consider the simplest case, that is, the case where there is only one curve l enclosed by L (Figure 7.39).

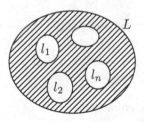

Fig. 7.38

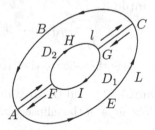

Fig. 7.39

We divide D into two parts, D_1 and D_2, by curves AF and CG. By Green's Theorem on D_1 and D_2

$$\iint\limits_{D_1} \left(\frac{\partial Q}{\partial x} - \frac{\partial P}{\partial y}\right) dx\,dy = \int_{\overrightarrow{AEC}} + \int_{\overrightarrow{CG}} + \int_{\overrightarrow{GIF}} + \int_{\overrightarrow{FA}} P\,dx + Q\,dy,$$

$$\iint\limits_{D_2} \left(\frac{\partial Q}{\partial x} - \frac{\partial P}{\partial y}\right) dx\,dy = \int_{\overrightarrow{AF}} + \int_{\overrightarrow{FHG}} + \int_{\overrightarrow{GC}} + \int_{\overrightarrow{CBA}} P\,dx + Q\,dy.$$

Adding the corresponding terms of the above two equations, we have

$$\iint\limits_{D} \left(\frac{\partial Q}{\partial x} - \frac{\partial P}{\partial y}\right) dx\,dy = \oint_{L} + \oint_{l} P\,dx + Q\,dy.$$

Green's Theorem can be applied to find the area of a planar region. If L is a closed curve on the xy-plane, then the area inside L is

$$A = \oint_{L} x\,dy = -\oint_{L} y\,dx = \frac{1}{2}\oint_{L} x\,dy - y\,dx.$$

In fact, the first integral is obtained by letting $P = 0$ and $Q = x$ in Green's Theorem. The second integral is obtained by letting $P = -y$ and $Q = 0$ in Green's Theorem. The third integral is obtained by averaging the first and the second integrals.

Example 1.

$$\oint_{\frac{x^2}{a^2}+\frac{y^2}{b^2}=1} (x+y)\,dx - (x-y)\,dy = -2\iint\limits_{\frac{x^2}{a^2}+\frac{y^2}{b^2}\leqslant 1} dx\,dy = -2\pi ab.$$

Example 2.

$$\oint_{x^2+y^2=1} y^2\,dx = -2\iint\limits_{x^2+y^2\leqslant 1} y\,dx\,dy = 0.$$

Example 3.

$$\oint_{L} 2xy\,dx + x^2\,dy = \iint\limits_{\substack{0\leqslant y\leqslant x \\ 0\leqslant x\leqslant 1}} (2x - 2x)\,dx\,dy = 0,$$

where L is the sides of the triangle bounded by $y = 0$, $x = 1$ and $y = x$.

Example 4. Find the area of the ellipse.

The area of the ellipse with the parametric equations

$$x = a\cos t, \quad y = b\sin t, \quad 0 \leqslant t \leqslant 2\pi$$

is

$$A = -\oint_{L} y\,dx = -\int_{0}^{2\pi} b\sin t\,d(a\cos t) = ab\int_{0}^{2\pi} \sin^2 t\,dt = \pi ab.$$

7.4.2 *Gauss's Theorem, Divergence*

Gauss's Theorem and Stokes' Theorem are generalizations of Green's Theorem in three dimensional space. Gauss's Theorem establishes the connection between surface integrals and triple integrals.

Gauss's Theorem: Suppose V is a closed region bounded by a closed surface Σ in the three dimensional space. Suppose also that the functions $P(x,y,z)$, $Q(x,y,z)$ and $R(x,y,z)$ have continuous first order partial derivatives on V. Then we have

$$\oiint_{\Sigma_{\text{outside}}} P\,dy\,dz + Q\,dz\,dx + R\,dx\,dy = \iiint_V \left(\frac{\partial P}{\partial x} + \frac{\partial Q}{\partial y} + \frac{\partial R}{\partial z} \right) dV,$$

where the normal vectors of Σ_{outside} points outward.

Proof: Suppose that V is a cylinder with generating lines parallel to the z-axis, and that both the top and the bottom surfaces have explicit representation and that the projection of V to the xy-plane is D (Figure 7.40). Suppose the bottom surface is

$$\Sigma_1 : z = z_1(x,y), \quad (x,y) \in D,$$

the top surface is

$$\Sigma_2 : z = z_2(x,y), \quad (x,y) \in D,$$

and the side of the cylinder is Σ_3. Then we have

$$\iiint_V \frac{\partial R}{\partial z}\,dx\,dy\,dz = \iint_D dx\,dy \int_{z_1(x,y)}^{z_2(x,y)} \frac{\partial R}{\partial z}\,dz$$

$$= \iint_D R(x,y,z_2(x,y))\,dx\,dy - \iint_D R(x,y,z_1(x,y))\,dx\,dy$$

$$= \iint_{\Sigma_{2\text{up}}} R\,dx\,dy + \iint_{\Sigma_{1\text{down}}} R\,dx\,dy,$$

where $\Sigma_{2\text{up}}$ represents the surface Σ_2 with the orientation such that the normal vectors of Σ_2 point upward, and $\Sigma_{1\text{down}}$ represents the surface Σ_1 with the orientation such that the normal vectors of Σ_1 point downward. By property (2) in Section 7.3.2, we have

$$\iint_{\Sigma_{3\text{out}}} R\,dx\,dy = 0.$$

Therefore,

$$\iiint\limits_{V} \frac{\partial R}{\partial z}\, dx\, dy\, dz = \iint\limits_{\Sigma_{2\mathrm{up}}} R\, dx\, dy + \iint\limits_{\Sigma_{1\mathrm{down}}} R\, dx\, dy + \iint\limits_{\Sigma_{3\mathrm{out}}} R\, dx\, dy$$

$$= \oiint\limits_{\Sigma_{\mathrm{out}}} R\, dx\, dy.$$

Similar results can be proven for

$$\iiint\limits_{V} \frac{\partial Q}{\partial y}\, dx\, dy\, dz,$$

and

$$\iiint\limits_{V} \frac{\partial P}{\partial x}\, dx\, dy\, dz.$$

Now we can conclude that Gauss's Theorem holds for the simpler region V.

In general, for any three dimensional closed region bounded by a closed surface, we partition it into several closed simpler regions as V, and Gauss's Theorem holds for each of the simpler regions. By adding the corresponding integral terms together, the conclusion that Gauss's Theorem holds for any closed region follows.

As is the case for Green's Theorem, there is a general form of Gauss's Theorem:

Suppose closed surfaces $\Sigma_1, \cdots, \Sigma_n$ are enclosed in the closed surface Σ, and they form the boundary of the region V (Figure 7.41). Also, we assume that the functions P, Q and R have continuous first order partial derivatives on V. Then

$$\iiint\limits_{V} \left(\frac{\partial P}{\partial x} + \frac{\partial Q}{\partial y} + \frac{\partial R}{\partial z} \right) dV$$

$$= \oiint\limits_{\Sigma_{\mathrm{outside}}} + \oiint\limits_{\Sigma_{1\mathrm{inside}}} + \cdots + \oiint\limits_{\Sigma_{n\mathrm{inside}}} P\, dy\, dz + Q\, dz\, dx + R\, dx\, dy.$$

Gauss's Theorem can also be used to calculate volumes. Suppose Σ is a closed surface, and V is the volume of the region inside Σ. Then

$$V = \oiint\limits_{\Sigma_{\mathrm{outside}}} x\, dy\, dz = \oiint\limits_{\Sigma_{\mathrm{outside}}} y\, dz\, dx = \oiint\limits_{\Sigma_{\mathrm{outside}}} z\, dx\, dy$$

$$= \frac{1}{3} \oiint\limits_{\Sigma_{\mathrm{outside}}} x\, dy\, dz + y\, dz\, dx + z\, dx\, dy.$$

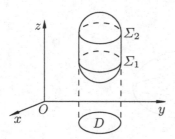

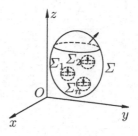

Fig. 7.40   Fig. 7.41

The physical significance of the integral

$$\oiint_{\Sigma_{\text{outside}}} P \, dy \, dz + Q \, dz \, dx + R \, dx \, dy$$

is the flux of the vector field $\vec{v} = P\vec{i} + Q\vec{j} + R\vec{k}$ from Σ_{inside} to Σ_{outside}.

Now we study the physical significance of the integral

$$\iiint_V \left(\frac{\partial P}{\partial x} + \frac{\partial Q}{\partial y} + \frac{\partial R}{\partial z} \right) dV.$$

Consider a fluid velocity field $\vec{v}$, and assume that Σ is a closed surface with unit outward normal vector $\vec{n}$. Then the flux of $\vec{v}$ is

$$\oiint_{\Sigma} \vec{v} \cdot \vec{n} \, dS.$$

If the flux is not zero, then the amount of fluid that flows into Σ and the amount of fluid that flows out from Σ are different. This implies that there are some positive or negative "sources" inside Σ.

In order to study the distribution of these sources in the vector field, we introduce the concept of divergence: Suppose $\vec{v}$ is a vector field and M is a point in the domain of $\vec{v}$ (Figure 7.42). Consider a closed surface Σ in the domain of $\vec{v}$ that contains the point M. Let V be the region inside Σ (the volume of this region is also denoted as V), and let the unit outward normal vector of Σ be $\vec{n}$. The *divergence* of the vector field $\vec{v}$ at M is defined as:

$$(\text{div } \vec{v})_M = \lim_{\Sigma \to M} \frac{\oiint_{\Sigma} \vec{v} \cdot \vec{n} \, dS}{V}.$$

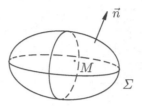

Fig. 7.42

The divergence of a vector field describes the intensity of the distribution of sources in the vector field. The divergence of $\vec{v}$ is

$$\operatorname{div} \vec{v} = \frac{\partial P}{\partial x} + \frac{\partial Q}{\partial y} + \frac{\partial R}{\partial z}.$$

In fact, by Gauss's theorem,

$$\oiint_{\Sigma} \vec{v} \cdot \vec{n} \, dS = \oiint_{\Sigma_{\text{out}}} P \, dy \, dz + Q \, dz \, dx + R \, dx \, dy$$

$$= \iiint_{V} \left(\frac{\partial P}{\partial x} + \frac{\partial Q}{\partial y} + \frac{\partial R}{\partial z} \right) dV.$$

By the Mean-Value Theorem for multiple integrals, the right hand side of the above equation is

$$\left(\frac{\partial P}{\partial x} + \frac{\partial Q}{\partial y} + \frac{\partial R}{\partial z} \right)_{\widetilde{M}} \cdot V,$$

where $\widetilde{M}$ is a point in V. Dividing by V on both sides of the equation and let Σ contract to the point M (of course at this time $\widetilde{M} \to M$). Then we have

$$(\operatorname{div} \vec{v})_M = \lim_{\Sigma \to M} \frac{\displaystyle\oiint_{\Sigma} \vec{v} \cdot \vec{n} \, dS}{V} = \left(\frac{\partial P}{\partial x} + \frac{\partial Q}{\partial y} + \frac{\partial R}{\partial z} \right)_M.$$

Gauss's theorem can now also be stated as follows: Suppose $\vec{v}$ is a vector field, V is a closed region bounded by a closed surface Σ in the domain of the vector field. Suppose also that $\vec{n}$ is the unit exterior normal vector. Then we have

$$\oiint_{\Sigma} \vec{v} \cdot \vec{n} \, dS = \iiint_{V} \operatorname{div} \vec{v} \, dV.$$

If the boundary of V consists of closed surfaces $\Sigma, \Sigma_1, \cdots, \Sigma_n$, where $\Sigma_1, \cdots, \Sigma_n$ are enclosed in Σ, and they do not contain each other, $\vec{n}$ is the unit exterior normal vector of Σ and unit interior normal vectors of $\Sigma_1, \cdots, \Sigma_n$ (Figure 7.41), then

$$\oiint_{\Sigma} + \oiint_{\Sigma_1} + \cdots + \oiint_{\Sigma_n} \vec{v} \cdot \vec{n}\, dS = \iiint_V \operatorname{div} \vec{v}\, dV.$$

It is easy to verify that if α, β are constants and φ is a function, then we have

(i) $\operatorname{div}(\alpha \vec{v}_1 + \beta \vec{v}_2) = \alpha \operatorname{div} \vec{v}_1 + \beta \operatorname{div} \vec{v}_2$,

(ii) $\operatorname{div} \varphi \vec{v} = \varphi \operatorname{div} \vec{v} + \vec{v} \cdot \mathbf{grad}\, \varphi$.

Divergence can be represented by the Nabla operator ∇:

$$\operatorname{div} \vec{v} = \nabla \cdot \vec{v} = \vec{i} \cdot \frac{\partial \vec{v}}{\partial x} + \vec{j} \cdot \frac{\partial \vec{v}}{\partial y} + \vec{k} \cdot \frac{\partial \vec{v}}{\partial z}.$$

Example 1. Suppose O is a fixed point, $\vec{r} = \overrightarrow{OM}$ and $r = |\vec{r}|$. Compute $\operatorname{div} r^n \vec{r}$.

Solution: Let O be the origin and the coordinate of the point M be (x, y, z). Then we have

$$\operatorname{div} \vec{r} = \nabla \cdot \vec{r} = \frac{\partial x}{\partial x} + \frac{\partial y}{\partial y} + \frac{\partial z}{\partial z} = 3.$$

On the other hand

$$\nabla r^n = nr^{n-1} \nabla r = nr^{n-1} \cdot \frac{r}{r} = nr^{n-2}\vec{r}.$$

By (ii), we have

$$\nabla \cdot r^n \vec{r} = 3r^n + \vec{r} \cdot (nr^{n-2}\vec{r}) = (3+n)r^n.$$

Especially, when $n = -3$, we have

$$\nabla \cdot \frac{\vec{r}}{r^3} = 0.$$

Another important operator is the Laplace operator Δ:

$$\Delta = \frac{\partial^2}{\partial x^2} + \frac{\partial^2}{\partial y^2} + \frac{\partial^2}{\partial z^2}.$$

Thus

$$\Delta = \nabla^2 = \nabla \cdot \nabla.$$

Hence

$$\operatorname{div} \mathbf{grad}\, u = \Delta u.$$

Example 2. The physics significance of divergence of electric fields.

Consider an electric charge q at a point O. The electric field generated by q at the point M is

$$\vec{E} = q\frac{\overrightarrow{OM}}{|OM|^3} = \frac{q\vec{r}}{r^3},$$

where $\vec{r} = \overrightarrow{OM}$.

The field is not defined at the point O. By Example 1, we have

$$\text{div } \vec{E} = 0$$

except at point O.

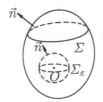

Fig. 7.43

Suppose Σ is a closed surface that contains the point O. Consider a sphere Σ_ε that contains the point O with radius ε inside Σ (Figure 7.43). If $\vec{n}$ is the unit normal vector of Σ and Σ_ε, then by Gauss's Theorem, we have

$$\oiint_\Sigma - \oiint_{\Sigma_\varepsilon} \vec{E} \cdot \vec{n}\, dS = \iiint_V \text{div } \vec{E}\, dV = 0.$$

Since $\vec{n} = \dfrac{\vec{r}}{r}$ on Σ_ε, we have

$$\oiint_\Sigma \vec{E} \cdot \vec{n}\, dS = \oiint_{\Sigma_\varepsilon} \vec{E} \cdot \vec{n}\, dS = \oiint_{\Sigma_\varepsilon} q\frac{\vec{r}}{r^3} \cdot \frac{\vec{r}}{r}\, dS$$

$$= \oiint_{\Sigma_\varepsilon} \frac{q}{r^2}\, dS = \frac{q}{\varepsilon^2} \oiint_{\Sigma_\varepsilon} dS = \frac{q}{\varepsilon^2} 4\pi\varepsilon^2 = 4\pi q.$$

If Σ contains several electric charges $q_1, \cdots, q_n$, and $\vec{E}_i$ is the electric field generated by q_i, $(i = 1, \cdots, n)$, then the electric field in Σ is $\vec{E} = \vec{E}_1 + \cdots + \vec{E}_n$. Thus,

$$\oiint_{\Sigma} \vec{E} \cdot \vec{n}\, dS = \sum_{i=1}^{n} \oiint_{\Sigma_i} \vec{E}_i \cdot \vec{n}\, dS = 4\pi \sum_{i=1}^{n} q_i.$$

If the electric charges are distributed continuously inside Σ, and the charge density is ρ, then we have

$$\iiint_{V} \operatorname{div} \vec{E}\, dV = \oiint_{\Sigma} \vec{E} \cdot \vec{n}\, dS = 4\pi \iiint_{V} \rho\, dV.$$

Hence,

$$(\operatorname{div} \vec{E})_M = \lim_{\Sigma \to M} \frac{1}{V} \oiint_{\Sigma} \vec{E} \cdot \vec{n}\, dS$$

$$= \lim_{\Sigma \to M} \frac{4\pi}{V} \iiint_{V} \rho\, dV = 4\pi\rho(M).$$

Therefore, the divergence of the electric field is the density of charges.

7.4.3   *Stokes' Theorem, and The Curl of a Vector Field*

Green's theorem establishes the connection between line integrals and double integrals on the plane. Gauss's theorem establishes the connection between surface integrals and triple integrals in space. Stokes' theorem establishes the connection between line integrals and surface integrals in space.

Stokes' Theorem: Suppose Σ is a spatial surface bounded by a closed curve L (Figure 7.44). If the functions P, Q, R have continuous first order partial derivatives, then

$$\oint_{L} P\, dx + Q\, dy + R\, dz = \iint_{\Sigma_B} \left(\frac{\partial R}{\partial y} - \frac{\partial Q}{\partial z} \right) dy\, dz$$

$$+ \left(\frac{\partial P}{\partial z} - \frac{\partial R}{\partial x} \right) dz\, dx + \left(\frac{\partial Q}{\partial x} - \frac{\partial P}{\partial y} \right) dx\, dy,$$

where the direction of L and the side B of Σ are chosen so that when one travel along L, B is on the left hand side.

Proof: As in the proof of Green's Theorem and Gauss's Theorem, we treat the case where the surface is explicitly expressed first. Suppose the explicit expression of the surface Σ is

$$z = z(x,y), \quad (x,y) \in D,$$

where D is a closed region with boundary L^* on xy-plane (Figure 7.45).

Suppose the parametric equations of the boundary L of the surface Σ are

$$x = x(t), \quad y = y(t), \quad z = z(t) = z(x(t), y(t)), \quad \alpha \leqslant t \leqslant \beta,$$

and the parameter increases in the direction of the curve in the graph.

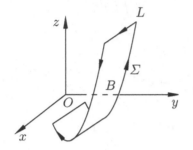

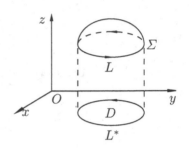

Fig. 7.44   Fig. 7.45

Obviously, the parametric equations of L^* is

$$x = x(t), \quad y = y(t), \quad \alpha \leqslant t \leqslant \beta.$$

Thus

$$\oint_L P(x,y,z)\,dx + Q(x,y,z)\,dy$$

$$= \int_\alpha^\beta (P(x(t),y(t),z(x(t),y(t)))x'(t)$$

$$+ Q(x(t),y(t),z(x(t),y(t)))y'(t))\,dt$$

$$= \oint_{L^*} P(x,y,z(x,y))\,dx + Q(x,y,z(x,y))\,dy.$$

Moreover,

$$\oint_L R\,dz = \int_\alpha^\beta R\,z'\,dt = \int_\alpha^\beta R(z_x'\,x' + z_y'\,y')\,dt$$

$$= \oint_{L^*} Rz_x'\,dx + Rz_y'\,dy.$$

Applying the Green's theorem to the sum of above two equations, we have

$$
\oint_L P\,dx + Q\,dy + R\,dz = \oint_{L^*} (P + Rz'_x)\,dx + (Q + Rz'_y)\,dy
$$

$$
= \iint_D \left[\frac{\partial}{\partial x}(Q + Rz'_y) - \frac{\partial}{\partial y}(P + Rz'_x) \right] dx\,dy
$$

$$
= \iint_D \left[\left(\frac{\partial Q}{\partial z} - \frac{\partial R}{\partial y} \right) z'_x + \left(\frac{\partial R}{\partial x} - \frac{\partial P}{\partial z} \right) z'_y + \left(\frac{\partial Q}{\partial x} - \frac{\partial P}{\partial y} \right) \right] dx\,dy.
$$

On the other hand, consider $z = z(x, y)$ as a parametric equation of Σ with parameters x and y. By the calculation formula of parametric surface integrals of the second kind in Section 7.3.3, we have

$$
\iint_{\Sigma_B} \left(\frac{\partial R}{\partial y} - \frac{\partial Q}{\partial z} \right) dy\,dz + \left(\frac{\partial P}{\partial z} - \frac{\partial R}{\partial x} \right) dz\,dx + \left(\frac{\partial Q}{\partial x} - \frac{\partial P}{\partial y} \right) dx\,dy
$$

$$
= \iint_D
\begin{vmatrix}
\dfrac{\partial R}{\partial y} - \dfrac{\partial Q}{\partial z} & \dfrac{\partial P}{\partial z} - \dfrac{\partial R}{\partial x} & \dfrac{\partial Q}{\partial x} - \dfrac{\partial P}{\partial y} \\[2mm]
\dfrac{\partial x}{\partial x} & \dfrac{\partial y}{\partial x} & \dfrac{\partial z}{\partial x} \\[2mm]
\dfrac{\partial x}{\partial y} & \dfrac{\partial y}{\partial y} & \dfrac{\partial z}{\partial y}
\end{vmatrix}
dx\,dy
$$

$$
= \iint_D \left[\left(\frac{\partial Q}{\partial z} - \frac{\partial R}{\partial y} \right) z'_x + \left(\frac{\partial R}{\partial x} - \frac{\partial P}{\partial z} \right) z'_y + \left(\frac{\partial Q}{\partial x} - \frac{\partial P}{\partial y} \right) \right] dx\,dy.
$$

From the above two equations, we can see that the Stokes' theorem is true for surfaces with explicit expressions.

In the general case, we can partition the surface such that each piece in the partition has an explicit expression. As in Figure 7.46, the doted line divides Σ into Σ_1, Σ_2, and both pieces have explicit expressions. Apply Stokes' theorem on each piece respectively. Since integrals on the doted line are canceled due to the opposite directions, Stokes' theorem holds for general surfaces.

The Stokes' theorem can also be written in the following form:

$$
\oint_L P\,dx + Q\,dy + R\,dz
$$

$$
= \iint_{\Sigma_B}
\begin{vmatrix}
dy\,dz & dz\,dx & dx\,dy \\[1mm]
\dfrac{\partial}{\partial x} & \dfrac{\partial}{\partial y} & \dfrac{\partial}{\partial z} \\[1mm]
P & Q & R
\end{vmatrix}
= \iint_{\Sigma_B}
\begin{vmatrix}
\cos\alpha & \cos\beta & \cos\gamma \\[1mm]
\dfrac{\partial}{\partial x} & \dfrac{\partial}{\partial y} & \dfrac{\partial}{\partial z} \\[1mm]
P & Q & R
\end{vmatrix}
dS,
$$

where $\cos\alpha$, $\cos\beta$, $\cos\gamma$ are direction cosines of the normal vector of Σ that points outward from the side B. The integral over L is in the direction where B is on the left hand side.

Example 1. Compute the circulation of the vector field $\vec{v} = y^2\vec{i} + z^2\vec{j} + x^2\vec{k}$ along the triangle $ABCA$, where A, B, C are on the positive side of x-axis, y-axis, z-axis, and $OA = OB = OC = 1$ (Figure 7.47).

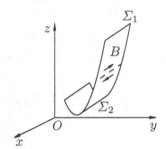
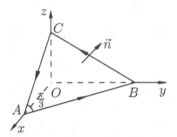

Fig. 7.46   Fig. 7.47

Solution: The surface Σ bounded by the triangle $ABCA$ can be expressed by equations $x + y + z = 1$, $x \geqslant 0$, $y \geqslant 0$, $z \geqslant 0$. The unit outward normal vector of Σ is $\left(\dfrac{1}{\sqrt{3}}, \dfrac{1}{\sqrt{3}}, \dfrac{1}{\sqrt{3}}\right)$. By Stokes' Theorem, we have

$$\oint_{ABCA} y^2\,dx + z^2\,dy + x^2\,dz = \iint_{\Sigma} \begin{vmatrix} \dfrac{1}{\sqrt{3}} & \dfrac{1}{\sqrt{3}} & \dfrac{1}{\sqrt{3}} \\[2mm] \dfrac{\partial}{\partial x} & \dfrac{\partial}{\partial y} & \dfrac{\partial}{\partial z} \\[2mm] y^2 & z^2 & x^2 \end{vmatrix} dS$$

$$= -\frac{2}{\sqrt{3}} \iint_{\Sigma} (x + y + z)\,dS = -\frac{2}{\sqrt{3}} \iint_{\Sigma} dS$$

$$= -\frac{2}{\sqrt{3}} \cdot \frac{1}{2} \cdot \sqrt{2} \cdot \sqrt{2} \cdot \sin\frac{\pi}{3} = -1.$$

Obviously, $\oint P\,dx + Q\,dy + R\,dz$ is the circulation of the vector field $v = P\vec{i} + Q\vec{j} + R\vec{k}$. Now we interpret the physical significance of the expression on the right hand side of the Stokes' Theorem.

Consider a point M on a plane in the three dimensional space. Let $\vec{n}$ be a unit vector perpendicular to the plane and with its initial point at M. We draw a closed curve L on the plane such that L encloses M. Consider the unit tangent vectors $\vec{t}$ at each point of L such that the vector $\vec{n}$ is on the left side of $\vec{t}$. Suppose there are two vector fields $\vec{v}_1$ and $\vec{v}_2$ with the same magnitude $|\vec{v}_1| = |\vec{v}_2|$, and $\vec{v}_1$ turns towards $\vec{t}$ more than $\vec{v}_2$ (Figure 7.48). This implies that the circulation of $\vec{v}_1$ along L is greater than the circulation of $\vec{v}_2$ along L. In other words, $\vec{v}_1$ rotates more than $\vec{v}_2$ along L if we observe $\vec{v}_1$ and $\vec{v}_2$ from the perspective of $\vec{n}$. If the magnitude of $\vec{v}_2$ is greater than the magnitude of $\vec{v}_1$, then it is possible that the circulation of $\vec{v}_2$ is greater than the circulation of $\vec{v}_1$ along L. In other words, the rotation of $\vec{v}_2$ is greater than the rotation of $\vec{v}_1$ if we observe the motion of $\vec{v}_1$ and $\vec{v}_2$ from the perspective of $\vec{n}$ (Figure 7.49).

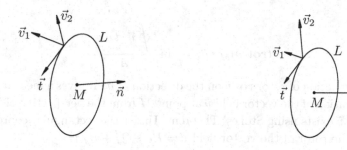

Fig. 7.48   Fig. 7.49

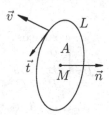

Fig. 7.50

This motivates us to define a new concept, the rotor (or curl), that describes the rotation at every point in every direction of a vector field. This concept provides a field theory interpretation of the right hand side of Stokes' Theorem.

The *rotor* (**rot** $\vec{v}$) of a vector field at a point M is a vector with the following definition:

Let $\vec{n}$ be a unit vector with initial point M. Consider a plane passing through M and is perpendicular to $\vec{n}$. Suppose L is a closed curve that encloses M. Let A be the area inside of L. At every point of L, draw a tangent vector $\vec{t}$ such that the vector $\vec{n}$ is on the left side (Figure 7.50). The circulation is

$$\oint P\,dx + Q\,dy + R\,dz = \oint \vec{v}\cdot\vec{t}\,ds$$

Define

$$(\textbf{rot } \vec{v})_M \cdot \vec{n} = \lim_{L\to M} \frac{\oint \vec{v}\cdot\vec{t}\,ds}{A}.$$

as the projection of rotor **rot** $\vec{v}$ on the direction of $\vec{n}$. It gives a measurement of the rotation of the vector field $\vec{v}$ at point M from the perspective of $\vec{n}$. We show **rot** $\vec{v}$ exists using Stokes' Theorem. Under the rectangular coordinate system, the rotor of the vector field $\vec{v} = P\vec{i} + Q\vec{j} + R\vec{j}$ is

$$\textbf{rot } \vec{v} = \begin{vmatrix} \vec{i} & \vec{j} & \vec{k} \\ \dfrac{\partial}{\partial x} & \dfrac{\partial}{\partial y} & \dfrac{\partial}{\partial z} \\ P & Q & R \end{vmatrix} = \left(\frac{\partial R}{\partial y} - \frac{\partial Q}{\partial z}\right)\vec{i} + \left(\frac{\partial P}{\partial z} - \frac{\partial R}{\partial x}\right)\vec{j} + \left(\frac{\partial Q}{\partial x} - \frac{\partial P}{\partial y}\right)\vec{k}.$$

In fact, let $\vec{n} = \cos\alpha\vec{i} + \cos\beta\vec{j} + \cos\gamma\vec{k}$, and Σ be the plane region inside of L. By Stokes Theorem and the Mean-Value Theorem of Integrals, we have

$$\oint_L \vec{v}\cdot\vec{t}\,ds = \iint_\Sigma \begin{vmatrix} \cos\alpha & \cos\beta & \cos\gamma \\ \dfrac{\partial}{\partial x} & \dfrac{\partial}{\partial y} & \dfrac{\partial}{\partial z} \\ P & Q & R \end{vmatrix} dS$$

$$= \begin{vmatrix} \cos\alpha & \cos\beta & \cos\gamma \\ \dfrac{\partial}{\partial x} & \dfrac{\partial}{\partial y} & \dfrac{\partial}{\partial z} \\ P & Q & R \end{vmatrix}_{\tilde{M}} \cdot A,$$

where $\tilde{M}$ is a point on Σ. Thus,

$$(\mathbf{rot}\ \vec{v})_M \cdot \vec{n} = \lim_{L \to M} \frac{\oint \vec{v} \cdot \vec{t}\, ds}{A} = \begin{vmatrix} \cos\alpha & \cos\beta & \cos\gamma \\ \dfrac{\partial}{\partial x} & \dfrac{\partial}{\partial y} & \dfrac{\partial}{\partial z} \\ P & Q & R \end{vmatrix}_M = \begin{vmatrix} \vec{i} & \vec{j} & \vec{k} \\ \dfrac{\partial}{\partial x} & \dfrac{\partial}{\partial y} & \dfrac{\partial}{\partial z} \\ P & Q & R \end{vmatrix} \cdot \vec{n}.$$

By letting $\vec{n}$ equal to $\vec{i}$, $\vec{j}$ and $\vec{k}$ respectively, we can get the coordinate components of $(\mathbf{rot}\ \vec{v})_M$:

$$\frac{\partial R}{\partial y} - \frac{\partial Q}{\partial z}, \quad \frac{\partial P}{\partial z} - \frac{\partial R}{\partial x}, \quad \frac{\partial Q}{\partial x} - \frac{\partial P}{\partial y}.$$

Hence, the Stokes theorem can also be stated as: Suppose $\vec{v}$ is a vector, Σ is a surface in the field with curved boundary L, $\vec{n}$ is the unit normal vector of Σ, $\vec{t}$ is the unit tangent vector of L, and $\vec{n}$ is on the left of $\vec{t}$. Then

$$\oint_L \vec{v} \cdot \vec{t}\, ds = \iint_\Sigma (\mathbf{rot}\ v) \cdot \vec{n}\, ds.$$

Also, it is easy to verify that if α, β are constants, and φ is a function, then

(1) $\mathbf{rot}\,(\alpha \vec{v_1} + \beta \vec{v_2}) = \alpha\, \mathbf{rot}\ \vec{v_1} + \beta\, \mathbf{rot}\ \vec{v_2}$;

(2) $\mathbf{rot}\ \varphi \vec{v} = \varphi\, \mathbf{rot}\ \vec{v} + \mathbf{grad}\,\varphi \times \vec{v}$;

(3) $\mathrm{div}\,(\vec{v_1} \times \vec{v_2}) = \vec{v_2} \cdot \mathbf{rot}\ \vec{v_1} - \vec{v_1} \cdot \mathbf{rot}\ \vec{v_2}$;

(4) $\mathbf{rot}\,(\vec{v_1} \times \vec{v_2}) = (\vec{v_2} \cdot \nabla)\vec{v_1} - (\vec{v_1} \cdot \nabla)\vec{v_2} + \vec{v_1}\,\mathrm{div}\ \vec{n}_2 - \vec{v_2}\,\mathrm{div}\ \vec{v_1}$;

(5) $\mathbf{rot}\,\mathbf{rot}\ \vec{v} = \nabla(\nabla \cdot \vec{v}) - \nabla^2 \vec{v}$.

Also

$$\mathbf{rot}\,\mathbf{grad}\ \vec{u} = 0, \quad \mathrm{div}\,\mathbf{rot}\ \vec{v} = 0.$$

We leave the verifications of the above equations to our readers.

With the notation of Nabla operation ∇, the rotor of a vector field can be written as

$$\mathbf{rot}\ \vec{v} = \nabla \times \vec{v} = \vec{i} \times \frac{\partial \vec{v}}{\partial x} + \vec{j} \times \frac{\partial \vec{v}}{\partial y} + \vec{k} \times \frac{\partial \vec{v}}{\partial z}.$$

Example 2. If O is a fixed point, and $\vec{r} = \overrightarrow{OM}$, then $\mathbf{rot}\ \vec{r} = 0$. If $\vec{v}$ is a constant vector, $r = |\overrightarrow{OM}|$, then

$$\mathbf{rot}\ r\vec{v} = r\,\mathbf{rot}\ \vec{v} + \mathbf{grad}\,r \times \vec{v}$$

$$= 0 + \frac{\vec{r}}{r} \times \vec{v} = \frac{\vec{r} \times \vec{v}}{r}.$$

Example 3. A rigid body rotates about an axis that passes the point O, and with the angular velocity $\vec{\omega}$. For any point M on this rigid body, it has a velocity $\vec{v} = \vec{\omega} \times \overrightarrow{OM} = \vec{\omega} \times \vec{r}$ at any moment. Since $\vec{\omega}$ is a constant vector, we have

$$\mathbf{rot}\ \vec{v} = \mathbf{rot}\ (\vec{\omega} \times \vec{r})$$
$$= (\vec{r} \cdot \nabla)\vec{\omega} - (\vec{\omega} \cdot \nabla)\vec{r} + \vec{\omega}\,\mathrm{div}\,\vec{r} - \vec{r}\,\mathrm{div}\,\vec{\omega}$$
$$= -(\vec{\omega} \cdot \nabla)\vec{r} + \vec{\omega}\,\mathrm{div}\,\vec{r} = -\vec{\omega} + 3\vec{\omega} = 2\vec{\omega}.$$

Again, this shows that rotor is a measurement of the rotation of a vector field.

Exercises 7.4

1. Find the following line integrals by applying Green's theorem.

(1) $\oint_C xy^2\,dy - x^2 y\,dx$, where C is the circle $x^2 + y^2 = a^2$, and the direction of the line integral around C is in the counterclockwise direction.

(2) $\oint_C (x + y)\,dx - (x - y)\,dy$, where C is the ellipse

$$\frac{x^2}{a^2} + \frac{y^2}{b^2} = 1,$$

and the direction of the line integral around C is in the counterclockwise direction.

(3) $\oint_C e^x[(1 - \cos y)\,dx - (y - \sin y)\,dy]$, where C is the boundary of the region $0 < x < \pi$, $0 < y < \sin x$, and the direction of the line integral around C is in the counterclockwise direction.

(4) $\int_C (e^x \sin y - my)\,dx + (e^x \cos y - m)\,dy$, where C is the upper half of the circle $x^2 + y^2 = ax$, from the point $A(a, 0)$ to the origin $O(0, 0)$.

(5) $\int_C (xy + x + y)\,dx + (xy + x - y)\,dy$,
(i) where C is the ellipse

$$\frac{x^2}{a^2} + \frac{y^2}{b^2} = 1,$$

and the direction of the line integral around C is in the counterclockwise direction.
(ii) where C is the circle $x^2 + y^2 = ax$, and the direction of the line integral around C is in the counterclockwise direction.

2. Show that $\int_c (yx^3 + e^y)\,dx + (xy^3 + xe^y - 2y)\,dy = 0$, if C is a closed curve that is symmetric in both the x-axis and the y-axis.

3. Evaluate $\oint_C \dfrac{x\,dy - y\,dx}{x^2 + y^2}$, where C is a simple closed curve which does not pass the origin. (Study two situations: the origin is inside of C, and the origin is outside of C.)

4. Evaluate the following integrals by Gauss's Theorem.

   (1) $\displaystyle\iint_S x^2\,dy\,dz + y^2\,dz\,dx + z^2\,dx\,dy$, where S is the outside surface of the boundary of the cube $0 \leqslant x \leqslant a,\, 0 \leqslant y \leqslant a,\, 0 \leqslant z \leqslant a$.

   (2) $\displaystyle\iint_S x^3\,dy\,dz + y^3\,dz\,dx + z^3\,dx\,dy$, where S is the outside surface of the sphere $x^2 + y^2 + z^2 = a^2$.

   (3) $\displaystyle\iint_S yz\,dx\,dy + zx\,dy\,dz + xy\,dz\,dx$, where S is the outside surface of the region bounded by the cylinder $x^2 + y^2 = R^2$, and the planes $z = 0,\, z = H,\, (H > 0)$ in the first octant.

   (4) $\displaystyle\iint_S xz\,dx\,dy + xy\,dy\,dz + yz\,dz\,dx$, where S is the outside surface of the tetrahedron bounded by $x = 0,\, y = 0,\, z = 0,\, x + y + z = 1$.

   (5) $\displaystyle\iint_S (x-y)\,dx\,dy + (y-z)\,dy\,dz + (z-x)\,dz\,dx$, where S is the outside surface of the cylinder $x^2 + y^2 = 1,\, z = 0,\, z = 3$.

5. Evaluate the following integrals by Stokes' Theorem.

   (1) $\displaystyle\oint_C y\,dx + z\,dy + x\,dz$, where C is a circle:

   $$x^2 + y^2 + z^2 = a^2 \text{ and } x + y + z = 0$$

   and the direction of the line integral around C is in the counterclockwise direction from the perspective of the positive z-axis.

   (2) $\displaystyle\oint_C (y + z)\,dx + (z + x)\,dy + (x + y)\,dz$, where C is an ellipse:

   $$x = a\sin^2 t,\ y = 2a\sin t\cos t,\ z = a\cos^2 t,\ (0 \leqslant t \leqslant \pi),$$

   and the line integral around C follows the direction of parameter increases.

   (3) $\displaystyle\oint_C (y - z)\,dx + (z - x)\,dy + (x - y)\,dz$, where C is an ellipse:

   $$x^2 + y^2 = a^2 \text{ and } \frac{x}{a} + \frac{z}{b} = 1,\ (a > 0,\, b > 0),$$

and the direction of line integral around C is in the counterclockwise direction from the perspective of the positive z-axis.

(4) $\oint_C y\,dx + z\,dy + x\,dz$, where C is the circle:

$$x^2 + y^2 + z^2 = a^2 \text{ and } z = \frac{a}{2},$$

and the direction of the line integral around C is in the counterclockwise direction from the perspective of the positive z-axis.

6. Compute the divergence and the rotor of the following vector fields.

   (1) $\vec{A} = (y^2 + z^2)\vec{i} + (z^2 + x^2)\vec{j} + (x^2 + y^2)\vec{k}$;

   (2) $\vec{C} = x^2 yz\vec{i} + xy^2 z\vec{j} + xyz^2\vec{k}$.

7. Compute:

   (1) $\operatorname{div}(\mathbf{grad}(x^2 + y^2 + z^2))$;

   (2) $\operatorname{div}(\mathbf{grad} f(r))$, where $r = \sqrt{x^2 + y^2 + z^2}$.

8. Let $\vec{r} = x\vec{i} + y\vec{j} + z\vec{k}$, $r = |\vec{r}|$ and $\vec{a}$, $\vec{b}$, $\vec{c}$ be constant vectors. Compute the rotor of the following vector fields.

   (1) $f(\vec{r})\vec{c}$;   (3) $f(r)\vec{r}$;   (5) $\vec{c} \times \vec{r}$;

   (2) $\vec{r}$;   (4) $(\vec{r} \cdot \vec{a})\vec{b}$;   (6) $\vec{c} \times f(r)\vec{r}$.

9. Let $\vec{r} = x\vec{i} + y\vec{j} + z\vec{k}$, $r = |\vec{r}|$ and $\vec{c}$ a constant vector. Compute the divergence of the following vector fields.

   (1) $\vec{r}$;

   (2) $\dfrac{\vec{r}}{r}$;   (3) $\dfrac{\vec{r}}{r^3}$;

   (4) $\vec{c} \times \vec{r}$;   (5) $f(r)\vec{c}$;   (6) $\nabla f(\vec{r})$.

10. Find the rotor of the following vector fields.

    (1) The electric field generated by a point electric charge.

    (2) The velocity field generated by the rotation of a rigid body about a fixed axis with a constant angular velocity.

11. Find the area of the regions having as their boundaries the following closed curves (Use line integrals).

    (1) The ellipse $x = a\cos t$, $y = b\sin t$, $(0 \leqslant t \leqslant 2\pi)$;

    (2) The asteroid $x = a\cos^3 t$, $y = a\sin^3 t$, $(0 \leqslant t \leqslant 2\pi)$;

    (3) The cycloid $x = a(t - \sin t)$, $y = a(1 - \cos t)$, $(0 \leqslant t \leqslant 2\pi)$ and the x-axis.

7.5 Total Differentials and Line Integrals

7.5.1 *Line Integrals that are Independent of Paths*

Suppose A and B are two points in a region Σ, and l is a curve which lays inside of Σ with end points A and B. If the value of a line integral

$$\int_l P\,dx + Q\,dy + R\,dz$$

depends only on the initial point A and the end point B of the curve, and is independent of the choice of l, then such a line integral is said to be *independent of the path.*

We start with the planar case. Assume that Σ is a simply connected region, that is, any closed curve can continuously contract to a point in this region. In other words, this region has no "holes" in it.

Suppose P and Q are continuous in Σ with continuous partial derivatives. We want to prove that the integral

$$\int_{(A)}^{(B)} P(x,y)\,dx + Q(x,y)\,dy$$

is independent of the path if and only if $\dfrac{\partial P}{\partial y} = \dfrac{\partial Q}{\partial x}$. This condition is called the *exact differential condition.* First, we show that

$$\int_{(A)}^{(B)} P(x,y)\,dx + Q(x,y)\,dy$$

is independent of the path if and only if

$$\int_l P(x,y)\,dx + Q(x,y)\,dy = 0$$

for any closed curve l inside Σ.

Let l_1 and l_2 be two curves from A to B (Figure 7.51), and

$$\int_{l_1} P\,dx + Q\,dy = \int_{l_2} P\,dx + Q\,dy.$$

Then we have

$$\int_{l_1} P\,dx + Q\,dy - \int_{l_2} P\,dx + Q\,dy = 0.$$

Let l be a closed curve consisting of l_1 and l_2. Then

$$\int_l P\,dx + Q\,dy = 0.$$

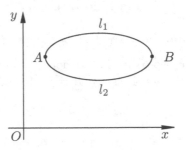

Fig. 7.51

This implies that such an integral is zero when l is any closed curve in the domain.

On the other hand, if such an integral is zero for any closed curve inside Σ, then the integral is independent of paths.

Now we show that a line integral is zero for any closed curve inside Σ if and only if $\dfrac{\partial P}{\partial y} = \dfrac{\partial Q}{\partial x}$.

By Green's theorem, if l is a closed curve, then

$$\int_l P\, dx + Q\, dy = \iint_\Sigma \left(\frac{\partial Q}{\partial x} - \frac{\partial P}{\partial y} \right) dx\, dy.$$

Thus, the condition of the integral is zero for any closed curve is equivalent to

$$\iint_\Sigma \left(\frac{\partial Q}{\partial x} - \frac{\partial P}{\partial y} \right) dx\, dy = 0$$

for any region Σ.

If $\dfrac{\partial Q}{\partial x} - \dfrac{\partial P}{\partial y} \neq 0$, then there exists a point (x_0, y_0) such that $\dfrac{\partial Q}{\partial x} - \dfrac{\partial P}{\partial y} \neq 0$, without loss of generosity, we assume that $\dfrac{\partial Q}{\partial x} - \dfrac{\partial P}{\partial y} > 0$ at this point.

Since $\dfrac{\partial Q}{\partial x}$ and $\dfrac{\partial P}{\partial y}$ are continuous, $\dfrac{\partial Q}{\partial x} - \dfrac{\partial P}{\partial y} \geqslant c$ for $c > 0$ in a small circle Σ_0 with the center (x_0, y_0). This implies that

$$\iint_{\Sigma_0} \left(\frac{\partial Q}{\partial x} - \frac{\partial P}{\partial y} \right) dx\, dy \geqslant c\Sigma_0 > 0,$$

and this contradicts with the assumption. Therefore, if the line integral is independent of the path, then

$$\frac{\partial Q}{\partial x} = \frac{\partial P}{\partial y}.$$

The inverse statement is also true.

Hence, if $\dfrac{\partial Q}{\partial x} = \dfrac{\partial P}{\partial y}$ is true in Σ, then we can define

$$\int_{(x_0,y_0)}^{(x,y)} P\,dx + Q\,dy = U(x,y).$$

Fixing the variable y and consider $U(x,y)$ as a function of x, we have

$$U(x+\Delta x, y) - U(x,y) = \int_{(x_0,y_0)}^{(x+\Delta x, y)} P\,dx + Q\,dy - \int_{(x_0,y_0)}^{(x,y)} P\,dx + Q\,dy.$$

By the assumption that the integral is independent of path, we can choose the path from (x_0, y_0) to $(x+\Delta x, y)$ to be the path from (x_0, y_0) to (x, y) followed by the path from (x, y) to $(x+\Delta x, y)$. Then

$$U(x+\Delta x, y) - U(x,y) = \int_{(x,y)}^{(x+\Delta x, y)} P\,dx + Q\,dy = \int_x^{x+\Delta x} P\,dx.$$

It follows that

$$\frac{\partial U}{\partial x} = \lim_{\Delta x \to 0} \frac{U(x+\Delta x, y) - U(x,y)}{\Delta x} = P(x,y).$$

Similarly,

$$\frac{\partial U}{\partial y} = Q(x,y).$$

Therefore

$$dU = \frac{\partial U}{\partial x}\,dx + \frac{\partial U}{\partial y}\,dy = P\,dx + Q\,dy.$$

This implies that if $\dfrac{\partial P}{\partial y} = \dfrac{\partial Q}{\partial x}$, then $P\,dx + Q\,dy$ is a total differential of a function $U(x,y)$.

Since the opposite conclusion is also true, we can say that the differential form $P\,dx + Q\,dy$ is the total differential of a function U if and only if

$$\frac{\partial P}{\partial y} = \frac{\partial Q}{\partial x}.$$

The function U is given by

$$U(x,y) = \int_{(x_0,y_0)}^{(x,y)} P\,dx + Q\,dy + C,$$

where C is the integral constant.

The assumption of simply connectedness for the region Σ is indispensable. For instance, If Σ is the region $x^2 + y^2 > 0$, and the integral curve is $x^2 + y^2 = 1$, then

$$\oint \frac{x\,dy + y\,dx}{x^2 + y^2} = \int_0^{2\pi} (\cos^2 t + \sin^2 t)\,dt = 2\pi \neq 0.$$

But the following equations hold on Σ

$$\frac{\partial}{\partial x}\frac{x}{x^2 + y^2} = \frac{y^2 - x^2}{(y^2 + x^2)^2} = \frac{\partial}{\partial y}\frac{-y}{x^2 + y^2}.$$

This is because the region $x^2 + y^2 > 0$ is not simply connected. There is a hole at the origin, and the inside of the integration curve $x^2 + y^2 = 1$ includes the origin.

Example: Solve the differential equation

$$(x + y + 1)\,dx + (x - y^2 + 3)\,dy = 0.$$

Solution: Let $P = x + y + 1$, $Q = x - y^2 + 3$. Since

$$\frac{\partial Q}{\partial x} = 1 = \frac{\partial P}{\partial y},$$

there exits a function u such that

$$du = (x + y + 1)\,dx + (x - y^2 + 3)\,dy.$$

Thus, we have $u = C$ (where C is a constant) from $du = 0$. On the other hand, we choose the integral curve as the polygonal line OBM (Figure 7.52), then

$$u(x, y) = \int_{0,0}^{x,y} (x + y + 1)\,dx + (x - y^2 + 3)\,dy$$

$$= \int_{OB} + \int_{BM} (x + y + 1)\,dx + (x - y^2 + 3)\,dy$$

$$= \int_0^x (x + 1)\,dx + \int_0^y (x - y^2 + 3)\,dy$$

$$= \frac{1}{2}x^2 + x + xy - \frac{1}{3}y^3 + 3y.$$

Therefore, the solution of this differential equation is

$$3x^2 + 6x + 6xy - 2y^2 + 18y = C.$$

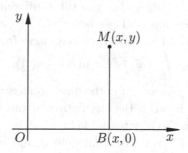

Fig. 7.52

The line integral $\int P\,dx + Q\,dy + R\,dz$ is independent of path if and only if

$$\frac{\partial R}{\partial y} = \frac{\partial Q}{\partial z}, \quad \frac{\partial P}{\partial z} = \frac{\partial R}{\partial x}, \quad \frac{\partial Q}{\partial x} = \frac{\partial P}{\partial y}.$$

hold in the integral region.

This conclusion can be obtained the same way as the corresponding conclusion in two dimensions by applying Stokes' theorem instead of Green's theorem.

7.5.2 *Potential Fields*

Suppose $\vec{v} = \vec{v}(M)$ is a vector field in the three-dimensional region V. If there exists a function $u = u(M)$ such that $\vec{v} = \mathbf{grad}\, u$ in V, then u is said to be a *potential function* of $\vec{v}$, and $\vec{v}$ is a *potential field*.

Let $\vec{v} = P\vec{i} + Q\vec{j} + R\vec{k}$. Since

$$\mathbf{grad}\, u = \frac{\partial u}{\partial x}\vec{i} + \frac{\partial u}{\partial y}\vec{j} + \frac{\partial u}{\partial z}\vec{k},$$

that is

$$P = \frac{\partial u}{\partial x}, \quad Q = \frac{\partial u}{\partial y}, \quad R = \frac{\partial u}{\partial z},$$

we have

$$\frac{\partial R}{\partial y} = \frac{\partial Q}{\partial z}, \quad \frac{\partial P}{\partial z} = \frac{\partial R}{\partial x}, \quad \frac{\partial Q}{\partial x} = \frac{\partial P}{\partial y}.$$

This implies that the line integral is independent of path, and there exists a function u such that

$$P\,dx + Q\,dy + R\,dz = du.$$

In other words, $P\,dx + Q\,dy + R\,dz$ is a total differential.

By the study in Section 7.5.1, we have:

(i) If $\vec{v}$ is a potential field, and L_{AB} is a curve from A to B in V, then

$$\int_{L_{AB}} \vec{v} \cdot \vec{t}\,ds = u(B) - u(A),$$

where $\vec{t}$ is the unit tangent vector in the direction towards B (Section 7.2.6).

(ii) On the other hand, if the circulation of the vector field $\vec{v}$ is independent of path, then $\vec{v}$ is a potential field.

(iii) A vector field $\vec{v}$ is a potential field if and only if the circulation around all closed curves in the field are zero.

(iv) A vector field $\vec{v}$ is a potential field if and only if $\vec{v}$ is an irrotational field (that is $\mathbf{rot}\ \vec{v} = 0$). This can be considered as a criterion for checking for a potential field.

Example 1. Suppose O is a fixed point, $\vec{r} = \overrightarrow{OM}$, $r = |\overrightarrow{OM}|$. Prove that the central field $\vec{v} = \varphi(r)\vec{r}$ is a potential field.

Proof:
$$\mathbf{rot}\ \vec{v} = \varphi(r)\,\mathbf{rot}\ \vec{r} + \mathbf{grad}\ \varphi(r) \times \vec{r} = \vec{0} + \varphi'(r)\frac{\vec{r}}{r} \times \vec{r} = \vec{0}.$$

Example 2. Suppose there is an electric charge q at a point O, generating an electric field

$$\vec{E} = q\frac{\overrightarrow{OM}}{|OM|^3} = q\frac{\vec{r}_M}{r_M^3}.$$

This is a central field. By Example 1, it is a potential field. Find its potential function.

Solution: In Figure 7.53, F is a fixed point, $L_{M'M}$ is on the spherical surface, M' is on the extension line of the ray OF. We have

$$u(M) = \int_F^M \vec{E} \cdot \vec{t}\,ds = \int_{FM'} \vec{E} \cdot \vec{t}\,ds + \int_{L_{M'M}} \vec{E} \cdot \vec{t}\,ds$$

$$= \int_{FM'} \vec{E} \cdot \vec{t}\,ds = \int_{FM'} q\frac{\vec{r}_p}{r_p^3} \cdot \frac{\vec{r}_p}{r_p}\,ds$$

$$= q\int_{FM'} \frac{ds}{r_p^2} = q\int_{r_F}^{r_{M'}} \frac{dr_p}{r_p^2}.$$

(Here we use the fact that $s = r_p - r_F$, so $ds = dr_p$.)

Since $r_{M'} = r_M$, we have

$$u(M) = -\frac{q}{r_M} + \frac{q}{r_F} = -\frac{q}{r_M} + C.$$

In physics, it is customary to call $U = \dfrac{q}{r_M}$ the potential function of E. And we have $\vec{E} = -\mathbf{grad}\ U$.

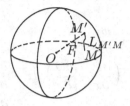

Fig. 7.53

7.5.3 *Solenoidal Vector Fields*

Suppose there is a vector field $\vec{v} = \vec{v}(M)$ in a closed region V. If there exists a vector $\vec{b}$ such that $\vec{v} = \mathbf{rot}\ \vec{b}$, then $\vec{v}$ is said to be a *rotor field* (or *curl field*) of $\vec{b}$, and $\vec{b}$ is said to be a *vector potential* of $\vec{v}$. A vector field $\vec{v}$ has a vector potential (that is, $\vec{v}$ is a curl field) if and only if div $\vec{v} = 0$ (that is, $\vec{v}$ is a *source-free field*).

In fact, if $\vec{v}$ has the vector potential $\vec{b}$, then

$$\text{div}\ \vec{v} = \text{div}\ \mathbf{rot}\ \vec{b} = 0.$$

On the other hand, if div $\vec{v} = 0$, then the vector potential $\vec{b}$ of $\vec{v}$ can be found in the following way:

Assume that $\vec{v} = P\vec{i} + Q\vec{j} + R\vec{k}$, and $\vec{b} = b_1\vec{i} + b_2\vec{j} + b_3\vec{k}$. We have

$$\begin{vmatrix} \vec{i} & \vec{j} & \vec{k} \\ \dfrac{\partial}{\partial x} & \dfrac{\partial}{\partial y} & \dfrac{\partial}{\partial z} \\ b_1 & b_2 & b_3 \end{vmatrix} = P\vec{i} + Q\vec{j} + R\vec{k}.$$

This is equivalent to solving the following equations

$$\frac{\partial b_3}{\partial y} - \frac{\partial b_2}{\partial z} = P, \quad \frac{\partial b_1}{\partial z} - \frac{\partial b_3}{\partial x} = Q, \quad \frac{\partial b_2}{\partial x} - \frac{\partial b_1}{\partial y} = R.$$

Let $b_3 = 0$. Then from the first equation, we get

$$b_2(x, y, z) = -\int_{z_0}^{z} P(x, y.z)\, dz,$$

and from the second equation, we get

$$b_1(x, y, z) = \int_{z_0}^{z} Q(x, y.z)\, dz + f(x, y).$$

Substituting these two results into the third equation. Since div $\vec{v} = 0$, we have

$$R(x, y, z) = \frac{\partial b_2}{\partial x} - \frac{\partial b_1}{\partial y}$$

$$= -\int_{z_0}^{z} \left(\frac{\partial P}{\partial x} + \frac{\partial Q}{\partial y} \right) dz - \frac{\partial f}{\partial y} = \int_{z_0}^{z} \frac{\partial R}{\partial z} dz - \frac{\partial f}{\partial y}$$

$$= R(x, y, z) - R(x, y, z_0) - \frac{\partial f}{\partial y}.$$

Choose

$$f(x, y) = -\int_{y_0}^{y} R(x, y, z_0) \, dy.$$

We get

$$\vec{b} = \left[\int_{z_0}^{z} Q \, dz - \int_{y_0}^{y} R(x, y, z_0) \, dy \right] \vec{i} - \int_{z_0}^{z} P \, dz \vec{j}$$

such that $\vec{v} = \mathbf{rot}\, \vec{b}$.

Notice that $\mathbf{rot}\, \vec{a} = \mathbf{rot}\, \vec{b}$ if and only if $\vec{a} = \vec{b} + \mathbf{grad}\, u$. Thus, the vector potential is not unique. The difference between any two vector potentials of a vector field is a gradient.

Example: Suppose $\vec{v} = (xy + 1)\vec{i} + z\vec{j} - yz\vec{k}$. Prove that $\vec{v}$ is a curl field and find its vector potential.

Solution: Since div $\vec{v} = y + 0 - y = 0$, we conclude that $\vec{v}$ is a curl field.

In order to find the vector potential, we need to solve the following equations

$$\frac{\partial b_3}{\partial y} - \frac{\partial b_2}{\partial z} = xy + 1, \quad \frac{\partial b_1}{\partial z} - \frac{\partial b_3}{\partial x} = z, \quad \frac{\partial b_2}{\partial x} - \frac{\partial b_1}{\partial y} = -yz.$$

Let $b_3 = 0$. By the first equation, we have

$$b_2 = -\int_0^z (xy + 1) \, dz = -(xy + 1)z.$$

By the second equation, we have

$$b_1 = \int_0^z z \, dz + f(x, y) = \frac{z^2}{2} + f(x, y).$$

Substituting these two results into the third equation, we have

$$-yz = \frac{\partial b_2}{\partial x} - \frac{\partial b_1}{\partial y} = -yz - \frac{\partial f}{\partial y}.$$

Choosing $f = 0$, we get

$$\vec{b} = \frac{1}{2}z^2\vec{i} - (xy + 1)z\vec{j}.$$

We have shown that curl fields are source-free fields. Now we prove that a vector field $\vec{v}$ is a source-free field if and only if the flux of $\vec{v}$ is zero on any closed surface.

In fact, suppose div $\vec{v} = 0$, Σ is a closed surface, and V is the region inside Σ. By Gauss's theorem (Section 7.4.2) we have

$$\oiint_\Sigma \vec{v} \cdot \vec{n} \, dS = \iiint_V \operatorname{div} \vec{v} \, dV = 0.$$

On the other hand, suppose the flux of $\vec{v}$ on any closed surface is zero, M is an arbitrary point inside of the closed surface Σ, and V is the region inside of Σ (the volume of this region is also denoted as V).

By Gauss's theorem and the Mean-Value Theorem for multiple integrals, we have

$$0 = \oiint_\Sigma \vec{v} \cdot \vec{n} \, dS = \iiint_V \operatorname{div} \vec{v} \, dV = (\operatorname{div} \vec{v})_{\widetilde{M}} \cdot V,$$

where $\widetilde{M}$ is a point in V. Thus, $(\operatorname{div} \vec{v})_{\widetilde{M}} = 0$.

The point $\widetilde{M}$ will approach to M and eventually coincide with M when Σ contract to M. It follows that $(\operatorname{div} \vec{v})_M = 0$. The field is a source-free field since M is an arbitrary point in the field.

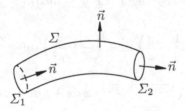

Fig. 7.54

Suppose $\vec{v}$ is a source-free field (Figure 7.54). Consider a pipe-shaped surface Σ composed of streamlines. We denote Σ_1 as the sectional plane on one end of the pipe, and Σ_2 as the sectional plane on the other end of the pipe. If $\vec{n}$ is the unit normal vector of this surface, with the direction towards inside of the pipe for $\vec{n}$ on Σ_1, and with the direction towards outside of the pipe for $\vec{n}$ on Σ_2, then

$$-\iint_{\Sigma_1} \vec{v} \cdot \vec{n} \, dS + \iint_{\Sigma} \vec{v} \cdot \vec{n} \, dS + \iint_{\Sigma_2} \vec{v} \cdot \vec{n} \, dS = 0.$$

Since $\vec{v} \cdot \vec{n} = 0$ on Σ, we have

$$\iint_{\Sigma_1} \vec{v} \cdot \vec{n} \, dS = \iint_{\Sigma_2} \vec{v} \cdot \vec{n} \, dS.$$

If $\vec{v}$ is considered as a velocity field, then this equation means that the volume of fluid flowing into the pipe from one side of the pipe is equal to the volume of fluid flowing out of the pipe from the other side of the pipe. Hence, source-free field is also called *solenoidal vector field*.

Exercises 7.5

1. Verify that the following integrals are independent of path, and find their values.

   (1) $\displaystyle\int_{(0,1)}^{(2,3)} (x+y) \, dx + (x-y) \, dy$;

   (2) $\displaystyle\int_{(2,1)}^{(1,2)} \frac{y \, dx - x \, dy}{x^2}$;   (3) $\displaystyle\int_{(1,0)}^{(6,8)} \frac{x \, dx - y \, dy}{\sqrt{x^2+y^2}}$;

   (4) $\displaystyle\int_{(-2,-1)}^{(3,0)} (x^4 + 4xy^3) \, dx + (6x^2y^2 - 5y^4) \, dy$;

   (5) $\displaystyle\int_{(1,1,1)}^{(2,2,2)} \left(1 - \frac{1}{y} + \frac{y}{z}\right) dx + \left(\frac{x}{z} + \frac{x}{y^2}\right) dy - \frac{xy}{z^2} \, dz$;

   (6) $\displaystyle\int_{(x_1,y_1,z_1)}^{(x_2,y_2,z_2)} f(\sqrt{x^2 + y^2 + z^2})(x \, dx + y \, dy + z \, dz)$,
   where f has continuous partial derivatives.

2. Verify whether the following equations are total differential equations, and solve these equations.

   (1) $x \, dy + y \, dx = 0$;   (2) $2xy \, dx + x^2 \, dy = 0$;

   (3) $e^y \, dx + (xe^y - 2y) \, dy = 0$;

   (4) $(1 + x\sqrt{x^2 + y^2}) \, dx + (-1 + \sqrt{x^2 + y^2})y \, dy = 0$.

3. Show that the field
$$\vec{a} = yz(2x + y + z)\vec{i} + zx(2y + z + x)\vec{j} + xy(2z + x + y)\vec{k}$$
is a potential field, and find its potential function.

4. Show that the central force field $\vec{F} = f(r)\dfrac{\vec{r}}{r}$ is a potential field, and find its potential function.

5. Determine whether the following vector fields are potential fields, source-free (zero-divergence) fields, or harmonic fields (potential field and source-free field).

(1) $\vec{a} = (xy + y^2)\vec{i} + x^2 y\vec{j}$;

(2) $\vec{a} = (2x + y)\vec{i} + (4y + x + 2z)\vec{j} + (2y - 6z)\vec{k}$;

(3) The electrostatic field generated by a point charge;

(4) The magnetic field generated by a straight conductor wire with infinite length;

(5) Gravitational field.

6. Verify that the following fields are source-free fields.

(1) $\vec{a} = z\vec{i} + x\vec{j} + y\vec{k}$; (2) $\vec{a} = xy\vec{i} - y^2\vec{j} + yz\vec{k}$;

(3) $\vec{a} = (z - y)\vec{i} + (x - z)\vec{j} + (y - z)\vec{k}$.

7.6 Exterior Differential Forms

7.6.1 *Exterior Products, Exterior Differential Forms*

In the last several sections, we have seen some differential forms when we studied line integrals, surface integrals and volume integrals. For instance, in the line integral
$$\int A\,dx + B\,dy + C\,dz,$$
$$\omega = A\,dx + B\,dy + C\,dz$$
is a first degree expression of differentials dx, dy, dz. We call it a first degree differential form.

In the surface integral
$$\iint P\,dy\,dz + Q\,dz\,dx + R\,dx\,dy,$$
$P\,dy\,dz + Q\,dz\,dx + R\,dx\,dy$ is a second degree expression of differentials dx, dy, dz. We call it a second degree differential form.

In the volume integral

$$\iiint H \, dx \, dy \, dz,$$

$H \, dx \, dy \, dz$ is a third degree expression of differentials dx, dy, dz. We call it a third degree differential form.

In the three dimensional space, these are all the possible differential forms. We also consider a function f as a zero degree form. Notice that in all the differential forms above, dx, dy, dz appear at most once in each term. We don't see terms involving $dx \, dx$, $dy \, dz \, dy$, $\cdots$.

Also, we have three formulas that connect line integrals, surface integrals and volume integrals:

(1) Green's Theorem

$$\oint_L P \, dx + Q \, dy = \iint_D \left(\frac{\partial Q}{\partial x} - \frac{\partial P}{\partial y} \right) dx \, dy,$$

where D is a closed region bounded by L, P and Q are functions on D with continuous first order partial derivatives.

(2) Gauss's Theorem

$$\oiint_{\Sigma_{\text{outside}}} P \, dy \, dz + Q \, dz \, dx + R \, dx \, dy = \iiint_V \left(\frac{\partial P}{\partial x} + \frac{\partial Q}{\partial y} + \frac{\partial R}{\partial z} \right) dx \, dy \, dz,$$

where V is a closed three-dimension region bounded by the closed surface Σ, P, Q and R are functions on V with continuous first order partial derivatives.

(3) Stokes' Theorem

$$\oint_L P \, dx + Q \, dy + R \, dz = \iint_\Sigma \left(\frac{\partial R}{\partial y} - \frac{\partial Q}{\partial z} \right) dy \, dz$$

$$+ \left(\frac{\partial P}{\partial z} - \frac{\partial R}{\partial x} \right) dz \, dx + \left(\frac{\partial Q}{\partial x} - \frac{\partial P}{\partial y} \right) dx \, dy,$$

where the closed curve L is the boundary of the surface Σ, P, Q and R are functions on Σ with continuous first order partial derivatives.

Suppose u is a scalar field, and $\vec{v} = P\vec{i} + Q\vec{j} + R\vec{k}$ is a vector field. Then

$$\mathbf{grad}\, u = \nabla u = \frac{\partial u}{\partial x}\vec{i} + \frac{\partial u}{\partial y}\vec{j} + \frac{\partial u}{\partial z}\vec{k},$$

$$\mathbf{rot}\, \vec{v} = \nabla \times \vec{v} = \begin{vmatrix} \vec{i} & \vec{j} & \vec{k} \\ \dfrac{\partial}{\partial x} & \dfrac{\partial}{\partial y} & \dfrac{\partial}{\partial z} \\ P & Q & R \end{vmatrix},$$

$$\mathrm{div}\, \vec{v} = \nabla \cdot \vec{v} = \frac{\partial P}{\partial x} + \frac{\partial Q}{\partial y} + \frac{\partial R}{\partial z}$$

are the mathematical aspects of gradient, curl(rotor) and divergence.

What is the connection between above three formulas? What is the Newton-Leibniz formula in higher-dimensions?

To discover the internal connections among the three formulas and to answer the above two questions, we need to introduce the concept of exterior differential forms.

The integral domains of line and surface integrals of second kind that we have studied are oriented. A curve L with end points A and B has two orientations. One is from A to B, the other is from B to A. The value of the line integral along L from A to B and the value of the line integral along L from B to A differ by a sign. This is similar to the following formula in definite integrals

$$\int_a^b f(x)\, dx = -\int_b^a f(x)\, dx.$$

A surface is orientable if a normal vector of the surface moves from a point continuously and its direction remains unchanged when it gets back to the starting point. In other words, the surface has two sides, called inside and outside, and the corresponding normal vectors, inward-pointing normal, and outer-pointing normal.

Notice that there exist surfaces with only one side. When a normal vector of such a surface moves from a point of the surface continuously and get back to the starting point, the direction of the vector may change to the opposite side. To have a clear view of this kind of surface, we take a paper strip with vertices A, B, C, D, twist one short edge before glueing the two short edges together. That is, let A coincide with C, and B coincide with D as in Figure 7.55 and Figure 7.56. We can see that this surface has only one side, and it is a non-orientable surface. We do not study non-orientable surfaces in this book.

The values of surface integrals on an orientable surface differ by a sign if a function is integrated over a surface with opposite orientations.

Recall the definition of double integrals: Suppose a function $f(x, y)$ is defined on D, the integral of $f(x, y)$ on D is

$$\iint_D f(x, y)\, dA = \lim \sum f(\xi_i, \eta_i)\, \Delta A_i,$$

where ΔA_i are area elements.

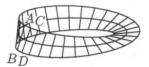

Fig. 7.55   Fig. 7.56

Thus, if

$$x = x(u, v), \quad y = y(u, v),$$

then

$$dA = dx\, dy = \left| \frac{\partial(x, y)}{\partial(u, v)} \right| du\, dv,$$

$$\iint_D f(x, y)\, dx\, dy = \iint_{D'} f(x(u, v), y(u, v)) \left| \frac{\partial(x, y)}{\partial(u, v)} \right| du\, dv.$$

Since the orientation of the area elements is not considered, we have to take the absolute value of the Jacobian determinant. If the orientation of the area elements is considered, then we don't have to take the absolute value of the Jacobian determinant

$$\iint_D f(x, y)\, dx\, dy = \iint_{D'} f(x(u, v), y(u, v)) \frac{\partial(x, y)}{\partial(u, v)}\, du\, dv,$$

where D is an oriented region and D' is the region under the inverse transformation of

$$x = x(u, v), y = y(u, v),$$

of D, and it is also orientated. Hence

$$dx\, dy = \frac{\partial(x, y)}{\partial(u, v)}\, du\, dv = \begin{vmatrix} \dfrac{\partial x}{\partial u} & \dfrac{\partial x}{\partial v} \\ \dfrac{\partial y}{\partial u} & \dfrac{\partial y}{\partial v} \end{vmatrix}\, du\, dv.$$

From this formula, we can see that

(i) If we take $y = x$, then

$$dx\, dx = \begin{vmatrix} \dfrac{\partial x}{\partial u} & \dfrac{\partial x}{\partial v} \\ \dfrac{\partial x}{\partial u} & \dfrac{\partial x}{\partial v} \end{vmatrix}\, du\, dv = 0.$$

(ii) If we switch the position of x and y, then

$$dy\, dx = \frac{\partial(y, x)}{\partial(u, v)}\, du\, dv = \begin{vmatrix} \dfrac{\partial y}{\partial u} & \dfrac{\partial y}{\partial v} \\ \dfrac{\partial x}{\partial u} & \dfrac{\partial x}{\partial v} \end{vmatrix}\, du\, dv = - \begin{vmatrix} \dfrac{\partial x}{\partial u} & \dfrac{\partial x}{\partial v} \\ \dfrac{\partial y}{\partial u} & \dfrac{\partial y}{\partial v} \end{vmatrix}\, du\, dv.$$

Therefore, $dy\, dx \neq dx\, dy$. In other words, the product of dx and dy will change its sign if the order of dx and dy is switched.

A product of differentials with these characteristics is called the *exterior product of differentials*. It is denoted as $dx \wedge dy$ to distinguish it from the ordinary product of differentials. With this notation, we have

$$dx \wedge dx = 0,$$
$$dx \wedge dy = -dy \wedge dx.$$

Of course, we can deduce $dx \wedge dx = 0$ from $dx \wedge dy = -dy \wedge dx$.

An *exterior differential form* consists of products of functions and exterior products of differentials. If P, Q, R, A, B, C and H are functions of variables x, y and z, then

$$P\, dx + Q\, dy + R\, dz$$

is called a first degree exterior differential form (or 1-form) and it is the same as the ordinary differential form.

The expression

$$A\, dx \wedge dy + B\, dy \wedge dz + C\, dz \wedge dx$$

is called a second degree exterior differential form (or 2-form), and the expression

$$H \, dx \wedge dy \wedge dz$$

is called a third degree exterior differential form (or 3-form). The functions P, Q, R, A, B, C and H are called the coefficients of the corresponding differential forms.

For instance,

$$\sin y \, dx + \tan z \, dy + \cos x \, dz$$

is a first degree exterior differential form;

$$z \, dx \wedge dy + x \, dy \wedge dz + y \, dz \wedge dx$$

is a second degree exterior differential form; and

$$e^{x+y+z} \, dx \wedge dy \wedge dz$$

is a third degree exterior differential form.

The exterior product $\lambda \wedge \mu$ of any two exterior differential forms λ and μ can be defined as the sum of the exterior products of the terms of λ and μ.

Let A, B, C, E, F, G, P, Q and R be functions of variables x, y and z, and

$$\lambda = A \, dx + B \, dy + C \, dz,$$
$$\mu = E \, dx + F \, dy + G \, dz,$$
$$\nu = P \, dy \wedge dz + Q \, dz \wedge dx + R \, dx \wedge dy.$$

Then

$$\begin{aligned}
\lambda \wedge \mu &= (A \, dx + B \, dy + C \, dz) \wedge (E \, dx + F \, dy + G \, dz) \\
&= AE \, dx \wedge dx + BE \, dy \wedge dx + CE \, dz \wedge dx \\
&\quad + AF \, dx \wedge dy + BF \, dy \wedge dy + CF \, dz \wedge dy \\
&\quad + AG \, dx \wedge dz + BG \, dy \wedge dz + CG \, dz \wedge dz.
\end{aligned}$$

Since

$$dx \wedge dx = dy \wedge dy = dz \wedge dz = 0,$$
$$dy \wedge dx = -dx \wedge dy, \ dz \wedge dy = -dy \wedge dz, \ dx \wedge dz = -dz \wedge dx,$$

we have

$$\begin{aligned}
\lambda \wedge \mu &= (BG - CF) \, dy \wedge dz + (CE - AG) \, dz \wedge dx \\
&\quad + (AF - BE) \, dx \wedge dy.
\end{aligned}$$

Similarly,

$$\lambda \wedge \nu = (A\,dx + B\,dy + C\,dz) \wedge (P\,dy \wedge dz + Q\,dz \wedge dx + R\,dx \wedge dy)$$
$$= AP\,dx \wedge dy \wedge dz + BP\,dy \wedge dy \wedge dz + CP\,dz \wedge dy \wedge dz$$
$$+ AQ\,dx \wedge dz \wedge dx + BQ\,dy \wedge dz \wedge dx + CQ\,dz \wedge dz \wedge dx$$
$$+ AR\,dx \wedge dx \wedge dy + BR\,dy \wedge dx \wedge dy + CR\,dz \wedge dx \wedge dy.$$

Since

$$dy \wedge dy \wedge dz = 0,$$
$$dz \wedge dy \wedge dz = -dy \wedge dz \wedge dz = 0,$$
$$dx \wedge dz \wedge dx = -dx \wedge dx \wedge dz = 0,$$
$$dy \wedge dz \wedge dx = -dy \wedge dx \wedge dz = (-1)^2 dx \wedge dy \wedge dz = dx \wedge dy \wedge dz,$$
$$dz \wedge dz \wedge dx = 0,$$
$$dx \wedge dx \wedge dy = 0,$$
$$dy \wedge dx \wedge dy = -dy \wedge dy \wedge dx = 0,$$
$$dz \wedge dx \wedge dy = -dx \wedge dz \wedge dy = (-1)^2 dx \wedge dy \wedge dz = dx \wedge dy \wedge dz,$$

we have

$$\lambda \wedge \nu = (AP + BQ + CR)\,dx \wedge dy \wedge dz.$$

It follows that, the exterior product of exterior differential forms satisfies the distributive law and the associative law:

If λ, μ and ν are exterior differential forms, then

(1) $(\lambda + \mu) \wedge \nu = \lambda \wedge \nu + \mu \wedge \nu; \quad \lambda \wedge (\mu + \nu) = \lambda \wedge \mu + \lambda \wedge \nu.$

(2) $\lambda \wedge (\mu \wedge \nu) = (\lambda \wedge \mu) \wedge \nu.$

Obviously, the exterior product of exterior differential forms does not satisfy the commutative law, instead, it satisfies the equation

(3) $\mu \wedge \lambda = (-1)^{pq} \lambda \wedge \mu,$

where λ is an exterior differential form of degree p and μ is an exterior differential form of degree q. We leave the verification of this property to our readers.

Actually, we have encountered an operation which follows the same rules as the exterior product of exterior differential forms, the cross product of vectors in Chapter 5:

$$\vec{a} \times \vec{a} = \vec{0}, \quad \vec{a} \times \vec{b} = -\vec{b} \times \vec{a}$$

for any vectors $\vec{a}$ and $\vec{b}$. So the calculation of the exterior products of differential forms shouldn't be unfamiliar for our readers.

7.6.2 *Exterior Differentiation, Poincaré Lemma and its Inverse*

The exterior differential operator d is defined as follows:

For a function f, which is considered as an exterior differential form of degree zero,

$$df = \frac{\partial f}{\partial x}\,dx + \frac{\partial f}{\partial y}\,dy + \frac{\partial f}{\partial z}\,dz.$$

This is the operation of total differential. For an exterior differential form of the first degree

$$\omega = P\,dx + Q\,dy + R\,dz,$$

we define

$$d\omega = dP \wedge dx + dQ \wedge dy + dR \wedge dz.$$

That is, calculating the exterior product of the exterior differentials of P, Q and R. Since

$$dP = \frac{\partial P}{\partial x}\,dx + \frac{\partial P}{\partial y}\,dy + \frac{\partial P}{\partial z}\,dz,$$

$$dQ = \frac{\partial Q}{\partial x}\,dx + \frac{\partial Q}{\partial y}\,dy + \frac{\partial Q}{\partial z}\,dz,$$

$$dR = \frac{\partial R}{\partial x}\,dx + \frac{\partial R}{\partial y}\,dy + \frac{\partial R}{\partial z}\,dz,$$

we have

$$d\omega = \left(\frac{\partial P}{\partial x}\,dx + \frac{\partial P}{\partial y}\,dy + \frac{\partial P}{\partial z}\,dz\right) \wedge dx + \left(\frac{\partial Q}{\partial x}\,dx + \frac{\partial Q}{\partial y}\,dy + \frac{\partial Q}{\partial z}\,dz\right) \wedge dy$$

$$+ \left(\frac{\partial R}{\partial x}\,dx + \frac{\partial R}{\partial y}\,dy + \frac{\partial R}{\partial z}\,dz\right) \wedge dz$$

$$= \frac{\partial P}{\partial x}\,dx \wedge dx + \frac{\partial P}{\partial y}\,dy \wedge dx + \frac{\partial P}{\partial z}\,dz \wedge dx + \frac{\partial Q}{\partial x}\,dx \wedge dy$$

$$+ \frac{\partial Q}{\partial y}\,dy \wedge dy + \frac{\partial Q}{\partial z}\,dz \wedge dy + \frac{\partial R}{\partial x}\,dx \wedge dz$$

$$+ \frac{\partial R}{\partial y}\,dy \wedge dz + \frac{\partial R}{\partial z}\,dz \wedge dz.$$

Since

$$dx \wedge dx = dy \wedge dy = dz \wedge dz = 0,$$

$$dy \wedge dx = -dx \wedge dy, \ \ dz \wedge dy = -dy \wedge dz, \ \ dx \wedge dz = -dz \wedge dx,$$

we have

$$d\omega = \left(\frac{\partial R}{\partial y} - \frac{\partial Q}{\partial z}\right) dy \wedge dz + \left(\frac{\partial P}{\partial z} - \frac{\partial R}{\partial x}\right) dz \wedge dx$$
$$+ \left(\frac{\partial Q}{\partial x} - \frac{\partial P}{\partial y}\right) dx \wedge dy.$$

Similarly, for an exterior differential form of the second degree

$$\omega = A\,dy \wedge dz + B\,dz \wedge dx + C\,dx \wedge dy,$$

we define

$$d\omega = dA \wedge dy \wedge dz + dB \wedge dz \wedge dx + dC \wedge dx \wedge dy.$$

Substitute the expressions of dA, dB and dC into the above equation, we have

$$d\omega = \left(\frac{\partial A}{\partial x} + \frac{\partial B}{\partial y} + \frac{\partial C}{\partial z}\right) dx \wedge dy \wedge dz.$$

For an exterior differential form of the third degree

$$\omega = H\,dx \wedge dy \wedge dz,$$

we define

$$d\omega = dH \wedge dx \wedge dy \wedge dz.$$

Since

$$dH = \frac{\partial H}{\partial x}\,dx + \frac{\partial H}{\partial y}\,dy + \frac{\partial H}{\partial z}\,dz,$$

according to the properties of exterior products, we have

$$d\omega = \left(\frac{\partial H}{\partial x}\,dx + \frac{\partial H}{\partial y}\,dy + \frac{\partial H}{\partial z}\,dz\right) \wedge dx \wedge dy \wedge dz$$
$$= \frac{\partial H}{\partial x}\,dx \wedge dx \wedge dy \wedge dz + \frac{\partial H}{\partial y}\,dy \wedge dx \wedge dy \wedge dz$$
$$+ \frac{\partial H}{\partial z}\,dz \wedge dx \wedge dy \wedge dz = 0,$$

by the fact that at least two differentials are the same in each term. Therefore, the exterior differential of any third degree exterior differential forms in three dimensional space is zero.

Now we can see that, in calculation, the exterior differential operator d works the same way as the differential operator except that the exterior products are applied between the factors.

Poincaré Lemma: If ω is an exterior differential form, and its coefficients have continuous second order partial derivatives, then $dd\omega = 0$.

Proof: There are four kind of exterior differential forms in three dimensional space. The exterior differential form of degree zero, which is the function itself, the exterior differential forms of the first, the second and the third degree:

$$\omega_1 = P\,dx + Q\,dy + R\,dz,$$
$$\omega_2 = A\,dx \wedge dy + B\,dy \wedge dz + C\,dz \wedge dx,$$
$$\omega_3 = H\,dx \wedge dy \wedge dz.$$

We now verify the lemma for each one of them.

For exterior differential form of zero degree $\omega = f$, we have

$$d\omega = df = \frac{\partial f}{\partial x}\,dx + \frac{\partial f}{\partial y}\,dy + \frac{\partial f}{\partial z}\,dz,$$

$$dd\omega = ddf = d\left(\frac{\partial f}{\partial x}\right) \wedge dx + d\left(\frac{\partial f}{\partial y}\right) \wedge dy + d\left(\frac{\partial f}{\partial z}\right) \wedge dz$$

$$= \left(\frac{\partial^2 f}{\partial x^2}\,dx + \frac{\partial^2 f}{\partial x \partial y}\,dy + \frac{\partial^2 f}{\partial x \partial z}\,dz\right) \wedge dx$$

$$+ \left(\frac{\partial^2 f}{\partial y \partial x}\,dx + \frac{\partial^2 f}{\partial y^2}\,dy + \frac{\partial^2 f}{\partial y \partial z}\,dz\right) \wedge dy$$

$$+ \left(\frac{\partial^2 f}{\partial z \partial x}\,dx + \frac{\partial^2 f}{\partial z \partial y}\,dy + \frac{\partial^2 f}{\partial z^2}\,dz\right) \wedge dz.$$

Since

$$dx \wedge dx = dy \wedge dy = dz \wedge dz = 0,$$
$$dy \wedge dx = -dx \wedge dy, \ \ dz \wedge dy = -dy \wedge dz, \ \ dx \wedge dz = -dz \wedge dx,$$

we have

$$dd\omega = \left(\frac{\partial^2 f}{\partial y \partial x} - \frac{\partial^2 f}{\partial x \partial y}\right) dx \wedge dy + \left(\frac{\partial^2 f}{\partial z \partial y} - \frac{\partial^2 f}{\partial y \partial z}\right) dy \wedge dz$$

$$+ \left(\frac{\partial^2 f}{\partial x \partial z} - \frac{\partial^2 f}{\partial z \partial x}\right) dz \wedge dx.$$

By the assumption that the second order partial derivatives of f are continuous, we have:

$$\frac{\partial^2 f}{\partial x \partial y} = \frac{\partial^2 f}{\partial y \partial x}, \quad \frac{\partial^2 f}{\partial y \partial z} = \frac{\partial^2 f}{\partial z \partial y}, \quad \frac{\partial^2 f}{\partial z \partial x} = \frac{\partial^2 f}{\partial x \partial z}.$$

Thus, $ddf = 0$.

Similarly, for exterior differential forms of the first degree ω_1,

$$d\omega_1 = dP \wedge dx + dQ \wedge dy + dR \wedge dz$$

$$= \left(\frac{\partial R}{\partial y} - \frac{\partial Q}{\partial z}\right) dy \wedge dz + \left(\frac{\partial P}{\partial z} - \frac{\partial R}{\partial x}\right) dz \wedge dx$$

$$+ \left(\frac{\partial Q}{\partial x} - \frac{\partial P}{\partial y}\right) dx \wedge dy.$$

Hence,

$$ddomega_1 = \left(\frac{\partial^2 R}{\partial x \partial y} - \frac{\partial^2 Q}{\partial x \partial z} + \frac{\partial^2 P}{\partial y \partial z} - \frac{\partial^2 R}{\partial y \partial x}\right.$$

$$\left. + \frac{\partial^2 Q}{\partial z \partial x} - \frac{\partial^2 P}{\partial z \partial y}\right) dx \wedge dy \wedge dz = 0.$$

For exterior differential forms of the second degree ω_2,

$$d\omega_2 = dA \wedge dy \wedge dz + dB \wedge dz \wedge dx + dC \wedge dx \wedge dy.$$

This is an exterior differential form of the third degree (3-form), so we have $dd\omega_2 = 0$ by our previous discussions.

It has been proved that $d\omega_3 = 0$ for the exterior differential form of the third degree $\omega_3 = H \, dx \wedge dy \wedge dz$ (3-form). The equation $dd\omega_3 = 0$ follows.

On the other hand, the inverse result of Poincaré lemma is also true: If ω is an exterior differential form of degree p and $d\omega = 0$, then there exists an exterior differential form α of degree $p \quad 1$ such that $\omega = d\alpha$.

In fact, we have proved this result in sections 7.5.2 and 7.5.3. We now give a more detailed proof:

For a 3-form $\omega_3 = H \, dx \wedge dy \wedge dz$, let

$$\alpha_0 = \int_0^x H(t, y, z) \, dt.$$

Then it can be verified directly that $\alpha_2 = \alpha_0 \, dy \wedge dz$ is the solution of the equation $\omega_3 = d\alpha_2$.

For a 2-form $\omega_2 = A \, dy \wedge dz + B \, dz \wedge dx + C \, dx \wedge dy$, since $d\omega_2 = 0$, we have

$$d\omega_2 = \left(\frac{\partial A}{\partial x} + \frac{\partial B}{\partial y} + \frac{\partial C}{\partial z}\right) dx \wedge dy \wedge dz = 0.$$

Thus

$$\frac{\partial A}{\partial x} + \frac{\partial B}{\partial y} + \frac{\partial C}{\partial z} = 0.$$

Obviously,

$$d\left(\int_0^y A(x,t,z)\,dt\,dz\right) = A(x,y,z)\,dy \wedge dz$$

$$+ \int_0^y \frac{\partial A(x,t,z)}{\partial x}\,dt\,dx \wedge dz,$$

$$d\left(\int_0^y -C(x,t,z)\,dt\,dx\right) = -C(x,y,z)\,dy \wedge dx$$

$$- \int_0^y \frac{\partial C(x,t,z)}{\partial z}\,dt\,dz \wedge dx.$$

Therefore,

$$d\left(\int_0^y A(x,t,z)\,dt\,dz - \int_0^y C(x,t,z)\,dt\,dx\right)$$

$$= A(x,y,z)\,dy \wedge dz + C(x,y,z)\,dx \wedge dy$$

$$- \int_0^y \left(\frac{\partial A(x,t,z)}{\partial x} + \frac{\partial C(x,t,z)}{\partial z}\right)dt\,dz \wedge dx$$

$$= A(x,y,z)\,dy \wedge dz + C(x,y,z)\,dx \wedge dy + \int_0^y \frac{\partial B(x,t,z)}{\partial t}\,dt\,dz \wedge dx$$

$$= A(x,y,z)\,dy \wedge dz + C(x,y,z)\,dx \wedge dy$$

$$+ B(x,y,z)\,dz \wedge dx - B(x,0,z)\,dz \wedge dx.$$

Let

$$\alpha_1 = \int_0^y A(x,t,z)\,dt\,dz - \int_0^y C(x,t,z)\,dt\,dx + \int_0^y B(x,0,z)\,dt\,dx.$$

Then $d\alpha_1 = \omega_2$.

For a 1-form $\omega_1 = P\,dx + Q\,dy + R\,dz$, since $d\omega_1 = 0$, we have

$$d\omega_1 = \left(\frac{\partial R}{\partial y} - \frac{\partial Q}{\partial z}\right)dy \wedge dz + \left(\frac{\partial P}{\partial z} - \frac{\partial R}{\partial x}\right)dz \wedge dx$$

$$+ \left(\frac{\partial Q}{\partial x} - \frac{\partial P}{\partial y}\right)dx \wedge dy = 0.$$

Thus

$$\frac{\partial R}{\partial y} = \frac{\partial Q}{\partial z}, \quad \frac{\partial P}{\partial z} = \frac{\partial R}{\partial x}, \quad \frac{\partial Q}{\partial x} = \frac{\partial P}{\partial y}.$$

Let $f(x,y,z) = \int_0^x P(t,y,z)\,dt$. Then we have

$$df = P(x,y,z)\,dx + \int_0^x \frac{\partial P(t,y,z)}{\partial y}\,dt\,dy + \int_0^x \frac{\partial P(t,y,z)}{\partial z}\,dt\,dz$$

$$= P(x,y,z)\,dx + \int_0^x \frac{\partial Q(t,y,z)}{\partial t}\,dt\,dy + \int_0^x \frac{\partial R(t,y,z)}{\partial t}\,dt\,dz$$

$$= P(x,y,z)\,dx + Q(x,y,z)\,dy + R(x,y,z)\,dz$$

$$- Q(0,y,z)\,dy - R(0,y,z)\,dz.$$

Therefore, if we let
$$\alpha = \int_0^x P(t, y, z)\, dt + \int_0^y Q(0, t, z)\, dt + \int_0^z R(0, 0, t)\, dt,$$
then we have $d\alpha = \omega_1$. This proves the inverse of the Poincaré lemma.

In the process of the above proof, we assumed that the regions involved are rectangular parallelepiped. Without this assumption, there is no guarantee of the existence of these integrals. However, we can reduce this assumption to a weaker form as follows:

(1) For an 1-form ω_1, if any closed curve in the region can continuously contract to a point in the region, and if $d\omega_1 = 0$, then there exists a 0-form α such that $d\alpha = \omega_1$.

(2) For a 2-form ω_2, if the region is a star-shaped region (that is, there exists a point $\overline{P_0}$ in the region, such that for any point P of the region, the line segment $\overline{PP_0}$ lay entirely inside of this region), and if $d\omega_2 = 0$, then there is a 1-form α_1, such that $d\alpha_1 = \omega_2$.

We omit the proofs of these results.

7.6.3 *Mathematical Meaning of Gradient, Divergence and Curl*

Now we can restate the results in the previous sections using exterior differential forms to gain some more clarified statements

First, let's explore the relationship between the exterior differential forms and gradient, divergence and curl.

For the 0-form $\omega = f$, the exterior derivative is
$$d\omega = df = \frac{\partial f}{\partial x}\, dx + \frac{\partial f}{\partial y}\, dy + \frac{\partial f}{\partial z}\, dz.$$
The gradient of f is
$$\mathbf{grad}\ f = \frac{\partial f}{\partial x}\vec{i} + \frac{\partial f}{\partial y}\vec{j} + \frac{\partial f}{\partial z}\vec{k}.$$
The similarity between these two expressions is obvious.

For the 1-form $\omega_1 = P\, dx + Q\, dy + R\, dz$, the exterior derivative is
$$d\omega_1 = \left(\frac{\partial R}{\partial y} - \frac{\partial Q}{\partial z}\right) dy \wedge dz + \left(\frac{\partial P}{\partial z} - \frac{\partial R}{\partial x}\right) dz \wedge dx$$
$$+ \left(\frac{\partial Q}{\partial x} - \frac{\partial P}{\partial y}\right) dx \wedge dy$$
$$= \begin{vmatrix} dy \wedge dz & dz \wedge dx & dx \wedge dy \\ \dfrac{\partial}{\partial x} & \dfrac{\partial}{\partial y} & \dfrac{\partial}{\partial z} \\ P & Q & R \end{vmatrix}.$$

The curl of vector $\vec{u} = (P, Q, R) = P\vec{i} + Q\vec{j} + R\vec{k}$ is

$$\mathbf{rot}\, u = \left(\frac{\partial R}{\partial y} - \frac{\partial Q}{\partial z}\right)\vec{i} + \left(\frac{\partial P}{\partial z} - \frac{\partial R}{\partial x}\right)\vec{j} + \left(\frac{\partial Q}{\partial x} - \frac{\partial P}{\partial y}\right)\vec{k}$$

$$= \begin{vmatrix} \vec{i} & \vec{j} & \vec{k} \\ \frac{\partial}{\partial x} & \frac{\partial}{\partial y} & \frac{\partial}{\partial z} \\ P & Q & R \end{vmatrix}.$$

We can also see the similarity between these two expressions.

For the 2-form $\omega_2 = A\, dy \wedge dz + B\, dz \wedge dx + C\, dx \wedge dy$, the exterior derivative is

$$d\omega_2 = \left(\frac{\partial A}{\partial x} + \frac{\partial B}{\partial y} + \frac{\partial C}{\partial z}\right) dx \wedge dy \wedge dz.$$

The divergence of the vector $\vec{v}$ is

$$\operatorname{div} \vec{v} = \frac{\partial A}{\partial x} + \frac{\partial B}{\partial y} + \frac{\partial C}{\partial z}.$$

Also, notice the similarity between these two expressions.

A question is raised from this point of view: Is there any other "forms"? Since the exterior differential of a 3-form is zero, the answer of this question is "no" for the three dimensional space. The corresponding relationship between exterior differential forms and gradient, divergence and curl can be listed as follows:

The Degree of Exterior Differential Forms	Vector Operators
0	Gradient
1	Curl (Rotor)
2	Divergence

Furthermore, from the point of view of field theory, the Poincaré lemma condition $dd\omega = 0$ can be interpreted as follows:

If ω is an exterior differential form of degree zero, that is, $\omega = f$, then $ddf = 0$ implies that

$$\mathbf{rot}\,\mathbf{grad}\, f = 0.$$

If ω is an exterior differential form of degree one, that is, $\omega_1 = P\, dx + Q\, dy + R\, dz$, then $dd\omega_1 = 0$ implies that

$$\operatorname{div} \mathbf{rot}\, \vec{u} = 0,$$

where $\vec{u} = (P, Q, R)$.

The inverse of the Poincaré lemma can also be interpreted from the point of view of field theory:

From Section 7.5, we know that: $\vec{v}$ is a potential field if and only if $\vec{v}$ is an irrotational field. That is, $\vec{v} = \mathbf{grad}\ f$ if and only if $\mathbf{rot}\ \vec{v} = 0$. This is the Poincaré lemma and its inverse, if $d\omega = 0$, then $\omega = d\alpha$. That is: if the exterior differential of an exterior differential form of degree one is zero, then this exterior differential form must be an exterior differential of a function.

Also, $\vec{v}$ is a curl field if and only if $\vec{v}$ is a solenoidal field. That is, $\vec{v} = \mathbf{rot}\ \vec{b}$ if and only if $\mathrm{div}\ \vec{v} = 0$. This is the Poincaré lemma and its inverse, if $d\omega = 0$, then $\omega = d\alpha$. That is: if the exterior differential of an exterior differential form of degree two is zero, then this exterior differential form must be an exterior differential of an exterior differential form of degree one.

7.6.4 Fundamental Theorem of Calculus in Several Variables (Stokes' Theorem)

The Fundamental Theorem of Calculus in several variables can be clearly stated using exterior differential forms.

Let's start from the Green's Theorem

$$\oint_L P\,dx + Q\,dy = \iint_D \left(\frac{\partial Q}{\partial x} - \frac{\partial P}{\partial y}\right) dx\,dy.$$

Denote $\omega_1 = P\,dx + Q\,dy$. Then ω_1 is an 1-form, and

$$d\omega_1 = \left(\frac{\partial Q}{\partial x} - \frac{\partial P}{\partial y}\right) dx \wedge dy.$$

Since the curve L is orientated, Green's Theorem can be written as

$$\oint \omega_1 = \iint d\omega_1.$$

Similarly, Gauss's theorem is

$$\oiint_{\Sigma_{\text{outside}}} P\,dy\,dz + Q\,dz\,dx + R\,dx\,dy = \iiint_V \left(\frac{\partial P}{\partial x} + \frac{\partial Q}{\partial y} + \frac{\partial R}{\partial z}\right) dx\,dy\,dz.$$

Since Σ is orientated, the expression

$$P\,dy\,dz + Q\,dz\,dx + R\,dx\,dy$$

can be considered as a 2-form

$$\omega_2 = P\,dy \wedge dz + Q\,dz \wedge dx + R\,dx \wedge dy.$$

Then

$$d\omega_2 = \left(\frac{\partial P}{\partial x} + \frac{\partial Q}{\partial y} + \frac{\partial R}{\partial z} \right) dx \wedge dy \wedge dz.$$

Thus, Gauss's theorem can be written as

$$\oiint \omega_2 = \iiint d\omega_2.$$

The Stokes' Theorem states

$$\oint_L P \, dx + Q \, dy + R \, dz = \iint_\Sigma \left(\frac{\partial R}{\partial y} - \frac{\partial Q}{\partial z} \right) dy \, dz$$

$$+ \left(\frac{\partial P}{\partial z} - \frac{\partial R}{\partial x} \right) dz \, dx + \left(\frac{\partial Q}{\partial x} - \frac{\partial P}{\partial y} \right) dx \, dy.$$

Since L and Σ are orientated, the expression

$$\omega = P \, dx + Q \, dy + R \, dz$$

can be considered as an 1-form. Then

$$d\omega = \left(\frac{\partial R}{\partial y} - \frac{\partial Q}{\partial z} \right) dy \wedge dz + \left(\frac{\partial P}{\partial z} - \frac{\partial R}{\partial x} \right) dz \wedge dx$$

$$+ \left(\frac{\partial Q}{\partial x} - \frac{\partial P}{\partial y} \right) dx \wedge dy.$$

Thus, the Stokes' Theorem can be written as

$$\oint \omega = \iint d\omega.$$

Therefore, Green's Theorem, Gauss's Theorem and Stokes' Theorem can be expressed by one equation, that is

$$\int_{\partial \Sigma} \omega = \int_\Sigma d\omega,$$

where ω is an exterior differential form and $d\omega$ is the exterior differential of ω, Σ is the closed region of the integral of $d\omega$, $\partial \Sigma$ is the boundary of Σ, and the type of integral $\int$ corresponds to the dimension of the region.

From the above study, we can also conclude that there are no more integral equations that connect the region and its boundary besides Green's formula, Gauss's formula and Stokes' formula, since the exterior derivative of exterior differential forms of degree three is zero in the three dimensional space.

The formula

$$\int_{\partial \Sigma} \omega = \int_\Sigma d\omega$$

can be generalized to higher dimensional spaces, and also to manifolds (We will not cover manifolds in this book).

The exterior differential forms and the integrals play the same roles as the derivative and integral in the Fundamental Theorem of Calculus:

$$\int_a^b \frac{d}{dx} f(x)\, dx = f(x,y)\Big|_a^b = f(b) - f(a).$$

Here Σ is the interval $[a, b]$, $\partial\Sigma$ is the boundary of Σ, which consists of the two points a and b, and $d\omega$ is $\dfrac{df(x)}{dx}\, dx$.

The formula

$$\int_{\partial\Sigma} \omega = \int_\Sigma d\omega,$$

is also called the **Stokes' Theorem**. It is the Fundamental Theorem of Calculus in higher dimensional spaces.

We summarize what we learned in the following chart.

Degree of the exterior differential form	Space	Formula
0	Straight line	Newton-Leibniz
1	Planar region	Green's
1	Surface in Space	Stokes'
2	Region in Space	Gauss's

This section is an intuitive instruction of the exterior differential forms. We hope our reader can have a deep and profound understanding of the essential theorems of calculus, and be prepared for the further studies of calculus of several variables.

Exercises 7.6

1. Compute the following exterior product.
   (1) $(x\, dx + 7x^2\, dy) \wedge (y\, dx - \sin 3x\, dy + dz)$;
   (2) $(5\, dx + 3\, dy) \wedge (3\, dx + 2\, dy)$;
   (3) $(6\, dx \wedge dy + 27\, dx \wedge dz) \wedge (dx + dy + dz)$.

2. Compute the following exterior differentials.
   (1) $d(\cos y\, dx - \sin x\, dy)$;
   (2) $d(2xy\, dx + x^2\, dy)$;
   (3) $d(6z\, dx \wedge dy - xy\, dx \wedge dz)$.

3. Suppose λ is an exterior differential form of degree p and μ is an exterior differential form of degree q. Prove the following equations.

(1) $\mu \wedge \lambda = (-1)^{pq} \lambda \wedge \mu$;

(2) $d(\mu \wedge \lambda) = d\mu \wedge \lambda + (-1)^q \mu \wedge d\lambda$.

4. Suppose $\omega = \sum a_{ij}\, dx_i \wedge dx_j$, $a_{ij} = -a_{ji}$. Show that

$$d\omega = \frac{1}{3} \sum_{i,j,k} \left(\frac{\partial a_{ij}}{\partial x_k} - \frac{\partial a_{kj}}{\partial x_i} + \frac{\partial a_{ki}}{\partial x_j} \right) dx_k \wedge dx_i \wedge dx_j.$$

Chapter 8

Some Applications of Calculus in Several Variables

Calculus in several variables has extensive applications in many different areas. We provide a little enlightenment about it by studying a few examples in this chapter.

8.1 Taylor Expansions and Extremal Problems

8.1.1 *Taylor Expansions of Functions in Several Variables*

Just as for functions in one variable, there are Taylor expansions for functions in several variables. We study Taylor expansions of functions in two variables here. Taylor expansions of functions in more than two variables can be obtained in a similar way.

Suppose $u = f(x, y)$ is a function defined on a planar region D, and it has continuous partial derivatives of order $n + 1$ in both variables. Let $M_0(x_0, y_0)$ be a point in D, and h, k be two positive numbers with their values small enough that the line segment between $(x_0 + h, y_0 + k)$ and M_0 lays entirely inside D. Define $\varphi(t) = f(a + th, b + tk)$, then we have

$$\varphi(0) = f(a, b), \quad \varphi(1) = f(a + h, b + k).$$

The *Taylor expansion* of $\varphi(t)$ at $t = 0$ gives us (refer to Section 3.3)

$$\varphi(1) = \varphi(0) + \frac{\varphi'(0)}{1!} + \frac{\varphi''(0)}{2!} + \cdots + \frac{\varphi^{(n)}(0)}{n!} + \frac{\varphi^{(n+1)}(\theta)}{(n+1)!},$$

where $0 < \theta < 1$.

On the other hand,

$$\frac{d\varphi}{dt} = \frac{\partial f}{\partial x} h + \frac{\partial f}{\partial y} k = \left(h \frac{\partial}{\partial x} + k \frac{\partial}{\partial y} \right) f.$$

Therefore,

$$\varphi^{(p)}(t) = \frac{d^p \varphi}{dt^p} = \left(h \frac{\partial}{\partial x} + k \frac{\partial}{\partial y} \right)^p f = \sum_{r=0}^{p} c_p^r h^r k^{p-r} \frac{\partial^p f}{\partial x^r \partial y^{p-r}},$$

and

$$\varphi^{(p)}(0) = \left(h \frac{\partial}{\partial x} + k \frac{\partial}{\partial y} \right)^p f(x,y) \Big|_{\substack{x=a \\ y=b}},$$

$$\varphi^{(n+1)}(\theta) = \left(h \frac{\partial}{\partial x} + k \frac{\partial}{\partial y} \right)^{n+1} f(x,y) \Big|_{\substack{x=a+\theta k \\ y=b+\theta k}}.$$

Substituting the above results into the original equation, we get the Taylor expansion of $f(x,y)$ at the point (a,b):

$$\begin{aligned}
f(a+h, b+k) = f(a,b) &+ \left(h \frac{\partial}{\partial x} + k \frac{\partial}{\partial y} \right) f(x,y) \Big|_{\substack{x=a \\ y=b}} \\
&+ \frac{1}{2!} \left(h \frac{\partial}{\partial x} + k \frac{\partial}{\partial y} \right)^2 f(x,y) \Big|_{\substack{x=a \\ y=b}} \\
&+ \cdots + \frac{1}{n!} \left(h \frac{\partial}{\partial x} + k \frac{\partial}{\partial y} \right)^n f(x,y) \Big|_{\substack{x=a \\ y=b}} \\
&+ \frac{1}{(n+1)!} \left(h \frac{\partial}{\partial x} + k \frac{\partial}{\partial y} \right) f(x,y)^{n+1} \Big|_{\substack{x=a+\theta k \\ y=b+\theta k}},
\end{aligned}$$

where $0 < \theta < 1$.

Obviously, Taylor expansion of a function at a point is determined by the properties of the function in a small neighborhood of the point, and is not related to the properties of the function in other places.

8.1.2 *Extremal Problems of Functions in Several Variables*

We start this section with functions in two variables. Suppose $f(x,y)$ is defined on a region D and (a,b) is a point inside D. If there exits a neighborhood of (a,b), such that $f(a,b) \geqslant f(x,y)$ for all (x,y) in this neighborhood, then $f(a,b)$ is called the maximum value of f in this neighborhood. The minimum value of f in a neighborhood of a point of its domain is defined similarly. Obviously, the maximum value and the minimum value that we study are local properties of a function.

For instance, the values of function $f(x,y) = 3x^2 + 4y^2$ are greater than zero at all points around $(0,0)$, and $f(x,y) = 0$ at the point $(0,0)$. Therefore, $f(x,y)$ has a minimum value at point $(0,0)$.

The value of function $f(x, y) = xy$ at $(0, 0)$ is zero, but it is neither a maximum nor a minimum. In any neighborhood of $(0, 0)$, one can find a point where f is positive, and one can also find a point where f is negative. The graph of $z = xy$ is a hyperbolic paraboloid and its shape in a neighborhood of $(0, 0, 0)$ is a saddle.

Assume that $f(x, y)$ has continuous partial derivatives of degree n, where n is as high as we need in our study.

Consider the function

$$\varphi(t) = f(a + \lambda t, b + \mu t).$$

It is a function of t whenever λ and μ are not zero simultaneously. If $f(x, y)$ has a maximum value at $x = a, y = b$, then $\varphi(t)$ has a maximum value at $t = 0$. Thus, $\varphi'(0) = 0$, that is

$$\varphi'(0) = \left[\lambda \frac{\partial f}{\partial x} + \mu \frac{\partial f}{\partial y} \right]_{\substack{x=a \\ y=b}} = 0.$$

By choosing $\lambda = 1$, $\mu = 0$, and then $\lambda = 0$, $\mu = 1$, we get

$$\frac{\partial f}{\partial x} \bigg|_{\substack{x=a \\ y=b}} = \frac{\partial f}{\partial y} \bigg|_{\substack{x=a \\ y=b}} = 0.$$

Hence, $f(x, y)$ has a maximum or minimum at an interior point (a, b) of D if the value of every partial derivative of f is zero at the point.

Furthermore,

$$\varphi''(0) = \left(\lambda^2 \frac{\partial^2 f}{\partial x^2} + 2\lambda\mu \frac{\partial^2 f}{\partial x \partial y} + \mu^2 \frac{\partial^2 f}{\partial y^2} \right) \bigg|_{\substack{x=a \\ y=b}}.$$

If $\varphi''(0) < 0$ for all λ and μ (λ and μ are not zero simultaneously), then, $\varphi(t)$ has a maximum at $t = 0$, that is, $f(x, y)$ has a maximum at (a, b).

Denote

$$\frac{\partial^2 f}{\partial x^2} \bigg|_{\substack{x=a \\ y=b}} = A, \qquad \frac{\partial^2 f}{\partial x \partial y} \bigg|_{\substack{x=a \\ y=b}} = B, \qquad \frac{\partial^2 f}{\partial y^2} \bigg|_{\substack{x=a \\ y=b}} = C,$$

then the question becomes: For what kind of numbers A, B, C, the inequality

$$A\lambda^2 + 2B\lambda\mu + C\mu^2 < 0$$

is true for all λ and μ that are not zero simultaneously?

Choose $\lambda = 1$, $\mu = 0$, we get $A < 0$. Since

$$A\lambda^2 + 2B\lambda\mu + C\mu^2 = A \left(\lambda + \frac{B}{A}\mu \right)^2 + \left(C - \frac{B^2}{A} \right)\mu^2,$$

we also need $C - \dfrac{B^2}{A} < 0$. Otherwise, we can choose $\mu = 1$, $\lambda = -\dfrac{B}{A}$, and

$$A\lambda^2 + 2B\lambda\mu + C\mu^2 \geqslant 0.$$

On the other hand, if $A < 0$, $C - \dfrac{B^2}{A} < 0$, then

$$A\lambda^2 + 2B\lambda\mu + C\mu^2 < 0$$

when λ and μ are not zero simultaneously. Hence, if the inequalities

$$\frac{\partial^2 f}{\partial x^2} < 0, \quad \left(\frac{\partial^2 f}{\partial x\partial y}\right)^2 - \left(\frac{\partial^2 f}{\partial x^2}\right)\left(\frac{\partial^2 f}{\partial y^2}\right) < 0,$$

hold at the point $x = a, y = b$, then $f(x, y)$ reaches its maximum at the point.

Similarly, if the inequalities

$$\frac{\partial^2 f}{\partial x^2} > 0, \quad \left(\frac{\partial^2 f}{\partial x\partial y}\right)^2 - \left(\frac{\partial^2 f}{\partial x^2}\right)\left(\frac{\partial^2 f}{\partial y^2}\right) < 0,$$

hold at the point $x = a, y = b$, then $f(x, y)$ has a minimum value at the point.

If

$$\left(\frac{\partial^2 f}{\partial x\partial y}\right)^2 - \left(\frac{\partial^2 f}{\partial x^2}\right)\left(\frac{\partial^2 f}{\partial y^2}\right) > 0$$

at $x = a$, $y = b$, then f has no extremal value at this point.

In fact, with this assumption, the equation $A\lambda^2 + 2B\lambda\mu + C\mu^2 = 0$ has real solutions

$$\lambda = \frac{-B \pm \sqrt{B^2 - AC}}{A}\mu.$$

Therefore, $\varphi''(0) = 0$ for λ and μ satisfying this equation. We can find λ, μ such that $\varphi''(0) > 0$. We can also find λ, μ such that $\varphi''(0) < 0$. That is, f has no extremal values.

If

$$\left(\frac{\partial^2 f}{\partial x\partial y}\right)^2 - \left(\frac{\partial^2 f}{\partial x^2}\right)\left(\frac{\partial^2 f}{\partial y^2}\right) = 0$$

at the point $x = a, y = b$, then whether the function has a maximum value or a minimum value cannot be determined outright, and we need to examine the values of $\varphi'''(0)$, $\varphi^{(4)}$, etc.

Hence, the process of finding the extremal values of $f(x, y)$ on the region D is: First, solve the system of equations

$$f'_x(x, y) = 0, \quad f'_y(x, y) = 0.$$

If a, b is one of solutions of the above system of equations, then let

$$\frac{\partial^2 f(a,b)}{\partial x^2} = A, \quad \frac{\partial^2 f(a,b)}{\partial x \partial y} = B, \quad \frac{\partial^2 f(a,b)}{\partial y^2} = C.$$

If $B^2 - AC < 0$, and $A < 0$, then $f(x,y)$ has a maximum value at the point (a,b). If $B^2 - AC < 0$, and $A > 0$, then $f(x,y)$ has a minimum value at the point (a,b). If $B^2 - AC > 0$, then the value of $f(x,y)$ at the point (a,b) is neither a maximum nor a minimum. If $B^2 - AC = 0$, then a further study about the values of $\varphi'''(0)$ and $\varphi^{(4)}$ is needed.

Example 1. Find the extremal points of the function $f(x,y) = x^3 - y^3 + 3x^2 + 3y^2 - 9x$.

Solution: By solving the system of equations

$$f'_x(x,y) = 3x^2 + 6x - 9 = 0, \quad f'_y(x,y) = -3y^2 + 6x = 0,$$

we get the following solutions

$$M_1(1,0), \quad M_2(1,2), \quad M_3(-3,0), \quad M_4(-3,2).$$

Then we calculate the second order partial derivatives:

$$A = f''_{xx}(x,y) = 6x + 6, \quad B = f''_{xy}(x,y) = 0, \quad C = f''_{yy}(x,y) = -6y + 6.$$

At the point $M_1(1,0)$, $A = 12$, $B = 0$, $C = 6$. Since $A = 12 > 0$ and $B^2 - AC = -72 < 0$, $M_1(1,0)$ is a minimum point.

At the point $M_2(1,2)$, $A = 12$, $B = 0$, $C = -6$. Since $A = 12 > 0$ and $B^2 - AC = 72 > 0$, $M_2(1,2)$ is not an extremal point.

At the point $M_3(-3,0)$, $A = -12$, $B = 0$, $C = 6$. Since $A = -12 < 0$ and $B^2 - AC = 72 > 0$, $M_3(-3,0)$ is not an extremal point.

At the point $M_4(-3,2)$, $A = -12$, $B = 0$, $C = -6$. Since $A = -12 < 0$ and $B^2 - AC = -72 < 0$, $M_4(-3,2)$ is a maximum point.

Example 2. Find the maximum value of the function $u = \sin x + \sin y - \sin(x + y)$, on the region $x \geqslant 0$, $y \geqslant 0$, $x + y \leqslant 2\pi$.

Solution: By solving

$$\begin{cases} \dfrac{\partial u}{\partial x} = \cos x - \cos(x + y) = 0, \\ \dfrac{\partial u}{\partial y} = \cos y - \cos(x + y) = 0, \end{cases}$$

we get $x = y = \dfrac{2\pi}{3}$ and $u = \dfrac{3\sqrt{3}}{2}$.

Since $u = 0$ on the boundary of the region $x = 0$, $y = 0$ and $x + y = 2\pi$, the function has a maximum at $\left(\dfrac{2\pi}{3}, \dfrac{2\pi}{3}\right)$.

Example 3. The *method of least squares*.

Suppose there are $n(> 2)$ linear equations

$$a_i x + b_i y = d_i \quad (1 \leqslant i \leqslant n),$$

where all a_i, b_i, d_i are constants.

Suppose

$$\begin{vmatrix} a_1 & b_1 \\ a_2 & b_2 \end{vmatrix} \neq 0.$$

The solution of the first two equations may not necessarily satisfy the rest of the equations. We want to find the values of x and y such that the value of

$$w = \sum_{i=1}^{n} \delta_i^2 = \sum_{i=1}^{n} (a_i x + b_i y - d_i)^2$$

is a minimum.

By the method of finding the minimum value that we studied earlier, we solve the system of equations

$$\frac{\partial w}{\partial x} = 2 \sum_{i=1}^{n} a_i (a_i x + b_i y - d_i) = 0,$$

$$\frac{\partial w}{\partial y} = 2 \sum_{i=1}^{n} b_i (a_i x + b_i y - d_i) = 0.$$

Let

$$(a, b) = \sum_{i=1}^{n} a_i b_i, \quad (a, a) = \sum_{i=1}^{n} a_i^2, \quad \cdots.$$

Then the above system of equations becomes

$$(a, a)x + (a, b)y = (a, d),$$
$$(a, b)x + (b, b)y = (b, d).$$

Since

$$\begin{vmatrix} (a, a) & (a, b) \\ (a, b) & (b, b) \end{vmatrix} = \sum_{1 \leqslant i < j \leqslant n} \begin{vmatrix} a_i & b_i \\ a_j & b_j \end{vmatrix}^2 \geqslant \begin{vmatrix} a_1 & b_1 \\ a_2 & b_2 \end{vmatrix}^2 > 0,$$

the solution of this system of equations exist.

It is easy to verify that w has a minimum at this solution point.

8.1.3 Conditional Extremum Problems

In this section, we study the extremum problem of a function $u = f(x, y, z)$ under the condition that the variables x, y satisfy the equation $g(x, y, z) = 0$.

Consider x, y as independent variables and z as a function of x, y defined by $g(x, y, z) = 0$.

The conditions for u to have an extremum are

$$\frac{\partial u}{\partial x} = \frac{\partial f}{\partial x} + \frac{\partial f}{\partial z}\frac{\partial z}{\partial x} = 0, \quad \frac{\partial u}{\partial y} = \frac{\partial f}{\partial y} + \frac{\partial f}{\partial z}\frac{\partial z}{\partial y} = 0.$$

Also, from $g(x, y, z) = 0$, we have

$$\frac{\partial g}{\partial x} + \frac{\partial g}{\partial z}\frac{\partial z}{\partial x} = 0, \quad \frac{\partial g}{\partial y} + \frac{\partial g}{\partial z}\frac{\partial z}{\partial y} = 0.$$

Now we eliminate $\dfrac{\partial z}{\partial x}$, $\dfrac{\partial z}{\partial y}$ from these four equations, and obtain that

$$\frac{\partial f}{\partial x}\frac{\partial g}{\partial z} - \frac{\partial f}{\partial z}\frac{\partial g}{\partial x} = 0, \quad \frac{\partial f}{\partial y}\frac{\partial g}{\partial z} - \frac{\partial f}{\partial z}\frac{\partial g}{\partial y} = 0.$$

Solving x, y, z from the above two equations and the equation $g(x, y, z) = 0$, we get all the possible extremal points.

These two equations can also be written as

$$\frac{\dfrac{\partial f}{\partial x}}{\dfrac{\partial g}{\partial x}} = \frac{\dfrac{\partial f}{\partial y}}{\dfrac{\partial g}{\partial y}} = \frac{\dfrac{\partial f}{\partial z}}{\dfrac{\partial g}{\partial z}}(= -\lambda).$$

That is, there exists a λ such that

$$\frac{\partial f}{\partial x} + \lambda\frac{\partial g}{\partial x} - 0, \quad \frac{\partial f}{\partial y} + \lambda\frac{\partial g}{\partial y} - 0, \quad \frac{\partial f}{\partial z} + \lambda\frac{\partial g}{\partial z} = 0.$$

If we consider x, y, z, λ as unknowns, then x, y, z, λ can be solved from these three equations and the equation $g = 0$. The function f has conditional extrema only at those points that are solutions of this system of equations.

On the other hand, consider the function

$$w = f(x, y, z) + \lambda g(x, y, z).$$

The extremum points of this function satisfy the following equations:

$$\frac{\partial w}{\partial \lambda} = g(x, y, z) = 0,$$

$$\frac{\partial w}{\partial x} = \frac{\partial f}{\partial x} + \lambda\frac{\partial g}{\partial x} = 0, \quad \frac{\partial w}{\partial y} = \frac{\partial f}{\partial y} + \lambda\frac{\partial g}{\partial y} = 0,$$

$$\frac{\partial w}{\partial z} = \frac{\partial f}{\partial z} + \lambda\frac{\partial g}{\partial z} = 0.$$

This is called *Lagrange multiplier method*.

Example 1. Find the shortest distance from the point (a, b) to the line $Ax + By + C = 0$.

Solution: Suppose the distance between (a, b) and a point (x, y) on the line is r. We have

$$r^2 = (x - a)^2 + (y - b)^2.$$

Let $w = (x - a)^2 + (y - b)^2 + \lambda(Ax + By + C)$. Then by solving the system of equations

$$\frac{\partial w}{\partial x} = 2(x - a) + \lambda A = 0,$$

$$\frac{\partial w}{\partial y} = 2(y - b) + \lambda B = 0,$$

$$\frac{\partial w}{\partial \lambda} = Ax + By + C = 0,$$

we get

$$x = a - \frac{1}{2}\lambda A, \quad y = b - \frac{1}{2}\lambda B, \quad Ax + By + C = 0.$$

Therefore,

$$\lambda = \frac{2(Aa + Bb + C)}{A^2 + B^2}.$$

Since $r \to \infty$ as $x^2 + y^2 \to \infty$, the shortest distance exists. Hence, the distance between (a, b) and (x_0, y_0) is the shortest, here

$$x_0 = a - \frac{A(Aa + Bb + C)}{A^2 + B^2} = \frac{aB^2 - bAB - AC}{A^2 + B^2},$$

$$y_0 = b - \frac{B(Aa + Bb + C)}{A^2 + B^2} = \frac{bA^2 + aAB + BC}{A^2 + B^2}.$$

The shortest distance is

$$\sqrt{(x_0 - a)^2 + (y_0 - b)^2} = \frac{|Aa + Bb + C|}{\sqrt{A^2 + B^2}}.$$

The Lagrange multiplier method can be generalized to functions of more variables. The result is similar to that for functions of three variables except there are more equations. We will not go into more details here, just give an example.

Example 2. Write a positive number a as the sum of n positive numbers such that the product of these n positive numbers is the greatest.

Solution: Suppose $a = x_1 + x_2 + \cdots + x_n$ where $x_1 > 0$, $x_2 > 0, \cdots, x_n > 0$. We would like to find the maximum value of the function
$$f(x_1, x_2, \cdots, x_n) = x_1 x_2 \cdots x_n.$$
Let
$$F(x_1, x_2, \cdots, x_n) = x_1 x_2 \cdots x_n + \lambda(x_1 + x_2 + \cdots + x_n - a).$$
Solve the following system of equations:
$$\frac{\partial F}{\partial x_1} = x_2 x_3 \cdots x_n + \lambda = 0,$$
$$\frac{\partial F}{\partial x_2} = x_1 x_3 \cdots x_n + \lambda = 0,$$
$$\cdots$$
$$\frac{\partial F}{\partial x_n} = x_1 x_2 \cdots x_{n-1} + \lambda = 0,$$
$$\frac{\partial F}{\partial \lambda} = x_1 + x_2 + \cdots + x_n - a = 0,$$
we obtain $x_1 = x_2 = \cdots = x_n$ from the first n equations. Substituting this result into the last equation, we get that $x_1 = x_2 = \cdots = x_n = \dfrac{a}{n}$. Thus, we have the only possible extremum point.

Since from the problem statement there exists the maximum value, this is the maximum point. That is, each of the positive number is $\dfrac{a}{n}$ and the maximum value is $\left(\dfrac{a}{n}\right)^n$.

From the result above, we can get an important inequality:
Since
$$x_1 x_2 \cdots x_n \leqslant \left(\frac{a}{n}\right)^n = \left(\frac{x_1 + x_2 + \cdots + x_n}{n}\right)^n,$$
we have
$$\sqrt[n]{x_1 x_2 \cdots x_n} \leqslant \frac{x_1 + x_2 + \cdots + x_n}{n}.$$
That is, the geometric mean of any n positive numbers is less than or equal to the arithmetic mean of the n numbers.

What we just studied is the Lagrange multiplier method with one constraint. If there are more than one constraints, for instance, to find the extremum points of the function $u = f(x_1, x_2, \cdots, x_n)$ under $m(< n)$ conditions
$$g_1(x_1, x_2, \cdots, x_n) = 0, \quad g_2(x_1, x_2, \cdots, x_n) = 0,$$
$$\cdots$$
$$g_m(x_1, x_2, \cdots, x_n) = 0,$$

we can use a similar method. Let

$$w = f + \lambda_1 g_1 + \lambda_2 g_2 + \cdots + \lambda_m g_m,$$

where $\lambda_1, \cdots, \lambda_1$ are undetermined coefficients. Then, we can find the possible extremum point by solving the following system of equations

$$\frac{\partial w}{\partial \lambda_1} = g_1 = 0,$$

$$\cdots$$

$$\frac{\partial w}{\partial \lambda_m} = g_m = 0,$$

$$\frac{\partial w}{\partial x_1} = \frac{\partial f}{\partial x_1} + \lambda_1 \frac{\partial g_1}{\partial x_1} + \cdots + \lambda_m \frac{\partial g_m}{\partial x_1} = 0,$$

$$\cdots$$

$$\frac{\partial w}{\partial x_n} = \frac{\partial f}{\partial x_n} + \lambda_1 \frac{\partial g_1}{\partial x_n} + \cdots + \lambda_m \frac{\partial g_m}{\partial x_m} = 0.$$

Example 3. Find the length of the major and miner axes of the ellipse

$$\frac{x^2}{a^2} + \frac{y^2}{b^2} + \frac{z^2}{c^2} = 1, \quad Ax + By + Cz = 0.$$

Solution: Since the center of this ellipse is the origin, the problem becomes finding the maximum and the minimum of the function

$$u = x^2 + y^2 + z^2$$

under the conditions

$$g_1(x, y, z) = \frac{x^2}{a^2} + \frac{y^2}{b^2} + \frac{z^2}{c^2} - 1 = 0,$$

$$g_2(x, y, z) = Ax + By + Cz = 0.$$

Let

$$w = x^2 + y^2 + z^2 + \lambda_1 g_1 + \lambda_2 g_2,$$

By solving the following system of equations

$$2x + 2\lambda_1 \frac{x}{a^2} + \lambda_2 A = 0,$$

$$2y + 2\lambda_1 \frac{y}{b^2} + \lambda_2 B = 0,$$

$$2z + 2\lambda_1 \frac{z}{c^2} + \lambda_2 C = 0,$$

$$\frac{x^2}{a^2} + \frac{y^2}{b^2} + \frac{z^2}{c^2} - 1 = 0,$$

$$Ax + By + Cz = 0,$$

we obtain that (multiplying x, y, z with the first three equations respectively, then adding the three equations together)

$$r^2 + \lambda_1 = 0, \quad (r^2 = x^2 + y^2 + z^2).$$

Thus,

$$x = \frac{1}{2}\lambda_2 \frac{Aa^2}{r^2 - a^2}, \quad y = \frac{1}{2}\lambda_2 \frac{Bb^2}{r^2 - b^2}, \quad z = \frac{1}{2}\lambda_2 \frac{Cc^2}{r^2 - c^2}.$$

Multiplying A, B, C with these three equations respectively and then adding them together, we have

$$\frac{A^2a^2}{r^2 - a^2} + \frac{B^2b^2}{r^2 - b^2} + \frac{C^2c^2}{r^2 - c^2} = 0.$$

The maximum and the minimum of u can be obtained by solving this equation for r^2.

Exercises 8.1

1. Find the Taylor expansion of the function

$$f(x, y) = 2x^2 - xy - y^2 - 6x - 3y + 5$$

   in a neighborhood of the point $A(1, -2)$.

2. By using the Taylor expansion in two variables, show that when $|x|$, $|y|$ are sufficiently small, the following approximation holds

$$\frac{\cos x}{\cos y} \approx 1 - \frac{1}{2}(x^2 - y^2).$$

3. Find the Taylor expansion of the function $f(x, y) = x^y$ in a neighborhood of the point $A(1, 1)$. Write out up to second order terms.

4. Find the Taylor expansion of the function $f(x, y) = \sqrt{1 - x^2 - y^2}$ at the point $(0, 0)$. Write out up to fourth order terms.

5. Find the Taylor expansion of the function $f(x, y) = \sin(x^2 + y^2)$ at the point $(0, 0)$.

6. Find the extremum values of the following functions.

   (1) $z = x^2 + (y - 1)^2$;   (2) $z = x^2 - xy + y^2 - 2x + y$;

   (3) $z = x^3 + y^3 - 3xy$;

   (4) $z = xy + \dfrac{50}{x} + \dfrac{20}{y}$, $(x > 0, y > 0)$.

7. Find the extremum values of the implicit function z with variables x, y.

   (1) $x^2 + y^2 + z^2 - 2x - 2y - 4z - 10 = 0$;

   (2) $x^2 + y^2 + z^2 - xz - yz + 2x + 2y + 2z - 2 = 0$.

8. Find the conditional extremum values of the following functions.

   (1) $u = xy$, if $x + y = 1$;

   (2) $u = x^2 + y^2$, if $\dfrac{x}{a} + \dfrac{y}{b} = 1$;

   (3) $u = x + y + z$, if $\dfrac{1}{x} + \dfrac{1}{y} + \dfrac{1}{z} = 1$, $(x > 0, y > 0, z > 0)$;

   (4) $u = xyz$, if $x^2 + y^2 + z^2 = 1$, $x + y + z = 0$.

9. Find the isosceles triangle inscribed in the ellipse $x^2 + 3y^2 = 12$ with its base parallel to the major axis that has the largest area.

10. An ellipse is the intersection of the plane $x+y+z = 1$ and the paraboloid $z = x^2 + y^2$. Find the shortest and the longest distances from the origin to the ellipse.

11. Find a point on the plane $3x - 2z = 0$ such that the sum of the squares of the distances between the point and $A(1,1,1)$ and $B(2,3,4)$ is the smallest.

12. Find the right triangle with hypotenuse l that has the longest perimeter.

13. A bath tub has the shape of a half cylinder with surface area S. Find the dimensions of the bath tub such that it has the largest volume.

14. Find the shortest distance between the parabola $y = x^2$ and the line $x - y - 2 = 0$.

15. Suppose a rectangle has perimeter $2p$. Find the dimensions of the rectangle such that it gives the largest volume of the solid that is generated by revolving the rectangle about its one side.

16. A tent consists of two parts. The lower part is a cylinder with hight H and base radius R. The upper part is a cone with hight h. Let V_0 be the volume of the tent. Show that the tent has the smallest surface area when $R = \sqrt{5}H$ and $h = 2H$.

17. The sum of all sides of a parallelepiped is $12a$. Find the dimension of this parallelepiped such that it has maximum volume.

18. Find a point on the xy-plane such that the sum of the squares of the distances between the point and the three lines $x = 0$, $y = 0$, $x + 2y - 16 = 0$ is minimum.

8.2 Examples of Applications in Physics

8.2.1 *Barycenter, Moment of Inertia and Gravitational Force*

To find the barycenter of an object V, we partition V into n small blocks V_i. Let $M_i(\xi_i, \eta_i, \zeta_i)$ be an arbitrary point in V_i. If $\rho(x, y, z)$ is the density at each point of V, then the mass of V_i is approximately equal to $\rho(\xi_i, \eta_i, \zeta_i)\Delta V_i$, where ΔV_i is the volume of V_i.

Consider these n small blocks as n particles $M_i(\xi_i, \eta_i, \zeta_i)$ (the mass of M_i is $\rho(M_i)\Delta V_i$).

The coordinates of the barycenter of this system of particles are:

$$
\begin{cases}
\bar{x} = \dfrac{\sum\limits_i \xi_i \rho(\xi_i, \eta_i, \zeta_i)\Delta V_i}{\sum\limits_i \rho(\xi_i, \eta_i, \zeta_i)\Delta V_i}, \\[2em]
\bar{y} = \dfrac{\sum\limits_i \eta_i \rho(\xi_i, \eta_i, \zeta_i)\Delta V_i}{\sum\limits_i \rho(\xi_i, \eta_i, \zeta_i)\Delta V_i}, \\[2em]
\bar{z} = \dfrac{\sum\limits_i \zeta_i \rho(\xi_i, \eta_i, \zeta_i)\Delta V_i}{\sum\limits_i \rho(\xi_i, \eta_i, \zeta_i)\Delta V_i}.
\end{cases}
$$

Take the limit as n approaches to infinity, we get the coordinates of the barycenter of V:

$$
x_0 = \lim \frac{\sum\limits_i \xi_i \rho(\xi_i, \eta_i, \zeta_i)\Delta V_i}{\sum\limits_i \rho(\xi_i, \eta_i, \zeta_i)\Delta V_i} = \frac{\iiint\limits_V x\rho(x, y, z)\, dx\, dy\, dz}{\iiint\limits_V \rho(x, y, z)\, dx\, dy\, dz},
$$

$$
y_0 = \lim \frac{\sum\limits_i \eta_i \rho(\xi_i, \eta_i, \zeta_i)\Delta V_i}{\sum\limits_i \rho(\xi_i, \eta_i, \zeta_i)\Delta V_i} = \frac{\iiint\limits_V y\rho(x, y, z)\, dx\, dy\, dz}{\iiint\limits_V \rho(x, y, z)\, dx\, dy\, dz},
$$

$$z_0 = \lim \frac{\sum_i \zeta_i \rho(\xi_i, \eta_i, \zeta_i) \Delta V_i}{\sum_i \rho(\xi_i, \eta_i, \zeta_i) \Delta V_i} = \frac{\iiint\limits_V z \rho(x, y, z) \, dx \, dy \, dz}{\iiint\limits_V \rho(x, y, z) \, dx \, dy \, dz}.$$

If the object V has uniform density, then ρ is a constant, and $M = \rho V$ (where V is the volume of the object). Thus, the coordinates of the barycenter are:

$$x_0 = \frac{1}{V} \iiint\limits_V x \, dx \, dy \, dz, \quad y_0 = \frac{1}{V} \iiint\limits_V y \, dx \, dy \, dz,$$

$$z_0 = \frac{1}{V} \iiint\limits_V z \, dx \, dy \, dz.$$

If an object is an uniform density thin slice D (bounded), then the coordinates of its barycenter are

$$x_0 = \frac{1}{A} \iint\limits_D x \, dx \, dy, \quad y_0 = \frac{1}{A} \iint\limits_D y \, dx \, dy,$$

where A is the area of D.

To find the moment of inertia, similarly, we consider V as n particles, then the moment of inertia of this system of particles with respect to the x-axis is

$$\sum_i (\eta_i^2 + \zeta_i^2) \rho(\xi_i, \eta_i, \zeta_i) \Delta V_i.$$

Take the limit as n approaches to infinity, we get the moment of inertia of V with respect to the x-axis

$$I_x = \iiint\limits_V (y^2 + z^2) \rho(x, y, z) \, dx \, dy \, dz.$$

Similarly, the moment of inertia of V with respect to the y-axis and the z-axis are:

$$I_y = \iiint\limits_V (x^2 + z^2) \rho(x, y, z) \, dx \, dy \, dz,$$

$$I_z = \iiint\limits_V (x^2 + y^2) \rho(x, y, z) \, dx \, dy \, dz.$$

The moment of inertia of V with respect to origin is:

$$I_0 = \iiint\limits_V (x^2 + y^2 + z^2)\rho(x, y, z)\, dx\, dy\, dz.$$

Example 1. Find the barycenter of the ball $x^2 + y^2 + z^2 \leqslant 2az$. Assume that the density at each point in the ball is inversely proportional to the distance from the origin. (Figure 8.1)

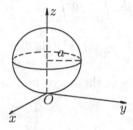

Fig. 8.1

Solution: The barycenter of the ball is on the z-axis, and its density is

$$\rho = \frac{k}{\sqrt{x^2 + y^2 + z^2}}$$

where k is the proportional constant.

The z coordinate of the barycenter is

$$z_0 = \frac{\displaystyle\iiint\limits_V \frac{kz}{\sqrt{x^2 + y^2 + z^2}}\, dx\, dy\, dz}{\displaystyle\iiint\limits_V \frac{k}{\sqrt{x^2 + y^2 + z^2}}\, dx\, dy\, dz}.$$

Denote the numerator of the above fraction as M_{xy}, and the denominator as m. Then $z_0 = \dfrac{M_{xy}}{m}$.

Under spherical coordinates

$$x = r\sin\theta\cos\varphi, \quad y = r\sin\theta\sin\varphi, \quad z = r\cos\theta,$$

the sphere is

$$0 \leqslant r \leqslant 2a\cos\theta,$$

where $0 \leqslant \varphi \leqslant 2\pi,\ 0 \leqslant \theta \leqslant \dfrac{\pi}{2}$. Thus, the mass of the sphere is

$$m = \int_0^{2\pi} d\varphi \int_0^{\frac{\pi}{2}} d\theta \int_0^{2a\cos\theta} \frac{k}{r} |r^2 \sin\theta|\, dr$$

$$= 2\pi k \int_0^{\frac{\pi}{2}} 2a^2 \cos^2\theta \sin\theta\, d\theta = \frac{4}{3}\pi ka^2.$$

Also,

$$M_{xy} = \int_0^{2\pi} d\varphi \int_0^{\frac{\pi}{2}} d\theta \int_0^{2a\cos\theta} \frac{k}{r} r\cos\theta |r^2 \sin\theta|\, dr$$

$$= 2\pi k \int_0^{\frac{\pi}{2}} \frac{8}{3}a^3 \cos^4\theta \sin\theta\, d\theta = \frac{16}{15}\pi ka^3.$$

Therefore, the coordinates of the barycenter are:

$$x_0 = y_0 = 0, \quad z_0 = \frac{\dfrac{16}{15}\pi ka^3}{\dfrac{4}{3}\pi ka^2} = \frac{4}{5}a.$$

Example 2. Find the moment of inertia of a cylinder with respect to its axis. Assume the cylinder has uniform density.

Solution: Let ρ be the density, R be the radius of the base of the cylinder, l be the hight of the cylinder. Choose a coordinate system so that the z-axis is the axis of the cylinder, and V is the inside of the cylinder bounded by $x^2 + y^2 = R^2$, $z = 0$, $z = l$. The moment of inertia of the cylinder with respect to the z-axis is

$$I_z = \iiint_V \rho(x^2 + y^2)\, dx\, dy\, dz = \int_0^l dz \iint_{x^2+y^2 \leqslant R^2} \rho(x^2 + y^2)\, dx\, dy$$

$$= \rho \int_0^l dz \int_0^{2\pi} d\varphi \int_0^R r^3\, dr = 2\pi l\rho\frac{R^4}{4} = \frac{1}{2}MR^2,$$

where M is the mass of the cylinder.

Example 3. Find the moment of inertia of a sphere rotating about its diameter. Assume the sphere has uniform density, and radius R.

Solution: Let the center of the sphere be the origin, and the diameter be on the z-axis. Then V is the inside of the sphere $x^2 + y^2 + z^2 = R^2$. The moment of inertia is

$$I_z = \iiint\limits_{V} \rho(x^2 + y^2)\, dx\, dy\, dz = \int_{-R}^{R} dz \iint\limits_{x^2+y^2 \leqslant R^2-z^2} \rho(x^2 + y^2)\, dx\, dy$$

$$= \rho \int_{-R}^{R} dz \int_{0}^{2\pi} d\varphi \int_{0}^{\sqrt{R^2-z^2}} r^3\, dr = \frac{1}{2}\pi\rho \int_{-R}^{R} (R^2 - z^2)^2\, dz$$

$$= \pi R^5 \rho \int_{0}^{\frac{\pi}{2}} \cos^5 t\, dt = \frac{8}{15}\pi R^5 \rho = \frac{2}{5} M R^2,$$

where M is the mass of the ball.

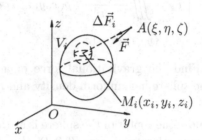

Fig. 8.2

Now we study the gravitational force.

Assume there is a particle $A(\xi, \eta, \zeta)$ with mass m outside of an object V with density $\rho = \rho(x, y, z)$ (Figure 8.2). To find the gravitational force F of V to A, we only need to find the projections of $\vec{F}$ on the three coordinate axes. Partition V into n small blocks $V_1, V_2, \cdots, V_n$. Denote the volume of V_i as ΔV_i. Let $M_i(x_i, y_i, z_i)$ be an arbitrary point in V_i. Suppose the mass of V_i, $\rho \Delta V_i$, is concentrated on the point M_i. The gravitational force ΔF_i of V_i to A is approximately equal to

$$|\Delta \vec{F}_i| = k \frac{m \rho \Delta V_i}{r_i^2},$$

where $r_i^2 = (x_i - \xi)^2 + (y_i - \eta)^2 + (z_i - \zeta)^2$, and k is a proportional constant. The direction of $\Delta \vec{F}_i$ is the same as the direction of $\overrightarrow{AM_i}$.

Without loss of generosity, assume the ratio k is equal to 1. The unit vector with the same direction of $\overrightarrow{AM_i}$ is

$$\left(\frac{x_i - \xi}{r_i}, \frac{y_i - \eta}{r_i}, \frac{z_i - \zeta}{r_i} \right).$$

Thus, $\Delta\vec{F_i}$ is

$$\frac{m\rho\Delta V_i}{r_i^2}\left(\frac{x_i-\xi}{r_i},\frac{y_i-\eta}{r_i},\frac{z_i-\zeta}{r_i}\right).$$

Calculating the limit of the sum of $\Delta\vec{F_i}$ over all small blocks of the partition of V, then taking the projections of $\vec{F}$ on the coordinate axes, we get

$$\begin{cases} F_x = \iiint\limits_V \dfrac{m(x-\xi)}{r^3}\rho(x,y,z)\,dx\,dy\,dz, \\[2mm] F_y = \iiint\limits_V \dfrac{m(y-\eta)}{r^3}\rho(x,y,z)\,dx\,dy\,dz, \\[2mm] F_z = \iiint\limits_V \dfrac{m(z-\zeta)}{r^3}\rho(x,y,z)\,dx\,dy\,dz, \end{cases}$$

where $r^2 = (x-\xi)^2 + (y-\eta)^2 + (z-\zeta)^2$.

Example 4. Find the gravitational force of a sphere to a particle A. Assume that the sphere has uniform density and radius R, the mass of the particle is 1.

Solution: Suppose the center of the sphere is at the origin, and the particle A is on the z-axis with coordinates $(0,0,l)$, $(l > R)$. If the density $\rho = 1$, then

$$F_x = \iiint\limits_V \frac{x}{r^3}\,dx\,dy\,dz, \qquad F_y = \iiint\limits_V \frac{y}{r^3}\,dx\,dy\,dz,$$

$$F_z = \iiint\limits_V \frac{z-l}{r^3}\,dx\,dy\,dz,$$

where $r^2 = x^2 + y^2 + (z-l)^2$; V is the sphere $x^2 + y^2 + z^2 \leqslant R^2$.

Since the sphere is symmetric with respect to the z-axis, it is easy to verify that $F_x = F_y = 0$. In fact, since

$$\int_0^{2\pi} \cos\varphi\,d\varphi = 0,$$

we have

$$F_x = \int_{-R}^{R} dz \iint\limits_{x^2+y^2\leqslant R^2-z^2} \frac{x}{r^3}\,dx\,dy$$

$$= \int_{-R}^{R} dz \int_0^{2\pi} d\varphi \int_0^{\sqrt{R^2-z^2}} \frac{\rho^2\cos\varphi}{[\rho^2+(z-l)^2]^{\frac{3}{2}}}\,d\rho$$

$$= \int_{-R}^{R} dz \int_0^{2\pi} \cos\varphi\,d\varphi \int_0^{\sqrt{R^2-z^2}} \frac{\rho^2}{[\rho^2+(z-l)^2]^{\frac{3}{2}}}\,d\rho = 0.$$

Similarly, $F_y = 0$. For the projection on the z-axis, we have

$$F_z = \int_{-R}^{R} (z - l)\, dz \iint\limits_{x^2+y^2 \leqslant R^2-z^2} \frac{dx\, dy}{[x^2 + y^2 + (z - l)^2]^{\frac{3}{2}}}.$$

Since

$$\iint\limits_{x^2+y^2 \leqslant R^2-z^2} \frac{dx\, dy}{[x^2 + y^2 + (z - l)^2]^{\frac{3}{2}}}$$

$$= \int_0^{2\pi} d\varphi \int_0^{\sqrt{R^2-z^2}} \frac{\rho}{[\rho^2 + (z - l)^2]^{\frac{3}{2}}}\, d\rho$$

$$= 2\pi \left(\frac{1}{l - z} - \frac{1}{\sqrt{R^2 - 2lz + l^2}} \right),$$

we have

$$F_z = 2\pi \int_{-R}^{R} \left[-1 - \frac{z - l}{\sqrt{R^2 - 2lz + l^2}} \right] dz = -\frac{4\pi R^3}{3l^2}.$$

The mass of the sphere is $M = \frac{4}{3}\pi R^3$, and the mass of the particle A is $m = 1$. Therefore, the gravitational force is

$$F_z = -\frac{Mm}{l^2}.$$

Therefore, the gravitational force generated by a uniform density sphere to an outside point is the same as that generated by a point at the center of the sphere with the same mass.

In astrophysics, the gravitational force between two celestial bodies is considered as the gravitational force between two particles that are at the centers of the celestial bodies with the corresponding masses.

8.2.2 *Complete System of Equations of Fluid Dynamics*

Suppose V is the region of a fluid velocity vector field that is a solenoidal field. That is, there are no sources or sinks in V. The density of fluids may change under different time and position, that is, the fluid is compressible.

Let S be the boundary of V. Then the amount of fluid flowing out in a unit time is

$$Q = \iint\limits_{S} \rho \vec{v} \cdot \vec{n}\, dS,$$

where $\rho = \rho(x, y, z, t)$ is the density of the fluid, $\vec{n}$ is the outward normal unit vector, and $\vec{v} = (u, v, w)$ is the velocity of the fluid.

On the other hand, the change of density in time interval dt is $\dfrac{\partial \rho}{\partial t}\, dt$, the change of mass $\rho\, dV$ of the volume element dV is $\dfrac{\partial \rho}{\partial t}\, dt\, dV$. Thus, the total change of fluid volume of V is

$$dt \iiint\limits_V \frac{\partial \rho}{\partial t}\, dV.$$

This is the amount of fluid flowing into V in time interval dt.

The amount of fluid flowing out of V in an unit time interval is

$$Q = -\iiint\limits_V \frac{\partial \rho}{\partial t}\, dV.$$

Hence,

$$\iint\limits_S \rho \vec{v} \cdot \vec{n}\, dS + \iiint\limits_V \frac{\partial \rho}{\partial t}\, dV = 0.$$

By Gauss's formula, we have

$$\iiint\limits_V \left(\operatorname{div} \rho \vec{v} + \frac{\partial \rho}{\partial t} \right) dV = 0.$$

This equation is true for any part of V. Therefore, by the Mean-Value Theorem of integrals, we get the continuity equation of fluid dynamics

$$\operatorname{div} \rho \vec{v} + \frac{\partial \rho}{\partial t} = 0.$$

Now we study the motion equation of inviscid flow.

The movement of an object is determined by the external forces and the internal forces that act on it. Suppose the external force is in direct proportion to the mass. Let $\vec{F}$ be a force that acts on an object with unit mass. Then the force acting on the volume element dV is $\rho\, dV \vec{F}$.

The internal force is the pressure on the boundary S of V. Let $(\cos \lambda,\ \cos \mu,\ \cos \nu)$ be the direction cosine of the outward normal of S. Then the projection of the force acting on the area element dS is

$$-p \cos \lambda\, dS, \quad -p \cos \mu\, dS, \quad -p \cos \nu\, dS,$$

where p is the pressure on a unit area. Thus, the force acting on V is

$$\left(-\iint\limits_S p \cos \lambda\, dS, \ -\iint\limits_S p \cos \mu\, dS, \ -\iint\limits_S p \cos \nu\, dS \right).$$

By Gauss's formula, that is

$$\left(-\iiint\limits_V \frac{\partial p}{\partial x}\, dV, \ -\iiint\limits_V \frac{\partial p}{\partial y}\, dV, \ -\iiint\limits_V \frac{\partial p}{\partial z}\, dV \right).$$

This implies that the force acting on dV is

$$-\left(\frac{\partial p}{\partial x}\, dV, \frac{\partial p}{\partial y}\, dV, \frac{\partial p}{\partial z}\, dV \right) = -dV\, \mathbf{grad}\, p.$$

By Newton's second law

$$\rho\, dV \vec{a} = \rho\, dV \vec{F} - dV\, \mathbf{grad}\, p,$$

where $\vec{a} = \dfrac{d\vec{v}}{dt} = \dfrac{d^2 \vec{r}}{dt^2}$ is the acceleration. Therefore

$$\frac{d^2 \vec{r}}{dt^2} = \vec{a} = \vec{F} - \frac{1}{\rho}\, \mathbf{grad}\, p.$$

This is the motion equation of inviscid flow.

Since

$$\vec{v} = \frac{d\vec{r}}{dt} = (u, v, w),$$

the position (x, y, z) and velocity (u, v, w) are related to t, we have

$$\frac{d^2 x}{dt^2} = \frac{du}{dt} = \frac{\partial u}{\partial t} + \frac{\partial u}{\partial x}\frac{dx}{dt} + \frac{\partial u}{\partial y}\frac{dy}{dt} + \frac{\partial u}{\partial z}\frac{dz}{dt}$$

$$= \frac{\partial u}{\partial t} + \frac{\partial u}{\partial x}u + \frac{\partial u}{\partial y}v + \frac{\partial u}{\partial z}w,$$

$$\frac{d^2 y}{dt^2} = \frac{\partial v}{\partial t} + \frac{\partial v}{\partial x}u + \frac{\partial v}{\partial y}v + \frac{\partial v}{\partial z}w,$$

$$\frac{d^2 z}{dt^2} = \frac{\partial w}{\partial t} + \frac{\partial w}{\partial x}u + \frac{\partial w}{\partial y}v + \frac{\partial w}{\partial z}w.$$

If $\vec{F} = (F_x, F_y, F_z)$, then the motion equation can be written as

$$\begin{cases} \dfrac{\partial u}{\partial t} + \dfrac{\partial u}{\partial x}u + \dfrac{\partial u}{\partial y}v + \dfrac{\partial u}{\partial z}w = F_x - \dfrac{1}{\rho}\dfrac{\partial p}{\partial x}, \\[2mm] \dfrac{\partial v}{\partial t} + \dfrac{\partial v}{\partial x}u + \dfrac{\partial v}{\partial y}v + \dfrac{\partial v}{\partial z}w = F_y - \dfrac{1}{\rho}\dfrac{\partial p}{\partial y}, \\[2mm] \dfrac{\partial w}{\partial t} + \dfrac{\partial w}{\partial x}u + \dfrac{\partial w}{\partial y}v + \dfrac{\partial w}{\partial z}w = F_z - \dfrac{1}{\rho}\dfrac{\partial p}{\partial z}. \end{cases}$$

This is called *Euler's equation*.

This formula can also be written as

$$\frac{\partial \vec{v}}{\partial t} + \left(u\frac{\partial}{\partial x} + v\frac{\partial}{\partial y} + w\frac{\partial}{\partial z} \right)\vec{v} = \vec{F} - \frac{1}{\rho}\,\mathbf{grad}\,p.$$

That is

$$\frac{\partial \vec{v}}{\partial t} + (v \cdot \nabla)\vec{v} = \vec{F} - \frac{1}{\rho}\,\mathbf{grad}\,p.$$

In summary, the velocity $\vec{v} = (u, v, w)$, the pressure p and the density ρ satisfy

$$\begin{cases} \dfrac{\partial p}{\partial t} + \mathrm{div}(\rho\vec{v}) = 0 \quad \text{(Continuity equation)}, \\[2mm] \dfrac{\partial \vec{v}}{\partial t} + (\vec{v} \cdot \nabla)\vec{v} = \vec{F} - \dfrac{1}{\rho}\,\mathbf{grad}\,p \quad \text{(Motion equation)}. \end{cases}$$

There are five unknown functions and four equations. The remaining equation is the relation between the pressure p and the density ρ, called the equation of state

$$p = f(\rho).$$

These equations together form the *complete system of equations of fluid dynamics*.

8.2.3   *Propagation of Sound*

We now apply the equations of fluid dynamics to the propagation of sound.

(i) Suppose that the propagation of sound is adiabatic, that is, the equation of state is

$$\frac{p}{p_0} = \left(\frac{\rho}{\rho_0} \right)^{\gamma}, \quad \gamma = \frac{c_p}{c_v},$$

where ρ_0 is the initial density, p_0 is the initial pressure, c_p is the specific heat at constant pressure and c_v is the specific heat at constant volume.

(ii) The vibration of gas is very small, so the high degree terms in the expressions of velocity, the gradient of density can be omitted.

Consider the relative change of the density

$$s = s(x, y, z, t) = \frac{\rho - \rho_0}{\rho_0}.$$

We have

$$\rho = (1 + s)\rho_0.$$

Thus,

$$\frac{1}{\rho}\operatorname{grad} p = \frac{1}{\rho_0}(1+s)^{-1}\operatorname{grad} p \approx \frac{1}{\rho_0}\operatorname{grad} p,$$

$$\operatorname{div}\rho\vec{v} = \vec{v}\operatorname{grad}\rho + \rho\operatorname{div}\vec{v} \approx \rho_0\operatorname{div}\vec{v}.$$

The equations of fluid dynamics can be written as

$$\begin{cases} \dfrac{\partial\vec{v}}{\partial t} = \vec{F} - \dfrac{1}{\rho_0}\operatorname{grad} p, \\[2mm] \dfrac{\partial\rho}{\partial t} = -\rho\operatorname{div}\vec{v}, \\[2mm] p = p_0(1+s)^\gamma \approx p_0(1+\gamma s). \end{cases}$$

Let $a^2 = \gamma\dfrac{p_0}{\rho_0}$. We obtain

$$\begin{cases} \dfrac{\partial\vec{v}}{\partial t} = \vec{F} - a^2\operatorname{grad} s, \\[2mm] \dfrac{\partial s}{\partial t} = -\operatorname{div}\vec{v}. \end{cases}$$

Taking the partial derivative of the second equation with respect to t, and exchanging the order of derivatives to eliminate v, we have

$$-\frac{\partial^2 s}{\partial t^2} = \frac{\partial}{\partial t}\operatorname{div}\vec{v} = \operatorname{div}(-a^2\operatorname{grad} s + \vec{F}) = \operatorname{div}\vec{F} - a^2\Delta s.$$

From this we get the wave equation

$$\frac{\partial^2 s}{\partial t^2} = a^2\Delta s - \operatorname{div}\vec{F}.$$

If there is no external force, that is, $\vec{F} = 0$, then

$$\frac{\partial^2 s}{\partial t^2} = a^2\Delta s = a^2\left(\frac{\partial^2 s}{\partial x^2} + \frac{\partial^2 s}{\partial y^2} + \frac{\partial^2 s}{\partial z^2}\right).$$

This is the *wave equation* that is usually seen elsewhere.

Notice that s describes the compression and expansion of the medium. The equation represents the law of propagation of sound, and $\operatorname{div}\vec{F}$ is the sound source.

8.2.4 Heat Exchange

A temperature field $\varphi(x,y,z,t)$ on an object V is a function of positions and time.

The vector $-k\operatorname{grad}\varphi$ is called heat flux vector, where $k > 0$ is a specific heat coefficient. The negative sign means the heat flows to the place with lower temperature. The direction of **grad** φ is the direction that the temperature changes the most rapidly.

Let S be the boundary surface of V. The heat that passes through a surface element dS in time dt is directly proportional to $dt\,dS$ and $\dfrac{\partial \varphi}{\partial \vec{n}}$, where $\vec{n}$ is the normal of S. That is

$$\Delta Q = k\,dt\,dS\frac{\partial \varphi}{\partial \vec{n}} = -k\,dt\,dS\,\mathbf{grad}\,\varphi \cdot \vec{n}.$$

Thus, the total amount of heat that passes through S is

$$-dt \iint\limits_{S} k\,\mathbf{grad}\,\varphi \cdot \vec{n}\,dS.$$

If the field is solenoidal, then the amount of heat released from V that passes through the boundary surface S in a unit time interval is

$$Q = - \iint\limits_{S} k\,\mathbf{grad}\,\varphi \cdot \vec{n}\,dS.$$

By Gauss's formula, the amount of heat that V absorbed is

$$\iiint\limits_{V} \mathrm{div}(k\,\mathbf{grad}\,\varphi)\,dV.$$

Now we calculate the heat of V another way. The temperature increases

$$d\varphi = \frac{\partial \varphi}{\partial t}\,dt$$

in time dt. Therefore, the heat that dV needs is

$$C\,d\varphi\rho\,dV = C\frac{\partial \varphi}{\partial t}\,dt\rho\,dV,$$

where C is the thermal capacity, ρ is the density of the object. The amount of heat that the entire V absorbed in time dt is

$$dt \iiint\limits_{V} C\rho\frac{\partial \varphi}{\partial t}\,dV.$$

The amount of heat that the entire V absorbed in a unit time is

$$\iiint\limits_{V} C\rho\frac{\partial \varphi}{\partial t}\,dV.$$

Thus,

$$\iiint\limits_{V} \left[C\rho\frac{\partial \varphi}{\partial t} - \mathrm{div}\,(k\,\mathbf{grad}\,\varphi)\right]\,dV = 0$$

for any part of V. Therefore,

$$C\rho\frac{\partial\varphi}{\partial t} = \text{div}\,(k\,\mathbf{grad}\,\varphi).$$

This is the *heat equation*.

In homogeneous medium, let $a^2 = \dfrac{k}{C\rho}$, then we get the equation

$$\frac{\partial\varphi}{\partial t} = a^2\Delta\varphi = a^2\left(\frac{\partial^2\varphi}{\partial x^2} + \frac{\partial^2\varphi}{\partial y^2} + \frac{\partial^2\varphi}{\partial z^2}\right).$$

If the function φ is independent of time t, then φ satisfies the Laplace equation

$$\Delta\varphi = 0.$$

If the source of heat exists, then we have

$$\iiint\limits_V\left[C\rho\frac{\partial\varphi}{\partial t} - \text{div}\,(k\,\mathbf{grad}\,\varphi)\right]dV = \iiint\limits_V e\,dV,$$

where $e = e(x, y, z, t)$ is the heat-intensity that is distributed in V continuously.

The right hand side of the above equation represents the heat released from V in an unit time. Therefore,

$$C\rho\frac{\partial\varphi}{\partial t} - \text{div}\,(k\,\mathbf{grad}\,\varphi) = e.$$

In homogeneous medium, we have

$$\frac{\partial\varphi}{\partial t} = a^2\Delta\varphi + \frac{e}{C\rho}.$$

Exercises 8.2

1. Find the mass of a thin slice in the shape of an ellipse

$$\frac{x^2}{a^2} + \frac{y^2}{b^2} \leqslant 1$$

with density $\rho = \dfrac{x^2}{a^2} + \dfrac{y^2}{b^2}$.

2. A disk has radius a. The density at each point is directly proportional to the distance between the point and the center with the proportionality coefficient 1. Cut away a small disk with radius $\dfrac{a}{2}$ that is inscribed in the larger disk. Find the coordinates of the barycenter of the remaining disk.

3. A thin slice is bounded by two concentric circles of radii R and r, $(R > r)$. The density at each point is inversely proportional to the distance between the point and the center. The density on the circle with radius r is 1. Find the mass of this figure.

4. An object is bounded by two concentric spheres of radii R and r, $(R > 1 > r)$. The density of the material at each point is inversely proportional to the distance between the point and the center. The density on the sphere with radius 1 is k. Find the mass of this figure.

5. Find the coordinates of the barycenter of an object with even density, that is bounded by the surfaces
$$\frac{x^2}{a^2} + \frac{y^2}{b^2} = \frac{z^2}{c^2} \text{ and } z = c.$$

6. Find the coordinates of the barycenter of an object with even density, that is bounded by the paraboloid $z = x^2 + y^2$ and $z = 1$.

7. The density at each point of the ball $x^2 + y^2 + z^2 \leqslant 2az$ is inversely proportional to the distance between the point and the origin. Find the coordinates of the barycenter.

8. Find the moment of inertia of the ball $x^2 + y^2 + z^2 \leqslant a^2$ of density 1, rotating about the coordinate axes.

9. Find the moment of inertia of the following objects with density 1.

   (1) A rod with mass m and length l.

   (i) The axis of rotation is perpendicular to the rod at its mid-point.

   (ii) The axis of rotation is perpendicular to the rod at one end.

   (2) A fine ring with mass m and radius R.

   (i) The axis of rotation is perpendicular to the ring at its center.

   (ii) The axis of rotation is a diameter of the ring.

   (3) A thin disk with mass m and radius R.

   (i) The axis of rotation is perpendicular to the disk at its center.

   (ii) The axis of rotation is a diameter of the disk.

   (4) A ball with mass m and radius R. The axis of rotation is a tangent line the ball.

10. Find the gravitational force of a truncated cone
$$z = \sqrt{x^2 + y^2}, \ (0 < a \leqslant z \leqslant b)$$
with uniform density ρ to a particle with mass m at the vertex of the cone.

11. A spherical shell has even density ρ and radius R. Find the gravitational force of the shell to a particle of mass m at a point with distance a from the center of the shell. Study the situations that the particle is inside of the shell and outside of the shell.

12. A spherical shell has even density ρ and radius R. It rotates about its diameter with the angular velocity ω. Find its moment of inertia and the rotational kinetic energy.

13. Prove that the moment of inertia of an object rotating about any axis of rotation is $Md^2 + I_c$, where M is the mass of the object, d is the distance between the axis of rotation and the barycenter C, I_c is the moment of inertia of the object rotating about an axis that passes the center of gravity and is parallel to the given axis.

Chapter 9

The ε-δ Definitions of Limits

9.1 The ε-N Definition of Limits of Number Sequences

9.1.1 *Definition of Limits of Number Sequences*

In this chapter, we re-state the contents of the previous chapters in a more precise language. Let's start from the concept of limits.

An infinite number sequence

$$a_1, a_2, \cdots, a_n, \cdots,$$

has a limit a if the terms approach a fixed number a as n increases without boundary. That is, for any small positive number, there exists a term in the sequence, such that the absolute value of the difference between each term after this term and a is less than this small positive number. The number a is called the limit of this infinite number sequence, and is denoted as

$$\lim_{n \to \infty} a_n = a.$$

In other words, for any small positive number ε, all terms a_n in the sequence is in the interval $(a - \varepsilon, a + \varepsilon)$ (Figure 9.1) when n is large enough.

The full definition of limit of number sequences is:

For a number sequence a_n, if there is a fixed number a, such that for any small positive number ε, there exits a positive integer $N = N(\varepsilon)$ that depends on ε, such that $|a_n - a| < \varepsilon$ if $n > N$, then a is the limit of the number sequence a_n, and is denoted as $\lim_{n \to \infty} a_n = a$ or $\lim a_n = a$ or $a_n \to a$ as $n \to \infty$. There are only finite many a_n outside of the interval $(a - \varepsilon, a + \varepsilon)$. (Figure 9.2)

This is called the ε-N *definition of limits.*

Example 1. Show that $\lim \dfrac{1}{n^\alpha} = 0, (\alpha > 0)$.

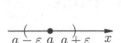

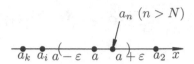

Fig. 9.1   Fig. 9.2

Proof: For any given $\varepsilon > 0$, the inequality

$$\left| \frac{1}{n^\alpha} - 0 \right| = \frac{1}{n^\alpha} < \varepsilon$$

is true if $n^\alpha > \dfrac{1}{\varepsilon}$, or $n > \left(\dfrac{1}{\varepsilon} \right)^{\frac{1}{\alpha}}$. Thus, we only need to find a positive

integer $N > \left(\dfrac{1}{\varepsilon} \right)^{\frac{1}{\alpha}}$, (for instance, we can choose $N = \left[\left(\dfrac{1}{\varepsilon} \right)^{\frac{1}{\alpha}} \right] + 1$, where

$\left[\left(\dfrac{1}{\varepsilon} \right)^{\frac{1}{\alpha}} \right]$ is the integer part of the number inside the square brackets) then
we have

$$\left| \frac{1}{n^\alpha} - 0 \right| < \frac{1}{N^\alpha} < \varepsilon$$

if $n > N$.

Example 2. Show that $\lim a^{\frac{1}{n}} = 1$, $(a > 1)$.

Proof: Since $a > 1$, we have $a^{\frac{1}{n}} > 1$. Let $\alpha_n = a^{\frac{1}{n}} - 1$. Then $\alpha_n > 0$.
Thus,

$$a = (1 + \alpha_n)^n = 1 + n\alpha_n + \cdots + \alpha_n^n > n\alpha_n.$$

It follows that $\alpha_n < \dfrac{a}{n}$. Hence, for any $\varepsilon > 0$, let $N > \dfrac{a}{\varepsilon}$. Then

$$\left| a^{\frac{1}{n}} - 1 \right| = \alpha_n < \frac{a}{N} < \varepsilon$$

if $n > N$.

If a number sequence α_n has a limit a, then we say that α_n is *convergent*, and that α_n converges to a. Otherwise, we say that the sequence α_n is *divergent*.

9.1.2 *Properties of Limits of Number Sequences*

Now we can prove some basic properties of number sequences by the ε-N definition. If

$$\lim_{n \to \infty} a_n = a, \ \lim_{n \to \infty} b_n = b,$$

then

$$\lim_{n \to \infty} (a_n \pm b_n) = a \pm b, \quad \lim_{n \to \infty} (a_n \cdot b_n) = a \cdot b,$$

$$\lim_{n \to \infty} \frac{a_n}{b_n} = \frac{a}{b}, (b \neq 0).$$

To proof the first equation, we start from the fact that

$$|(a_n \pm b_n) - (a \pm b)| = |(a_n - a) \pm (b_n - b)| \leqslant |a_n - a| + |b_n - b|.$$

For any $\varepsilon > 0$, there exists a positive number N_1, such that

$$|a_n - a| < \frac{\varepsilon}{2}$$

if $n > N_1$. There also exists a positive number N_2, such that

$$|b_n - b| < \frac{\varepsilon}{2}$$

if $n > N_2$. Thus,

$$|(a_n \pm b_n) - (a \pm b)| \leqslant |a_n - a| + |b_n - b| < \frac{\varepsilon}{2} + \frac{\varepsilon}{2} = \varepsilon.$$

if $n > N = \max(N_1, N_2)$. This proves the first equation. The rest of equations can be proved in a similar way.

From

$$\lim_{n \to \infty} (a_n \cdot b_n) = a \cdot b$$

we have that, if k is a constant, then

$$\lim_{n \to \infty} ka_n = ka.$$

More properties can be proved using the ε-N definition:

1. If a number sequence is convergent, then it has only one limit.

In fact, suppose there are two limits a and b ($a \neq b$). If we choose $\varepsilon = \dfrac{|a - b|}{3}$, then the intervals $(a-\varepsilon, a+\varepsilon)$ and $(b-\varepsilon, b+\varepsilon)$ are disjoint. Since a is a limit of a_n, there are only finitely many terms of a_n in $(b - \varepsilon, b + \varepsilon)$. Since b is also a limit of a_n, this is impossible.

If there exists a positive number M, such that $|a_n| \leqslant M, (n = 1, 2, \cdots)$, then we say that a_n is bounded. If there exists a number A, such that

$a_n \leqslant A$, $(n = 1, 2, \cdots)$, then A is said to be an upper bound of the number series. If there exists a number B, such that $a_n \geqslant B$, $(n = 1, 2, \cdots)$, then B is said to be a lower bound of the number series. If a number sequence has both an upper bound and a lower bound, then it is called a *bounded sequence*.

2. If a number sequence is convergent, then it is bounded.

In fact, suppose a number sequence a_n is convergent and has limit a. Let $\varepsilon = 1$. Then there exists a natural number N such that

$$|a_n| = |a_n - a + a| \leqslant |a_n - a| + |a| < 1 + |a|$$

if $n > N$.

Let $M = \max(|a_1|, |a_2|, \cdots, |a_n|, |a| + 1)$. Then obviously $|a_n| \leqslant M$ for all $n = 1, 2, \cdots$, and the conclusion follows.

3. If $\lim a_n = a$, $\lim b_n = b$, and $a_n \leqslant b_n$, $(n = 1, 2, \cdots)$, then $a \leqslant b$.

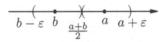

Fig. 9.3

In fact, suppose $a > b$. Let $\varepsilon = \dfrac{a - b}{2}$. Then there exists a natural number N such that

$$b_n < \frac{a + b}{2} < a_n$$

if $n > N$. This contradicts the assumption.

Notice that even if the assumption is $a_n < b_n$, the conclusion is still $a \leqslant b$, not necessarily $a < b$. For instance, the sequence $\dfrac{1}{n} > 0$, $(n = 1, 2, \cdots)$, but

$$\lim 0 = 0 = \lim \frac{1}{n}.$$

4. If $b_n \leqslant a_n \leqslant c_n$, and $\lim b_n = \lim c_n = a$, then $\lim a_n = a$.

In fact, by the assumption, for any $\varepsilon > 0$, there exists a natural number N, such that

$$a - \varepsilon < b_n < a + \varepsilon, \quad a - \varepsilon < c_n < a + \varepsilon,$$

if $n > N$. Thus, we have

$$a - \varepsilon < b_n \leqslant a_n \leqslant c_n < a + \varepsilon$$

if $n > N$. That is, $a - \varepsilon < a_n < a + \varepsilon$ if $n > N$.

Example 1. Find the limit $\lim\limits_{n\to\infty} \dfrac{2n^3 + 3n^2 + 4}{5n^3 + 6n + 7}$.

Solution: Since

$$\frac{2n^3 + 3n^2 + 4}{5n^3 + 6n + 7} = \frac{2 + 3\dfrac{1}{n} + 4\dfrac{1}{n^3}}{5 + 6\dfrac{1}{n^2} + 7\dfrac{1}{n^3}}$$

approaches

$$\frac{2 + 0 + 0}{5 + 0 + 0} = \frac{2}{5}$$

as $n \to \infty$, the limit is $\dfrac{2}{5}$.

Example 2. Find the limit $\lim\limits_{n\to\infty} \left(\dfrac{1}{n^2} + \dfrac{2}{n^2} + \cdots + \dfrac{n}{n^2} \right)$.

Solution:

$$\lim_{n\to\infty} \left(\frac{1}{n^2} + \frac{2}{n^2} + \cdots + \frac{n}{n^2} \right) = \lim_{n\to\infty} \frac{n(n+1)}{2n^2} = \lim_{n\to\infty} \left(\frac{1}{2} + \frac{1}{2n} \right) = \frac{1}{2}.$$

Example 3. If $a > 0$, then $\lim\limits_{n\to\infty} a^{\frac{1}{n}} = 1$.

Proof: The proof is the same as the Example 2 in section 9.1.1 for $a > 1$. The conclusion follows immediately for $a = 1$.

We only need to discuss the case $0 < a < 1$. Since

$$a^{\frac{1}{n}} = \frac{1}{\left(\dfrac{1}{a}\right)^{\frac{1}{n}}},$$

and $\dfrac{1}{a} > 1$, therefore, $\left(\dfrac{1}{a}\right)^{\frac{1}{n}} \to 1$ as $n \to \infty$, we have $a^{\frac{1}{n}} \to 1$ as $n \to \infty$.

Therefore, $a^{\frac{1}{n}} = \sqrt[n]{a} \to 1$ as $n \to \infty$.

Example 4. Find $\lim\limits_{n\to\infty} \sqrt[n]{n}$.

Solution: Since

$$(1+\lambda)^n = 1 + n\lambda + \frac{n(n-1)}{2}\lambda^2 + \cdots + \lambda^n$$

for $\lambda > 0$, we have

$$(1+\lambda)^n > \frac{n(n-1)}{2}\lambda^2.$$

Let $\lambda = \sqrt[n]{n} - 1$, we have

$$n > \frac{n(n-1)}{2}(\sqrt[n]{n} - 1)^2.$$

Thus

$$0 < \sqrt[n]{n} - 1 < \frac{\sqrt{2}}{\sqrt{n-1}}$$

if $n \geqslant 2$. The limit of the term on the right hand side is

$$\lim_{n\to\infty} \frac{\sqrt{2}}{\sqrt{n-1}} = 0.$$

It follows that $\lim\limits_{n\to\infty} (\sqrt[n]{n} - 1) = 0$, and $\lim\limits_{n\to\infty} \sqrt[n]{n} = 1$.

9.1.3 *Criteria for the Existence of Limits*

What kind of number sequences is convergent? What is the limit of a convergent sequence? Of course, we can always follow the definition of the limit and try some numbers, but this method seems aimless. We introduce two criteria of convergence in this section.

A sequence a_n is said to be *monotonically increasing* if $a_n \leqslant a_{n+1}$ for all $n = 1, 2, \cdots$. Similarly, if $a_n \geqslant a_{n+1}$ for all $n = 1, 2, \cdots$, then it is said to be *monotonically decreasing*.

1. If a monotonically increasing (decreasing) sequence has an upper (a lower) bound, then the sequence is convergent.

The proof of this criterion is not as straightforward as its statement.

We regard all real numbers as infinite decimals. A real number a can be written as an integral part α_0 and a decimal part $0.\alpha_1\alpha_2\cdots\alpha_n\cdots$, where each α_i, $(1 \leqslant i < \infty)$, is one of the ten digits $0, 1, 2, \cdots, 9$. If there are only finitely many digits (there are infinitely many zeros after a certain digit) in this expression, or if the digits become periodic after a certain digit, then this real number is a *rational number*. It can be written as $\dfrac{p}{q}$, where p and

q are integers. If the digits of an infinite decimal never have any kind of repetition, then this infinite decimal is called an *irrational number*.

Suppose $a_1 \leqslant a_2 \leqslant \cdots \leqslant a_n \leqslant a_{n+1} \leqslant \cdots$ is a monotonically increasing sequence with an upper bound M. That is,

$$a_n \leqslant M, n = 1, 2, \cdots.$$

Since every real number a_n is an infinite decimal, a_n has an upper bound. The integer parts of a_n, $n = 1, 2, \cdots$ are the same after a certain term. We name this integer α_0. The first digit of a_n, $n = 1, 2, \cdots$ after decimal point also are the same after a certain term, and we name it α_1. Continue this process, we get an infinite decimal

$$a = \alpha_0.\alpha_1\alpha_2 \cdots \alpha_n \cdots.$$

Now we show that a is the limit of the sequence a_n.

In fact, for any $\varepsilon > 0$, we can find a positive integer k, such that

$$a - \varepsilon < \alpha_0.\alpha_1\alpha_2 \cdots \alpha_k.$$

Since $\alpha_0.\alpha_1\alpha_2 \cdots \alpha_k$ is not an upper bound of a_n, there exists a term a_N, such that

$$a - \varepsilon < \alpha_0.\alpha_1\alpha_2 \ldots \alpha_k < a_N \leqslant a.$$

By the assumption that a_n is monotonically increasing, we have

$$a - \varepsilon < a_N \leqslant a_n < a + \varepsilon$$

if $n > N$. Thus, a_n converges to a.

The conclusion that If a monotonically decreasing sequence has a lower bound, then the sequence is convergent can be proved in a similar way.

Example 1. Suppose $\alpha \geqslant 2$ and

$$a_n = 1 + \frac{1}{2^\alpha} + \frac{1}{3^\alpha} + \cdots + \frac{1}{n^\alpha}, \quad (n = 1, 2, \cdots).$$

Prove that the sequence a_n is convergent.

Proof: Obviously, a_n is monotonically increasing, so we only need to show the existence of its upper bound. In fact,

$$\begin{aligned}
a_n &< 1 + \frac{1}{2^2} + \frac{1}{3^2} + \cdots + \frac{1}{n^2} \\
&\leqslant 1 + \frac{1}{2 \cdot 1} + \frac{1}{3 \cdot 2} + \cdots + \frac{1}{n \cdot (n-1)} \\
&= 1 + \left(1 - \frac{1}{2}\right) + \left(\frac{1}{2} - \frac{1}{3}\right) + \cdots + \left(\frac{1}{n-1} - \frac{1}{n}\right) \\
&= 2 - \frac{1}{n} < 2, \quad (n = 1, 2, \cdots).
\end{aligned}$$

Example 2. Prove that the limit of the sequence

$$a_n = \sqrt{2 + \sqrt{2 + \cdots + \sqrt{2}}}, \quad \text{(n-level nested radical)}$$

exists.

Proof: When $n = 1$, $a_1 = \sqrt{2} < 2$. If $a_k < 2$, then $a_{k+1} = \sqrt{2 + a_k} < 2$. Thus, by mathematical induction, $a_n < 2$. Since a_n is monotonically increasing, it is convergent.

The next example is important.

Example 3. Prove that the limit of the sequence

$$a_n = \left(1 + \frac{1}{n}\right)^n$$

exists.

Proof: First, we show that a_n is monotonically increasing,

$$a_n = \left(1 + \frac{1}{n}\right)^n = 1 + \sum_{k=1}^{n} \frac{n(n-1)\cdots(n-k+1)}{k!} \frac{1}{n^k}$$

$$= 1 + \sum_{k=1}^{n} \frac{1}{k!}\left(1 - \frac{1}{n}\right)\cdots\left(1 - \frac{k-1}{n}\right)$$

$$< 1 + \sum_{k=1}^{n} \frac{1}{k!}\left(1 - \frac{1}{n+1}\right)\left(1 - \frac{2}{n+1}\right)\cdots\left(1 - \frac{k-1}{n+1}\right)$$

$$< 1 + \sum_{k=1}^{n+1} \frac{1}{k!}\left(1 - \frac{1}{n+1}\right)\left(1 - \frac{2}{n+1}\right)\cdots\left(1 - \frac{k-1}{n+1}\right)$$

$$= a_{n+1}.$$

Now we show that a_n has an upper bound.

$$a_n = 1 + \sum_{k=1}^{n} \frac{1}{k!}\left(1 - \frac{1}{n}\right)\left(1 - \frac{2}{n}\right)\cdots\left(1 - \frac{k-1}{n}\right)$$

$$< 1 + \sum_{k=1}^{n} \frac{1}{k!} < 1 + 1 + \sum_{k=2}^{n} \frac{1}{k(k-1)}$$

$$= 2 + \left(1 - \frac{1}{n}\right) < 3.$$

We have learned (Section 1.3.3) that the limit of a_n is

$$e = 2.718281828459045\cdots.$$

The monotonicity of a sequence is a necessary condition in this criterion.

If a sequence not only has an upper bound but also has a lower bound, then we say that it is bounded, and it has a convergent subsequence. This is the next criterion.

2. Bolzano–Weierstrass Criterion

There is a convergent subsequence in any bounded sequence.

The proof is not difficult: Since a_n is bounded, there exists an interval $[A, B]$ such that all terms of the sequences are in the interval. Bisecting $[A, B]$ into two subintervals, one of the subintervals must contains infinitely many terms of the sequence. If we denote this subinterval as $[A_1, B_1]$, then

$$B_1 - A_1 = \frac{1}{2}(B - A).$$

Bisecting $[A_1, B_1]$ into two subintervals, one of the subinterval must contain infinitely many terms of the sequence. Repeating the same process, we get a sequence of intervals each of which is in the preceding interval. The length of each interval is half of the length of the preceding interval. The length of the kth interval is

$$B_k - A_k = \frac{1}{2^k}(B - A).$$

This length tends to zero as k tends to infinity. Thus, we obtain two monotonic sequences

$$A_1 \leqslant A_2 \leqslant A_3 \leqslant \cdots \leqslant A_k \leqslant A_{k+1} \leqslant \cdots,$$
$$B_1 \geqslant B_2 \geqslant B_3 \geqslant \cdots \geqslant B_k \geqslant B_{k+1} \geqslant \cdots.$$

Obviously, both sequences are bounded and therefore have limits. Since $B_k - A_k \to 0$, the two limits are the same. If we denote this limit as a, then

$$\lim_{k \to \infty} A_k = \lim_{k \to \infty} B_k = a.$$

The interval $[A_k, B_k]$ contains infinitely many terms of a_n. Let a_{n_k} be an arbitrary one of them. Since

$$A_k \leqslant a_{n_k} \leqslant B_k,$$

and

$$\lim_{k \to \infty} A_k = \lim_{k \to \infty} B_k = a,$$

we have

$$\lim_{k \to \infty} a_{n_k} = a.$$

Hence, we can choose a subsequence a_{n_k} from the sequence a_n, such that the limit of a_{n_k} exists.

From the Bolzano-Weierstrass criterion, we can get the next important criterion.

3. Cauchy Criterion

Suppose a_n is a number sequence. If for any small positive number ε, there exists a natural number $N = N(\varepsilon)$ (N varies with ε) such that

$$|a_n - a_m| < \varepsilon$$

for all n, m greater than N, then the limit of the sequence a_n exists.

In fact, by the assumption, for any $\varepsilon > 0$, there exists an N, such that

$$|a_n - a_m| < \varepsilon$$

for all n, m greater than N.

For a fixed m, we have

$$a_m - \varepsilon < a_n < a_m + \varepsilon$$

if $n > N$.

This implies that there are infinitely many terms in $[a_m - \varepsilon, a_m + \varepsilon]$.

By Bolzano-Weierstrass criterion, we can find a subsequence a_{n_k}, such that

$$\lim_{k \to \infty} a_{n_k} = a.$$

To prove a_n also tends to a, we can choose n_k large enough, such that

$$|a_{n_k} - a| < \varepsilon.$$

Let $n_k > N$. Then we have

$$|a_n - a_{n_k}| < \varepsilon$$

if $n > N$. Hence,

$$|a_n - a| \leqslant |a_n - a_{n_k}| + |a_{n_k} - a| < 2\varepsilon.$$

if $n > N$.

It follows that $\lim_{n \to \infty} a_n = a$.

Cauchy criterion is a necessary and sufficient condition of the existence of the limit of a number sequence.

We have shown that if a sequence satisfies the condition in Cauchy criterion, then it is convergent. On the other hand, if a sequence a_n is convergent, then it satisfies the condition in Cauchy criterion.

Indeed, If a_n is convergent, then by definition, for any $\varepsilon > 0$, there exists an N, such that

$$|a_n - a| < \frac{\varepsilon}{2}$$

if $n > N$.

Thus, for any m and n greater than N, we have

$$|a_m - a_n| \leqslant |a_m - a| + |a_n - a| < \frac{\varepsilon}{2} + \frac{\varepsilon}{2} = \varepsilon.$$

Since the actual value of the limit of the sequence is not involved in Cauchy criterion, it is easy to use. Also, we can regard Cauchy criterion as a definition of the existence of the limit of a sequence.

Example 4. Prove that the sequence

$$a_n = \frac{\sin 1}{2} + \frac{\sin 2}{2^2} + \cdots + \frac{\sin n}{2^n}$$

is convergent.

Proof: Since

$$|a_{n+p} - a_n| = \left| \frac{\sin(n+1)}{2^{n+1}} + \cdots + \frac{\sin(n+p)}{2^{n+p}} \right|$$

$$\leqslant \frac{1}{2^{n+1}} + \cdots + \frac{1}{2^{n+p}}$$

$$= \frac{1}{2^{n+1}} \frac{1 - \dfrac{1}{2^p}}{1 - \dfrac{1}{2}} = \frac{1}{2^n}\left(1 - \frac{1}{2^p}\right) < \frac{1}{2^n}$$

is true for all positive integer p, for any $\varepsilon > 0$, there must exists $N(\varepsilon)$, such that $\dfrac{1}{2^n} < \varepsilon$ whenever $n > N$. Thus, $|a_{n+p} - a_n| < \varepsilon$ is true for any positive integer p. That is, $|a_m - a_n| < \varepsilon$ if m and n are greater than N. By Cauchy criterion, the limit of the sequence a_n exists.

The sign of the value of difference

$$a_{n+1} - a_n = \frac{\sin(n+1)}{2^{n+1}}$$

can be different depending on the value of n. This implies the sequence is not monotone, and the first criterion in this section does not apply. It is also not easy to verify by the definition since the value of the limit of a_n is not obvious. But it is not hard to determine the existence of the limit by Cauchy criterion.

Cauchy criterion can also be used to decide whether a sequence is divergent.

Example 5. Prove that

$$a_n = 1 + \frac{1}{2} + \frac{1}{3} + \cdots + \frac{1}{n}$$

is divergent.

Proof: Since

$$|a_{n+p} - a_n| = \frac{1}{n+1} + \cdots + \frac{1}{n+p} > \frac{p}{n+p},$$

we have

$$|a_{2n} - a_n| > \frac{n}{n+n} = \frac{1}{2}$$

if we choose $p = n$.

Therefore, if we choose $\varepsilon = \frac{1}{2}$, then

$$|a_{2n} - a_n| > \varepsilon$$

is always true for any large n. The sequence does not satisfy the convergent condition in Cauchy criterion. Therefore, the sequence a_n diverges.

Exercises 9.1

1. Prove the following by the ε-N definition of limit of number sequences.

(1) $\displaystyle\lim_{n\to\infty} \frac{n+1}{n} = 1$;

(2) $\displaystyle\lim_{n\to\infty} (-1)^n (0.1)^n = 0$;

(3) $\displaystyle\lim_{n\to\infty} \frac{\sin n}{n} = 0$;

(4) $\displaystyle\lim_{n\to\infty} \frac{n^2 + 1}{2n^2 + 1} = \frac{1}{2}$;

(5) $\displaystyle\lim_{n\to\infty} \frac{3}{1 + 2^n} = 0$;

(6) $\displaystyle\lim_{n\to\infty} \frac{1}{1 + \sqrt{n+1}} = 0$.

2. Find the following limits.

(1) $\displaystyle\lim_{n\to\infty} \frac{1000n}{n^2 + 1}$;

(2) $\displaystyle\lim_{n\to\infty} \frac{4n^2 - 5n + 9}{6n^2 - 2n + 1}$;

(3) $\displaystyle\lim_{n\to\infty} \frac{\sqrt{n} + 1000}{n - 1}$;

(4) $\displaystyle\lim_{n\to\infty} \frac{\sqrt[3]{n^2} \sin n!}{n + 1}$;

(5) $\displaystyle\lim_{n\to\infty} \frac{(-2)^n + 3^n}{(-2)^{n+1} + 3^{n+1}}$;

(6) $\displaystyle\lim_{n\to\infty} \left(\frac{1 + 2 + \cdots + n}{n + 2} - \frac{n}{2} \right)$;

(7) $\displaystyle\lim_{n\to\infty} \left(\frac{1^2}{n^3} + \frac{2^2}{n^3} + \cdots + \frac{(n-1)^2}{n^3} \right)$;

(8) $\displaystyle\lim_{n\to\infty} \frac{\sqrt{n^2 - n + 1}}{2n + 1}$;

(9) $\displaystyle\lim_{n\to\infty} \frac{1 + a + a^2 + \cdots + a^n}{1 + b + b^2 + \cdots + b^n}$, $(|a| < 1, |b| < 1)$;

(10) $\lim\limits_{n\to\infty} (\sqrt{n+1} - \sqrt{n})$;

(11) $\lim\limits_{n\to\infty} (\sqrt{(n+a)(n+b)} - n)$;

(12) $\lim\limits_{n\to\infty} \sqrt[3]{n}(\sqrt[3]{n+1} - \sqrt[3]{n})$;

(14) $\lim\limits_{n\to\infty} n(\sqrt{n^4+4} - n^2)$;

(13) $\lim\limits_{n\to\infty} \left(\sqrt[3]{1 + \dfrac{1}{n}} - 1 \right) n$;

(15) $\lim\limits_{n\to\infty} \sum\limits_{k=1}^{n} \dfrac{1}{\sqrt{n^2+k^2}}$;

(16) $\lim\limits_{n\to\infty} \left| \dfrac{1}{n} - \dfrac{2}{n} + \dfrac{3}{n} - \cdots + (-1)^{n-1}\dfrac{n}{n} \right|$;

(17) $\lim\limits_{n\to\infty} \left(\dfrac{1}{1\cdot 2} + \dfrac{1}{2\cdot 3} + \cdots \dfrac{1}{n(n+1)} \right)$;

(18) $\lim\limits_{n\to\infty} \left(\dfrac{1}{2} + \dfrac{3}{2^2} + \cdots + \dfrac{2n-1}{2^2} \right)$;

(19) $\lim\limits_{n\to\infty} (\sqrt{2} \cdot \sqrt[4]{2} \cdot \sqrt[8]{2} \cdots \sqrt[2^n]{2})$.

3. Prove the following equations.

(1) $\lim\limits_{n\to\infty} \dfrac{n}{2^n} = 0$;

(2) $\lim\limits_{n\to\infty} \dfrac{n^2}{a^n} = 0, \quad (a > 1)$;

(3) $\lim\limits_{n\to\infty} nq^n = 0, \quad (|q| < 1)$;

(4) $\lim\limits_{n\to\infty} \dfrac{a^n}{n!} = 0$;

(5) $\lim\limits_{n\to\infty} \dfrac{\log_a n}{n} = 0, \quad (a > 1)$;

(6) $\lim\limits_{n\to\infty} \left(\dfrac{1}{(n+1)^2} + \dfrac{1}{(n+2)^2} + \cdots + \dfrac{1}{(2n)^2} \right) = 0$;

(7) $\lim\limits_{n\to\infty} \dfrac{1}{2} \cdot \dfrac{3}{4} \cdots \dfrac{2n-1}{2n} = 0$; Hint: $\dfrac{1}{2} \cdot \dfrac{3}{4} \cdots \dfrac{2n-1}{2n} < \dfrac{1}{\sqrt{2n+1}}$.

4. Suppose $\lim\limits_{n\to\infty} a_n = a$. Prove $\lim\limits_{n\to\infty} (a_n - a_{n-1}) = 0$.

5. Suppose $\lim\limits_{n\to\infty} a_n = a$. Prove $\lim\limits_{n\to\infty} |a_n| = |a|$. Does the converse hold?

6. Suppose $\lim\limits_{n\to\infty} a_n = a$ and $a > b$. Prove that $a_n > b$ for n large enough.

7. Prove that if $\lim\limits_{n\to\infty} a_{2n+1} = a$, $\lim\limits_{n\to\infty} a_{2n} = a$, then $\lim\limits_{n\to\infty} a_n = a$.

8. Suppose $|b_n| \leqslant M$, $(n = 1, 2, \cdots)$ and $\lim\limits_{n\to\infty} a_n = 0$. Show that $\lim\limits_{n\to\infty} a_n b_n = 0$.

9. If $\lim\limits_{n\to\infty} a_n = a$, does the equation $\lim\limits_{n\to\infty} \dfrac{a_{n+1}}{a_n} = 1$ hold?

10. Show that $\lim\limits_{n\to\infty} \sqrt[n]{a_1^n + a_2^n + \cdots + a_m^n} = \max(a_1, a_2, \cdots, a_m)$, where $a_i \geqslant 0$, $(i = 1, 2, \cdots, m)$.

11. (1) Suppose $\lim\limits_{n\to\infty} a_n = a$. Prove $\lim\limits_{n\to\infty} \dfrac{a_1 + a_2 + \cdots + a_n}{n} = a$.

(2) Find the following limits.

(i) $\displaystyle\lim_{n\to\infty}\frac{1+\dfrac{1}{2}+\cdots+\dfrac{1}{n}}{n}$;

(ii) $\displaystyle\lim_{n\to\infty}\frac{1+\sqrt{2}+\sqrt[3]{3}+\cdots+\sqrt[n]{n}}{n}$.

12. (1) Suppose $\displaystyle\lim_{n\to\infty} a_n = a$, $(a_n > 0,\ n = 1, 2, \cdots)$. Prove that $\displaystyle\lim_{n\to\infty}\sqrt[n]{a_1 a_2 \cdots a_n} = a$.

Hint: $\dfrac{n}{\dfrac{1}{a_1} + \dfrac{1}{a_2} + \cdots + \dfrac{1}{a_n}} \leqslant \sqrt[n]{a_1 a_2 \cdots a_n} \leqslant \dfrac{a_1 + a_2 + \cdots + a_n}{n}$.

(2) Suppose $\displaystyle\lim_{n\to\infty}\dfrac{a_{n+1}}{a_n} = a$, $(a_n > 0,\ n = 1, 2, \cdots)$. Prove that $\displaystyle\lim_{n\to\infty}\sqrt[n]{a_n} = a$.

13. Prove the following equations.

(1) $\displaystyle\lim_{n\to\infty}\frac{1}{\sqrt[n]{n!}} = 0$;

(2) $\displaystyle\lim_{n\to\infty}\frac{n}{\sqrt[n]{n!}} = e$.

14. Show that the following sequences are convergent.

(1) $a_n = \left(1 - \dfrac{1}{2}\right)\left(1 - \dfrac{1}{2^2}\right)\cdots\left(1 - \dfrac{1}{2^n}\right)$;

(2) $a_n = \dfrac{1}{3+1} + \dfrac{1}{3^2+1} + \cdots + \dfrac{1}{3^n+1}$;

(3) $a_n = \dfrac{10}{1}\cdot\dfrac{11}{3}\cdots\dfrac{n+9}{2n-1}$.

15. Show that $\displaystyle\lim_{n\to\infty}\frac{n!}{n^n} = 0$.

16. Suppose $c > 0$,

$$a_1 = \sqrt{c},\quad a_2 = \sqrt{c+\sqrt{c}},\quad\cdots,\quad a_n = \sqrt{c+\sqrt{c+\cdots\sqrt{c}}}$$

(n-level nested radical). Prove that $\displaystyle\lim_{n\to\infty} a_n = \dfrac{\sqrt{4c+1}+1}{2}$.

17. Suppose a_n is a monotonically increasing sequence and b_n is a monotonically decreasing sequence, and $a_n < b_n$ for $n = 1, 2, \ldots$. Show that if $\displaystyle\lim_{n\to\infty}(b_n - a_n) = 0$, then $\displaystyle\lim_{n\to\infty} a_n = \lim_{n\to\infty} b_n$.

18. Suppose $a > 0$,

$$a_0 > 0,\quad a_{n+1} = \frac{1}{2}\left(a_n + \frac{a}{a_n}\right)$$

for $n = 0, 1, \cdots$. Prove $\displaystyle\lim_{n\to\infty} a_n = \sqrt{a}$.

19. Suppose $b_n = \displaystyle\sum_{k=1}^{n-1} |a_{k+1} - a_k|$ is bounded. Prove that a_n converges.

20. Determine whether the following sequences are convergent by Cauchy criterion.

(1) $a_n = \dfrac{\cos 1}{1 \cdot 2} + \dfrac{\cos 2}{2 \cdot 3} + \cdots + \dfrac{\cos n}{n(n+1)}$;

(2) $a_n = \dfrac{a\cos 2 + b\sin 2}{2(2 + \sin 2!)} + \dfrac{a\cos 3 + b\sin 3}{3(3 + \sin 3!)} + \cdots + \dfrac{a\cos n + b\sin n}{n(n + \sin n!)}$;

(3) $a_n = a_0 + \dfrac{a_1}{10} + \dfrac{a_2}{10^2} + \cdots + \dfrac{a_n}{10^n}$, where $|a_n| < 10$ for $k = 1, 2, \cdots, n$;

(4) $a_n = \dfrac{1}{1^k} + \dfrac{1}{2^k} + \cdots + \dfrac{1}{n^k}$, $(0 < k < 1)$.

9.2 The ε-δ Definition of Continuity of Functions

9.2.1 *Limits of Functions*

Suppose $f(x)$ is a function defined in a neighborhood of $x = a$. That is, there exists a number $d > 0$ such that $f(x)$ has well defined function values at every point x that satisfies $0 < |x - a| < d$.

If for any $\varepsilon > 0$, there exists a $\delta > 0$ such that

$$|f(x) - A| < \varepsilon$$

whenever $0 < |x - a| < \delta$, then we say that $f(x)$ has *limit* A at a and we denote this limit as

$$\lim_{x \to a} f(x) = A.$$

If for any $\varepsilon > 0$, there exists a $\delta > 0$ such that

$$|f(x) - A| < \varepsilon$$

whenever $0 < x - a < \delta$, then we say that $f(x)$ has a *right limit* A at a and we denote this right limit as

$$\lim_{x \to a+0} f(x) = A. \quad (f(a + 0) = A)$$

The *left limit* of $f(x)$ at a can be defined in a similar way and we denote this left limit as

$$\lim_{x \to a-0} f(x) = A. \quad (f(a - 0) = A).$$

The function $f(x)$ has a limit at a if and only if both its right limit and left limit at a exist and are equal. This is called the ε-δ *definition of limits of functions*.

Example 1. Prove $\lim\limits_{x\to 0}(1+x)^{\frac{1}{m}} = 1$, where m is a positive integer.

Proof: Suppose $|x| < 1$. For $x \neq 0$, we have

$$1 - |x| < (1+x)^{\frac{1}{m}} < 1 + |x|.$$

That is,

$$|(1+x)^{\frac{1}{m}} - 1| < |x|.$$

Hence, for any $\varepsilon > 0$, we choose $\delta = \min(\varepsilon, 1)$, then

$$|(1+x)^{\frac{1}{m}} - 1| < \varepsilon$$

if $0 < |x| < \delta$. Therefore,

$$\lim_{x\to 0}(1+x)^{\frac{1}{m}} = 1.$$

If for any $E > 0$, there exists a $\delta > 0$, such that $|f(x)| > E$ for all $0 < |x - a| < \delta$, then we say that $f(x)$ tends to infinity as x tends to a. We denote this limit as

$$\lim_{x\to a} f(x) = \infty.$$

Example 2. The function $\cot x$ tends to $+\infty$ as x tends to $0+$; the function $\cot x$ tends to $-\infty$ as x tends to $0-$.

If for any $\varepsilon > 0$, there exists a positive number K, such that

$$|f(x) - A| < \varepsilon$$

for all $x > K$, then we define

$$\lim_{x\to +\infty} f(x) = A.$$

The following limits can be defined in similar ways:

$$\lim_{x\to +\infty} f(x) = +\infty, \quad \lim_{x\to +\infty} f(x) = -\infty,$$

$$\lim_{x\to -\infty} f(x) = A, \quad \lim_{x\to -\infty} f(x) = +\infty,$$

$$\lim_{x\to -\infty} f(x) = -\infty.$$

From the following discussion, we can see the close relationship between the limit of a function and the limit of a number sequence.

Suppose $x_1, x_2, \cdots, x_n, \cdots$ is a number sequence that tends to a as n tends to infinity. Let

$$f(x_n) = y_n.$$

If $\lim\limits_{x \to a} f(x) = A$, then

$$\lim_{n \to \infty} f(x_n) = \lim_{n \to \infty} y_n = A.$$

On the other hand, if for any sequence x_n that tends to a, all the corresponding y_n tends to A, then

$$\lim_{x \to a} f(x) = A.$$

In fact, if $\lim\limits_{x \to a} f(x)$ does not exist, or it exists but is not equal to A, then by the definition of the limit, there is an ε, such that for any small δ

$$|f(x) - A| > \varepsilon$$

for some x in $0 < |x - a| < \delta$.

Let δ_n be a sequence that tends to zero and x_n be a sequence of numbers in $0 < |x - a| < \delta_n$ such that

$$|f(x_n) - A| > \varepsilon.$$

Then x_n is a sequence that tends to a. But $f(x_n)$ does not tend to A as $n \to \infty$. This contradicts the assumption. Therefore, we must have

$$\lim_{x \to a} f(x) = A.$$

Cauchy criterion for the limit of a function: The necessary and sufficient condition for the existence of a finite limit of a function $f(x)$ as $x \to a$ is: for any $\varepsilon > 0$, there exists a $\delta > 0$ such that

$$|f(x) - f(x')| < \varepsilon$$

for any x and x' that satisfy

$$0 < |x - a| < \delta, \quad 0 < |x' - a| < \delta.$$

We leave the proof of this result to our readers. Also, we suggest our readers try to state the corresponding theorem when $a = \infty$ or the limit is infinity.

The arithmetic of the limits of functions are similar to that of the limits of sequences.

If the limits on the right hand side exist and are finite, then the following equations are true:

$$\lim_{x \to a} (f(x) \pm g(x)) = \lim_{x \to a} f(x) \pm \lim_{x \to a} g(x),$$

$$\lim_{x \to a} (f(x) \cdot g(x)) = \lim_{x \to a} f(x) \cdot \lim_{x \to a} g(x),$$

$$\lim_{x \to a} \frac{f(x)}{g(x)} = \frac{\lim_{x \to a} f(x)}{\lim_{x \to a} g(x)}, \quad (\lim_{x \to a} g(x) \neq 0).$$

These results can be verified by the ε-δ definition. The processes of verifications are similar to that of the corresponding results of the limits of number sequences in Section 9.1.2.

Example 3. Prove $\lim\limits_{x \to \infty} \left(1 + \dfrac{1}{x}\right)^x = e$.

Proof: We have already shown in Section 9.1.3, Example 3 that

$$\lim_{n \to \infty} \left(1 + \frac{1}{n}\right)^n = e.$$

For $x \to +\infty$, let n be the largest integer that is not greater than x. That is

$$n \leqslant x < n + 1.$$

We have that

$$1 + \frac{1}{n+1} < 1 + \frac{1}{x} \leqslant 1 + \frac{1}{n}.$$

Thus,

$$\left(1 + \frac{1}{n+1}\right)^n < \left(1 + \frac{1}{x}\right)^x < \left(1 + \frac{1}{n}\right)^{n+1},$$

and

$$\lim_{n \to \infty} \left(1 + \frac{1}{n+1}\right)^n \leqslant \lim_{x \to +\infty} \left(1 + \frac{1}{x}\right)^x \leqslant \lim_{n \to \infty} \left(1 + \frac{1}{n}\right)^{n+1}.$$

Since

$$\lim_{n \to \infty} \left(1 + \frac{1}{n+1}\right)^n = \lim_{n \to \infty} \frac{\left(1 + \frac{1}{n}\right)^{n+1}}{\left(1 + \frac{1}{n+1}\right)} = \frac{e}{1} = e,$$

$$\lim_{n \to \infty} \left(1 + \frac{1}{n}\right)^{n+1} = \lim_{n \to \infty} \left(1 + \frac{1}{n}\right)^n \left(1 + \frac{1}{n}\right) = e \cdot 1 = e,$$

we have

$$\lim_{x \to +\infty} \left(1 + \frac{1}{x}\right)^x = e.$$

For $x \to -\infty$, let $y = -x$. Then $y \to +\infty$ as $x \to -\infty$. Assume that $y > 1$, we have

$$\left(1 + \frac{1}{x}\right)^x = \left(1 + \frac{1}{y}\right)^{-y} = \left(\frac{y}{y-1}\right)^y$$

$$= \left(1 + \frac{1}{y-1}\right)^{y-1} \left(1 + \frac{1}{y-1}\right).$$

Hence,

$$\lim_{x \to -\infty} \left(1 + \frac{1}{x}\right)^x = \lim_{y \to +\infty} \left(1 + \frac{1}{y-1}\right)^{y-1} \left(1 + \frac{1}{y-1}\right) = e \cdot 1 = e.$$

Therefore

$$\lim_{x \to \infty} \left(1 + \frac{1}{x}\right)^x = e.$$

Now we study two kinds of limits: the infinitesimal and the infinite.

If the limit of a function $\alpha(x)$ is 0 as x tends to x_0 (which can be a finite number or the infinity), then we say that $\alpha(x)$ is an *infinitesimal* as x tends to x_0. If the limit of a function $\alpha(x)$ is infinity as x tends to x_0 (which can be a finite number or the infinity), then we say that $\alpha(x)$ is an *infinite* as x tends to x_0. (Translator's note: The word "infinite" is used here to represent the concept of the opposite of "infinitesimal".)

Suppose $\alpha(x)$, $\beta(x)$ are infinitesimals as x tends to x_0, and $\beta(x) \neq 0$. If

$$\lim_{x \to x_0} \frac{\alpha(x)}{\beta(x)} = A,$$

then, if $A \neq 0$, $\alpha(x)$ and $\beta(x)$ are called *infinitesimals of the same order* as x tends to x_0.

If $A = 1$, then $\alpha(x)$ and $\beta(x)$ are called *equivalent infinitesimals* as x tends to x_0. We use the notation $\alpha(x) \sim \beta(x) \, (x \to x_0)$.

If $A = 0$, then $\alpha(x)$ is called a *higher order infinitesimal* of $\beta(x)$ as x tends to x_0. We use the notation $\alpha(x) = o(\beta(x))(x \to x_0)$.

Suppose $u(x)$, $v(x)$ are infinites as x tends to x_0, and $v(x) \neq 0$. If

$$\lim_{x \to x_0} \frac{u(x)}{v(x)} = A, \quad \text{(a finite number)},$$

then, if $A \neq 0$, $u(x)$ and $v(x)$ are called the *infinites of the same order* as x tends to x_0.

If $A = 1$, then $u(x)$ and $v(x)$ are called *equivalent infinites* as x tends to x_0.

If $A = 0$, then $v(x)$ is called a *higher order infinite* of $u(x)$ as x tends to x_0.

More generally, if the limit of $\alpha(x)$ is α and the limit of $\beta(x)$ is β as x tends to x_0, then the limit of $\alpha(x) - \alpha$ is zero and the limit of $\beta(x) - \beta$ is also zero. Therefore we can compare $\alpha(x) - \alpha$ and $\beta(x) - \beta$. If

$$\lim_{x \to x_0} \frac{\alpha(x) - \alpha}{\beta(x) - \beta} = A,$$

and $A \neq 0$, then $\alpha(x) - \alpha$ and $\beta(x) - \beta$ are called infinitesimals of the same order as x tends to x_0.

We can define $\alpha(x) - \alpha = o(\beta(x) - \beta)$ $(x \to x_0)$ similarly.

Example 4. Compare the order of the infinitesimals $\alpha(x) = x$, and $\beta(x) = \sin x$ as $x \to 0$.

Solution: Since

$$\lim_{x \to 0} \frac{\sin x}{x} = 1,$$

$\alpha(x)$ and $\beta(x)$ are equivalent infinitesimals. That is,

$$\sin x \sim x \quad (x \to 0).$$

Example 5. Compare the order of the infinitesimals $\alpha(x) = x$ and $\gamma(x) = 1 - \cos x$ as $x \to 0$.

Solution: Since

$$\lim_{x \to 0} \frac{1 - \cos x}{x} = 0,$$

$\gamma(x)$ is a higher order infinitesimal of $\alpha(x)$. That is,

$$1 - \cos x = o(x) \quad (x \to 0).$$

Next, we study the limit of functions of two variables.

Let $f(x, y)$ be a function of two variables that is defined on a region D. Suppose $p_0(x_0, y_0)$ is an accumulation point of D.

If for any given $\varepsilon > 0$, there exists a $\delta > 0$, such that

$$|f(x, y) - A| < \varepsilon$$

for any point $p(x, y)$ in D that satisfies the inequality

$$0 < \sqrt{(x - x_0)^2 + (y - y_0)^2} < \delta$$

or

$$0 < |x - x_0| < \delta, \ 0 < |y - y_0| < \delta,$$

then A is called the limit of $f(x, y)$ as $p(x, y) \to p_0(x_0, y_0)$. We denote this limit as

$$\lim_{\substack{x \to x_0 \\ y \to y_0}} f(x, y) = A \quad \text{or} \quad \lim_{p \to p_0} f(p) = A.$$

Although the definition of the limit of functions with two variables is similar to the definition of the limit of functions with one variable, notice that the function $f(x, y)$ tends to A when $p(x, y)$ tends to $p_0(x_0, y_0)$ in every way. Thus, if the function $f(x, y)$ tends to a fixed value only when $p(x, y)$ tends to $p_0(x_0, y_0)$ in a special way (for instance, along a specific line or curve), then we can not reach the conclusion that the limit of $f(x, y)$ exists as $p(x, y) \to p_0(x_0, y_0)$. If $f(x, y)$ tends to different values when $p(x, y)$ tends to $p_0(x_0, y_0)$ in different ways, then we can claim that the limit of $f(x, y)$ does not exists as $p(x, y) \to p_0(x_0, y_0)$.

Example 6. Prove $\lim\limits_{\substack{x \to 0 \\ y \to 0}} \dfrac{x^2 y}{x^2 + y^2} = 0$.

Proof: Since $(x - y)^2 \geqslant 0$, we have

$$\left| \frac{2xy}{x^2 + y^2} \right| \leqslant 1.$$

Thus,

$$\left| \frac{x^2 y}{x^2 + y^2} \right| \leqslant \frac{|x|}{2} \left| \frac{2xy}{x^2 + y^2} \right| \leqslant \frac{|x|}{2}.$$

Therefore, for any $\varepsilon > 0$, let $\delta = 2\varepsilon$, then

$$\left| \frac{x^2 y}{x^2 + y^2} - 0 \right| < \varepsilon$$

when $0 < |x| < \delta$ and $0 < |y| < \delta$.

This proves that

$$\lim_{\substack{x \to 0 \\ y \to 0}} \frac{x^2 y}{x^2 + y^2} = 0.$$

Example 7. Prove $\lim\limits_{\substack{x \to 0 \\ y \to 0}} \dfrac{xy}{x^2 + y^2}$ does not exists (Section 6.1.1).

Proof: Let $y = kx$. Then

$$\frac{xy}{x^2 + y^2} = \frac{xkx}{x^2 + k^2 x^2} = \frac{k}{1 + k^2}.$$

Thus, the limit of the function $f(x, y)$ is $\dfrac{k}{1 + k^2}$ when (x, y) tends to $(0, 0)$ along the straight line $y = kx$. This limit will be different if the slope of the line k changes. Therefore

$$\lim_{\substack{x \to 0 \\ y \to 0}} \frac{xy}{x^2 + y^2}$$

does not exists.

In a similar way, it is not hard to define the limit of a function with n ($n \geqslant 3$) variables by ε-δ definition.

9.2.2 *Definition of Continuous Functions*

We define the continuity of a function by ε-δ definition.

Suppose $f(x)$ is a function defined near a point x_0. If

$$f(x_0 - 0) = f(x_0) = f(x_0 + 0),$$

then we say that $f(x)$ is *continuous* at $x = x_0$.

If $f(x_0) = f(x_0 + 0)$, then $f(x)$ is called *right continuous*. If $f(x_0) = f(x_0 - 0)$, then $f(x)$ is called *left continuous*.

A point at which a function is not continuous is called a *discontinuous point*.

The above definition can also be stated as:

Suppose a function $f(x)$ is defined near x_0 (including x_0). If for any $\varepsilon > 0$, there is a $\delta > 0$ such that $|f(x) - f(x_0)| < \varepsilon$ if $|x - x_0| < \delta$, then we say that $f(x)$ is continuous at x_0.

The definitions of right continuity and left continuity of a function can be stated in a similar way by ε-δ definition.

If $f(x)$ is continuous at every point of an open interval (a, b) (a closed interval $[a, b]$), then we say that $f(x)$ is continuous on the open interval (a, b) (the closed interval $[a, b]$).

The definition shows that the continuity of a function at a point x_0 is a "local" property. it is only related to the nature of the function near x_0.

We verify the following properties by ε-δ definition.

1. If functions $f(x)$ and $g(x)$ are continuous at a point x_0, then

$$f(x) \pm g(x), \quad f(x) \cdot g(x), \quad \frac{f(x)}{g(x)}$$

are also continuous at x_0. (Assume that $g(x_0) \neq 0$.)

2. Suppose $f(x)$ is continuous at a point x_0, and $f(x_0) = y_0$. If $\varphi(y)$ is continuous at $y = y_0$, then $\varphi(f(x))$ is also continuous at $x = x_0$.

In fact, since $\varphi(y)$ is continuous at $y = y_0$, for any $\varepsilon > 0$, there exists a $\sigma > 0$ such that $|\varphi(y) - \varphi(y_0)| < \varepsilon$ if $|y - y_0| < \sigma$. Since $f(x)$ is continuous at $x = x_0$, there exists a $\delta > 0$ such that $|f(x) - y_0| < \sigma$ if $|x - x_0| < \delta$. Thus, we have

$$|\varphi(f(x)) - \varphi(f(x_0))| = |\varphi(y) - \varphi(y_0)| < \varepsilon.$$

Hence, $\varphi(f(x))$ is continuous at $x = x_0$.

3. Suppose a continuous function $f(x)$ is strictly monotone on an interval I. If the range of $f(x)$ is J, then the inverse function $x = f^{-1}(y)$ is also continuous on J.

The proof is not difficult. Let $y_0 = f(x_0)$ be an arbitrary point in J. Since $f^{-1}(y)$ is strictly monotone, by the first criterion in Section 9.1.3, $f^{-1}(y_0 - 0)$ and $f^{-1}(y_0 + 0)$ exist and are finite. Now we need to show that

$$f^{-1}(y_0 - 0) = f^{-1}(y_0 + 0) = f^{-1}(y_0).$$

Let

$$x_0 = f^{-1}(y_0), \quad x_1 = f^{-1}(y_0 - 0), \quad x_2 = f^{-1}(y_0 + 0).$$

Since $f(x)$ is continuous, we have

$$y_0 = \lim_{y \to y_0 - 0} y = \lim_{y \to y_0 - 0} f[f^{-1}(y)] = \lim_{x \to x_1} f(x) = f(x_1),$$

$$y_0 = \lim_{y \to y_0 + 0} y = \lim_{y \to y_0 + 0} f[f^{-1}(y)] = \lim_{x \to x_2} f(x) = f(x_2).$$

Therefore,

$$x_1 = f^{-1}(y_0) = x_0, \quad x_2 = f^{-1}(y_0) = x_0$$

by the monotonicity of $f(x)$.

A function is called an *elementary function* if it is a power function, an exponential function, a trigonometric function, the inverse of a trigonometric function, a logarithmic function, a combination of these functions by a finite number of arithmetic operations (addition, subtraction, multiplication, division) and root extractions, or a finite composition of these functions.

4. All elementary functions are continuous in their domains.

We have already applied this result in previous chapters, now we prove it.

(1) The continuity of trigonometric functions.

The function $\sin x$ is continuous on the real number axis.

For any real number x_0,

$$|\sin x - \sin x_0| = 2 \left|\cos \frac{x + x_0}{2}\right| \left|\sin \frac{x - x_0}{2}\right|$$

$$\leqslant 2 \left|\sin \frac{x - x_0}{2}\right| \leqslant 2 \left|\frac{x - x_0}{2}\right| = |x - x_0|.$$

Thus,

$$\lim_{x \to x_0} \sin x = \sin x_0.$$

That is, $\sin x$ is continuous on the real number axis.

Since

$$\cos x = \sin\left(\frac{\pi}{2} - x\right),$$

the function $\cos x$ is also continuous on the real number axis. Since

$$\tan x = \frac{\sin x}{\cos x}, \quad \sec x = \frac{1}{\cos x}, \quad \cot x = \frac{\cos x}{\sin x}, \quad \csc x = \frac{1}{\sin x},$$

the functions $\tan x$ and $\sec x$ are continuous except at the points where $\cos x = 0$, and $\cot x$ and $\csc x$ are continuous except at the points where $\sin x = 0$.

From the third result, the inverse functions of trigonometric functions are continuous on their domains. For instance, $\sin x$ is strictly monotonically increasing on $\left[-\frac{\pi}{2}, \frac{\pi}{2}\right]$. By 3, the function $\arcsin x$ is continuous in $[-1, 1]$.

The continuity of $\arccos x$ and $\arctan x$ can be discussed in a similar way. Since trigonometric functions are periodic functions, the inverse functions of trigonometric functions are defined on one period.

(2) The continuity of exponential functions and logarithmic functions.

Let $a > 1$. If we can show that

$$\lim_{x \to 0} a^x = 1,$$

then for any x_0,

$$\lim_{x \to x_0} a^x = \lim_{x \to x_0} a^{x_0 + (x - x_0)} = a^{x_0} \lim_{x \to x_0} a^{x - x_0} = a^{x_0} \lim_{y \to 0} a^y = a^{x_0}.$$

Thus, a^x is a continuous function.

In Example 2 of Section 9.1.1, we have proved that

$$\lim_{n \to \infty} a^{\frac{1}{n}} = 1, \quad (a > 1).$$

For any x that satisfies $0 < x < 1$, there exists a positive integer n, such that

$$\frac{1}{n+1} < x < \frac{1}{n}.$$

Thus,

$$a^{\frac{1}{n+1}} < a^x < a^{\frac{1}{n}}.$$

Hence,

$$\lim_{n\to\infty} a^{\frac{1}{n+1}} \leqslant \lim_{x\to 0} a^x \leqslant \lim_{n\to\infty} a^{\frac{1}{n}}.$$

That is, $\lim\limits_{x\to 0} a^x = 1$. Therefore, a^x is a continuous function when $a > 1$. The function a^x is also a continuous function when $0 < a < 1$. This can be proved in a similar way.

If $a > 1$, then a^x is a strictly monotonically increasing and continuous function. If $0 < a < 1$, then a^x is a strictly monotonically decreasing and continuous function. By property 3, $\log_a x$, $(a > 0, a \neq 0, x > 0)$ is also a continuous function. The functions e^x and $\ln x$ are both strictly monotonically increasing and continuous functions.

(3) The continuity of power function x^μ $(x > 0)$.

Since e^x and $\ln x$ are continuous functions, by property 2, the composite function

$$e^{\mu \ln x} = x^\mu$$

is also a continuous function. (Where μ is an arbitrary real number.)

The finite composition and finite arithmetic operations of these basic elementary functions are also continuous. Thus, all elementary functions are continuous.

9.2.3 *Properties of Continuous Functions*

In Section 1.1.2, we listed the following properties of continuous functions, and proved some important theorems, such as the Mean-Value Theorem of Integrals, by these properties. Now we restate and prove these properties.

1. Suppose a function $f(x)$ is continuous on a closed interval $[a, b]$. If the values of $f(x)$ at a and b have different signs, then there exists a point c between a and b such that

$$f(c) = 0, \quad (a < c < b).$$

In other words, if a continuous curve extends from one side of the x-axis to the other side, then the curve must intersect with the x-axis.

Indeed, bisecting $[a, b]$, if $f\left(\dfrac{a+b}{2}\right) = 0$, then the conclusion is true.
Otherwise, in one of the two intervals $\left[a, \dfrac{a+b}{2}\right]$ or $\left[\dfrac{a+b}{2}, b\right]$, the values
of $f(x)$ have different signs at the two end points. We denote this interval
as $[a_1, b_1]$.

Suppose $f(a_1 < 0)$, $f(b_1 > 0)$. Repeat the same process and
we obtain a series of intervals each of which contains the next:
$[a_2, b_2]$, $[a_3, b_3]$, $\cdots$, $[a_n, b_n]$, $\cdots$, where

$$b_n - a_n = \frac{b-a}{2^n},$$

and $f(a_2) < 0, \cdots, f(a_n) < 0, \cdots, f(b_2) > 0, \cdots, f(b_n) > 0, \cdots$.
Since

$$a_1 \leqslant a_2 \leqslant \cdots \leqslant a_n \leqslant \cdots,$$

$$b_1 \geqslant b_2 \geqslant \cdots \geqslant b_n \geqslant \cdots,$$

and $b_n - a_n \to 0$ as $n \to \infty$, we have

$$\lim a_n = \lim b_n = c.$$

Also because

$$f(c) = \lim_{n\to\infty} f(a_n) \leqslant 0, \quad f(c) = \lim_{n\to\infty} f(b_n) \geqslant 0,$$

it follows that

$$f(c) = 0.$$

2. With the same assumptions as above, if $f(a) = A$, $f(b) = B$, then
for any number C between A and B, there exists a point $x = c$, such that
$f(c) = C$.

This is a corollary of above property. Consider the function $f(x) - C$,
and the conclusion follows.

3. If a function $f(x)$ is continuous on a closed interval $[a, b]$, then $f(x)$
is bounded. In other words, there exist two constants m and M, such that

$$m \leqslant f(x) \leqslant M, \quad (a \leqslant x \leqslant b).$$

In fact, if $f(x)$ is not bounded on $[a, b]$, then for any natural number n,
there exists an x_n in $[a, b]$ such that

$$|f(x_n)| \geqslant n.$$

By Bolzano-Weierstrass criterion (Section 9.1.3), we can find a subsequence $|x_{n_k}|$ from the sequence $|x_n|$, such that

$$x_{n_k} \to x_0, (k \to \infty).$$

Obviously, $a \leqslant x_0 \leqslant b$. By the continuity of the function,

$$f(x_{n_k}) \to f(x_0),$$

and this is not possible.

The Mean-Value Theorem for Integrals in Section 1.1.2 follows from the above properties 2 and 3.

The smallest number M that makes the following inequality true is called the *least upper bound* (or *supremum*) of $f(x)$

$$m \leqslant f(x) \leqslant M,$$

and the largest number m that makes this inequality true is called the *greatest lower bound* (or *infimum*).

4. With the same assumptions as above properties, $f(x)$ attains its infimum and supremum in $[a, b]$. In other words, there exist x_0 and x_1 in $[a, b]$, such that the maximum value of $f(x)$ and the minimum value of $f(x)$ are $f(x_0)$ and $f(x_1)$ respectively.

The proof is not difficult. Let M be the supremum. By property 3, M is finite. Assume that $f(x) < M$ for all x, and let

$$\varphi(x) = \frac{1}{M - f(x)}.$$

This is a well defined continuous function (the denominator is not zero). By property 3, it is bounded. In other words, there exists $\mu > 0$ such that $\varphi(x) \leqslant \mu$. Thus,

$$f(x) \leqslant M - \frac{1}{\mu}.$$

This implies that the supremum of $f(x)$ can be a number $M - \dfrac{1}{\mu}$ which is less than M. This contradicts the definition of the supremum. Thus, there exists an x_0, such that $f(x_0) = M$.

Similarly, there exists an x_1 such that $f(x_1) = m$.

9.2.4   *Uniform Continuity of Functions*

If a function $f(x)$ is continuous on every point of an interval (a, b) (or $[a, b]$), then we say that $f(x)$ is continuous on (a, b) (or $[a, b]$). In other words, for any $\varepsilon > 0$ and any point x_0 of (a, b) (or $[a, b]$), there exists a $\delta(\varepsilon, x_0) > 0$ such that

$$|f(x) - f(x_0)| < \varepsilon$$

if $|x - x_0| < \delta$. In this definition, the number δ is not only related to ε, but also related to x_0. For different point x_0, the number δ is different. If there exists a δ that is not related to any point in (a, b) (or $[a, b]$), then the function is *uniformly continuous* on (a, b) (or $[a, b]$).

The precise definition of uniformly continuous functions is:

Suppose a function $f(x)$ is defined on an interval (a, b) (or $[a, b]$). If for any $\varepsilon > 0$, there exists $\delta(\varepsilon) > 0$ (δ varies only with ε), such that

$$|f(x) - f(x_0)| < \varepsilon$$

if $|x - x_0| < \delta$, for every point x_0, then $f(x)$ is uniformly continuous on (a, b) (or $[a, b]$).

This definition can also be stated as:

Suppose a function $f(x)$ is defined on an interval (a, b) (or $[a, b]$). If for any $\varepsilon > 0$, there exists $\delta(\varepsilon) > 0$, such that

$$|f(x') - f(x'')| < \varepsilon$$

if $|x' - x''| < \delta$ for any two points x' and x'' in (a, b) (or $[a, b]$), then $f(x)$ is uniformly continuous on (a, b) (or $[a, b]$).

Example: The function $f(x) = \dfrac{1}{x}$ is continues at every point of the open interval $(0, 1)$, but it is not uniformly continuous on $(0, 1)$.

Proof: Let $\varepsilon = 1$. For any $\delta > 0$, and any positive integer n that is large enough, there are two points $x' = \dfrac{1}{n}$ and $x'' = \dfrac{1}{n+1}$ in $(0, 1)$, such that

$$|f(x') - f(x'')| = |n - (n + 1)| = 1$$

although

$$|x' - x''| = \frac{1}{n} - \frac{1}{n+1} = \frac{1}{n(n+1)} < \delta.$$

This implies that, for $\varepsilon = 1$, there is no such δ as in the original definition. Therefore, $f(x) = \dfrac{1}{x}$ is not uniformly continuous on $(0, 1)$.

For closed intervals, we have the following result:

If $f(x)$ is continues on a closed interval $[a, b]$, then it is uniformly continuous on $[a, b]$.

Indeed, if $f(x)$ is not uniformly continuous on $[a, b]$, then there exists $\varepsilon > 0$, for any positive integer n, there exist x_n, x'_n in $[a, b]$ such that $|x_n - x'_n| < \dfrac{1}{n}$ but

$$|f(x_n) - f(x'_n)| \geq \varepsilon.$$

Since x_n is bounded, by Bolzano-Weierstrass criterion in Section 9.1.3, there exists a subsequence x_{n_k} that is convergent.

Assume $x_{n_k} \to x_0$. Then x_0 is in $[a, b]$. On the other hand, $x'_{n_k} \to x_0$ since $|x_{n_k} - x'_{n_k}| < \dfrac{1}{n_k} \to 0$ as $k \to \infty$. Thus, from

$$|f(x_{n_k}) - f(x'_{n_k})| \geq \varepsilon$$

and the continuity of $f(x)$, let $k \to \infty$, we have $0 \geq \varepsilon$. And this is a contradiction.

Exercises 9.2

1. Prove the following equations by ε-δ definition.

    (1) $\lim\limits_{x \to 1} x^3 = 1$;

    (2) $\lim\limits_{x \to 0} x^2 \sin \dfrac{1}{x} = 0$;

    (3) $\lim\limits_{h \to 0} \dfrac{(x + h)^2 - x^2}{h} = 2x$;

    (4) $\lim\limits_{x \to 0} \sqrt{2x + 1} = 1$.

2. Suppose $f(x) = \dfrac{\sin x}{x}$. Find $f(0 + 0)$, $f(0 - 0)$.

3. Suppose

$$f(x) = \begin{cases} x^2, & x \geq 2, \\ -ax, & x < 2. \end{cases}$$

    (1) Find $f(2+0)$, $f(2-0)$. (2) For what value of a does $\lim\limits_{x \to 2} f(x)$ exist?

4. Prove the following:

    (1) If $\lim\limits_{x \to x_0} f(x) = a$, then $f(x)$ is bounded when x is close enough to x_0;

    (2) If $\lim\limits_{x \to x_0} f(x) = a$, $(a > 0)$, then $f(x) > 0$ when x is close enough to x_0.

5. Prove that the following limits do not exist.

   (1) $\lim\limits_{x \to \infty} \sin x$;

   (2) $\lim\limits_{x \to 0} \cos \dfrac{1}{x}$.

6. Find the following limits.

   (1) $\lim\limits_{x \to 0} \dfrac{x^2 - 1}{2x^2 - x - 1}$;

   (2) $\lim\limits_{x \to 1} \dfrac{x^2 - 1}{2x^2 - x - 1}$;

   (3) $\lim\limits_{x \to \infty} \dfrac{x^2 - 1}{2x^2 - x - 1}$;

   (4) $\lim\limits_{x \to 0} \dfrac{(1 + x)(1 + 2x)(1 + 3x) - 1}{x}$;

   (5) $\lim\limits_{x \to 0} \dfrac{(1 + x)^5 - (1 + 5x)}{x^2 + x^5}$;

   (6) $\lim\limits_{x \to 0} \dfrac{(1 + mx)^n - (1 + nx)^m}{x^2}$, ($m$, n are positive integers);

   (7) $\lim\limits_{x \to 1} \dfrac{x^m - 1}{x^n - 1}$, ($m$, n are positive integers);

   (8) $\lim\limits_{x \to 1} \dfrac{x + x^2 + \cdots + x^n - n}{x - 1}$, ($n$ is a positive integer);

   (9) $\lim\limits_{x \to +\infty} (\sqrt{x + 2} - \sqrt{x + 1})$;

   (10) $\lim\limits_{x \to +\infty} \dfrac{\sqrt{x + \sqrt{x + \sqrt{x}}}}{\sqrt{x + 1}}$;

   (11) $\lim\limits_{x \to 0} \dfrac{\sqrt{x + 1} - (x + 1)}{\sqrt{x + 1} - 1}$;

   (12) $\lim\limits_{x \to +\infty} (\sin \sqrt{x + 1} - \sin \sqrt{x})$;

   (13) $\lim\limits_{x \to +\infty} (\sqrt{(x + a)(x + b)} - x)$;

   (14) $\lim\limits_{x \to 0} \dfrac{\sqrt{x + 1} - \sqrt{1 - x}}{\sqrt[3]{1 + x} - \sqrt[3]{1 - x}}$;

   (15) $\lim\limits_{x \to 0} \dfrac{\sqrt[m]{1 + \alpha x} - \sqrt[n]{1 + \beta x}}{x}$, ($m$, n are positive integers, α, β are real numbers);

   (16) $\lim\limits_{x \to 0} \dfrac{\sqrt[m]{1 + \alpha x} \, \sqrt[n]{1 + \beta x} - 1}{x}$, ($m$, n are positive integers, α, β are real numbers).

7. Prove the following equations.

   (1) $\lim\limits_{x \to 0} \dfrac{(1 + x)^r - 1}{x} = r$, ($r$ is a rational number);

   (2) $\lim\limits_{x \to 0} \dfrac{\sqrt[m]{1 + p(x)} - 1}{x} = \dfrac{a_1}{m}$, where $p(x) = a_1 x + a_2 x^2 + \cdots + a_n x^n$, m is a positive integer.

8. Find constants a, b such that the following equations hold.

(1) $\lim\limits_{x \to \infty} \left(\dfrac{x^2 + 1}{x + 1} - ax - b \right) = 0$;

(2) $\lim\limits_{x \to +\infty} (\sqrt{x^2 - x + 1} - ax - b) = 0$;

(3) $\lim\limits_{x \to -\infty} (\sqrt{x^2 - x + 1} - ax - b) = 0$.

9. Suppose $R(x) = \dfrac{a_0 x^n + a_1 x^{n-1} + \cdots + a_n}{b_0 x^m + b_1 x^{m-1} + \cdots + b_m}$, where $a_0 \neq 0$, $b_0 \neq 0$, m, n are positive integers. Show that

$$\lim\limits_{x \to +\infty} R(x) = \begin{cases} \infty, & \text{if } n > m, \\ \dfrac{a_0}{b_0}, & \text{if } n = m, \\ 0, & \text{if } n < m. \end{cases}$$

10. Find the following limits.

(1) $\lim\limits_{x \to \infty} \left(1 - \dfrac{1}{x} \right)^x$;

(2) $\lim\limits_{x \to 0} (1 - 2x)^{\frac{1}{x}}$;

(3) $\lim\limits_{x \to \infty} \left(\dfrac{x^2 + 1}{x^2 - 1} \right)^{x^2}$;   (4) $\lim\limits_{x \to \infty} \left(\dfrac{1 + x}{2 + x} \right)^{\frac{1 - x^2}{1 - x}}$;

(5) $\lim\limits_{x \to 0} \dfrac{\sin 5x}{x}$;

(6) $\lim\limits_{x \to \pi} \dfrac{\sin mx}{\sin nx}$, $(m, n$ are positive integers$)$;

(7) $\lim\limits_{x \to 0} \dfrac{\tan x - \sin x}{\sin^3 x}$;   (8) $\lim\limits_{x \to 0} \dfrac{\sin(\sin x)}{x}$;

(9) $\lim\limits_{x \to a} \dfrac{\cos x - \cos a}{x - a}$;

(10) $\lim\limits_{x \to 0} \dfrac{\sin(x^m)}{\sin^n x}$, $(m, n$ are positive integers$)$.

11. Prove the following equations.

(1) $\lim\limits_{x \to +\infty} \dfrac{x^k}{a^x} = 0$, $(a > 1, k > 0)$;   (2) $\lim\limits_{x \to +\infty} \dfrac{\ln x}{x^k} = 0$, $(k > 0)$;

(3) $\lim\limits_{n \to +\infty} \cos \dfrac{x}{2} \cos \dfrac{x}{2^2} \cdots \cos \dfrac{x}{2^n} = \dfrac{\sin x}{x}$.

12. Show that the functions $e^{2x} - e^x$ and $\sin 2x - \sin x$ are equivalent infinitesimals as $x \to 0$.

13. Find the order of infinitesimals for the following functions as $x \to 0$.

(1) $e^x - \cos x$;

(2) $2x - 3x^3 + x^{10}$;

(3) $\sqrt{1 + x^2} - 1$;

(4) $\sin(\sqrt{1 + x} - 1)$.

14. Compare the order of infinites for the following pairs of functions as $x \to +\infty$.

(1) 2^x and 3^x;

(2) $\sqrt{1 + x^2}$ and $x - 1$;

(3) $\sqrt{x^3 + x + 1}$ and $\sqrt{x + \sin^2 x}$;

(4) $\ln(x + \sqrt{x^2 + 1})$ and $\sqrt{x^2 + x + 1}$.

15. Prove the following results.

(1) $x^2 \sin \dfrac{1}{x} = o(x)$, $(x \to 0)$;

(2) If $a_0 \neq 0$, then $a_0 x^n + a_1 x^{n-1} + \cdots + a_n$ and x^n are infinites of the same order when $x \to \infty$;

(3) For what value of A do we have $\dfrac{\arctan x}{1 + x^2} \sim \dfrac{A}{x^2}$, $(x \to +\infty)$?

16. Suppose $\alpha(x)$, $\beta(x)$, $\alpha_1(x)$, $\beta_1(x)$ are infinitesimals as $x \to x_0$. Suppose also that $\alpha(x) \sim \alpha_1(x)$, $x \to x_0$, $\beta(x) \sim \beta_1(x)$, $x \to x_0$, and that the limit $\lim\limits_{x \to x_0} \dfrac{\alpha_1(x)}{\beta_1(x)}$ exists. Prove that $\lim\limits_{x \to x_0} \dfrac{\alpha(x)}{\beta(x)} = \lim\limits_{x \to x_0} \dfrac{\alpha_1(x)}{\beta_1(x)}$.

17. Find the following limits.

(1) $\lim\limits_{x \to 0} \dfrac{\sin mx}{\sin nx}$;

(2) $\lim\limits_{x \to 0} \dfrac{\sqrt{1 + x + x^2} - 1}{\sin 2x}$;

(3) $\lim\limits_{x \to 0} \dfrac{\sqrt{1 + x^2} - 1}{1 - \cos x}$;

(4) $\lim\limits_{x \to 0} \dfrac{\tan(\tan x)}{x}$.

18. Prove the continuity of the following functions by ε-δ definition.

(1) $f(x) = x^2$, $x \in (-\infty, +\infty)$; (2) $f(x) = \sqrt{x}$, $x \in [0, +\infty)$;

(3) $f(x) = \dfrac{1}{x}$, $x \in (0, +\infty)$.

19. Study the continuity of the following functions.

(1) $f(x) = \begin{cases} x^2, & |x| \leqslant 1, \\ 0, & |x| > 1; \end{cases}$

(2) $f(x) = \begin{cases} x, & 0 \leqslant x < 1, \\ 1, & 1 \leqslant x < 3, \\ -x + 4, & 3 \leqslant x \leqslant 4; \end{cases}$

(3) $f(x) = \begin{cases} \dfrac{1}{1 + e^{\frac{1}{x-1}}}, & x \neq 1, \\ A, & x = 1; \end{cases}$

(4) $f(x) = \lim\limits_{n \to \infty} \dfrac{x + e^{nx}}{1 + x^2 e^{nx}}, \quad x \in (-\infty, +\infty);$

(5) $f(x) = \lim\limits_{n \to \infty} \dfrac{n^x - n^{-x}}{n^x + n^{-x}}, \quad x \in (-\infty, +\infty).$

20. Determine the values of a, b and c such that

$$f(x) = \begin{cases} -1, & x \leqslant -1, \\ ax^2 + bx + c, & |x| < 1, x \neq 0, \\ 0, & x = 0, \\ 1, & x \geqslant 1, \end{cases}$$

is continuous.

21. Assume that the function $f(x)$ is continuous at x_0. Prove that $|f(x)|$ is also continuous at x_0. Does the converse hold?

22. Study the continuity of functions $f(x) + g(x)$ and $f(x) \cdot g(x)$ in the following situations:

    (1) $f(x)$ is continuous at x_0 and $g(x)$ is not continuous at x_0;

    (2) neither $f(x)$, nor $g(x)$ is continuous at x_0.

23. Show that if the function $f(x)$ is continuous on $[a, b]$, and $x_1, x_2, \cdots, x_n$ are n points in $[a, b]$, then there exists a point ξ in (a, b), such that

$$f(\xi) = \frac{1}{n}[f(x_1) + f(x_2) + \cdots + f(x_n)].$$

24. Prove the following:

    (1) The equation $x^3 - 3x = 1$ has a root in $[1, 2]$;

    (2) The equation $x2^x = 1$ has a root in $[0, 1]$;

    (3) The equation

$$\frac{a_1}{x - \lambda_1} + \frac{a_2}{x - \lambda_2} + \frac{a_3}{x - \lambda_3} = 0,$$

(where $a_1, a_2, a_3 > 0$, and $\lambda_1 < \lambda_2 < \lambda_3$), has a root in each of the intervals (λ_1, λ_2), (λ_2, λ_3).

    (4) The equation $x^{2n+1} + a_1 x^{2n} + \cdots + a_{2n} x + a_{2n+1} = 0$ has at least one root. Here all coefficients are real.

25. Show that if $f(x)$ is continuous on $[a, b]$, and $f(x) \neq 0$, $(a \leqslant x \leqslant b)$, then $f(x)$ does not change signs on $[a, b]$.

26. Study the uniform continuity of the following functions on the given intervals.

    (1) $f(x) = x$, $x \in (-\infty, +\infty)$; (2) $f(x) = x^2$, $x \in (-\infty, +\infty)$;

    (3) $f(x) = \sin x$, $x \in (-\infty, +\infty)$;

    (4) $f(x) = \dfrac{1}{x}$, $x \in (a, 1)$, $(a > 0)$;

    (5) $f(x) = \sqrt{x}$, $x \in [1, +\infty)$; (7) $f(x) = \sqrt[3]{x}$, $x \in [0, +\infty)$.

    (6) $f(x) = \cos \dfrac{1}{x}$, $x \in (0, 1)$;

27. Show that the function $f(x) = \sin \dfrac{\pi}{x}$ is bounded and continuous on $(0, 1)$, but it is not uniformly continuous on the interval.

28. Show that the function $f(x) = \sin x^2$ is bounded and continuous on $(-\infty, +\infty)$, but it is not uniformly continuous on the interval.

29. Show that if a function $f(x)$ is continuous on $[a, +\infty)$, the limit $\lim\limits_{x \to +\infty} f(x)$ exists and is finite, then $f(x)$ is uniformly continuous on $[a, +\infty)$.

9.3   Existence of Definite Integrals

9.3.1   *Darboux Sums*

We studied the concept of the definite integral of a continuous function $f(x)$ on a closed interval $[a, b]$ in Section 1.2.2.

Suppose $f(x)$ is a continuous function defined on the interval $[a, b]$. Let T be an arbitrary partition of $[a, b]$:

$$a = x_0 < x_1 < x_2 < \cdots < x_{n-1} < x_n = b.$$

Consider the sum with respect to this partition

$$\sigma = \sum_{i=1}^{n} f(\xi_i) \Delta x_i,$$

where $\Delta x_i = x_i - x_{i-1}$ and ξ_i is an arbitrary point in the interval $[x_{i-1}, x_i]$. This is called the *integral sum* of $f(x)$ on the interval $[a, b]$. Obviously, this sum is related to the partition T and is also related to the choice of ξ_i. Let the length of the longest interval among $[x_{i-1}, x_i]$, $(i = 1, 2, \cdots, n)$ be

$\lambda(T) = \max_{1 \leqslant i \leqslant n} \Delta x_i$. If all the different integral sums tend to the same value I when $\lambda(T) \to 0$, then the value is called the *definite integral* of $f(x)$ on $[a, b]$.

We studied some examples in Section 1.2.1 that were about finding the area under certain curves by taking limits of integral sums with respect to different choices of partitions T and different choice of ξ. How do we know that these integral sums with respect to different partitions will tend to the same value when the length of the longest interval of each partition tends to zero? How do we know that for the same partition T but different choice of the points ξ_i in each small interval, the integral sums σ will also tend to the same value? This is the existence problem of definite integrals. When we studied the definite integrals in chapters 1, 6 and 7, we have assumed the existence of those integrals. Now we give a proof by ε-δ definition.

We only study this problem in one dimension, since the problem in two and three dimension are similar.

This definition of definite integrals was given by Riemann, and this kind of integrals is called *Riemann integrals*. For a function, if its Riemann integral exists on an interval, then the function is said to be *Riemann integrable*. The result above implies that continuous functions are Riemann integrable. Thus, what we need to prove is: For a continuous function $f(x)$ on $[a, b]$, the sum σ tends to the same limit when $\lambda(T) \to 0$ for any partition T of $[a, b]$, and any points ξ_i in the ith interval $[x_{i-1}, x_i]$ $(i = 1, 2, \cdots, n)$. For a partition T, the sum

$$\sigma = \sum_{i=1}^{n} f(\xi_i)\Delta x_i,$$

is called a *Riemann sum*. We use $S(f, T)$ to denote the sum σ in order to emphasize that it is related to the partition T.

Since $f(x)$ is continuous on the closed interval $[a, b]$, by property 4 in Section 9.2.3, $f(x)$ has the maximum value and the minimum value on any interval $[x_{i-1}, x_i]$ of the partition T:

$$[a, x_1], \cdots, [x_{i-1}, x_i], \cdots, [x_{n-1}; b].$$

Let $M_i(x)$ be the maximum value and $m_i(x)$ be the minimum value of $f(x)$ on $[x_{i-1}, x_i]$. Then the largest sum among all $S(f, T)$ is

$$S_1(f, T) = \sum_{i=1}^{n} M_i(x)\Delta x_i,$$

and the smallest sum is

$$S_0(f, T) = \sum_{i=1}^{n} m_i(x)\Delta x_i.$$

These two sums are called the *upper Darboux sum* and the *lower Darboux sum* of $f(x)$ on $[a, b]$ respectively. Obviously,

$$S_0(f, T) \leqslant S(f, T) \leqslant S_1(f, T).$$

9.3.2   *Integrability of Continuous Functions*

Now we study the integrability of continuous functions on closed intervals.

1. Suppose T_1 and T_2 are two partitions of the interval $[a, b]$. If all points of division of T_1 are points of division of T_2. Then

$$S_0(f, T_1) \leqslant S_0(f, T_2) \leqslant S_1(f, T_2) \leqslant S_1(f, T_1).$$

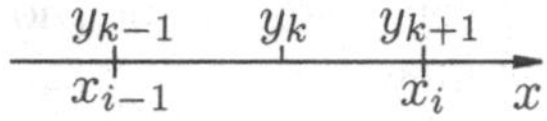

Fig. 9.4

We only prove the last inequality.

To simplify the problem, assuming the two partitions are as shown in Figure 9.4, where x_{i-1}, x_i are two points of division of T_1, $y_{k-1}(= x_{i-1})$, y_k, $y_{k+1}(= x_i)$ are points of division of T_2. Then

$$M_k(y)\Delta y_k + M_{k+1}(y)\Delta y_{k+1} \leqslant M_i(x)\Delta y_k + M_i(x)\Delta y_{k+1} = M_i(x)\Delta x_i.$$

Taking the summation with respect to k on the left hand side and taking the summation with respect to i on the right hand side of the above inequality, we get what we wanted to prove.

2. Suppose T_1 and T_2 are two different partitions of $[a, b]$. Combining all division points of T_1 with all division points of T_2, we get a new partition T_0. By result 1, we have

$$S_0(f, T_1) \leqslant S_0(f, T_0) \leqslant S_1(f, T_0) \leqslant S_1(f, T_2).$$

Thus, for any two partitions T_1 and T_2, we always have

$$S_0(f, T_1) \leqslant S_1(f, T_2).$$

Hence, an upper bound of $S_0(f, T)$ exists for every partition T. The smallest upper bound is called the *least upper bound* or *supremum* and is denoted by I_0. Also, by the above inequality, a lower bound of $S_1(f, T)$ exists for every partition T. The largest lower bound is called the *greatest lower bound* or *infimum* and is denoted by I_1. Obviously,

$$I_0 \leqslant I_1.$$

From the above discussion, we have

3. If for any partition T $(\lambda(T) \to 0)$, the limit

$$\lim_{\lambda(T) \to 0} [S_1(f, T) - S_0(f, T)] = 0,$$

then the integral of $f(x)$ on $[a, b]$ is $I = I_0 = I_1$.

This is the necessary and sufficient condition for a function $f(x)$ to be integrable on an interval $[a, b]$.

To prove the sufficiency of the condition, let $\lambda(T) \to 0$ in the inequality

$$0 < I_1 - I_0 \leqslant S_1(f, T) - S_0(f, T).$$

We get $I_1 = I_0$, denote this by I. Since

$$S_0(f, T) \leqslant S(f, T) \leqslant S_1(f, T),$$

and

$$I_1 \leqslant S_1(f, T), \quad I_0 \geqslant S_0(f, T), \quad I_1 = I_0 = I,$$

we have

$$S_0(f, T) - S_1(f, T) \leqslant S(f, T) - I \leqslant S_1(f, T) - S_0(f, T).$$

That is

$$|S(f, T) - I| \leqslant S_1(f, T) - S_0(f, T).$$

Let $\lambda(T) \to 0$, we get

$$\lim_{\lambda(T) \to 0} S(f, T) = I.$$

This implies that $f(x)$ is integrable on $[a, b]$, and the value of the integral is $I = I_0 = I_1$.

To prove the necessity of the condition, assume that $f(x)$ is integrable on $[a, b]$, and the value of the integral is I. That is

$$\lim_{\lambda(T) \to 0} S(f, T) = I.$$

For any $\varepsilon > 0$, there exists a $\delta > 0$ such that for any choice of ξ_i in $[x_{i-1}, x_i]$

$$I - \varepsilon < S(f, T) < I + \varepsilon$$

as $\lambda(T) < \delta$.

Since ξ_i is an arbitrary point in $[x_{i-1}, x_i]$, $I - \varepsilon$ and $I + \varepsilon$ are constants and is not related to ξ_i, from the above inequality, we have

$$I - \varepsilon \leqslant S_0(f, T) \leqslant S_1(f, T) \leqslant I + \varepsilon.$$

That is

$$0 \leqslant S_1(f, T) - S_0(f, T) \leqslant 2\varepsilon.$$

Thus

$$\lim_{\lambda(T) \to 0} (S_1(f, T) - S_0(f, T)) = 0.$$

4. Now we are ready to prove that a continuous function $f(x)$ on $[a, b]$ is integrable.

In fact, since $f(x)$ is continuous on $[a, b]$, by the discussion in Section 9.2.4, $f(x)$ is uniformly continuous on $[a, b]$. Thus, for any $\varepsilon > 0$, there exists a $\delta > 0$ such that

$$|f(t) - f(s)| < \frac{\varepsilon}{2(b - a)}$$

for all $t, s \in [a, b]$, and $|t - s| < \delta$. Thus, the above inequality is true on every interval $[x_{i-1}, x_i]$, $(i = 1, 2, \cdots, n)$ when $\lambda(T) < \delta$. Hence, the maximum value of $|f(t) - f(s)|$, which is denoted by $L_i(x)$, is not greater than $\dfrac{\varepsilon}{2(b - a)}$ for any t and s in $[x_{i-1}, x_i]$, $(i = 1, 2, \cdots, n)$.

Since

$$M_i(x) - m_i(x) = L_i(x),$$

we have

$$S_1(f, T) - S_0(f, T) = \sum_{i=1}^{n} (M_i(x) - m_i(x)) \Delta x_i = \sum_{i=1}^{n} L_i(x) \Delta x$$

$$\leqslant \sum_{i=1}^{n} \frac{\varepsilon}{2(b - a)} \Delta x_i = \frac{\varepsilon}{2(b - a)} \sum_{i=1}^{n} \Delta x_i$$

$$= \frac{\varepsilon}{2} < \varepsilon,$$

when $\lambda(T) < \delta$. Therefore,

$$\lim_{\lambda(T) \to 0} (S_1(f, T) - S_0(f, T)) = 0.$$

By property 3 above, $f(x)$ is integrable on $[a, b]$.

The property of being uniformly continuous is crucial in the process of the proof of the above result.

Similarly, we can prove that: if a function $f(x)$ is bounded and continuous on an interval $[a, b]$ except for finitely many discontinuous points, then the function is integrable.

Suppose $\alpha_1, \alpha_2, \cdots, \alpha_r$ are the discontinuous points of $f(x)$ on $[a, b]$, and they are listed in ascending order. Let ε be an arbitrarily small positive number such that intervals

$$(\alpha_i - \varepsilon, \alpha_i + \varepsilon), \quad (1 \leqslant i \leqslant r)$$

are disjoint from each other. (Figure 9.5)

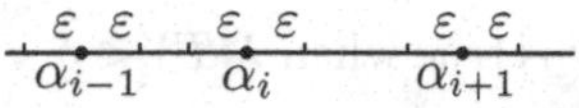

Fig. 9.5

For any partition T of $[a, b]$, there are two kinds of small intervals of T: The first kind are those intervals that are entirely contained in one of the small intervals $[a, \alpha_1 - \varepsilon]$, $[\alpha_{i-1} + \varepsilon, \alpha_i - \varepsilon]$ $(1 \leqslant i \leqslant r)$, and $[\alpha_r + \varepsilon, b]$. The second kind are those intervals that have common points with one of small intervals $(\alpha_i - \varepsilon, \alpha_i + \varepsilon)$ $(1 \leqslant i \leqslant r)$. In this way, the difference of the upper sum and the lower sum of $f(x)$ with respect to T can be written as

$$S_1(f, T) - S_0(f, T) = \sum_{i=1}^{n}(M_i(x) - m_i(x))\Delta x_i$$

$$= {\sum}' (M_i(x) - m_i(x))\Delta x_i + {\sum}'' (M_i(x) - m_i(x))\Delta x_i,$$

where the first sum on the right hand side of the above equation is the sum with respect to the first kind of intervals in the partition T, and the second sum on the right hand side of the above equation is the sum with respect to the second kind of intervals in partition T.

Since $f(x)$ is continuous on every small interval $[a, \alpha_1 - \varepsilon]$, $[\alpha_{i-1} + \varepsilon, \alpha_i - \varepsilon]$ $(1 \leqslant i \leqslant r)$, and $[\alpha_r + \varepsilon, b]$, for each interval of the first kind, there exists a $\delta > 0$ small enough such that the difference of the upper sum M_i and the lower sum m_i of $f(x)$ on every small interval (x_{i-1}, x_i) is less than ε when $\lambda(T) < \delta$. Thus

$$ {\sum}' (M_i(x) - m_i(x))\Delta x_i \leqslant \varepsilon \sum_{i=1}^{n} \Delta x_i = \varepsilon(b - a).$$

For the second kind of interval, since the sum of the length of all intervals is less than $(2\varepsilon + 2\delta)r$, the sum of the length of all intervals is less than $4r\varepsilon$ when $\delta < \varepsilon$.

The second sum

$$\sum{}'' (M_i(x) - m_i(x))\Delta x_i \leqslant (M - m)\sum{}'' \Delta x_i \leqslant 4(M - m)r\varepsilon,$$

where M and m are the supremum and the infimum of $f(x)$ on $[a, b]$. Thus,

$$S_1(f, T) - S_0(f, T) = \sum{}' (M_i(x) - m_i(x))\Delta x_i$$
$$+ \sum{}'' (M_i(x) - m_i(x))\Delta x_i$$
$$< (b - a)\varepsilon + 4(M - m)r\varepsilon = |(b - a) + 4(M - m)r|\varepsilon.$$

Hence, above inequality is true when $\lambda(T) < \delta < \varepsilon$ for any $\varepsilon > 0$. Therefore

$$\lim_{\lambda(T) \to 0} |S_1(f, T) - S_0(f, T)| = 0.$$

By property 3, the result follows.

Suppose $f(x)$ is a monotonic function defined on $[a, b]$. Without lose of generosity, assume $f(x)$ is a monotonically increasing function.

If $f(a) = f(b)$, then $f(x)$ is a constant function on $[a, b]$, and it is integrable.

Assuming that $f(b) > f(a)$. For any given $\varepsilon > 0$, let $\delta = \dfrac{\varepsilon}{f(b) - f(a)}$.

Then by the monotonicity of the function

$$S_1(f, T) - S_0(f, T) = \sum_{i=1}^{n} M_i(x)\Delta x_i - \sum_{i=1}^{n} m_i(x)\Delta x_i$$
$$= \sum_{i=1}^{n} (f(x_i) - f(x_{i-1}))\Delta x_i \leqslant \delta \sum_{i=1}^{n} (f(x_i) - f(x_{i-1}))$$
$$= \delta(f(b) - f(a)) < \varepsilon$$

when $\lambda(T) < \delta$.

Thus,

$$\lim_{\lambda(T) \to 0} [S_1(f, T) - S_0(f, T)] = 0.$$

Therefore by property 3, any monotonic function defined on a closed interval is integrable.

Now we give an example of a non-integrable function.

Let

$$D(x) = \begin{cases} 1, & \text{if } x \text{ is a rational number,} \\ 0, & \text{if } x \text{ is an irrational number.} \end{cases}$$

Obviously, it is bounded on the interval $[0, 1]$. For any partition T of $[0, 1]$, there are rational numbers and irrational numbers in every small interval $[x_{i-1}, x_i]$. Thus, $M_i = 1$, $m_i = 0$. Hence

$$S_1(D, T) - S_0(D, T) = 1$$

when $\lambda(T)$ tends to zero. Therefore, the function is not integrable by property 3.

Before we study the necessary and sufficient condition of whether a function is Riemann integrable, we need to classify infinite sets first.

Let N_+ be the set of all positive integers,

$$N_+ = \{1, 2, 3, \cdots\}.$$

This set contains infinitely many elements. Suppose a set A also consists of infinite many elements. If there is a one-to-one correspondence between A and N_+, then A is called a *countable set*. If A consists of finitely many elements, or is a countable set, then it is called an *at most countable set*. Otherwise, the set is an *uncountable set*.

The set of all rational numbers is a countable set. The set of all irrational numbers is an uncountable set. The set of all real numbers is also an uncountable set.

Suppose A is a set that consists of some real numbers. If for any given $\varepsilon > 0$, there exists an at most countable list of open intervals $\{I_n, n \in N_+\}$, such that any point of A must be an inner point of one of I_n (that is, $\{I_n, n \in N_+\}$ is an open covering of A), and

$$\sum_{n=1}^{\infty} |I_n| < \varepsilon$$

where $|I_n|$ is the length of the interval I_n, then A is called a *set of measure zero*. Obviously, the empty set is a set of measure zero. Also, if A is an "at most countable" set, then A is a set of measure zero.

In fact, suppose that A is a countable set

$$A = \{a_1, a_2, \cdots, a_n, \cdots\}.$$

For any $\varepsilon > 0$, consider intervals

$$I_n = \left(a_n - \frac{\delta}{2^{n+1}}, a_n - \frac{\delta}{2^{n+1}}\right), \quad n = 1, 2, \cdots$$

where $0 < \delta < \varepsilon$.

The set of all intervals $\{I_n, n \in N_+\}$ is an open covering of A, and

$$\sum_{n=1}^{\infty} |I_n| = \sum_{n=1}^{\infty} 2 \cdot \frac{\delta}{2^{n+1}} = \delta \sum_{n=1}^{\infty} \frac{1}{2^n} = \delta < \varepsilon.$$

Thus, A is a set of measure zero.

The conclusion is true obviously when A is a finite set.

On the other hand, if the length an interval is not zero, then this interval is not a set of measure zero.

In fact, consider an open interval (a, b), where $a < b$. If $\{I_n, n \in N_+\}$ is an open covering of (a, b), then

$$\sum_{n=1}^{\infty} |I_n| \geqslant b - a > 0.$$

Since $b - a$ is a fixed positive number, the interval (a, b) is not a set of measure zero.

With these preparations, we are now ready to state the following important theorem.

Lebesgue Theorem: Suppose a function $f(x)$ is bounded on a finite interval $[a, b]$. Then $f(x)$ is Riemann integrable if and only if the set of all points of discontinuity of $f(x)$ in $[a, b]$ is a set of measure zero.

In other words, a Riemann integrable function is an "almost continuous" function except on a set of measure zero.

The function $D(x)$ in the example that we just studied is not integrable since it is not continuous at all irrational numbers and the set of all irrational numbers is not a set of measure zero.

We will not prove this theorem in this book. Readers can find the proof of this theorem in any Real Analysis textbook. The class of Lebesgue integrable functions is an expansion of the class of Riemann integrable functions. Readers can find discussions of the subject of Lebesgue integrable functions in any Real Analysis textbook.

We believe our readers are able to do the following work by using ε-δ definition now:

(1) Define the derivative of a function.

(2) Define the continuity and uniform continuity of a function with several variables.

(3) Prove that continuous functions with several variables on a closed region are uniformly continuous.

(4) Prove that continuous functions with several variables on a closed region are Riemann integrable by Darboux sum.

(5) Define line integrals, surface integrals and partial derivatives of functions with several variables.

9.3.3 Generalization of the Concept of Definite Integrals (Improper Integrals)

In our study of the existence of Riemann integrals, we assumed that the function is continuous and the interval of integration is closed. Also, the integrand is bounded since a continuous function is bounded on a closed interval. Riemann integrals can be generalized to functions and integration regions that do not satisfy these restrictions.

For instance, consider calculating the area of the region bounded by the curve

$$y = \frac{1}{x^2}, \quad (x \geqslant 1)$$

and the lines $x = 1$, $y = 0$ (Figure 9.6). On the face of it, this area is equal to the value of the integral

$$\int_1^{+\infty} \frac{1}{x^2}\, dx.$$

However, according to the definition of Riemann integrals, this is meaningless. Thus, we need a new definition for this kind of integral.

Consider the region bounded by $y = \frac{1}{x^2}$, $x = 1$, $x = b$ and $y = 0$. The area of this region is (Figure 9.7)

$$\int_1^b \frac{1}{x^2}\, dx = -\frac{1}{x}\Big|_1^b = 1 - \frac{1}{b}.$$

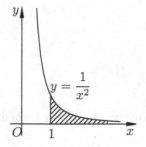

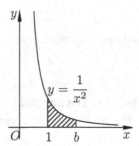

Fig. 9.6   Fig. 9.7

The area increases as b increases. The area of the region is the value of the above integral as $b \to \infty$. Thus, we define:

$$\int_1^{+\infty} \frac{1}{x^2}\, dx = \lim_{b \to +\infty} \int_1^b \frac{1}{x^2}\, dx = 1.$$

To state formally: Suppose $f(x)$ is continuous on the interval $[a, +\infty)$. If the limit of the integral

$$I(b) = \int_a^b f(x)\, dx$$

exists, as $b \to +\infty$, that is

$$\lim_{b \to +\infty} I(b) = \lim_{b \to +\infty} \int_a^b f(x)\, dx$$

is a finite number, then the infinite integral

$$\int_a^{+\infty} f(x)\, dx$$

is said to *converge*, and its value is defined to be the above limit

$$\int_a^{+\infty} f(x)\, dx = \lim_{b \to +\infty} \int_a^b f(x)\, dx.$$

Otherwise, the integral is said to *diverge*.

Similarly, if a function $f(x)$ is continuous on $(-\infty, b]$, and the limit of the integral

$$\int_a^b f(x)\, dx, \qquad \lim_{a \to -\infty} \int_a^b f(x)\, dx,$$

exists, then the infinite integral

$$\int_{-\infty}^b f(x)\, dx$$

is said to converge, and

$$\int_{-\infty}^b f(x)\, dx = \lim_{a \to -\infty} \int_a^b f(x)\, dx.$$

Otherwise, the integral is said to diverge.

If a function $f(x)$ is continuous on the interval $(-\infty, +\infty)$, then

$$\int_{-\infty}^{+\infty} f(x)\, dx = \lim_{\substack{a \to -\infty \\ b \to +\infty}} \int_a^b f(x)\, dx$$

can be defined in a similar way, if the limit of the right-hand-side integral exists.

Example 1. Show that the following integral converges:

$$\int_0^{+\infty} \frac{1}{1 + x^2}\, dx.$$

Proof: Since

$$\lim_{b\to+\infty} \int_0^b \frac{1}{1+x^2}\,dx = \lim_{b\to+\infty} \arctan b = \frac{\pi}{2},$$

the integral converges.

Example 2. Show that the integral

$$\int_a^{+\infty} \frac{1}{x^p}\,dx \quad (a>0)$$

converges if $p > 1$ and diverges if $p \leqslant 1$.

Proof: Since

$$\lim_{b\to+\infty} \int_a^b \frac{1}{x^p}\,dx = \lim_{b\to+\infty} \frac{1}{1-p}(b^{1-p} - a^{1-p}) = \frac{a^{1-p}}{p-1},$$

for $p > 1$, the integral converges.

Since

$$\lim_{b\to+\infty} \int_a^b \frac{1}{x^p}\,dx = \lim_{b\to+\infty} \frac{1}{1-p}(b^{1-p} - a^{1-p}) = +\infty$$

for $p < 1$, the integral diverges.

Since

$$\lim_{b\to+\infty} \int_a^b \frac{1}{x}\,dx = \lim_{b\to+\infty}(\ln b - \ln a) = +\infty$$

for $p = 1$, the integral diverges.

The Fundamental Theorem of Calculus holds for infinite integrals.

Suppose $f(x)$ is a continuous function and $F(x)$ is an antiderivative of $f(x)$. Let

$$F(+\infty) = \lim_{x\to+\infty} F(x), \quad F(-\infty) = \lim_{x\to-\infty} F(x).$$

Then

$$\int_a^{+\infty} f(x)\,dx = F(x)\Big|_a^{+\infty} = F(+\infty) - F(a),$$

$$\int_{-\infty}^b f(x)\,dx = F(x)\Big|_{-\infty}^b = F(b) - F(-\infty),$$

$$\int_{-\infty}^{+\infty} f(x)\,dx = F(x)\Big|_{-\infty}^{+\infty} = F(+\infty) - F(-\infty).$$

Example 3. Evaluate

$$\int_{-\infty}^{+\infty} \frac{1}{1+x^2}\, dx.$$

Solution:

$$\int_{-\infty}^{+\infty} \frac{1}{1+x^2}\, dx = \arctan x \Big|_{-\infty}^{+\infty} = \frac{\pi}{2} - \left(-\frac{\pi}{2}\right) = \pi.$$

Example 4. Evaluate

$$\int_{0}^{+\infty} e^{-x} \cos x \, dx.$$

Solution: Applying the method of integration by parts twice, we have

$$\int_{0}^{+\infty} e^{-x} \cos x \, dx = e^{-x} \sin x \Big|_{0}^{+\infty} + \int_{0}^{+\infty} e^{-x} \sin x \, dx$$

$$= \int_{0}^{+\infty} e^{-x} \sin x \, dx = -e^{-x} \cos x \Big|_{0}^{+\infty} - \int_{0}^{+\infty} e^{-x} \cos x \, dx$$

$$= 1 - \int_{0}^{+\infty} e^{-x} \cos x \, dx.$$

Thus,

$$\int_{0}^{+\infty} e^{-x} \cos x \, dx = \frac{1}{2}.$$

In a similar way, the restriction that the integrand is bounded on $[a, b]$ can also be removed.

For instance, consider the integral

$$\int_{0}^{1} \frac{1}{\sqrt{x}}\, dx.$$

According to the definition of Riemann integrals, it is meaningless, since $\frac{1}{\sqrt{x}} \to \infty$ as $x \to 0$. The value of the integral is the area under the curve in Figure 9.8. Consider the shaded area under the curve in Figure 9.9, that is

$$\int_{\varepsilon}^{1} \frac{1}{\sqrt{x}}\, dx = 2\sqrt{x}\,\Big|_{\varepsilon}^{1} = 2(1 - \sqrt{\varepsilon}).$$

The area gets larger when ε gets closer to zero. That is, the area in Figure 9.9 will tends to the area in Figure 9.8 when ε tends to zero:

$$\lim_{\varepsilon \to 0} \int_{\varepsilon}^{1} \frac{1}{\sqrt{x}}\, dx = 2.$$

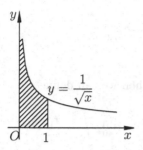

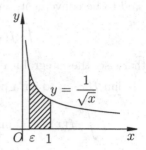

Fig. 9.8   Fig. 9.9

Naturally, we define

$$\int_0^1 \frac{1}{\sqrt{x}}\, dx = \lim_{\varepsilon \to 0} \int_\varepsilon^1 \frac{1}{\sqrt{x}}\, dx = 2.$$

Now we state this definition formally:

Suppose a function $f(x)$ is continuous on $(a, b]$, and $f(x) \to \infty$ as $x \to a + 0$ (the point a is called an *improper point*). If the limit

$$\lim_{\varepsilon \to 0^+} \int_{a+\varepsilon}^b f(x)\, dx$$

exists, then the integral

$$\int_a^b f(x)\, dx$$

is said to be *convergent*, and we define

$$\int_a^b f(x)\, dx = \lim_{\varepsilon \to 0^+} \int_{a+\varepsilon}^b f(x)\, dx.$$

Otherwise, the integral is *divergent*.

Similarly, suppose $f(x)$ is continuous on $[a, b)$, and $f(x) \to \infty$ as $x \to b - 0$. If the limit

$$\lim_{\varepsilon \to 0^+} \int_a^{b-\varepsilon} f(x)\, dx$$

exists, then the integral

$$\int_a^b f(x)\, dx$$

is said to be convergent, and we define

$$\int_a^b f(x)\, dx = \lim_{\varepsilon \to 0^+} \int_a^{b-\varepsilon} f(x)\, dx.$$

Otherwise, the integral is divergent.

If $\lim\limits_{x \to c} f(x) = \infty$ for a point c in $[a, b]$, then we can define

$$\int_a^b f(x)\, dx = \lim_{\varepsilon \to 0^+} \int_a^{c-\varepsilon} f(x)\, dx + \lim_{\varepsilon' \to 0^+} \int_{c+\varepsilon'}^b f(x)\, dx$$

if both limits on the right hand side exist.

The Fundamental Theorem of Calculus holds for integrals with an improper point:

If $F(x)$ is the antiderivative of $f(x)$, then

$$\int_a^b f(x)\, dx = F(b-0) - F(a) \quad \text{(if b is an improper point)};$$

$$\int_a^b f(x)\, dx = F(b) - F(a+0) \quad \text{(if a is an improper point)};$$

$$\int_a^b f(x)\, dx = F(b-0) - F(a+0) \quad \text{(if both a and b are improper points)}.$$

Example 5. Evaluate

$$\int_0^1 \ln x\, dx.$$

Solution: The function

$$\ln x \to -\infty$$

as $x \to 0^+$. Thus, $x = 0$ is an improper point. The integral can be evaluated as follows:

$$\int_0^1 \ln x\, dx = x \ln x \Big|_1^{0^+} - \int_0^1 dx = -1.$$

Example 6. Show that the integral

$$\int_0^a \frac{1}{x^p}\, dx \quad (a > 0)$$

converges if $p > 1$, and diverges if $p \geqslant 1$.

Proof: The point $x = 0$ is an improper point. We have

$$\int_0^a \frac{1}{x^p}\, dx = \frac{1}{1-p} x^{1-p} \Big|_{0^+}^a$$

if $p \neq 1$.

If $p < 1$, then $\lim\limits_{x \to 0+} x^{1-p} = 0$. Thus

$$\int_0^a \frac{1}{x^p}\, dx = \frac{1}{1-p} a^{1-p},$$

and the integral converges.

If $p > 1$, then

$$\lim\limits_{x \to 0+} x^{1-p} = +\infty,$$

and the integral diverges.

If $p = 1$, then

$$\int_0^a \frac{1}{x}\, dx = \ln x \Big|_{0+}^a = \infty,$$

and the integral diverges.

With the generalizations in this section, we can talk about integrals of unbounded functions and over infinite intervals. More detailed discussion about infinite integrals can be found in Chapter 10.

Exercises 9.3

1. Identify the supremum and the infimum of the following number sets.

(1) $A = \{a_1, a_2, \cdots, a_n\}$;

(2) $A = \left\{ \left(1 + \dfrac{1}{n}\right)^n \Big|_{n=1,2,\cdots} \right\}$;   (3) $A = \left\{ \sin \dfrac{\pi}{n} \Big|_{n=1,2,\cdots} \right\}$;

(4) $A = \left\{ \dfrac{(-1)^n}{n} \Big|_{n=1,2,\cdots} \right\}$;   (5) $A = \left\{ x \Big| \sin \dfrac{\pi}{x} = 0, x > 0 \right\}$;

(6) $A = \{x | \ln x < 0\}$;

(7) $A = \left\{ x \Big| x > (1 + \dfrac{1}{n})^n, n = 1, 2, \cdots \right\}$;

(8) $A = \left\{ x \Big| x < \dfrac{1}{n}, n = 1, 2, \cdots \right\}$.

2. Find the supremum and the infimum of the following functions.

(1) $f(x) = x^2$, $(-2 < x < 5)$;

(2) $f(x) = \dfrac{1}{1 + x^2}$, $(-\infty < x < +\infty)$;

(3) $f(x) = \dfrac{2x}{1 + x^2}$, $(0 < x < +\infty)$;

(4) $f(x) = \sin x$, $(0 < x < +\infty)$;

(5) $f(x) = 2^x$, $(-1 < x < 2)$;

(6) $f(x) = \sin x + \cos x$, $(0 \leqslant x \leqslant 2\pi)$.

3. Partition the interval $[-1, 4]$ into n subintervals of equal lengths. Let ξ_i be the mid-points of each interval respectively $(i = 0, 1, 2, \cdots, n-1)$. Find the Riemann sum of $f(x) = 1 + x$ on the interval.

4. Partition the domain into n equal subintervals, find the upper Darboux sum and the lower Darboux sum of the following functions on given intervals.

(1) $f(x) = x^3$, $(-2 \leqslant x \leqslant 3)$;

(2) $f(x) = \sqrt{x}$, $(0 \leqslant x \leqslant 1)$.

5. Evaluate $\displaystyle\int_0^T (v_0 + gt)\, dt$ by the definition of integral, where v_0 and g are constants.

6. Find the following limits by using definite integrals.

(1) $\displaystyle\lim_{n \to \infty} \frac{1}{n} \left(\sqrt{1 + \frac{1}{n}} + \sqrt{1 + \frac{2}{n}} + \cdots + \sqrt{1 + \frac{n}{n}} \right)$;

(2) $\displaystyle\lim_{n \to \infty} \left(\frac{1}{n+1} + \frac{1}{n+2} + \cdots + \frac{1}{n+n} \right)$;

(3) $\displaystyle\lim_{n \to \infty} \left(\frac{n}{n^2 + 1^2} + \frac{n}{n^2 + 2^2} + \cdots + \frac{n}{n^2 + n^2} \right)$.

7. Evaluate the following integrals.

(1) $\displaystyle\int_1^{+\infty} \frac{1}{x^4}\, dx$;

(2) $\displaystyle\int_0^{+\infty} e^{-ax}\, dx$, $(a > 0)$;

(3) $\displaystyle\int_{-\infty}^{+\infty} \frac{1}{x^2 + 2x + 2}\, dx$;

(4) $\displaystyle\int_2^{+\infty} \frac{1}{x^2 + x - 2}\, dx$;

(5) $\displaystyle\int_0^{+\infty} xe^{-x^2}\, dx$;

(6) $\displaystyle\int_0^{+\infty} e^{-x} \sin x\, dx$;

(7) $\displaystyle\int_0^{+\infty} x^n e^{-x}\, dx$, ($n$ is a positive integer);

(8) $\displaystyle\int_0^{+\infty} \frac{1}{(a^2 + x^2)^n}\, dx$, ($n$ is a positive integer);

(9) $\displaystyle\int_0^1 \ln x \, dx;$

(10) $\displaystyle\int_0^1 \frac{1}{\sqrt{1-x^2}} \, dx;$

(11) $\displaystyle\int_0^1 x \ln x \, dx;$

(12) $\displaystyle\int_0^1 \frac{1}{(2-x)\sqrt{1-x}} \, dx;$

(13) $\displaystyle\int_1^2 \frac{x}{\sqrt{x-1}} \, dx;$

(14) $\displaystyle\int_0^1 \frac{x^n}{\sqrt{1-x^2}} \, dx,$ (n is a positive integer);

(15) $\displaystyle\int_0^1 \frac{(1-x)^n}{\sqrt{x}} \, dx,$ (n is a positive integer);

(16) $\displaystyle\int_0^1 (\ln x)^n \, dx,$ (n is a positive integer).

Chapter 10

Infinite Series and Infinite Integrals

10.1 Number Series

10.1.1 *Basic Concepts*

A function defined by finitely many arithmetic operations of elementary functions is still an elementary function, but a function defined by infinitely many arithmetic operations of elementary functions is not necessarily an elementary function. In this chapter and the next chapter, we study the infinite arithmetic combinations of some very simple elementary functions: the power functions, and the trigonometric functions.

In this chapter, we study power series, which is an infinite sum of power functions. (The Taylor series is an infinite sum of power functions.)

In the next chapter, we study trigonometric series, which is an infinite sum of trigonometric functions. (The Fourier series is an infinite sum of trigonometric functions.)

More discussions about the relation between exponential functions and trigonometric functions can be found in Complex Analysis, where it is shown that the exponential function and trigonometric functions can be expressed in each other. So a separate treatment of an infinite sum of exponential functions is not necessary.

The derivative of an elementary function is also an elementary function, but the integral of an elementary function is not necessarily an elementary function. We will study the problem of integrals with parameters later in this chapter.

We start our study from number series.

A number series is a sum of infinitely many numbers a_n $(n = 1, 2, \cdots)$
$$a_1 + a_2 + \cdots + a_n + \cdots,$$
where a_n can be real numbers or complex numbers. We only consider real

numbers here. The sum

$$S_n = a_1 + a_2 + \cdots + a_n$$

is called a *partial sum of the series*. If there is a number S such that

$$S_n \to S \quad (n \to \infty),$$

then the series is *convergent*, and it converges to S. Otherwise it is *divergent*.

Example 1. The series

$$\sum_{n=0}^{\infty} r^n = \lim_{N \to \infty} \sum_{n=0}^{N} r^n = \lim_{N \to \infty} (1 + r + \cdots + r^N)$$

$$= \lim_{N \to \infty} \frac{1 - r^{N+1}}{1 - r} = \frac{1}{1 - r}$$

if $|r| < 1$, so it is convergent.

Example 2. The partial sum of the series

$$\sum_{n=1}^{\infty} \ln \frac{n+1}{n} = \ln \frac{2}{1} + \ln \frac{3}{2} + \cdots + \ln \frac{n+1}{n} + \cdots$$

is

$$S_n = (\ln 2 - \ln 1) + (\ln 3 - \ln 2) + \cdots + (\ln(n+1) - \ln n) = \ln(n+1),$$

and it approaches to infinity as $n \to \infty$. Thus, the series is divergent.

Example 3. The partial sum of the series

$$\sum_{n=0}^{\infty} (-1)^n = 1 - 1 + \cdots + (-1)^n + \cdots$$

is

$$S_n = \begin{cases} 1, & \text{if } n \text{ is an even number}, \\ 0, & \text{if } n \text{ is an odd number}. \end{cases}$$

It has no limit when $n \to \infty$. Thus, the series is divergent.

The partial sums of the series

$$a_1 + a_2 + \cdots + a_n + \cdots$$

is

$$S_1, S_2, \cdots, S_n, \cdots .$$

This is a sequence. The convergence or divergence of the series depends on the convergence or divergence of the sequence of the partial sums S_n.

On the other hand, suppose

$$b_1, b_2, \cdots, b_n, \cdots,$$

is a given sequence, we can create a series

$$b_1 + (b_2 - b_1) + \cdots + (b_n - b_{n-1}) + \cdots.$$

The partial sum S_n of the series is the nth term b_n. The convergence or divergence of this sequence depends on the convergence or divergence of the series. Because of this relationship between number sequences and number series, we can translate the theorems for sequences into theorems for series.

10.1.2 *Some Convergence Criteria*

It is not hard to write the *Cauchy criterion* for series due to its similarity to the Cauchy criterion for sequences.

An infinite series

$$a_1 + a_2 + \cdots + a_n + \cdots$$

is convergent if and only if for any $\varepsilon > 0$, there exists a positive integer $N(\varepsilon)$, such that

$$|S_n - S_m| = |a_{m+1} + a_{m+2} + \cdots + a_n| < \varepsilon$$

if $n > m > N$.

Since the limit of any bounded and monotonic sequence exists, we have the following *convergence criterion for series of positive terms*.

If $a_n \geqslant 0$ and $|S_n| \leqslant M$ for all n (where M is a positive number), then the series

$$a_1 + a_2 + \cdots + a_n + \cdots$$

is convergent.

Let $m = n - 1$ in the Cauchy criterion. We have $S_n - S_{n-1} = a_n$. Thus, if the series $a_1 + a_2 + \cdots + a_n + \cdots$ is convergent, then

$$\lim_{n \to \infty} a_n = 0.$$

This is a necessary condition of convergence of the series, but not a sufficient condition. For instance, the term of the series in Example 2

$$a_n = \ln \frac{n+1}{n}$$

tends to zero when n tends to infinity, but the series is divergent.

If the series

$$|a_1| + |a_2| + \cdots + |a_n| + \cdots$$

is convergent, then we say that the series

$$a_1 + a_2 + \cdots + a_n + \cdots$$

is *absolutely convergent*.

By Cauchy criterion and the fact that

$$|a_{m+1} + a_{m+2} + \cdots + a_n| \leqslant |a_{m+1}| + |a_{m+2}| + \cdots + |a_n|,$$

we conclude that if a series is absolutely convergent then it is convergent.

There is a more general result: If $b_n \geqslant 0$, and

$$b_1 + b_2 + \cdots + b_n + \cdots$$

is a convergent series, and $|a_n| \leqslant b_n$, then

$$a_1 + a_2 + \cdots + a_n + \cdots$$

is absolutely convergent.

Actually, the conclusion is true if $|a_n| \leqslant b_n$ when n is large enough.

More generally, we have the following criterion.

Comparison Criterion: Suppose $\sum\limits_{n=1}^{\infty} a_n$ and $\sum\limits_{n=1}^{\infty} b_n$ are two series of positive terms, (that is $a_n \geqslant 0$, $b_n \geqslant 0$). If $a_n \leqslant b_n$, then the following statements are true:

(1) If $\sum\limits_{n=1}^{\infty} b_n$ is convergent, then $\sum\limits_{n=1}^{\infty} a_n$ is convergent.

(2) If $\sum\limits_{n=1}^{\infty} a_n$ is divergent, then $\sum\limits_{n=1}^{\infty} b_n$ is divergent.

Root Test: Suppose the series $\sum\limits_{n=1}^{\infty} a_n$ satisfies

$$\lim_{n \to \infty} \sqrt[n]{|a_n|} = r.$$

Then the series is absolutely convergent if $r < 1$; the series is divergent if $r > 1$; and there is no definite conclusion if $r = 1$.

In fact, if $r < 1$, then let $b_n = \left(\dfrac{r+1}{2}\right)^n$, $\left(0 < \dfrac{r+1}{2} < 1\right)$ in the criterion by comparison. By Example 1, the series

$$\sum_{n=1}^{\infty} \left(\frac{r+1}{2}\right)^n$$

is convergent.

Since $\lim_{n\to\infty} \sqrt[n]{|a_n|} = r$, we have $\sqrt[n]{|a_n|} < \dfrac{r+1}{2}$, or $|a_n| < \left(\dfrac{r+1}{2}\right)^n$, when n is large enough. Therefore, the series $\sum_{n=1}^{\infty} a_n$ is absolutely convergent.

If $r > 1$, since $\lim_{n\to\infty} |a_n|^{\frac{1}{n}} = r > 1$, we have $\sqrt[n]{|a_n|} > 1$, that is $|a_n| > 1$ when n is large enough. Thus, a_n does not tend to zero as n approaches to infinity. Therefore, the series $\sum_{n=1}^{\infty} a_n$ is divergent.

If $r = 1$, we have examples of both convergent and divergent series.

In Example 4 below, we will show that the series $\sum_{n=1}^{\infty} \dfrac{1}{n^2}$ is convergent, and the series $\sum_{n=1}^{\infty} \dfrac{1}{n}$ is divergent, but both series satisfy $\lim_{n\to\infty} \sqrt[n]{|a_n|} = 1$, where a_n denotes the terms of series.

Ratio Test (d'Alembert Criterion): If the terms of a series $\sum_{n=1}^{\infty} a_n$ satisfy

$$\lim_{n\to\infty} \left| \frac{a_{n+1}}{a_n} \right| = r,$$

then the series is absolutely convergent if $r < 1$; the series is divergent if $r > 1$; and there is no definite conclusion if $r = 1$.

Indeed, if $r < 1$, for any $0 < \varepsilon < 1 - r$, there exists a positive integer $N(\varepsilon)$ such that

$$\left| \frac{a_{n+1}}{a_n} \right| \leqslant r + \varepsilon,$$

for $n > N$.

This implies that

$$|a_{n+1}| \leqslant (r + \varepsilon)|a_n|.$$

Thus, we have

$$|a_{N+p}| \leqslant (r + \varepsilon)^p |a_N|.$$

Therefore, $\displaystyle\sum_{n=1}^{\infty} a_n$ is absolutely convergent because $\displaystyle\sum_{p=1}^{\infty} |a_N|(r+\varepsilon)^p$ is convergent.

The case for $r > 1$ and $r = 1$ is analogous to the root test.

If $\displaystyle\sum_{n=1}^{\infty} a_n$ is a series of positive terms, and

$$\lim_{n\to\infty} a_n^{\frac{1}{n}} = r,$$

then the series converges if $r < 1$; the series diverges if $r > 1$; there is no definite conclusion if $r = 1$.

Similarly, If $\displaystyle\sum_{n=1}^{\infty} a_n$ is a series of positive terms, and

$$\lim_{n\to\infty} \frac{a_{n+1}}{a_n} = r,$$

then the series converges if $r < 1$; the series diverges if $r > 1$; there is no definite conclusion if $r = 1$.

These results can be verified by the same process.

Also, infinite integrals can be used to determine whether a series of positive terms is convergent.

If $f(x)$ is a non-negative and non-increasing continuous function on $[1, +\infty)$, then the series

$$\sum_{n=1}^{\infty} f(k) = f(1) + f(2) + \cdots + f(n) + \cdots$$

and the infinite integral

$$\int_1^{+\infty} f(x)\,dx$$

are convergent or divergent simultaneously.

Indeed, since $f(x)$ is non-increasing, we have

$$f(k+1) \leqslant f(x) \leqslant f(k)$$

for $k \leqslant x \leqslant k+1$. That is

$$f(k+1) = \int_k^{k+1} f(k+1)\,dx \leqslant \int_k^{k+1} f(x)\,dx \leqslant \int_k^{k+1} f(k)\,dx = f(k),$$

where $k = 1, 2, \cdots$.

Adding these inequalities from 1 to n, we obtain

$$\sum_{k=1}^{n} f(k+1) \leqslant \int_1^{n+1} f(x)\,dx \leqslant \sum_{k=1}^{n} f(k).$$

Since $f(x)$ is non-negative, the partial sum of the series

$$S_{n+1} = f(1) + \sum_{k=1}^{n} f(k+1) \leqslant f(1) + \int_{1}^{n+1} f(x)\,dx$$

$$\leqslant f(1) + \int_{1}^{+\infty} f(x)\,dx = M$$

is bounded if the infinite integral $\int_{1}^{+\infty} f(x)\,dx$ is convergent. Since the

limit of a bounded monotone increasing sequence exists, the series $\sum\limits_{n=1}^{\infty} f(k)$

is convergent.

Conversely, since

$$\int_{1}^{n+1} f(x)\,dx \leqslant \sum_{k=1}^{n} f(k),$$

the partial sum of the series is unbounded if the infinite integral is divergent. This implies that the series is divergent. Therefore, they are convergent or divergent simultaneously.

From the discussion above, we can see the connection between infinite series and infinite integrals. They are both sums of infinitely many terms, and the only difference is one is "discrete" and the other is "continuous". Their values can be estimated by each other.

Example 1. The series

$$\sum_{n=2}^{\infty} \frac{1}{(\ln n)^n}$$

is convergent by Cauchy criterion because

$$\lim_{n\to\infty} \sqrt[n]{a_n} = \lim_{n\to\infty} \frac{1}{\ln n} = 0 < 1.$$

Example 2. The series

$$\sum_{n=1}^{\infty} \frac{1}{2^n} \left(1 + \frac{1}{n}\right)^{n^2}$$

is divergent by Cauchy criterion because

$$\lim_{n\to\infty} \sqrt[n]{a_n} = \lim_{n\to\infty} \frac{1}{2} \left(1 + \frac{1}{n}\right)^n = \frac{e}{2} > 1.$$

Example 3. Prove that the series

$$\sum_{n=1}^{\infty} n! \left(\frac{2}{n}\right)^n$$

is convergent.

Proof: Since

$$\lim_{n \to \infty} \frac{a_{n+1}}{a_n} = \lim_{n \to \infty} \frac{(n+1)! \left(\dfrac{2}{n+1}\right)^{n+1}}{n! \left(\dfrac{2}{n}\right)^n}$$

$$= \lim_{n \to \infty} \frac{2}{\left(1 + \dfrac{1}{n}\right)^n} = \frac{2}{e} < 1,$$

the series is convergent by the ratio test.

Example 4. Prove that the series

$$\sum_{n=1}^{\infty} \frac{1}{n^\alpha} = 1 + \frac{1}{2^\alpha} + \frac{1}{3^\alpha} + \cdots + \frac{1}{n^\alpha} + \cdots$$

is convergent if $\alpha > 1$ and is divergent if $\alpha \leqslant 1$.

Proof: We cannot use Cauchy criterion and ratio test for this series since $r = 1$, but we can use the method of infinite integral.

Consider the function

$$f(x) = \frac{1}{x^\alpha}.$$

Since

$$\int_1^{+\infty} \frac{1}{x^\alpha}\, dx = \begin{cases} \dfrac{-1}{1-\alpha}, & \alpha > 1; \\ \infty, & \alpha \leqslant 1, \end{cases}$$

the series is convergent if $\alpha > 1$ and is divergent if $\alpha \leqslant 1$.

If $\alpha = 1$, it is the harmonic series

$$1 + \frac{1}{2} + \frac{1}{3} + \cdots + \frac{1}{n} + \cdots.$$

This is a divergent series.

10.1.3 *Conditionally Convergent Series*

If a convergent series is not absolutely convergent, then it is called *conditionally convergent*. For instance, the series

$$1 - \frac{1}{2} + \frac{1}{3} - \frac{1}{4} + \cdots$$

is not absolutely convergent. Since

$$\frac{1}{n} - \frac{1}{n+1} + \frac{1}{n+2} - \cdots + (-1)^p \frac{1}{n+p} \leqslant \frac{1}{n} - \left(\frac{1}{n+1} - \frac{1}{n+2} \right)$$

$$- \left(\frac{1}{n+3} - \frac{1}{n+4} \right) - \cdots - \left(\frac{1}{n+2[\frac{p}{2}]-1} - \frac{1}{n+2[\frac{p}{2}]} \right) \leqslant \frac{1}{n}$$

and

$$\frac{1}{n} - \frac{1}{n+1} + \frac{1}{n+2} - \cdots + (-1)^p \frac{1}{n+p} \geqslant \left(\frac{1}{n} - \frac{1}{n+1} \right)$$

$$+ \left(\frac{1}{n+2} - \frac{1}{n+3} \right) + \cdots + \left(\frac{1}{n+2[\frac{p-1}{2}]} - \frac{1}{n+2[\frac{p-1}{2}]+1} \right) \geqslant 0,$$

where $[x]$ represent the integer part of x. Thus, $S_n - S_{n+p} \to 0$ as $n \to \infty$. Therefore, the series is convergent by Cauchy criterion.

The above example can be generalized to another criterion.

Leibniz Criterion: If $0 < a_{n+1} \leqslant a_n$, and $\lim\limits_{n\to\infty} a_n = 0$, then the series $\sum\limits_{n=1}^{\infty} (-1)^n a_n$ is convergent.

In fact, since

$$0 \leqslant a_n - a_{n+1} + a_{n+2} - \cdots + (-1)^p a_{n+p} \leqslant a_n,$$

by Cauchy criterion, the series is convergent.

If the terms of a series are positive and negative alternately, then the series is called an *alternating series*. The Leibniz criterion is also called the criterion of convergence for alternating series.

A preparation is needed for further discussions.

Summation by Parts: If

$$S_k = a_1 + a_2 + \cdots + a_k$$

for $k = 1, 2, \cdots$, then

$$\sum_{k=1}^{n} a_k b_k = \sum_{k=1}^{n-1} S_k (b_k - b_{k+1}) + S_n b_n.$$

In fact, since $S_k - S_{k-1} = a_k$, we have

$$\sum_{k=1}^{n} a_k b_k = \sum_{k=1}^{n} b_k(S_k - S_{k-1}) = \sum_{k=1}^{n} b_k S_k - \sum_{k=1}^{n} b_k S_{k-1}$$

$$= \sum_{k=1}^{n} b_k S_k - \sum_{k=1}^{n-1} b_{k+1} S_k = \sum_{k=1}^{n-1} S_k(b_k - b_{k+1}) + S_n b_n.$$

Notice that summation by parts is a form of finite terms of integration by parts.

Abel Lemma: If $b_1 \geqslant b_2 \geqslant \cdots \geqslant b_n \geqslant 0$, and

$$m \leqslant S_k = \sum_{i=1}^{k} a_i \leqslant M$$

for $k = 1, 2, \cdots, n$, then

$$b_1 m \leqslant \sum_{k=1}^{n} a_k b_k \leqslant b_1 M.$$

Indeed, by summation by parts, we have

$$\sum_{k=1}^{n} a_k b_k = \sum_{k=1}^{n-1} S_k(b_k - b_{k+1}) + S_n b_n$$

$$\leqslant M \left(\sum_{k=1}^{n-1} (b_k - b_{k+1}) + b_n \right) = M b_1.$$

Similarly,

$$\sum_{k=1}^{n} a_k b_k \geqslant m b_1.$$

From Abel lemma, we have

Dirichlet Criterion: If b_k is monotonically decreasing and tends to zero, and the sun $S_k = a_1 + a_2 + \cdots + a_k$ is bounded, that is $|S_k| \leqslant M$ for $k = 1, 2, \cdots$, then the series

$$\sum_{k=1}^{\infty} a_k b_k$$

is convergent.

Indeed, since

$$-M \leqslant a_1 + a_2 + \cdots + a_n + a_{n+1} + \cdots + a_{n+p} \leqslant M,$$

that is

$$-M - S_n \leqslant a_{n+1} + \cdots + a_{n+p} \leqslant M - S_n$$

for $p = 1, 2, \cdots$, by Abel lemma, we have

$$-b_{n+1}(M + S_n) \leqslant a_{n+1}b_{n+1} + \cdots + a_{n+p}b_{n+p} \leqslant b_{n+1}(M - S_n).$$

Thus,

$$-2Mb_{n+1} \leqslant a_{n+1}b_{n+1} + \cdots + a_{n+p}b_{n+p} \leqslant 2Mb_{n+1}.$$

It follows that

$$|a_{n+1}b_{n+1} + \cdots + a_{n+p}b_{n+p}| \leqslant 2Mb_{n+1}.$$

Since $b_n \to 0$ as $n \to \infty$, we have $b_{n+1} < \dfrac{\varepsilon}{2M}$ when n is large enough. Hence, there exists an N, such that

$$|a_{n+1}b_{n+1} + \cdots + a_{n+p}b_{n+p}| \leqslant \varepsilon$$

as $n > N$ for any natural number p. Therefore, the series $\sum\limits_{k=1}^{\infty} a_k b_k$ is convergent by Cauchy criterion.

If we let $a_k = (-1)^{k-1}$ in above criterion, then

$$|S_k| = |a_1 + a_2 + \cdots + a_n| \leqslant 1.$$

Thus, the series

$$\sum_{k=1}^{\infty} (-1)^{k-1} b_k$$

is convergent if b_k is monotonically decreasing and tends to zero as k tends to infinity. This is Leibniz criterion.

Abel Criterion: If b_k is monotone and bounded, and the series $\sum\limits_{k=1}^{\infty} a_k$ is convergent, then the series $\sum\limits_{k=1}^{\infty} a_k b_k$ is convergent.

Indeed, the limit of b_k exists because b_k is monotone and bounded. We denote this limit as b. Let $b'_k = b_k - b$, if b_k is monotonically decreasing. Let $b'_k = b - b_k$, if b_k is monotonically increasing. In both situations, the sequence b'_k is monotonically decreasing and tends to 0. Since the series

$\sum\limits_{k=1}^{\infty} a_k$ is convergent, its partial sum S_k is bounded. Thus the series $\sum\limits_{k=1}^{\infty} a_k b_k'$ is convergent by Dirichlet criterion. Therefore,

$$\sum_{k=1}^{\infty} a_k b_k = \sum_{k=1}^{\infty} a_k (b_k' + b) = \sum_{k=1}^{\infty} a_k b_k' + b \sum_{k=1}^{\infty} a_k$$

in the first case (a similar equality holds in the second case), and the result follows.

Example 1. Study the convergence of the series $\sum\limits_{k=1}^{\infty} \dfrac{\cos nx}{n}$.

Solution: Let $b_n = \dfrac{1}{n}$. Then b_n is monotonically decreasing and tends to 0.

Let $a_n = \cos nx$. We have

$$\left| \sum_{n=1}^{N} a_n \right| = \left| \sum_{n=1}^{N} \cos nx \right| = \left| \frac{\sin\left(N + \dfrac{1}{2}\right)x - \sin\dfrac{1}{2}x}{2\sin\dfrac{x}{2}} \right| \leqslant \frac{1}{\left|\sin\dfrac{x}{2}\right|}.$$

Thus, $\sum\limits_{n=1}^{N} a_n$ is bounded when x is not a multiple of 2π. By Dirichlet criterion, the series is convergent if $x \neq 2k\pi$, $k = 0, \pm 1, \pm 2, \cdots$.

Example 2. Study the convergence of the series $\sum\limits_{n=1}^{\infty} \dfrac{\cos 3n}{n}\left(1 + \dfrac{1}{n}\right)^n$.

Solution: By the result of the last example, the series $\sum\limits_{k=1}^{\infty} \dfrac{\cos 3n}{n}$ is convergent. Since the sequence $b_n = \left(1 + \dfrac{1}{n}\right)^n$ is monotonically increasing and has a limit e, it is bounded. By Abel criterion, the series is convergent.

When we do addition of finite numbers, the value of the sum is the same if the order of the numbers is switched. This is not always true for the sum of infinitely many numbers. If a series is absolutely convergent, then this conclusion is correct. If a series is conditionally convergent, not only is the conclusion not true, but the series can be made to be convergent to any fixed number or can be made to be divergent by a rearrangement of its terms.

The value of the sum of an absolutely convergent series is the same when the terms are rearranged.

First, we show that this conclusion is true for series of positive terms. Suppose

$$\sum_{n=1}^{\infty} a_n = a_1 + a_2 + \cdots + a_n + \cdots$$

is a series of positive terms, it is convergent and the sum is S. Let

$$\sum_{n=1}^{\infty} a'_n = a'_1 + a'_2 + \cdots + a'_n + \cdots$$

be a series obtained by rearranging the terms of the above series arbitrarily. Since every term a'_k of the new series is a term of the original series, for any partial sum $S'_n = a'_1 + a'_2 + \cdots + a'_n$ of the new series, there exists a natural number m large enough such that each term $a'_1, a'_2, \cdots, a'_n$ is included in the partial sum of the original series $S_m = a_1 + a_2 + \cdots + a_m$. Thus, we have

$$S'_n \leqslant S_m \leqslant S.$$

This implies that S'_n has an upper bounded, and therefore the new series $\sum_{n=1}^{\infty} a'_n$ is convergent. Also, the sum of the new series

$$S' \leqslant S.$$

On the other hand, the original series can be considered as the rearrangement of the new series. The above discussion gives us

$$S \leqslant S'.$$

Therefore, $S' = S$.

Now we show that the conclusion is true for absolutely convergent series. Suppose a series $\sum_{n=1}^{\infty} a_n$ is absolutely convergent. We rewrite this series as the difference of two series of positive terms

$$\sum_{n=1}^{\infty} a_n = \sum_{n=1}^{\infty} b_n - \sum_{n=1}^{\infty} c_n,$$

where

$$b_n = \frac{|a_n| + a_n}{2}, c_n = \frac{|a_n| - a_n}{2}.$$

When the terms of the original series are rearranged, the terms of the two series of positive terms are also rearranged correspondingly, but their sums are the same by the above conclusion. Therefore the sum of an absolutely convergent series will be the same when the terms are rearranged.

If a series $\sum\limits_{n=1}^{\infty} a_n$ is conditionally convergent, then it can become a divergent series, or it can be made to converge to a fixed number S by rearrangement of its terms.

First, we show that the series can be made to be divergent by rearrangement of its terms. Since there are infinite many positive terms and infinite many negative terms in a conditionally convergent series (otherwise it is an absolutely convergent series), the series that consists of all the positive terms is divergent and the series that consists of all the negative terms is also divergent. To obtain a divergent series by rearranging the terms of the series, we take several positive terms in their original order in the series such that the sum of these terms is greater than 1. We then add the first negative term of the series after these positive terms.

Next we take some terms from the rest of the positive terms in the series in their original order such that the sum of these positive terms is greater than 1. We then add the second negative term of the series after these positive terms. Repeating the same process, every term in the original series will show up in the new series. Since there is always a segment of the series with indices larger than n with a sum greater than 1 no matter how large n is, the series is divergent according to Cauchy criterion.

Next, we try to obtain a series that converges to a fixed number S by rearranging the order of the terms of the series. Assume that $S \geqslant 0$. Take some positive terms in the series, in the original order, until the sum of these terms is barely greater than S. Take some negative terms in the series in the original order until the sum of these terms and those positive terms we have taken is barely less than S. Repeating the same process, we get a new series

$$b_1 + b_2 + \cdots + b_n + \cdots .$$

Obviously, the series $\sum\limits_{n=1}^{\infty} b_n$ is the series $\sum\limits_{n=1}^{\infty} a_n$ with a different arrangement of the terms.

Consider the sequence of the partial sums of $\sum\limits_{n=1}^{\infty} b_n$

$$\sigma_n = \sum_{k=1}^{n} b_k, \quad (n = 1, 2, \cdots).$$

Since the series $\sum_{n=1}^{\infty} a_n$ is convergent, $\lim_{n\to\infty} a_n = 0$. It follows that $\lim_{n\to\infty} b_n = 0$.

For any $\varepsilon > 0$, there exists an N, such that $|b_n| < \varepsilon$ for $n \geqslant N$. For a partial sum $\sigma_n (n \geqslant N)$, if one of σ_n and σ_{n-1} is greater than S and the other one is less than S, then we have

$$|\sigma_n - S| \leqslant |\sigma_n - \sigma_{n-1}| = |b_n| < \varepsilon.$$

If both σ_n and σ_{n-1} are greater than S or are less than S, then by the method of constructing the series $\sum_{n=1}^{\infty} b_n$, σ_n is closer to S than σ_{n-1}. Thus, in any situation, $|\sigma_n - S|$ is either less than ε or less than $|\sigma_{n-1} - S|$. Hence, $|\sigma_n - S| < \varepsilon$ if $n \geqslant N$. That is $\lim_{n\to\infty} \sigma_n = S$.

Exercises 10.1

1. Prove the following equations.

   (1) $\displaystyle\sum_{n=1}^{\infty} \frac{1}{(2n-1)(2n+1)} = \frac{1}{2}$;

   (2) $\displaystyle\sum_{n=1}^{\infty} (\sqrt{n+2} - 2\sqrt{n+1} + \sqrt{n}) = 1 - \sqrt{2}$;

   (3) $\displaystyle\sum_{n=1}^{\infty} \frac{1}{(3n-2)(3n+1)} = \frac{1}{3}$;   (4) $\displaystyle\sum_{n=1}^{\infty} \ln \frac{n(2n+1)}{(n+1)(2n-1)} = \ln 2$;

   (5) $\displaystyle\sum_{n=1}^{\infty} \frac{2n-1}{2^n} = 3$;

   (6) $\displaystyle\sum_{n=1}^{\infty} \frac{1}{n(n+m)} = \frac{1}{m}\left(1 + \frac{1}{2} + \cdots + \frac{1}{m}\right)$.

2. Find the sum of the following series.

   (1) $\displaystyle\sum_{n=1}^{\infty} \left(\frac{1}{2^n} + \frac{1}{3^n}\right)$;   (2) $\displaystyle\sum_{n=1}^{\infty} \frac{2^n + (-1)^n 3^n}{5^n}$.

3. Prove that the convergence of a series is not changed if a finitely many of its terms are altered.

4. Suppose the series $\displaystyle\sum_{n=1}^{\infty} a_n$ is convergent. Show that the series $\displaystyle\sum_{n=1}^{\infty} (a_n + a_{n+1})$ is also convergent. Is the converse true?

5. Prove that if the series $\displaystyle\sum_{n=1}^{\infty} a_n$ with positive terms is convergent, then the series $\displaystyle\sum_{n=1}^{\infty} a_n^2$ is also convergent. Show that the converse of this conclusion is not necessary true by finding an counterexample.

6. Prove that if series $\displaystyle\sum_{n=1}^{\infty} a_n^2$ and $\displaystyle\sum_{n=1}^{\infty} b_n^2$ are convergent, then series $\displaystyle\sum_{n=1}^{\infty} |a_n b_n|$, $\displaystyle\sum_{n=1}^{\infty} (a_n + b_n)^2$, and $\displaystyle\sum_{n=1}^{\infty} \frac{a_n}{n}$ are also convergent.

7. Prove that if $\displaystyle\lim_{n\to\infty} n a_n = a \neq 0$, then $\displaystyle\sum_{n=1}^{\infty} a_n$ is divergent.

8. Determine the convergence or divergence of the following series.

(1) $\displaystyle\sum_{n=1}^{\infty} \frac{1}{(3n^2+5)}$;

(2) $\displaystyle\sum_{n=1}^{\infty} \frac{n}{(n+1)}$;

(3) $\displaystyle\sum_{n=1}^{\infty} \left(\frac{2n}{(2n+1)} - \frac{2n-1}{2n} \right)$;

(4) $\displaystyle\sum_{n=1}^{\infty} \frac{1}{\sqrt[n]{n}}$;

(5) $\displaystyle\sum_{n=1}^{\infty} \frac{1}{n^{1+\frac{1}{n}}}$;

(6) $\displaystyle\sum_{n=2}^{\infty} \frac{1}{(\ln n)^{\ln n}}$;

(7) $\displaystyle\sum_{n=1}^{\infty} 2^n \sin \frac{\pi}{3^n}$;

(8) $\displaystyle\sum_{n=1}^{\infty} \frac{n+1}{n(n+2)}$;

(9) $\displaystyle\sum_{n=1}^{\infty} \arctan \frac{\pi}{4n}$;

(10) $\displaystyle\sum_{n=1}^{\infty} n \tan \frac{\pi}{2^{n+1}}$;

(11) $\displaystyle\sum_{n=2}^{\infty} \frac{n}{(\ln n)^n}$;

(12) $\displaystyle\sum_{n=1}^{\infty} \frac{n^2}{\left(1+\dfrac{1}{n}\right)^n}$;

(13) $\displaystyle\sum_{n=1}^{\infty} \frac{1000^n}{n!}$;

(14) $\displaystyle\sum_{n=1}^{\infty} \frac{2^n n!}{n^n}$;

(15) $\displaystyle\sum_{n=1}^{\infty} \frac{(n!)^2}{(2n)!}$;

(16) $\displaystyle\sum_{n=1}^{\infty} \frac{1}{3n} \left(\frac{n+1}{n} \right)^{n^2}$;

(17) $\displaystyle\sum_{n=1}^{\infty} \left(\frac{an}{n+1} \right)^n$, $(a > 0)$.

9. Determine the convergence or divergence of the following series.

(1) $\displaystyle\sum_{n=2}^{\infty} \frac{1}{n \ln^p n}$;

(3) $\displaystyle\sum_{n=1}^{\infty} \frac{1}{n(\ln n)^p (\ln \ln n)^q}$.

(2) $\displaystyle\sum_{n=3}^{\infty} \frac{1}{n \ln n \ln \ln \ln n}$;

10. Prove the following results.

(1) Suppose $a > 0$, $a_n > 0$ for $n = 1, 2, \cdots$. If $\dfrac{\ln \dfrac{1}{a_n}}{\ln n} \geqslant 1 + a$ for $n > N$, then the series $\displaystyle\sum_{n=1}^{\infty} a_n$ is convergent.

(2) If $\dfrac{\ln \dfrac{1}{a_n}}{\ln n} \leqslant 1$ for $n > N$, then the series $\displaystyle\sum_{n=1}^{\infty} a_n$ is divergent.

11. Show that the series $\displaystyle\sum_{n=1}^{\infty} \frac{1}{(\ln n)^{\ln n}}$ and the series $\displaystyle\sum_{n=1}^{\infty} \frac{1}{3^{\ln n}}$ are convergent by the result of Exercise 10.

12. Determine the convergence or divergence of the following series.

(1) $\displaystyle\sum_{n=1}^{\infty} \frac{(-1)^{\frac{n(n-1)}{2}}}{2^n}$;

(4) $\displaystyle\sum_{n=1}^{\infty} (-1)^n \frac{\sin^2 n}{n}$;

(2) $\displaystyle\sum_{n=1}^{\infty} (-1)^{n-1} \frac{\sqrt{n}}{n+1}$;

(5) $\displaystyle\sum_{n=1}^{\infty} \frac{(-1)^n}{\sqrt{n}}$;

(3) $\displaystyle\sum_{n=1}^{\infty} (-1)^n \left(\frac{2n+100}{3n+1} \right)^n$;

(6) $\displaystyle\sum_{n=1}^{\infty} (-1)^n \sin \frac{1}{n}$;

(7) $\displaystyle\sum_{n=1}^{\infty} \frac{\ln^{100} n}{n} \sin \frac{n\pi}{4}$.

13. Determine whether the following series are conditionally convergent or absolutely convergent.

(1) $\displaystyle\sum_{n=1}^{\infty} \frac{(-1)^{n-1}}{n^p}$;

(3) $\displaystyle\sum_{n=1}^{\infty} \frac{(-1)^{n-1}}{n^{p+\frac{1}{n}}}$;

(2) $\displaystyle\sum_{n=1}^{\infty} \frac{(-1)^{n-1}}{na^{2n}}$;

(4) $\displaystyle\sum_{n=1}^{\infty} \frac{(-1)^n}{x+n}$.

10.2   Function Series

10.2.1   *Infinite Sums*

Suppose

$$u_1(x), u_2(x), \cdots, u_n(x), \cdots$$

is a sequence of functions defined on an interval $[a, b]$. The sum

$$\sum_{n=1}^{\infty} u_n(x) = u_1(x) + u_2(x) + \cdots + u_n(x) + \cdots$$

is called a *function series* on $[a, b]$. For a fixed point x_0 in $[a, b]$,

$$\sum_{n=1}^{\infty} u_n(x_0) = u_1(x_0) + u_2(x_0) + \cdots + u_n(x_0) + \cdots$$

is a number series. If this series is convergent, then the function series $\sum_{n=1}^{\infty} u_n(x)$ is said to be convergent at the point x_0. Otherwise the function series is said to be divergent at the point.

If $\sum_{n=1}^{\infty} u_n(x)$ is convergent at every point of $[a, b]$, then it is said to be convergent on the interval $[a, b]$. The set of all points at which the series $\sum_{n=1}^{\infty} u_n(x)$ is convergent is called a convergence region of the series.

The sum of the series $\sum_{n=1}^{\infty} u_n(x)$ exists on the convergence region. We denote it as $S(x) = \sum_{n=1}^{\infty} u_n(x)$. For instance, the series

$$\sum_{n=1}^{\infty} x^n = 1 + x + x^2 + \cdots + x^n + \cdots$$

is a functions series on $(-\infty, +\infty)$, and it converges on $(-1, 1)$ to $S(x) = \dfrac{1}{1 - x}$. It is divergent on $(-\infty, -1]$ and $[1, +\infty)$.

We know the following: the sum of finitely many continuous functions is also a continuous function; the derivative of the sum of finitely many functions is equal to the sum of the derivatives of these functions; the integral of the sum of finitely many functions is equal to the sum of the integrals of these functions. Are these conclusions true for infinite series?

Example 1. On the interval $[0, 1]$, consider the series

$$x + \sum_{n=2}^{\infty} (x^n - x^{n-1}) = x + (x^2 - x) + \cdots .$$

The nth partial sum is $S_n(x) = x^n$, so the sum function

$$S(x) = \lim_{n \to \infty} S_n(x) = \begin{cases} 0, & 0 \leqslant x < 1; \\ 1, & x = 1. \end{cases}$$

It is not continuous at $x = 1$, but every term of the series

$$u_1(x) = x, \quad u_n(x) = x^n - x^{n-1}, \quad (n = 2, 3, \cdots)$$

is continuous on $[0, 1]$. Thus even if every term is a continuous function, the sum of the series is not necessarily a continuous function. The functions

$$u_1(x) = x, \quad u_n(x) = x^n - x^{n-1}, \quad (n = 2, 3, \cdots)$$

are differentiable on $[0, 1]$, but the sum of the series is not a differentiable function at $x = 1$. Thus even if every term is a differentiable function, the sum of the series is not necessarily a differentiable function.

Example 2. Consider the series

$$2xe^{-x^2} + (2 \cdot 2^2 xe^{-2^2 x^2} - 2 \cdot 1^2 xe^{-1^2 x^2}) + \cdots$$
$$+ (2n^2 xe^{-n^2 x^2} - 2(n-1)^2 xe^{-(n-1)^2 x^2}) + \cdots .$$

The partial sum is

$$S_n(x) = 2n^2 xe^{-n^2 x^2}.$$

Since

$$S(x) = \lim_{n \to \infty} S_n(x) = 0$$

on $[0, 1]$, we have

$$\int_0^1 S(x)\, dx = 0.$$

On the other hand, the termwise integral of the series is

$$\sum_{n=1}^{\infty} \int_0^1 [(2n^2 xe^{-n^2 x^2} - 2(n-1)^2 xe^{-(n-1)^2 x^2})]\, dx$$

$$= \sum_{n=1}^{\infty} [(1 - e^{-n^2}) - (1 - e^{-(n-1)^2})]$$

$$= \sum_{n=1}^{\infty} (e^{-(n-1)^2} - e^{-n^2}) = 1 \neq 0.$$

Thus, the integral of the sum function is not necessarily equal to the sum of the integrals of the function in each term of the series. What condition is needed in order to make this conclusion to be true? The answer is in the next section.

10.2.2 *Uniformly Convergent Sequences of Functions*

Since infinite sequences and infinite series can be converted to each other, infinite function sequences and infinite function series can also be converted to each other.

Let

$$a_1(x), a_2(x), \cdots, a_n(x), \cdots$$

be a sequence of functions that are defined on an interval $[a, b]$. Corresponding to an arbitrary point x in $[a, b]$, there is a sequence of numbers. If this sequence is convergent, then we say that the function sequence is convergent at the point. If the function sequence is convergent at every point in $[a, b]$, then we say that the function sequence is convergent on $[a, b]$. For a fixed point x in $[a, b]$, given $\varepsilon > 0$, there exists a natural number N such that

$$|a_n(x) - a(x)| < \varepsilon$$

if $n > N$. The number N is not only related to ε, but also related to x.

If there exists an N that is only related to ε, and is independent of x, such that

$$|a_n(x) - a(x)| < \varepsilon$$

if $n > N$, for all x in $[a, b]$, then we say that $a_n(x)$ is *uniformly convergent* to $a(x)$ on $[a, b]$.

Cauchy Criterion for Uniformly Convergent Sequence of Functions: A sequence of functions

$$a_1(x), a_2(x), \cdots, a_n(x), \cdots$$

is uniformly convergent if and only if for any $\varepsilon > 0$, there exists an N which is only related to ε, and independent of x, such that

$$|a_m(x) - a_n(x)| < \varepsilon$$

if $m, n > N$.

We show the necessity first. Suppose $a_n(x)$ is uniformly convergent to $a(x)$. For any given $\varepsilon > 0$, there exists an $N(\varepsilon)$, such that

$$|a_m(x) - a(x)| < \frac{\varepsilon}{2}, \quad |a_n(x) - a(x)| < \frac{\varepsilon}{2}$$

if $m, n > N$. Thus,

$$|a_m(x) - a_n(x)| \leqslant |a_m(x) - a(x)| + |a(x) - a_n(x)| < \frac{\varepsilon}{2} + \frac{\varepsilon}{2} = \varepsilon.$$

Now we show the sufficiency. The sequence $a_n(x)$ is convergent at any point x by the Cauchy criterion for number sequences. Let the limit function be $a(x)$. For any $\varepsilon > 0$, there exists an $N(\varepsilon)$ such that

$$|a_m(x) - a_n(x)| < \varepsilon$$

if $m, n > N$. Let $n \to \infty$ while m be fixed. Since $a_n(x) \to a(x)$, we have

$$|a_m(x) - a(x)| < \varepsilon$$

if $m > N$. This implies that the function sequence is uniformly convergent.

We also have the following conclusion:

The limit of an uniformly convergent sequence of continuous functions is also a continuous function.

Indeed, since the sequence of continuous functions $a_n(x)$ uniformly converges to $a(x)$, for any given $\varepsilon > 0$, there exists an $N(\varepsilon)$, such that

$$|a_n(x) - a(x)| < \frac{\varepsilon}{3}$$

if $n > N(\varepsilon)$. This N is independent of x.

Consider

$$|a(x + h) - a(x)| \leqslant |a_n(x + h) - a(x + h)|$$
$$+ |a_n(x) - a(x)| + |a_n(x + h) - a_n(x)|$$
$$< \frac{2\varepsilon}{3} + |a_n(x + h) - a_n(x)|.$$

Since $a_n(x)$ is continuous, for any $\varepsilon > 0$, there exists a $\delta > 0$ such that

$$|a_n(x + h) - a_n(x)| < \frac{\varepsilon}{3}$$

if $|h| < \delta$. Thus, for any given $\varepsilon > 0$, there exists a $\delta > 0$ such that

$$|a(x + h) - a(x)| < \varepsilon.$$

if $|h| < \delta$. This implies that $a(x)$ is a continuous function.

Example 1. The sequence of functions

$$a_n(x) = \frac{1}{1 + nx}, \quad 0 \leqslant x \leqslant 1, \quad n = 1, 2, \cdots$$

is convergent. The limit of this sequence is

$$a(x) = \begin{cases} 0, & \text{if } 0 < x \leqslant 1; \\ 1, & \text{if } x = 0. \end{cases}$$

Since $a(x)$ is not continuous, the series is not uniformly convergent on $[0, 1]$.

Example 2. The sequence of functions

$$a_n(x) = \frac{1}{n+x}, \quad 0 \leqslant x \leqslant 1, \quad n = 1, 2, \cdots$$

is convergent. The limit of this sequence is $a(x) = 0$.
Since

$$|a_n(x) - a(x)| = \frac{1}{n+x} < \frac{1}{n}$$

is true for all x in $[0, 1]$, for any $\varepsilon > 0$, let $N = \left[\frac{1}{\varepsilon}\right]$, then we have

$$|a_n(x) - a(x)| = \frac{1}{n} < \varepsilon$$

if $n > N$ for all x in $[0, 1]$. Obviously N is independent of x. Therefore, $a_n(x)$ is uniformly convergent on $[0, 1]$.

We also have the following conclusions:

If $a_n(x)$ is continuous on $[a, b]$, and uniformly converges to $a(x)$, then

$$\int_a^b a(x)\, dx = \lim_{n \to \infty} \int_a^b a_n(x)\, dx.$$

That is

$$\int_a^b \lim_{n \to \infty} a_n(x)\, dx = \lim_{n \to \infty} \int_a^b a_n(x)\, dx.$$

In other words, the order of the limit and the integral can be exchanged.

In fact, for any $\varepsilon > 0$, there exists a $N(\varepsilon)$ such that

$$|a_n(x) - a(x)| < \varepsilon$$

if $n > N$. Thus,

$$\left| \int_a^b a(x)\, dx - \int_a^b a_n(x)\, dx \right| \leqslant \int_a^b |a(x) - a_n(x)|\, dx < \varepsilon(b-a),$$

and the conclusion follows.

Example 3. Recall Example 2 in Section 10.2.1, the nth partial sum of the series is

$$S_n(x) = 2n^2 x e^{-n^2 x^2}, \quad 0 \leqslant x \leqslant 1.$$

Obviously, the limit of $S_n(x)$ is $S(x) = 0$, but the difference

$$\left| S_n\left(\frac{1}{n}\right) - S\left(\frac{1}{n}\right) \right| = 2n e^{-1}$$

does not tend to zero as $n \to \infty$. Thus, $S_n(x)$ does not converge uniformly to $S(x)$, and that is why:

$$1 = \lim_{n\to\infty} \int_0^1 2n^2 x e^{-n^2 x^2}\, dx \neq \int_0^1 S(x)\, dx = 0.$$

Now we study the condition of termwise differentiable sequence of functions.

If $a_n(x)$ is a sequence of differentiable functions that is convergent on $[a, b]$, and the sequence of derivatives $a'_n(x)$ is continuous and uniformly convergent on $[a, b]$, then the limit $a(x)$ of $a_n(x)$ is differentiable and

$$a'(x) = \lim_{n\to\infty} a'_n(x).$$

That is

$$\frac{d}{dx} \lim_{n\to\infty} a_n(x) = \lim_{n\to\infty} \frac{d}{dx} a_n(x).$$

In other words, the order of the limit and the derivative can be exchanged.

First, we show that $a_n(x)$ is uniformly convergent on $[a, b]$.

Since

$$a_n(x) = a_n(a) + \int_a^x a'_n(x)\, dx,$$

we have

$$|a_m(x) - a_n(x)| \leqslant |a_m(a) - a_n(a)| + \int_a^x |u'_m(x) - u'_n(x)|\, dx.$$

Since $a'_n(x)$ is uniformly convergent on $[a, b]$, for any given $\varepsilon > 0$, there exists an N_1 such that

$$|a'_m(x) - a'_n(x)| < \frac{\varepsilon}{2(b-a)}, \quad (a \leqslant x \leqslant b)$$

if $m, n > N_1$.

Since the sequence $a_n(x)$ converges at $x = a$. there exists an N_2 such that

$$|a_m(a) - a_n(a)| < \frac{\varepsilon}{2}$$

if $m, n > N_2$. Thus

$$|a_m(x) - a_n(x)| < \frac{\varepsilon}{2} + \frac{\varepsilon}{2(b-a)}(x-a) \leqslant \frac{\varepsilon}{2} + \frac{\varepsilon}{2} = \varepsilon$$

if $m, n > N = \max(N_1, N_2)$. This implies that $a_n(x)$ is uniformly convergent on $[a, b]$.

Suppose $b(x) = \lim_{n\to\infty} a'_n(x)$. By the result that the order of the operation of limit and the operation of integral can be exchanged, we have:

$$\int_a^x b(x)\,dx = \int_a^x \lim_{n\to\infty} a'_n(x)\,dx = \lim_{n\to\infty} \int_a^x a'_n(x)\,dx$$

$$= \lim_{n\to\infty} (a_n(x) - a_n(a)) = a(x) - a(a).$$

Differentiating both sides of above equation, we get

$$b(x) = \frac{d}{dx}a(x).$$

That is

$$a'(x) = \lim_{n\to\infty} a'_n(x).$$

10.2.3   *Uniformly Convergent Function Series*

Suppose a function series

$$u_1(x) + u_2(x) + \cdots + u_n(x) + \cdots$$

converges on $[a, b]$, that is, the sequence of partial sum

$$S_n(x) = \sum_{k=1}^n u_k(x)$$

converges to the sum function $S(x)$. If $S_n(x)$ uniformly converges to $S(x)$, then we say that the function series $\sum_{n=1}^{\infty} u_n(x)$ is uniformly convergent on $[a, b]$.

Since function series and function sequences can be converted to each other, the results of the last section can be restated as follows.

1. *Cauchy Criterion*: A function series $\sum_{n=1}^{\infty} u_n(x)$ is uniformly convergent on $[a, b]$ if and only if for any given $\varepsilon > 0$, there exists an N which is independent of x, such that

$$|u_{n+1}(x) + u_{n+2}(x) + \cdots + u_{n+l}(x)| < \varepsilon$$

if $n > N$, for all $l > 0$.

2. The sum function of a uniformly convergent series of continuous functions is also a continuous function.

3. A uniformly convergent series of continuous functions is a termwise integrable series. In other words, the order of the sum and the integration can be exchanged. That is:

$$\int_a^b \left(\sum_{n=1}^\infty u_n(x) \right) dx = \sum_{n=1}^\infty \int_a^b u_n(x)\, dx.$$

4. If every term of a function series is differentiable, every derivative function is continuous, and the series of derivative functions is uniformly convergent, then the series is termwise differentiable. That is, if a functional series of $\sum_{n=1}^\infty u_n(x)$ is convergent, the derivative function $u_n'(x)$ of each term is continuous and the function series $\sum_{n=1}^\infty u_n'(x)$ is uniformly convergent, then

$$\frac{d}{dx}\left(\sum_{n=1}^\infty u_n(x) \right) = \sum_{n=1}^\infty \frac{d}{dx} u_n(x) = \sum_{n=1}^\infty u_n'(x).$$

In other words, the order of the derivative $\dfrac{d}{dx}$ and the sum $\sum$ can be exchanged.

Now we study some criteria about how to determine whether a function series is uniformly convergent. These criteria are similar to the criteria in Sections 10.1.2 and 10.1.3.

Weierstrass Criterion: Suppose $\sum_{n=1}^\infty a_n$ is a convergent series of positive terms. If for any n that is large enough, and $a \leqslant x \leqslant b$

$$|u_n(x)| \leqslant a_n,$$

then the series $\sum_{n=1}^\infty u_n(x)$ is uniformly convergent on $[a, b]$.

Indeed, by Cauchy criterion for number series, for any given $\varepsilon > 0$, there exists an $N(\varepsilon)$ such that

$$a_{n+1} + a_{n+2} + \cdots + a_m < \varepsilon$$

for $m > n > N(\varepsilon)$. Thus,

$$|S_m(x) - S_n(x)| = |u_{n+1}(x) + u_{n+2}(x) + \cdots + u_m(x)|$$
$$< a_{n+1} + a_{n+2} + \cdots + a_m < \varepsilon.$$

By Cauchy criterion for function series, $\sum_{n=1}^{\infty} u_n(x)$ is uniformly convergent on $[a, b]$.

This result can be generalized to:

If $\sum_{n=1}^{\infty} v_n(x)$ is uniformly convergent on $[a, b]$, $v_n(x) \geqslant 0$, and

$$|u_n(x)| \leqslant v_n(x)$$

on $[a, b]$, then $\sum_{n=1}^{\infty} u_n(x)$ is also uniformly convergent on $[a, b]$.

Cauchy criterion for number series can be generalized to:

If there exists a constant $r(< 1)$ that is independent of x, such that

$$|u_n(x)|^{\frac{1}{n}} < r$$

on $[a, b]$ for n large enough, then $\sum_{n=1}^{\infty} u_n(x)$ is uniformly convergent on $[a, b]$.

Ratio test (d'Alembert criterion) for number series can be generalized to:

Suppose $u_n(x)$ is uniformly bounded. That is, there exists a constant M independent of x, such that

$$|u_n(x)| \leqslant M, \quad (n = 1, 2, \cdots).$$

If there exists a constant $r(< 1)$ independent of x, such that

$$\left| \frac{u_{n+1}(x)}{u_n(x)} \right| < r$$

on $[a, b]$ for n large enough, then $\sum_{n=1}^{\infty} u_n(x)$ is uniformly convergent on $[a, b]$.

Dirichlet criterion and Abel criterion for number series (Section 10.1.3) can be generalized to:

Dirichlet Criterion: Suppose the partial sum $B_n(x)$ of a function series $\sum_{n=1}^{\infty} b_n(x)$ is bounded on $[a, b]$. That is, there exists a constant M independent of x and n, such that

$$|B_n(x)| \leqslant M$$

at any point x in $[a, b]$. Moreover, suppose $a_n(x)$ is monotonically decreasing at any point x on $[a, b]$, and $a_n(x)$ tends to zero uniformly on $[a, b]$ as $n \to \infty$, then the series

$$\sum_{n=1}^{\infty} a_n(x)b_n(x)$$

is uniformly convergent on $[a, b]$.

Abel Criterion: Suppose $\sum_{n=1}^{\infty} b_n(x)$ is uniformly convergent on $[a, b]$, and $a_n(x)$ is a monotone sequence for every x. If there exists a constant K independent of n and x, such that

$$|a_n(x)| \leqslant K$$

for any point x in $[a, b]$, then the series

$$\sum_{n=1}^{\infty} a_n(x)b_n(x)$$

is uniformly convergent on $[a, b]$.

The proofs of these criteria are similar to the proofs of the corresponding criteria for number series.

Example: If a_n is a monotonically decreasing sequence and it tends to zero, then the series

$$\sum_{n=1}^{\infty} a_n \sin nx, \quad \sum_{n=1}^{\infty} a_n \cos nx$$

are uniformly convergent on any closed interval which does not include the points $2k\pi$, $(k = 0, \pm 1, \pm 2, \cdots)$.

Indeed, we have

$$\left| \sum_{n=1}^{\infty} \sin kx \right| = \left| \frac{\cos \frac{1}{2}x - \cos \left(n + \frac{1}{2} \right)x}{2 \sin \frac{1}{2}x} \right| \leqslant \frac{1}{\left| \sin \frac{x}{2} \right|}.$$

Since $\sin \frac{1}{2}x \neq 0$ on the interval above, there exists an upper bound that is independent of n and x. Thus, the series is uniformly convergent by Dirichlet criterion.

Now we can use the theory of series to prove some of the theorems that we have stated in earlier chapters.

10.2.4 *Existence Theorem of Implicit Functions*

The existence theorem of implicit functions was used in Section 6.2.2 without proof. Now we prove this theorem by the iteration method (successive approximation method) and uniform convergence.

The existence theorem of implicit functions: If a function $F(x, y)$ and $F'_y(x, y) = \dfrac{\partial}{\partial y} F(x, y)$ are continuous in the square

$$D : |x - x_0| \leqslant \varepsilon, \quad |y - y_0| \leqslant \varepsilon,$$

and

$$F(x_0, y_0) = 0, \quad F'_y(x_0, y_0) \neq 0,$$

then the implicit function

$$F(x, y) = 0$$

has a solution near (x_0, y_0). That is, there is a continuous function

$$y = f(x),$$

such that $y_0 = f(x_0)$ and

$$F(x, f(x)) = 0.$$

If $F'_x(x, y) = \dfrac{\partial}{\partial x} F(x, y)$ is also continuous in the square D, then the derivative of $y = f(x)$ exists and is continuous on the interval $(x_0 - \delta, x_0 + \delta)$, and

$$f'(x) = -\frac{\dfrac{\partial F}{\partial x}}{\dfrac{\partial F}{\partial y}}.$$

Proof: We rewrite the equation $F(x, y) = 0$ as

$$y = y_0 + \left[y - y_0 - \frac{F(x, y)}{F'_y(x_0, y_0)} \right] = y_0 + \varphi(x, y).$$

Then $\varphi(x_0, y_0) = 0$ and $\varphi'_y(x_0, y_0) = 0$.

Since φ and φ'_y are continuous in D, for any $0 < \lambda < 1$, we can choose ε small enough, such that

$$|\varphi'_y(x, y)| < \lambda$$

in D. Also, we can choose δ small enough such that

$$|\varphi(x, y_0)| < (1 - \lambda)\varepsilon$$

for $|x - x_0| \leqslant \delta$ $(\delta \leqslant \varepsilon)$.

Consider the region

$$D^* : |x - x_0| \leqslant \delta, \quad |y - y_0| \leqslant \varepsilon.$$

Substituting y with y_0 in the equation $y = y_0 + \varphi(x, y)$, we get a function of x

$$y_1 = y_1(x) = y_0 + \varphi(x, y_0).$$

Substituting y with y_1 in the equation $y = y_0 + \varphi(x, y)$, again we get a function of x

$$y_2 = y_2(x) = y_0 + \varphi(x, y_1).$$

Continue this process

$$y_3 = y_3(x) = y_0 + \varphi(x, y_2),$$

$$\cdots$$

$$y_n = y_n(x) = y_0 + \varphi(x, y_{n-1}),$$

$$\cdots$$

we obtain a sequence of functions

$$y_1(x), y_2(x), \cdots, y_n(x), \cdots.$$

First, we show that the range of $y_n(x)$ is $|y - y_0| < \varepsilon$ for every n by induction. Obviously,

$$|y_1 - y_0| = |\varphi(x, y_0)| < (1 - \lambda)\varepsilon < \varepsilon.$$

Suppose

$$|y_{n-1} - y_0| \leqslant \varepsilon.$$

Since

$$y_n - y_0 = \varphi(x, y_{n-1}),$$

and

$$|\varphi(x, y_{n-1})| \leqslant |\varphi(x, y_{n-1}) - \varphi(x, y_0)| + |\varphi(x, y_0)|.$$

By the Mean-Value Theorem, there exists a ξ between y_0 and y_{n-1} such that

$$|\varphi(x, y_{n-1}) - \varphi(x, y_0)| = |\varphi'_y(x, \xi)||y_{n-1} - y_0| < \lambda\varepsilon.$$

Also,

$$|\varphi(x, y_0)| < (1 - \lambda)\varepsilon.$$

Hence,

$$|y_n - y_0| < \lambda\varepsilon + (1 - \lambda)\varepsilon = \varepsilon.$$

Therefore, the conclusion follows.

Next, we show that $y_n(x)$ is a continuous function of x by induction. Obviously, $y_1(x)$ is continuous. Suppose $y_{n-1}(x)$ is continuous. Then the difference between $y_{n-1}(x)$ and $y_{n-1}(x')$ can be arbitrarily small when x and x' are sufficiently close. Since $\varphi(x, y)$ is a continuous function, the difference

$$y_n(x) - y_n(x') = \varphi(x, y_{n-1}(x)) - \varphi(x', y'_{n-1}(x'))$$

can be arbitrarily small. It follows that $y_n(x)$ is also continuous.

Now we study the convergence of the sequence

$$y_1, y_2, \cdots, y_n, \cdots.$$

Consider the series

$$y_0 + \sum_{n=1}^{\infty} (y_n - y_{n-1}).$$

By the Mean-Value Theorem,

$$|y_n - y_{n-1}| = |\varphi(x, y_{n-1}) - \varphi(x, y_{n-2})| < \lambda|y_{n-1} - y_{n-2}|.$$

Apply the Mean-Value Theorem multiple times, we have

$$|y_n - y_{n-1}| < \lambda|y_{n-1} - y_{n-2}|$$
$$< \lambda^2|y_{n-2} - y_{n-3}| < \cdots < \lambda^{n-1}(1 - \lambda)\varepsilon.$$

Since the geometric series

$$(1 - \lambda)\varepsilon \sum_{n=1}^{\infty} \lambda^{n-1}$$

is convergent, the series

$$y_0 + \sum_{n=1}^{\infty} (y_n - y_{n-1})$$

is uniformly convergent in $|x - x_0| \leqslant \delta$. Thus,

$$y = y(x) = \lim_{n \to \infty} y_n(x)$$

is continuous in $|x - x_0| \leqslant \delta$.

Since

$$y_n = y_0 + \varphi(x, y_{n-1}),$$

the function

$$y = y(x) = \lim_{n \to \infty} y_n(x)$$

is a solution of

$$y = y_0 + \varphi(x, y),$$

and it is also a solution of

$$F(x, y) = 0.$$

This solution is unique. In fact, if there exists another solution $\widetilde{y}$, then

$$\widetilde{y} = y_0 + \varphi(x, \widetilde{y})$$

Subtracting the equation

$$y = y_0 + \varphi(x, y)$$

we have

$$|y - \widetilde{y}| = |\varphi(x, y) - \varphi(x, \widetilde{y})| < \lambda |y - \widetilde{y}|.$$

It is impossible that $y \neq \widetilde{y}$ since $\lambda < 1$. Therefore the solution is unique.

Now we show that $y'(x)$ exists and is continuous.

By the assumption, $F(x, y)$ has continuous partial derivatives on

$$D^* : |x - x_0| \leqslant \delta, |y - y_0| \leqslant \varepsilon.$$

Thus,

$$\Delta F = F(x + \Delta x, y + \Delta y) - F(x, y) = \frac{\partial F}{\partial x} \Delta x + \frac{\partial F}{\partial y} \Delta y + o(\rho),$$

where $\rho = \sqrt{\Delta x^2 + \Delta y^2}$.

Suppose x and $x + \Delta x$ are in the interval $(x_0 - \delta, x_0 + \delta)$. Let

$$y = f(x), \quad y + \Delta y = f(x + \Delta x),$$

or

$$\Delta y = f(x + \Delta x) - f(x).$$

Obviously,

$$F(x + \Delta x, y + \Delta y) = 0.$$

It follows that

$$\Delta F = \frac{\partial F}{\partial x} \Delta x + \frac{\partial F}{\partial y} \Delta y + o(\rho) = 0.$$

Thus

$$\frac{\partial F}{\partial x}\Delta x + \frac{\partial F}{\partial y}\Delta y = o(\rho) = o(\sqrt{\Delta x^2 + \Delta y^2}),$$

or

$$\frac{\partial F}{\partial x} + \frac{\partial F}{\partial y}\frac{\Delta y}{\Delta x} = o\left(\sqrt{1 + \left(\frac{\Delta y}{\Delta x}\right)^2}\right) = o\left(1 + \left|\frac{\Delta y}{\Delta x}\right|\right).$$

Rewrite the above equation as

$$\frac{\partial F}{\partial x} + \frac{\partial F}{\partial y}\frac{\Delta y}{\Delta x} = \mu\left(1 \pm \frac{\Delta y}{\Delta x}\right),$$

where $\mu \to 0$ as $\rho \to 0$. Hence

$$\frac{\Delta y}{\Delta x} = -\frac{\dfrac{\partial F}{\partial x} - \mu}{\dfrac{\partial F}{\partial y} \mp \mu}.$$

Since $y = f(x)$ is continuous on the interval $(x_0 - \delta, x_0 + \delta)$, we have $\Delta y \to 0$ as $\Delta x \to 0$. It follows that $\rho \to 0$ and $\mu \to 0$.

However, $\dfrac{\partial F}{\partial x}$ and $\dfrac{\partial F}{\partial y}$ do not change as $\Delta x \to 0$. Therefore,

$$\lim_{\Delta x \to 0} \frac{\Delta y}{\Delta x} = f'(x) = -\frac{\dfrac{\partial F}{\partial x}}{\dfrac{\partial F}{\partial y}}.$$

This proves the existence theorem of implicit functions.

10.2.5 *Existence and Uniqueness Theorem of the Solution of Ordinary Differential Equations*

We proved the existence theorem of implicit functions by the iteration method (successive approximation method) and uniform convergence. In this section we prove the existence and uniqueness theorem of the solution of ordinary differential equations in a similar way.

The initial value problem of the second order homogeneous linear ordinary differential equations can be stated as follows:

$$\begin{cases} y'' + p(x)y' + q(x)y = 0, \\ y(x_0) = \alpha, \quad y'(x_0) = \beta. \end{cases} \tag{1}$$

In general, the solution of the equation can not be written as a simple expression except in some trivial cases. We will prove the existence and uniqueness of the solution of problem (1). And the process of the proof is the process of finding the solution.

Let $z = \dfrac{dy}{dx}$. Then (1) can be transformed into an equivalent problem

$$\begin{cases} \dfrac{dz}{dx} = -p(x)z - q(x)y, \\ \dfrac{dy}{dx} = z, \\ y(x_0) = \alpha, \\ z(x_0) = \beta. \end{cases}$$

More generally, we can consider the following problem

$$\begin{cases} \dfrac{dy_1}{dx} = a_{11}(x)y_1 + a_{12}(x)y_2, \\ \dfrac{dy_2}{dx} = a_{21}(x)y_1 + a_{22}(x)y_2, \\ y_1(x_0) = \alpha, \\ y_2(x_0) = \beta, \end{cases} \tag{2}$$

where $y_1(x)$, $y_2(x)$ are unknown functions, and a_{ij} are continuous on an interval $[a, b]$ that contains x_0.

Since $a_{ij}(x)$ are continuous on $[a, b]$, we can assume that

$$|a_{ij}(x)| \leqslant M, \quad i = 1, 2; \quad j = 1, 2.$$

on $[a, b]$. We need to show that the solution $y_1(x)$, $y_2(x)$ of problem (2) exists and is unique.

Integrating the system of differential equations (2) from x_0 to x, and using the initial conditions, we get that

$$\begin{cases} y_1(x) = \alpha + \displaystyle\int_{x_0}^{x} [a_{11}(t)y_1(t) + a_{12}(t)y_2(t)] \, dt, \\ y_2(x) = \beta + \displaystyle\int_{x_0}^{x} [a_{21}(t)y_1(t) + a_{22}(t)y_2(t)] \, dt. \end{cases} \tag{3}$$

The system (3) is equivalent to system (2). In fact, if $y_1(x)$, $y_2(x)$ satisfy (3), let $x = x_0$, then we have

$$y_1(x_0) = \alpha, \quad y_2(x_0) = \beta.$$

Differentiating system (3), we get the differential equations in (2). We now apply the iteration method on (3). Let

$$y_1^{(0)}(x) = \alpha, \quad y_2^{(0)}(x) = \beta.$$

Substituting $y_1^{(0)}(x)$, $y_2^{(0)}(x)$ for $y_1(x)$, $y_2(x)$ into the right hand side of (3), we get

$$y_1^{(1)}(x) = \alpha + \int_{x_0}^x [a_{11}(t)\alpha + a_{12}(t)\beta]\, dt,$$

$$y_2^{(1)}(x) = \beta + \int_{x_0}^x [a_{21}(t)\alpha + a_{22}(t)\beta]\, dt.$$

Substituting $y_1^{(1)}(x)$, $y_2^{(1)}(x)$ for $y_1(x)$, $y_2(x)$ into the right hand side of (3), we get

$$y_1^{(2)}(x) = \alpha + \int_{x_0}^x [a_{11}(t)y_1^{(1)} + a_{12}(t)y_2^{(1)}]\, dt,$$

$$y_2^{(2)}(x) = \beta + \int_{x_0}^x [a_{21}(t)y_1^{(1)} + a_{22}(t)y_2^{(1)}]\, dt.$$

Continuing this process of iteration, assume we already have $y_1^{(n)}$ and $y_2^{(n)}$, then we have

$$y_1^{(n+1)}(x) = \alpha + \int_{x_0}^x [a_{11}(t)y_1^{(n)} + a_{12}(t)y_2^{(n)}]\, dt,$$

$$y_2^{(n+1)}(x) = \beta + \int_{x_0}^x [a_{21}(t)y_1^{(n)} + a_{22}(t)y_2^{(n)}]\, dt.$$

Thus, we obtain two sequences of functions

$$y_1^{(0)}(x),\ y_1^{(1)}(x),\ y_1^{(2)}(x),\ \cdots,\ y_1^{(n)}(x),\ \cdots,$$

$$y_2^{(0)}(x),\ y_2^{(1)}(x),\ y_2^{(2)}(x),\ \cdots,\ y_2^{(n)}(x),\ \cdots.$$

All functions are continuous on $[a, b]$.

Now we show that the limit functions

$$y_1(x) = \lim_{n\to\infty} y_1^{(n)}(x),\quad y_2(x) = \lim_{n\to\infty} y_2^{(n)}(x)$$

exist, are continuous on $[a, b]$, and satisfy (3).

Consider the series

$$y_1^{(0)} + \sum_{n=0}^{\infty}(y_1^{(n+1)} - y_1^{(n)}) = \lim_{n\to\infty} y_1^{(n)}(x),$$

$$y_2^{(0)} + \sum_{n=0}^{\infty}(y_2^{(n+1)} - y_2^{(n)}) = \lim_{n\to\infty} y_2^{(n)}(x).$$

If these two series are uniformly convergent on $[a, b]$, then $y_1(x)$ and $y_2(x)$ are continuous.

To prove this, we let $A = \max\{|\alpha|, |\beta|\}$. Then

$$\left|y_1^{(1)} - y_1^{(0)}\right| = \left|\int_{x_0}^x [a_{11}(t)\alpha + a_{12}(t)\beta]\, dt\right|$$

$$\leqslant \left|\int_{x_0}^x [|a_{11}(t)\alpha| + |a_{12}(t)\beta|]\, dt\right|$$

$$\leqslant \left|\int_{x_0}^x 2MA\, dt\right| = 2MA|x - x_0|,$$

$$\left|y_2^{(1)} - y_2^{(0)}\right| = \left|\int_{x_0}^x [a_{21}(t)\alpha + a_{22}(t)\beta]\, dt\right|$$

$$\leqslant \left|\int_{x_0}^x [|a_{21}(t)\alpha| + |a_{22}(t)\beta|]\, dt\right|$$

$$\leqslant \left|\int_{x_0}^x 2MA\, dt\right| = 2MA|x - x_0|.$$

From these inequalities, we get

$$\left|y_1^{(2)} - y_1^{(1)}\right| = \left|\int_{x_0}^x [a_{11}(t)(y_1^{(1)} - y_1^{(0)}) + a_{12}(t)(y_2^{(1)} - y_2^{(0)})]\, dt\right|$$

$$\leqslant \left|\int_{x_0}^x [|a_{11}(t)|\,|y_1^{(1)} - y_1^{(0)}| + |a_{12}(t)|\,|y_2^{(1)} - y_2^{(0)}|]\, dt\right|$$

$$\leqslant \left|\int_{x_0}^x (2M)^2 A(t - x_0)\, dt\right| = \frac{(2M)^2 A}{2!}|x - x_0|^2,$$

$$\left|y_2^{(2)} - y_2^{(1)}\right| = \left|\int_{x_0}^x [a_{21}(t)(y_1^{(1)} - y_1^{(0)}) + a_{22}(t)(y_2^{(1)} - y_2^{(0)})]\, dt\right|$$

$$\leqslant \left|\int_{x_0}^x [|a_{21}(t)|\,|y_1^{(1)} - y_1^{(0)}| + |a_{22}(t)|\,|y_2^{(1)} - y_2^{(0)}|]\, dt\right|$$

$$\leqslant \left|\int_{x_0}^x (2M)^2 A(t - x_0)\, dt\right| = \frac{(2M)^2 A}{2!}|x - x_0|^2.$$

From these inequalities, we get

$$\left|y_1^{(3)} - y_1^{(2)}\right| \leqslant \frac{(2M)^3 A}{3!}|x - x_0|^3,$$

$$\left|y_2^{(3)} - y_2^{(2)}\right| \leqslant \frac{(2M)^3 A}{3!}|x - x_0|^3.$$

Repeating the process, we get that

$$\left|y_1^{(n+1)} - y_1^{(n)}\right| \leqslant \frac{(2M)^{n+1} A}{(n+1)!}|x - x_0|^{n+1} \leqslant A\frac{(2M)^{n+1}}{(n+1)!}(b - a)^{n+1},$$

$$\left|y_2^{(n+1)} - y_2^{(n)}\right| \leqslant \frac{(2M)^{n+1} A}{(n+1)!}|x - x_0|^{n+1} \leqslant A\frac{(2M)^{n+1}}{(n+1)!}(b - a)^{n+1}.$$

The series

$$A + \sum_{n=0}^{\infty} A \frac{(2M)^{n+1}}{(n+1)!} (b-a)^{n+1} = A e^{2M(b-a)}$$

is convergent. By Weierstrass criterion, the series

$$y_1^{(0)} + \sum_{n=0}^{\infty} (y_1^{(n+1)} - y_1^{(n)})$$

and

$$y_2^{(0)} + \sum_{n=0}^{\infty} (y_2^{(n+1)} - y_2^{(n)})$$

are uniformly convergent on $[a, b]$. It follows that, $y_1(x) = \lim_{n \to \infty} y_1^{(n)}(x)$ and $y_2(x) = \lim_{n \to \infty} y_2^{(n)}(x)$ are continuous functions on $[a, b]$.

Since the series is uniformly convergent, the order of the limit and the integration in the equations

$$y_1^{(n+1)}(x) = \alpha + \int_{x_0}^{x} [a_{11}(t)y_1^{(n)} + a_{12}(t)y_2^{(n)}]\, dt,$$

$$y_2^{(n+1)}(x) = \beta + \int_{x_0}^{x} [a_{21}(t)y_1^{(n)} + a_{22}(t)y_2^{(n)}]\, dt,$$

can be exchanged. Thus, we have

$$y_1(x) = \alpha + \int_{x_0}^{x} [a_{11}(t)y_1 + a_{12}(t)y_2]\, dt,$$

$$y_2(x) = \beta + \int_{x_0}^{x} [a_{21}(t)y_1 + a_{22}(t)y_2]\, dt.$$

This implies that $y_1(x)$ and $y_2(x)$ are solutions of (3).

The uniqueness of the solution:

Suppose there is another pair of continuous functions $z_1(x)$, $z_2(x)$ that is the solution. Let

$$u_1 = y_1 - z_1, \quad u_2 = y_2 - z_2.$$

Then

$$u_1 = \int_{x_0}^{x} [a_{11}(t)u_1 + a_{12}(t)u_2]\, dt,$$

$$u_2 = \int_{x_0}^{x} [a_{21}(t)u_1 + a_{22}(t)u_2]\, dt.$$

Since u_1 and u_2 are continuous on $[a, b]$, there exists an N such that

$$|u_1(x)| \leqslant N, \quad |u_2(x)| \leqslant N.$$

Thus,

$$\begin{aligned}
|u_1(x)| &= \left| \int_{x_0}^{x} [a_{11}(t)u_1 + a_{12}(t)u_2]\, dt \right| \\
&\leqslant \left| \int_{x_0}^{x} 2MN\, dt \right| = 2MN|x - x_0|,
\end{aligned}$$

$$\begin{aligned}
|u_2(x)| &= \left| \int_{x_0}^{x} [a_{21}(t)u_1 + a_{22}(t)u_2]\, dt \right| \\
&\leqslant \left| \int_{x_0}^{x} 2MN\, dt \right| = 2MN|x - x_0|.
\end{aligned}$$

It follows that

$$\begin{aligned}
|u_1(x)| &= \left| \int_{x_0}^{x} [a_{11}(t)u_1 + a_{12}(t)u_2]\, dt \right| \\
&\leqslant \left| \int_{x_0}^{x} (2M)^2 N(t - x_0)\, dt \right| = \frac{(2M)^2}{2!} N|x - x_0|^2,
\end{aligned}$$

$$\begin{aligned}
|u_2(x)| &= \left| \int_{x_0}^{x} [a_{21}(t)u_1 + a_{22}(t)u_2]\, dt \right| \\
&\leqslant \left| \int_{x_0}^{x} (2M)^2 N(t - x_0)\, dt \right| = \frac{(2M)^2}{2!} N|x - x_0|^2.
\end{aligned}$$

Repeating the process, we obtain

$$|u_1(x)| \leqslant \frac{(2M)^n}{n!} N|x - x_0|^n,$$
$$|u_2(x)| \leqslant \frac{(2M)^n}{n!} N|x - x_0|^n.$$

Since

$$\lim_{n \to \infty} \frac{(2M)^n}{n!} N|x - x_0|^n = 0,$$

we have $u_1 = 0$, $u_2 = 0$. This implies that $y_1 = z_1$, $y_2 = z_2$, and the uniqueness follows.

The statement of the *existence and uniqueness theorem of the solution for the system of ordinary differential equations*: If functions $a_{ij}(x)$, $(i, j =$

1, 2) are continuous on an interval $[a, b]$ that contains x_0, then the solution $y_1 = y_1(x)$, $y_2 = y_2(x)$ of the initial value problem

$$\begin{cases} \dfrac{dy_1}{dx} = a_{11}(x)y_1 + a_{12}(x)y_2, \\[2mm] \dfrac{dy_2}{dx} = a_{21}(x)y_1 + a_{22}(x)y_2, \\[2mm] y_1(x_0) = \alpha, \quad y_2(x_0) = \beta. \end{cases}$$

exists and is unique on $[a, b]$.

The existence and uniqueness of the solution of the above initial value problem implies the existence and uniqueness of the solution of the initial value problem of second order homogeneous linear ordinary differential equations.

Exercises 10.2

1. Determine the convergent regions for the following series of functions.

(1) $\displaystyle\sum_{n=1}^{\infty} \frac{n}{x^n}$;

(5) $\displaystyle\sum_{n=1}^{\infty} \frac{x^n}{1 + x^{2n}}$;

(2) $\displaystyle\sum_{n=1}^{\infty} ne^{-nx}$;

(6) $\displaystyle\sum_{n=1}^{\infty} \frac{(n + x)^n}{n^{n+x}}$;

(3) $\displaystyle\sum_{n=1}^{\infty} \frac{n-1}{n+1} \left(\frac{x}{3x + 1} \right)^n$;

(7) $\displaystyle\sum_{n=1}^{\infty} \frac{x^n}{n + y^n}$, $(y \geqslant 0)$;

(4) $\displaystyle\sum_{n=1}^{\infty} \left[\frac{x(x + n)}{n} \right]^n$;

(8) $\displaystyle\sum_{n=1}^{\infty} \frac{x^n y^n}{x^n + y^n}$, $(x > 0, y > 0)$.

2. Determine whether the following sequences of functions are uniformly convergent on the given intervals.

(1) $f_n(x) = x^n$ on (i) $0 \leqslant x \leqslant \dfrac{1}{2}$, and (ii) $0 \leqslant x \leqslant 1$;

(2) $f_n(x) = x^n - x^{n-1}$ on $0 \leqslant x \leqslant 1$;

(3) $f_n(x) = \dfrac{nx}{1 + n + x}$ on $0 \leqslant x \leqslant 1$;

(4) $f_n(x) = \dfrac{2nx}{1 + n^2 x^2}$ on (i) $0 \leqslant x \leqslant 1$ and (ii) $0 < x < +\infty$.

3. Determine whether the following series are uniformly convergent on the given intervals.

(1) $\displaystyle\sum_{n=1}^{\infty} \frac{\sin nx}{n^2}$ on $(-\infty < x < +\infty)$;

(2) $\displaystyle\sum_{n=1}^{\infty} \frac{1}{(x+n)(x+n+1)}$ on $(0 < x < +\infty)$;

(3) $\displaystyle\sum_{n=1}^{\infty} (-1)^{n-1} x^n$ on (i) $\left(-\dfrac{1}{2} \leqslant x \leqslant \dfrac{1}{2}\right)$ and (ii) $(-1 < x < 1)$;

(4) $\displaystyle\sum_{n=1}^{\infty} x^n e^{-nx}$ on $(0 \leqslant x < +\infty)$;

(5) $\displaystyle\sum_{n=1}^{\infty} \frac{(-1)^n}{x+2n}$ on $(-2 < x < +\infty)$;

(6) $\displaystyle\sum_{n=1}^{\infty} \frac{x}{1+n^4 x^2}$ on $(0 \leqslant x < +\infty)$;

(7) $\displaystyle\sum_{n=1}^{\infty} \frac{nx}{1+n^5 x^2}$ on $(-\infty < x < +\infty)$;

(8) $\displaystyle\sum_{n=1}^{\infty} \frac{n^2}{\sqrt{n!}}(x^n + x^{-n})$ on $\left(-\dfrac{1}{2} \leqslant |x| \leqslant \dfrac{1}{2}\right)$;

(9) $\displaystyle\sum_{n=1}^{\infty} 2^n \sin \frac{1}{3^n x}$ on $(0 < x < +\infty)$;

(10) $\displaystyle\sum_{n=1}^{\infty} \frac{(-1)^n}{n+\sin x}$ on $(0 \leqslant x \leqslant 2\pi)$;

(11) $\displaystyle\sum_{n=1}^{\infty} \frac{(-1)^n}{x+n}$ on $(0 < x < +\infty)$;

(12) $\displaystyle\sum_{n=1}^{\infty} \ln\left(1 + \frac{x}{n\ln^2 n}\right)$ on $|x| < a, a > 0$.

4. Determine the domain of the following functions and study their continuity.

(1) $f(x) = \displaystyle\sum_{n=1}^{\infty} \frac{x}{(1+x^2)^n}$; (2) $f(x) = \displaystyle\sum_{n=1}^{\infty} \frac{x+n(-1)^n}{x^2+n^2}$.

5. Suppose

$$f(x) = \sum_{n=1}^{\infty} \frac{x^n \cos \dfrac{n\pi}{x}}{(1+2x)^n}.$$

Find $\displaystyle\lim_{x \to 1} f(x)$ and $\displaystyle\lim_{x \to +\infty} f(x)$.

6. Prove that $f(x) = \sum_{n=1}^{\infty} ne^{-nx}$ is a continuous function for $x > 0$.

7. Prove that $f(x) = \sum_{n=1}^{\infty} \dfrac{\sin nx}{n^4}$ has continuous second-order derivative on $(-\infty, +\infty)$ and calculate it.

8. Suppose $f(x) = \sum_{n=1}^{\infty} ne^{-nx}$. Evaluate $\displaystyle\int_{\ln 2}^{\ln 3} f(x)\, dx$.

9. Suppose $f(x) = \sum_{n=1}^{\infty} \dfrac{\cos nx}{n^3}$. Evaluate $\displaystyle\int_{0}^{\pi} f(x)\, dx$.

10.3   Power Series and Taylor Series

10.3.1   *Convergence Radius of Power Series*

Power series is the simplest function series, also, it is the most important in applications. Since any partial sum of a power series is a polynomial, a convergent power series can be approximated by polynomials. It can be considered as polynomials of infinite degrees. They share a lot of similar properties.

For the power series

$$\sum_{n=1}^{\infty} a_n x^n = a_0 + a_1 x + \cdots + a_n x^n + \cdots$$

we have the following convergence theorem.

Abel Theorem: If the power series $\displaystyle\sum_{n=1}^{\infty} a_n x^n$ is convergent at a point $x = x_0$ ($x_0 \neq 0$), then it is absolutely convergent in the interval $|x| < |x_0|$.

Proof: For any point x in the interval $(-|x_0|, |x_0|)$, since $\displaystyle\sum_{n=1}^{\infty} a_n x^n$ is convergent, there exists an N such that $|a_n x_0^n| < 1$ if $n > N$. Thus

$$|a_n x^n| = |a_n x_0^n| \left(\frac{|x|}{|x_0|} \right)^n < \left(\frac{|x|}{|x_0|} \right)^n$$

if $n > N$.

By the assumption $\dfrac{|x|}{|x_0|} < 1$, the geometric series $\displaystyle\sum_{n=1}^{\infty} \left(\frac{|x|}{|x_0|} \right)^n$ is convergent. Thus, $\displaystyle\sum_{n=1}^{\infty} a_n x^n$ is convergent by the comparison criterion.

On the other hand, if the power series $\sum\limits_{n=1}^{\infty} a_n x^n$ is divergent at a point $x = x_0'$ $(x_0' \neq 0)$, then it is divergent at all points x that satisfy $|x| > |x_0|$.

Suppose the contrary. If there exists a point $|x_1| > |x_0'|$ such that the series is convergent at x_1, then the series must be convergent at $x = x_0'$ by Abel theorem, and this is a contradiction. Therefore, it is possible that a power series is convergent only at $x = 0$; it is possible that a power series is convergent on $(-\infty, +\infty)$; it is also possible that a power series is divergent at a finite point x_0'.

By Abel theorem, there exists a number R such that the series is convergent in $|x| < R$ and is divergent on $|x| > R$. The length of R does not exceed $|x_0'|$. This length of R is called the *radius of convergence* (convergence radius) of the series.

For the series $\sum\limits_{n=1}^{\infty} |a_n x^n|$, if the limit

$$\lim_{n \to \infty} \left| \frac{a_{n+1}}{a_n} \right| = l$$

exists, then

$$\lim_{n \to \infty} \left| \frac{a_{n+1} x^{n+1}}{a_n x^n} \right| = \lim_{n \to \infty} \left| \frac{a_{n+1}}{a_n} \right| |x| = l|x|.$$

By d'Alembert criterion, if $l|x| < 1$, that is, $|x| < \dfrac{1}{l}$, then the series $\sum\limits_{n=1}^{\infty} a_n x^n$ is absolutely convergent; if $l|x| > 1$, that is, $|x| > \dfrac{1}{l}$, then the series $\sum\limits_{n=1}^{\infty} a_n x^n$ is divergent. Thus,

$$R = \frac{1}{l} = \lim_{n \to \infty} \left| \frac{a_n}{a_{n+1}} \right|$$

is the convergence radius of the series $\sum\limits_{n=1}^{\infty} a_n x^n$.

Similarly, by Cauchy criterion, if the limit $\lim\limits_{n \to \infty} \sqrt[n]{|a_n|}$ exists, then

$$R = \frac{1}{\lim\limits_{n \to \infty} \sqrt[n]{|a_n|}}$$

is the convergence radius of the power series $\sum\limits_{n=0}^{\infty} a_n x^n$.

Example 1. The convergence radius of the series $\sum\limits_{n=0}^{\infty} x^n$ is 1 because

$$R = \frac{1}{l} = \lim_{n \to \infty} \left| \frac{a_n}{a_{n+1}} \right| = 1.$$

Example 2. Find the convergence radius of the series $\sum\limits_{n=0}^{\infty} n(2x)^n$.

Solution:

$$R = \frac{1}{\lim\limits_{n \to \infty} \sqrt[n]{|a_n|}} = \frac{1}{\lim\limits_{n \to \infty} 2\sqrt[n]{n}} = \frac{1}{2}.$$

Example 3. The convergence radius of the series $\sum\limits_{n=1}^{\infty} \frac{x^n}{n!}$ is infinity because

$$R = \lim_{n \to \infty} \left| \frac{a_n}{a_{n+1}} \right| = \lim_{n \to \infty} \frac{\dfrac{1}{n!}}{\dfrac{1}{(n+1)!}} = \infty.$$

Example 4. The convergence radius of the series $\sum\limits_{n=1}^{\infty} (nx)^n$ is 0, because

$$R = \frac{1}{\lim\limits_{n \to \infty} \sqrt[n]{|a_n|}} = \lim_{n \to \infty} \frac{1}{n} = 0.$$

If the limits $\lim\limits_{n \to \infty} \left| \dfrac{a_n}{a_{n+1}} \right|$ and $\lim\limits_{n \to \infty} \sqrt[n]{|a_n|}$ do not exist, then we can not say that the convergence radius of the series does not exist, we can only say that it can not be determined by these two methods.

Example 5. Find the convergence radius of the series $\sum\limits_{n=1}^{\infty} 2^n x^{2n}$.

Solution: In this power series, the coefficients of x with even powers are non-zero, and the coefficients of x with odd powers are zero. That is

$$a_{2n} = 2^n, \quad a_{2n+1} = 0.$$

The two limits above do not exists. Applying the d'Alembert criterion, we have that

$$\lim_{n \to \infty} \frac{2^{n+1}|x|^{2(n+1)}}{2^n |x|^{2n}} = 2|x|^2.$$

Thus, the series is convergent if $2|x|^2 < 1$ $\left(|x| < \dfrac{1}{\sqrt{2}}\right)$, and the series is divergent if $2|x|^2 > 1$ $\left(|x| > \dfrac{1}{\sqrt{2}}\right)$. Therefore, the convergence radius of the series is $\dfrac{1}{\sqrt{2}}$.

Remark: There is no certain conclusion about the convergence of the series at the two end points $x = -R$ and $x = R$ of the convergence interval. It depends on the actual series. For instance, the interval of convergence of the series

$$(1) \ \sum_{n=1}^{\infty} \frac{x^n}{n}, \quad (2) \ \sum_{n=1}^{\infty} \frac{x^n}{n^2}, \quad (3) \ \sum_{n=1}^{\infty} n x^n$$

are the same, they all converge on $(-1, 1)$, but series (1) is conditionally convergent at $x = -1$ and is divergent at $x = 1$. Series (2) is absolutely convergent at both end points. Series (3) is divergent at both end points.

10.3.2 *Properties of Power Series*

The theory of the function series in Section 10.2 can be applied to the power series.

Suppose the convergence radius of the series $\sum_{n=1}^{\infty} a_n x^n$ is R. It determines a sum function $S(x)$ in the interval $(-R, R)$, but the series is not necessary uniformly convergent on $(-R, R)$. For instance, the series

$$\sum_{n=0}^{\infty} x^n = 1 + x + x^2 + \cdots + x^n + \cdots$$

is not uniformly convergent in its interval of convergence. However, we have the following result:

Suppose the radius of convergence of the series $\sum_{n=0}^{\infty} a_n x^n$ is R. Then the series is uniformly convergent on $[-r, r]$ for any r that satisfies $0 < r < R$.
    Indeed, we have

$$|a_n x^n| = |a_n||x|^n \leqslant |a_n| r^n$$

for x in $[-r, r]$. Since $\sum_{n=1}^{\infty} |a_n| r^n$ is convergent, by Weierstrass criterion, the series is uniformly convergent on $[-r, r]$.

By the fact that power functions are continuous functions, we have the following result:

Suppose the convergence interval of the series $\sum\limits_{n=1}^{\infty} a_n x^n$ is $(-R, R)$. Then the sum function $S(x)$ of the series is continuous on $(-R, R)$.

To prove this conclusion, let x_0 be an arbitrary point in $(-R, R)$. Then $|x_0| < R$. The series is uniformly convergent on $[-r, r]$ for any number r between x_0 and R. Thus $S(x)$ is continuous at x_0. Thus, $S(x)$ is continuous on $(-R, R)$ since x_0 is an arbitrary point in $(-R, R)$.

Similarly, by the uniform convergence, derivatives of $S(x)$ of any order exist in the convergence interval. We have the following result.

Suppose the convergence radius of the series $\sum\limits_{n=0}^{\infty} a_n x^n$ is R. Then the sum function $S(x)$ is differentiable in the interval $(-R, R)$, and the derivative of $S(x)$ can be obtained by taking the derivative of the series termwise. That is

$$S'(x) = \sum_{n=1}^{\infty} n a_n x^{n-1}.$$

The radius of convergence of the series $S'(x)$ is also R.

First, we show that radius of convergence of the series $\sum\limits_{n=1}^{\infty} n a_n x^{n-1}$ is R.

Let x_1 be a point in $(-R, R)$. Then $|x_1| < R$. If r is a number that satisfies $|x_1| < r < R$, then the series $\sum\limits_{n=1}^{\infty} a_n r^n$ is convergent. It follows that $a_n r^{n-1}$ tends to zero as n tends to infinity. Thus, $|a_n r^{n-1}| < 1$ if n is sufficiently large. Hence,

$$\left| n a_n x_1^{n-1} \right| = \left| n a_n r^{n-1} \left(\frac{x_1}{r} \right)^{n-1} \right| < n \left(\frac{x_1}{r} \right)^{n-1}.$$

The series $\sum\limits_{n=1}^{\infty} n \left| \frac{x_1}{r} \right|^{n-1}$ is convergent because $\left| \frac{x_1}{r} \right| < 1$. Therefore $\sum\limits_{n=1}^{\infty} n |a_n x_1^{n-1}|$ is convergent by the comparison criterion.

Let x_2 be a point outside the interval $(-R, R)$. Then $|x_2| > R$. Let r' be a number that satisfies $|x_2| > r' > R$. Then we have

$$n |a_n r'^{n-1}| \geqslant |a_n r'^{n-1}| = \frac{1}{r'} |a_n r'^n|.$$

Since $\sum\limits_{n=1}^{\infty} a_n r'^n$ is divergent, $\sum\limits_{n=1}^{\infty} |a_n r'^n|$ is also divergent. Thus

$\sum\limits_{n=1}^{\infty} n|a_n r'^{n-1}|$ is divergent by comparison criterion. Hence $\sum\limits_{n=1}^{\infty} n a_n x_2^{n-1}$

is divergent (otherwise $\sum\limits_{n=1}^{\infty} n|a_n r'^{n-1}|$ would be convergent).

This shows that the convergence radius of the series $\sum\limits_{n=1}^{\infty} n a_n x^{n-1}$ is R. It follows that the series is uniformly convergent on the closed interval $[-r, r]$ (where $r < R$). Therefore, the order of the derivative and the sum can be exchanged, and

$$\frac{d}{dx} S(x) = \sum_{n=1}^{\infty} n a_n x^{n-1}$$

is true on $[-r, r]$. Since $|r - R|$ can be arbitrarily small, the above equation is also true on $(-R, R)$.

This conclusion can be applied repeatedly to obtain the next result:

Suppose the radius of convergence of the series $\sum\limits_{n=0}^{\infty} a_n x^n$ is R. Then the derivatives of any order of the sum function $S(x)$ exist in the interval $(-R, R)$, and

$$\frac{d^k}{dx^k} S(x) = \sum_{n=k}^{\infty} n(n-1) \cdots (n-k+1) a_n x^{n-k}, \quad (k = 1, 2, \cdots).$$

The convergence radius of this series is also R.

The next result is for the integral of the power series.

Suppose the convergence radius of the series $\sum\limits_{n=0}^{\infty} a_n x^n$ is R, and the sum function is $S(x)$. Then for any point x in $(-R, R)$, we have

$$\int_0^x S(x)\, dx = \int_0^x \left(\sum_{n=0}^{\infty} a_n x^n \right) dx = \sum_{n=0}^{\infty} \frac{a_n}{n+1} x^{n+1}.$$

The radius of convergence of this series is also R.

To show this result, suppose $x > 0$. Since the power series $\sum\limits_{n=0}^{\infty} a_n x^n$ is uniformly convergent on $[0, x]$, it is a termwise integrable series:

$$\int_0^x \left(\sum_{n=0}^{\infty} a_n x^n \right) dx = \sum_{n=0}^{\infty} \int_0^x a_n x^n\, dx = \sum_{n=0}^{\infty} \frac{a_n}{n+1} x^{n+1}.$$

Let the convergence radius of the series $\sum_{n=0}^{\infty} \frac{a_n}{n+1} x^{n+1}$ be R'. The deriva-

tive of $\sum_{n=0}^{\infty} \frac{a_n}{n+1} x^{n+1}$ is $\sum_{n=0}^{\infty} a_n x^n$ and its convergence radius is also R'. By the assumption, we have $R' = R$.

By these theorems, we can calculate sum functions for some power series, and we can also expand some elementary functions to power series.

Example 1. Find the sum function of the series $\sum_{n=0}^{\infty} \frac{x^n}{n!}$.

Solution: By Example 3 of Section 10.3.1, the convergence radius of the series $\sum_{n=0}^{\infty} \frac{x^n}{n!}$ is ∞. Suppose the sum function is $S(x)$. Differentiating the series termwise, we get that

$$S'(x) = \sum_{n=1}^{\infty} \frac{x^{n-1}}{(n-1)!} = \sum_{n=0}^{\infty} \frac{x^n}{n!} = S(x).$$

It is the differential equation

$$\frac{dS(x)}{dx} = S(x),$$

and the solution is

$$x = \ln S(x) + C.$$

Since $S(0) = 1$, let $x = 0$ in the above equation we get $C = 0$. Thus,

$$S(x) = e^x.$$

Therefore, we obtain the power series expansion of e^x:

$$e^x = 1 + x + \frac{x^2}{2!} + \cdots + \frac{x^n}{n!} + \cdots$$

for any real number x ($-\infty < x < \infty$).

Example 2. Find the power series expansion of $\ln(1+x)$ and $\arctan x$.

Solution: For x in $(-1, 1)$, we have

$$\frac{1}{1+x} = \sum_{n=0}^{\infty} (-1)^n x^n.$$

Integrating the series termwise from 0 to x, we obtain

$$\ln(1+x) = \sum_{n=0}^{\infty} \frac{(-1)^n}{n+1} x^{n+1} = \sum_{n=1}^{\infty} \frac{(-1)^{n-1}}{n} x^n,$$

where $-1 < x < 1$, and this is the power series expansion of $\ln(1 + x)$.

On the other hand, Integrating

$$\frac{1}{1 + x^2} = \sum_{n=0}^{\infty} (-1)^n x^{2n}$$

termwise, we get

$$\arctan x = \sum_{n=0}^{\infty} \frac{(-1)^n}{2n + 1} x^{2n+1},$$

where $-1 < x < 1$.

The following theorem is for the convergence of a power series $\sum_{n=0}^{\infty} a_n x^n$ at the end points $x = -R$ and $x = R$ of the convergence region.

Abel Theorem: Suppose the convergence radius of the series $\sum_{n=0}^{\infty} a_n x^n$ is R. If the series is convergent at $x = R$, then the sum function $S(x)$ is continuous on the left at $x = R$. That is

$$\lim_{x \to R^-} \sum_{n=0}^{\infty} a_n x^n = \sum_{n=0}^{\infty} a_n R^n.$$

Similarly, if the series is convergent at $x = -R$, then the sum function $S(x)$ is continuous on the right at $x = -R$. That is

$$\lim_{x \to -R^+} \sum_{n=0}^{\infty} a_n x^n = \sum_{n=0}^{\infty} a_n (-R)^n.$$

Proof: Suppose $\sum_{n=0}^{\infty} a_n x^n$ is convergent at $x = R$. It is uniformly convergent on $[0, R]$. In fact,

$$\sum_{n=0}^{\infty} a_n x^n = \sum_{n=0}^{\infty} a_n R^n \left(\frac{x}{R}\right)^n.$$

Since $\sum_{n=0}^{\infty} a_n R^n$ is convergent, and the sequence $\left(\frac{x}{R}\right)^n$ is monotone for every x in $[0, R]$, and is uniformly bounded:

$$\left(\frac{x}{R}\right)^n \leqslant 1, \quad (n = 1, 2, \cdots)$$

by the Abel criterion for function series in section 10.2, $\sum_{n=0}^{\infty} a_n x^n$ is uniformly convergent on $[0, R]$. It follows that $S(x)$ is continuous on the left at $x = R$. The other part of conclusion can be proved in a similar way.

In Example 2, the power series expansion of $\ln(1 + x)$, $\sum_{n=1}^{\infty}(-1)^n\dfrac{x^n}{n}$,
is convergent at the end point $x = 1$ of the convergence interval. The expansion of the series at this point is

$$1 - \frac{1}{2} + \frac{1}{3} - \frac{1}{4} + \cdots = \ln 2.$$

The power series expansion of the function $\arctan x$ is convergent at both end points $x = -1$ and $x = 1$ of the interval of convergence, and we get

$$1 - \frac{1}{3} + \frac{1}{5} - \frac{1}{7} + \cdots = \frac{\pi}{4}.$$

Now we consider power series in a generalized form

$$\sum_{n=0}^{\infty} a_n(x - a)^n,$$

where a is a fixed number. Performing the transformation $y = x - a$, this series becomes

$$\sum_{n=0}^{\infty} a_n y^n.$$

If the radius of convergence of this series is R, then the interval of convergence of $\sum_{n=0}^{\infty} a_n(x - a)^n$ is $(-R + a, R + a)$.

All conclusions about power series can be applied to this generalized form.

10.3.3   *Taylor Series*

In this section, we study the conditions under which a function can be expanded to a power series.

In Section 3.3, we showed that if the $(n + 1)$-order derivative of the function $y = f(x)$ exists at a point $x = x_0$, then the *Taylor expansion* of $f(x)$ at $x = x_0$ is:

$$f(x) = f(x_0) + f'(x_0)(x - x_0) + \frac{f''(x_0)}{2!}(x - x_0)^2 + \cdots$$

$$+ \frac{f'^{(n)}(x_0)}{n!}(x - x_0)^n + R_n(x),$$

where $R_n(x)$ is the *remainder of the Taylor expansion*, and

$$R_n(x) = \frac{f^{(n+1)}(\xi)}{(n + 1)!}(x - x_0)^{n+1},$$

where $\xi = x_0 + \theta(x - x_0)$, $0 < \theta < 1$.

Therefore, the condition for $f(x)$ to be able to be expanded to a power series in a neighborhood of x_0, $(x_0 - R, x_0 + R)$, is that there exist arbitrary order derivatives of $f(x)$ on $(x_0 - R, x_0 + R)$, and that $R_n(x)$ tends to zero as $n \to \infty$.

In other words, if there exist arbitrary order derivatives of $f(x)$ on $(x_0 - R, x_0 + R)$, then $f(x)$ can be expanded to a power series on $(x_0 - R, x_0 + R)$ if and only if

$$\lim_{n \to \infty} R_n(x) = 0.$$

Under this condition, the power series becomes

$$\sum_{n=0}^{\infty} \frac{f^{(n)}(x_0)}{n!} (x - x_0)^n.$$

This power series is called the *Taylor series* of $f(x)$ on $(x_0 - R, x_0 + R)$.

We must mention that, for a function $f(x)$ that has arbitrary order derivatives on $(x_0 - R, x_0 + R)$, there always exists the Taylor series

$$\sum_{n=0}^{\infty} \frac{f^{(n)}(x_0)}{n!} (x - x_0)^n.$$

However, this Taylor series is not necessarily convergent, or even if it is convergent, it does not necessarily converge to $f(x)$. The necessary and sufficient condition for this Taylor series to be convergent and to converge to $f(x)$ is

$$\lim_{n \to \infty} R_n(x) = 0.$$

For instance,

$$f(x) = \sum_{n=0}^{\infty} \frac{\sin(2^n x)}{n!}$$

has Taylor series that is divergent everywhere except at $x = 0$. Consider another function:

$$f(x) = \begin{cases} e^{-\frac{1}{x^2}}, & x \neq 0, \\ 0, & x = 0. \end{cases}$$

Since $f^{(n)}(0) = 0$, $(n = 1, 2, \cdots)$, although the Taylor series of this function is convergent, it is not convergent to $f(x)$ (except at $x = 0$).

A sufficient condition for the Taylor expansion of a function is the following:

If there exists a constant M, such that

$$|f^{(n)}(x)| \leqslant M$$

for every x in $(x_0 - R, x_0 + R)$ and all nature number n, then $f(x)$ can be expanded to a Taylor series in $(x_0 - R, x_0 + R)$.

We only need to show that, under the above assumption,

$$\lim_{n \to \infty} R_n(x) = 0$$

for x in $(x_0 - R, x_0 + R)$.

In fact,

$$|R_n(x)| = \left| \frac{f^{(n+1)}(x_0 + \theta(x - x_0))}{(n+1)!}(x - x_0)^{n+1} \right|$$

$$\leqslant \frac{M}{(n+1)!}|x - x_0|^{n+1} \leqslant M\frac{R^{n+1}}{(n+1)!}.$$

By Example 1 in Section 10.3.2, the series $\displaystyle\sum_{n=0}^{\infty} \frac{R^n}{n!}$ is convergent. It follows

that $\displaystyle\lim_{x \to \infty} \frac{R^{n+1}}{(n+1)!} = 0$. Thus

$$\lim_{x \to \infty} R_n(x) = 0$$

for x in $(x_0 - R, x_0 + R)$.

If $f(x)$ can be expanded to the Taylor series at $x = 0$,

$$f(0) + f'(0)x + \frac{f''(0)}{2!}x^2 + \cdots + \frac{f^{(n)}(0)}{n!}x^n + \cdots$$

then this power series is called the *Maclaurin series*.

The Maclaurin series of a function is a power series of x. Now we prove that, if $f(x)$ can be expanded to a power series, then the expansion is unique, and it is the Maclaurin series of $f(x)$.

In fact, if $f(x)$ can be expanded to a power series of x in a neighborhood of $x = 0$, $(-R, R)$, that is

$$f(x) = a_0 + a_1 x + a_2 x^2 + \cdots + a_n x^n + \cdots ,$$

then since the power series can be differentiated termwise in the region of convergence, we have

$$f'(x) = a_1 + 2a_2 x + \cdots + na_n x^{n-1} + \cdots ;$$
$$f''(x) = 2!a_2 + 3 \cdot 2a_3 x + \cdots + n(n-1)a_n x^{n-2} + \cdots ;$$
$$\cdots$$

$$f'^{(n)}(x) = n!a_n + (n+1)n(n-1)\cdots 2a_{n+1}x + \cdots .$$

Let $x = 0$. We get

$$a_0 = f(0), \quad a_1 = f'(0), \quad a_2 = \frac{f''(0)}{2!}, \cdots, \quad a_n = \frac{f^{(n)}(0)}{n!}, \cdots.$$

By the uniqueness of the power series expansion of a function, if $f(x)$ can be expanded to a power series of x, then this power series is the Maclaurin series of $f(x)$. On the other hand, if the Maclaurin series of $f(x)$ is convergent in a neighborhood of $x = 0$, it is not necessarily convergent to $f(x)$.

We learned in Section 10.3.2 that the function e^x can be expanded to the power series $\sum_{n=0}^{\infty} \frac{x^n}{n!}$.

By the above theorem, this result can be obtained in an easy way. We only need to show that $f^{(n)}(x)$ is bounded for all n when $|x| < R$, where R is an arbitrary number. It is obvious, since $f^{(n)}(x) = e^x$, we have $e^x < e^R$ when $|x| < R$. Thus, $\lim_{n \to \infty} R_n(x) = 0$. Therefore e^x can be expanded to the Maclaurin series $\sum_{n=0}^{\infty} \frac{x^n}{n!}$.

The Maclaurin series of $\sin x$ and $\cos x$ can be obtained in a similar way.

In fact,

$$(\sin x)^{(n)} = \sin(x + \frac{n\pi}{2}) \quad (n = 1, 2, \cdots).$$

Thus,

$$\frac{(\sin x)^{(n)}}{n!}\bigg|_{x=0} = \frac{1}{n!} \sin \frac{n\pi}{2} = \begin{cases} 0, & \text{if } n = 2k, \\ \dfrac{(-1)^k}{(2k+1)!}, & \text{if } n = 2k+1. \end{cases}$$

Since the inequality

$$|(\sin x)^{(n)}| = \left|\sin\left(x + \frac{n\pi}{2}\right)\right| \leqslant 1$$

is true for any x and n, we have

$$\sin x = \sum_{n=0}^{\infty} \frac{(-1)^k}{(2k+1)!} x^{2k+1} = x - \frac{x^3}{3!} + \frac{x^5}{5!} - \cdots$$

for all x.

The Maclaurin series of $\cos x$ can be obtained in a similar way:

$$\cos x = \sum_{n=0}^{\infty} \frac{(-1)^k}{(2k)!} x^{2k} = 1 - \frac{x^2}{2!} + \frac{x^4}{4!} - \cdots$$

for all x.

In order to obtain the Taylor expansion of the function $(1+x)^\alpha$ (where α is an arbitrary real number), we need to consider the remainder of the Taylor series in another form.

The remainder of the power series expansion of $f(x)$ at $x = x_0$ is

$$R_n(x) = f(x) - f(x_0) - \frac{f'(x_0)}{1!}(x - x_0) - \frac{f''(x_0)}{2!}(x - x_0)^2$$
$$- \cdots - \frac{f^{(n)}(x_0)}{n!}(x - x_0)^n.$$

Define an auxiliary function

$$\varphi(y) = f(x) - f(y) - \frac{f'(y)}{1!}(x - y) - \frac{f''(y)}{2!}(x - y)^2$$
$$- \cdots - \frac{f^{(n)}(y)}{n!}(x - y)^n,$$

and we have

$$\varphi(x_0) = R_n(x), \quad \varphi(x) = 0.$$

Since

$$\varphi'(y) = -f'(y) - \left[\frac{f''(y)}{1!}(x - y) - f'(y)\right]$$
$$- \left[\frac{f'''(y)}{2!}(x - y)^2 - \frac{f''(y)}{1!}(x - y)\right] - \cdots$$
$$- \left[\frac{f^{(n+1)}(y)}{n!}(x - y)^n - \frac{f^{(n)}(y)}{(n-1)!}(x - y)^{n-1}\right]$$
$$= -\frac{f^{(n+1)}(y)}{n!}(x - y)^n.$$

By the Mean-Value Theorem

$$\varphi(x) - \varphi(x_0) = (x - x_0)\varphi'(c),$$

where $c = x_0 + \theta(x - x_0)$, $0 < \theta < 1$. Thus,

$$R_n(x) = (x - x_0)\frac{f^{(n+1)}(x_0 + \theta(x - x_0))}{n!}(1 - \theta)^n(x - x_0)^n$$
$$= \frac{f^{(n+1)}(x_0 + \theta(x - x_0))}{n!}(1 - \theta)^n(x - x_0)^{n+1}.$$

Since

$$f^{(n)}(x) = \alpha(\alpha - 1)\cdots(\alpha - n + 1)(1 + x)^{\alpha - n},$$

we have

$$f^{(n)}(0) = \alpha(\alpha - 1) \cdots (\alpha - n + 1), \quad n = 1, 2, \cdots.$$

Thus, the Maclaurin series of $f(x) = (1 + x)^\alpha$ is

$$1 + \sum_{n=1}^{\infty} \frac{\alpha(\alpha - 1) \cdots (\alpha - n + 1)}{n!} x^n.$$

When α is a natural number or 0, this series only contains a finite number of terms, and is the binomial expansion.

If α is not a natural number or 0, then none of the terms in this series are 0. By the comparison criterion, the radius of convergence of this power series is 1. To show that this power series converges to $(1 + x)^\alpha$, we only need to show that $R_n(x) \to 0$ as $n \to \infty$. The remainder is

$$R_n(x) = \frac{f^{(n+1)}(\theta x)}{n!} (1 - \theta)^n x^{n+1}$$

$$= \frac{\alpha(\alpha - 1) \cdots (\alpha - n)(1 + \theta x)^{\alpha - n - 1}}{n!} (1 - \theta)^n x^{n+1}$$

$$= \left(\frac{(\alpha - 1)(\alpha - 2) \cdots [(\alpha - 1) - n + 1]}{n!} x^n \right)$$

$$\cdot (\alpha x (1 + \theta x)^{\alpha - 1}) \left(\frac{1 - \theta}{1 + \theta x} \right)^n.$$

The first factor on the right hand side is the general term of the Maclaurin series of $(1 + x)^{\alpha - 1}$. Since the Maclaurin series is convergent when $|x| < 1$, this factor tends to zero as $n \to \infty$. The second factor on the right hand side is bounded when $|x| < 1$.

Since $\dfrac{1 - \theta}{1 + \theta x} < 1$ when $|x| < 1$, the third factor on the right hand side is less than 1. Thus,

$$\lim_{n \to \infty} R_n(x) = 0.$$

Therefore,

$$(1 + x)^\alpha = 1 + \sum_{n=1}^{\infty} \frac{\alpha(\alpha - 1) \cdots (\alpha - n + 1)}{n!} x^n$$

for $|x| < 1$.

Especially, when $\alpha = -1$, $\alpha = \dfrac{1}{2}$, and $\alpha = -\dfrac{1}{2}$, we have

$$\frac{1}{1 + x} = 1 - x + x^2 - \cdots + (-1)^n x^n + \cdots,$$

where $-1 < x < 1$.

$$\sqrt{1+x} = 1 + \frac{1}{2}x - \frac{1}{2\cdot 4}x^2 + \frac{1\cdot 3}{2\cdot 4\cdot 6}x^3$$
$$- \cdots + (-1)^{n-1}\frac{(2n-3)!!}{(2n)!!}x^n + \cdots,$$

where $-1 \leqslant x \leqslant 1$.

$$\frac{1}{\sqrt{1+x}} = 1 - \frac{1}{2}x + \frac{1\cdot 3}{2\cdot 4}x^2 + \frac{1\cdot 3\cdot 5}{2\cdot 4\cdot 6}x^3$$
$$+ \cdots + (-1)^n\frac{(2n-1)!!}{(2n)!!}x^n + \cdots,$$

where $-1 < x \leqslant 1$.

It is easy to prove the following conclusion:

Suppose the convergence radius of the power series $\sum\limits_{n=0}^{\infty} a_n x^n$ is R_1 and the convergence radius of the power series $\sum\limits_{n=0}^{\infty} b_n x^n$ is R_2. Let $R = \min(R_1, R_2)$. Then in the interval $(-R, R)$, we have

$$\sum_{n=0}^{\infty} a_n x^n + \sum_{n=0}^{\infty} b_n x^n = \sum_{n=0}^{\infty} (a_n + b_n)x^n$$

and

$$\left(\sum_{n=0}^{\infty} a_n x^n\right)\left(\sum_{n=0}^{\infty} b_n x^n\right) = \sum_{n=0}^{\infty} c_n x^n,$$

where

$$c_n = \sum_{l=0}^{n} a_l b_{n-l} = a_0 b_n + a_1 b_{n-1} + \cdots + a_n b_0.$$

By this result, we can obtain the Taylor expansions for some functions. Consider the function $\dfrac{\ln(1-x)}{1-x}$. Since

$$\frac{1}{1-x} = \sum_{n=0}^{\infty} x^n, \quad \ln(1-x) = -\sum_{n=1}^{\infty} \frac{x^n}{n}$$

in $(-1, 1)$, by the multiplication rule for power series, we have

$$c_n = \sum_{l=0}^{n} a_l b_{n-l} = -\sum_{l=1}^{n} \frac{1}{l}.$$

Thus

$$\frac{\ln(1-x)}{1-x} = -\sum_{n=1}^{\infty} \left(1 + \frac{1}{2} + \cdots + \frac{1}{n}\right) x^n,$$

where $-1 < x < 1$.

The Maclaurin series of the following six elementary functions are often used:

$$e^x = \sum_{n=0}^{\infty} \frac{x^n}{n!}, \qquad (-\infty < x < \infty),$$

$$\sin x = \sum_{n=0}^{\infty} \frac{(-1)^n}{(2n+1)!} x^{2n+1}, \qquad (-\infty < x < \infty),$$

$$\cos x = \sum_{n=0}^{\infty} \frac{(-1)^n}{(2n)!} x^{2n}, \qquad (-\infty < x < \infty),$$

$$\ln(1+x) = \sum_{n=1}^{\infty} \frac{(-1)^{n-1}}{n} x^n, \qquad (-1 < x \leqslant 1),$$

$$\arctan x = \sum_{n=0}^{\infty} \frac{(-1)^n}{(2n+1)} x^{2n+1}, \qquad (-1 \leqslant x \leqslant 1),$$

$$(1+x)^\alpha = 1 + \sum_{n=1}^{\infty} \frac{\alpha(\alpha-1)\cdots(\alpha-n+1)}{n!} x^n, \qquad (-1 < x < 1).$$

10.3.4 *Applications of Power Series*

The infinite series is a powerful tool in mathematical analysis in the following studies:

(1) Proving the existence and uniqueness of the implicit function.

(2) Proving the existence and uniqueness of the solution of ordinary differential equations. Determining the solution function.

(3) Proving the existence and uniqueness of the solution of integral equations.

(4) Proving the existence and uniqueness of the solution of system of differential equations.

(5) Proving the existence and uniqueness of the solution of partial differential equations.

The infinite series is also a powerful tool in complex analysis (see another book by the author: "Concise Complex Analysis" (Revised Edition)).

In this section, we study some applications of power series:

(1) Approximate computation.

(2) Finding the approximative solutions of implicit functions.

(3) Finding the solutions for differential equations by power series.

Example 1. Find the approximative value of $\ln 2$ to within the accuracy of 10^{-4}.

Solution: In Section 10.3.2, we obtained

$$\ln 2 = 1 - \frac{1}{2} + \frac{1}{3} - \frac{1}{4} + \cdots + (-1)^{n-1}\frac{1}{n} + \cdots .$$

If we consider the partial sum of the first n terms, then the remainder satisfies

$$|r_n| < |a_{n+1}| = \frac{1}{n+1}.$$

We need to calculate at least $10,000$ terms in order to have $\dfrac{1}{n+1} < 10^4$.

There is a better way to do this calculation. Subtracting the following two equations

$$\ln(1+x) = x - \frac{x^2}{2} + \frac{x^3}{3} - \cdots + (-1)^{n-1}\frac{x^n}{n} + \cdots ,$$

$$\ln(1-x) = -x - \frac{x^2}{2} - \frac{x^3}{3} - \cdots - \frac{x^n}{n} - \cdots ,$$

we obtain that

$$\ln\frac{1+x}{1-x} = 2\left(x + \frac{1}{3}x^3 + \frac{1}{5}x^5 + \cdots + \frac{x^{2n-1}}{2n-1} + \cdots\right).$$

Let $x = \dfrac{1}{3}$. Then the above equation becomes

$$\ln 2 = 2\left[\frac{1}{3} + \frac{1}{3 \cdot 3^3} + \frac{1}{5 \cdot 3^5} + \frac{1}{7 \cdot 3^7} + \cdots\right].$$

The remainder of this series after the fourth term is

$$\frac{2}{9 \cdot 3^9} + \frac{2}{11 \cdot 3^{11}} + \frac{2}{13 \cdot 3^{13}} + \cdots$$

$$< \frac{2}{9 \cdot 3^9}\left(1 + \frac{1}{3^2} + \frac{1}{3^4} + \cdots\right) = \frac{1}{4 \cdot 3^9} = \frac{1}{78732} < 10^4.$$

Thus, we only need to calculate the first four terms.

$$\frac{1}{3} \approx 0.33333, \qquad \frac{1}{3 \cdot 3^3} \approx 0.01235,$$

$$\frac{1}{5 \cdot 3^5} \approx 0.00082, \qquad \frac{1}{7 \cdot 3^7} \approx 0.00007.$$

Therefore, $\ln 2 \approx 0.6931$ within the accuracy of 10^{-4}.

Let $x = \dfrac{1}{5}$. We can compute $\ln 3$. Let $x = \dfrac{1}{4}$. We can compute $\ln 5$. Also, since $\ln 4 = 2 \ln 2$, $\ln 6 = \ln 2 + \ln 3$, we can compile a table of logarithms this way.

Example 2. Compute the approximate value of the integral

$$\int_0^1 \frac{\sin x}{x}\, dx$$

to within the accuracy of 10^{-4}.

Solution: The Taylor series of the function $\dfrac{\sin x}{x}$ is

$$\frac{\sin x}{x} = 1 - \frac{x^2}{3!} + \frac{x^4}{5!} - \frac{x^6}{7!} + \cdots, \quad (-\infty < x < \infty).$$

Integrating this series termwise, we get

$$\int_0^1 \frac{\sin x}{x}\, dx = 1 - \frac{1}{3 \cdot 3!} + \frac{1}{5 \cdot 5!} - \frac{1}{7 \cdot 7!} + \cdots.$$

The difference is easy to estimate because the right hand side of the above equation is an alternating series. Since

$$\frac{1}{3 \cdot 3!} \approx 0.05556, \quad \frac{1}{5 \cdot 5!} \approx 0.00167,$$

$$\frac{1}{7 \cdot 7!} \approx 0.000028 < 10^{-4},$$

we only need to calculate the first three terms, that is

$$\int_0^1 \frac{\sin x}{x}\, dx \approx 1 - 0.05556 + 0.00167 \approx 0.9461.$$

In Section 8.1, we studied the Taylor expansion for functions of several variables. Following a similar discussion for functions of one variable, our readers can arrive at the necessary and sufficient condition for the existence of the Taylor series of functions of several variables.

Now we study how to solve for an implicit function by power series. Suppose

$$F(x, y) = 0$$

is an implicit function. It can be expanded to a power series with respect to $x - x_0$ and $y - y_0$ at the point (x_0, y_0). Since $F(x_0, y_0) = 0$ and $F_y'(x_0, y_0) \neq$

0, the implicit function $F(x, y) = 0$ can be written as (translator's note: we assume $x_0 = 0$, and $y_0 = 0$ here to simplify the calculation)

$$y = C_{10}x + C_{20}x^2 + C_{11}xy + C_{02}y^2 + C_{30}x^3$$
$$+ C_{21}x^2y + C_{12}xy^2 + C_{03}y^3 + \cdots .$$

If

$$y = a_1x + a_2x^2 + a_3x^3 + \cdots$$

satisfies $F(x, y) = 0$, then we have that

$$a_1x + a_2x^2 + a_3x^3 + \cdots$$
$$= C_{10}x + C_{20}x^2 + C_{11}x(a_1x + a_2x^2 + \cdots)$$
$$+ C_{02}(a_1x + a_2x^2 + \cdots)^2 + C_{30}x^3 + C_{21}x^2(a_1x + a_2x^2 + \cdots)$$
$$+ C_{12}x(a_1x + a_2x^2 + \cdots)^2 + C_{03}(a_1x + a_2x^2 + \cdots)^3 + \cdots .$$

Comparing the coefficients of corresponding terms on both side of the equation, we get

$$a_1 = C_{10}, \quad a_2 = C_{20} + C_{11}a_1 + C_{02}a_1^2,$$
$$a_3 = C_{11}a_2 + 2C_{02}a_1a_2 + C_{30} + C_{21}a_1 + C_{12}a_1^2 + C_{03}a_1^3,$$
$$\cdots .$$

and substituting earlier equations into later ones, we get

$$a_1 = C_{10}, \quad a_2 = C_{20} + C_{11}C_{10} + C_{02}C_{10}^2,$$
$$a_3 = (C_{11} + 2C_{02}C_{10})(C_{20} + C_{11}C_{10} + C_{02}C_{10}^2) + C_{30}$$
$$+ C_{21}C_{10} + C_{12}C_{10}^2 + C_{03}C_{10}^3 + \cdots ,$$
$$\cdots .$$

Thus, the solution is

$$y = a_1x + a_2x^2 + \cdots .$$

The power series is absolutely convergent at a neighborhood of $x = 0$. This method is called the *method of undetermined coefficients*.

Example 3. Find the approximate expression of the implicit function that is determined by the equation

$$xy - e^x + e^y = 0$$

in a neighborhood of the point $(0, 0)$.

Solution: Suppose the Taylor series of $y = y(x)$ at the point $(0,0)$ is

$$y(x) = \sum_{n=0}^{\infty} \frac{y^{(n)}(0)}{n!} x^n.$$

Taking the derivative of the equation, we get

$$y + xy' - e^x + e^y y' = 0.$$

Let $x = 0$ and $y(0) = 0$. we get $y'(0) = 1$. Continue the process, we have

$$y''(0) = -2, \quad y'''(0) = 12.$$

Thus,

$$y \approx y(0) + y'(0)x + \frac{y''(0)}{2!}x^2 + \frac{y'''(0)}{3!}x^3 = x - x^2 + 2x^3.$$

Sometimes, we can find the exact solution.

Example 4. Study the equation

$$y = a + x\varphi(y),$$

where $\varphi(y)$ can be expanded to a convergent power series in a neighborhood of the point $y = a$.

Solution: Taking derivatives of the equation with respect to x and a, we get

$$(1 - x\varphi'(y))\frac{\partial y}{\partial x} = \varphi(y), \quad (1 - x\varphi'(y))\frac{\partial y}{\partial a} = 1.$$

Thus,

$$\frac{\partial y}{\partial x} = \varphi(y)\frac{\partial y}{\partial a}.$$

By mathematical induction, we have that

$$\frac{\partial^n y}{\partial x^n} = \frac{\partial^{n-1}}{\partial a^{n-1}}\left(\varphi^n(y)\frac{\partial y}{\partial a}\right).$$

In fact, this equation is obviously true when $n = 1$. Assume the equation is true for $n = k$.

$$\frac{\partial^k y}{\partial x^k} = \frac{\partial^{k-1}}{\partial a^{k-1}}\left(\varphi^k(y)\frac{\partial y}{\partial a}\right).$$

Differentiating this equation with respect x, we get

$$\frac{\partial^{k+1}y}{\partial x^{k+1}} = \frac{\partial^{k-1}}{\partial a^{k-1}}\left(\varphi^k(y)\frac{\partial^2 y}{\partial a \partial x} + k\varphi^{k-1}(y)\frac{\partial\varphi}{\partial y}\frac{\partial y}{\partial x}\frac{\partial y}{\partial a}\right)$$

$$= \frac{\partial^{k-1}}{\partial a^{k-1}}\left(\varphi^k(y)\frac{\partial}{\partial a}\left(\varphi(y)\frac{\partial y}{\partial a}\right) + k\varphi^k(y)\frac{\partial\varphi}{\partial y}\left(\frac{\partial y}{\partial a}\right)^2\right)$$

$$= \frac{\partial^{k-1}}{\partial a^{k-1}}\left(\varphi^k(y)\frac{\partial\varphi}{\partial y}\left(\frac{\partial y}{\partial a}\right)^2 + \varphi^{k+1}(y)\frac{\partial^2 y}{\partial a^2} + k\varphi^k(y)\frac{\partial\varphi}{\partial y}\left(\frac{\partial y}{\partial a}\right)^2\right)$$

$$= \frac{\partial^{k-1}}{\partial a^{k-1}}\left((k+1)\varphi^k(y)\frac{\partial\varphi}{\partial y}\left(\frac{\partial y}{\partial a}\right)^2 + \varphi^{k+1}(y)\frac{\partial^2 y}{\partial a^2}\right)$$

$$= \frac{\partial^k}{\partial a^k}\left(\varphi^{k+1}(y)\frac{\partial y}{\partial a}\right).$$

Since $y = a$ when $x = 0$, we have

$$\frac{\partial^k y}{\partial x^k}\bigg|_{x=0} = \frac{\partial^{k-1}}{\partial a^{k-1}}(\varphi^k(a)).$$

Thus, the Taylor series of y is

$$y = a + x\varphi(a) + \frac{x^2}{2!}\frac{\partial}{\partial a}(\varphi^2(a)) + \cdots + \frac{x^n}{n!}\frac{\partial^{n-1}}{\partial a^{n-1}}(\varphi^n(a)) + \cdots.$$

This is called the *Lagrange series*.

Now we study how to solve differential equations by power series.

Consider a second-order homogeneous differential equation

$$y'' + p(x)y' + q(x)y = 0.$$

Assume that the power series of $p(x)$ and $q(x)$ are

$$p(x) = a_0 + a_1 x + \cdots + a_n x^n + \cdots = \sum_{k=0}^{\infty} a_k x^k,$$

$$q(x) = b_0 + b_1 x + \cdots + b_n x^n + \cdots = \sum_{k=0}^{\infty} b_k x^k.$$

If the solution of the equation is

$$y = \alpha_0 + \alpha_1 x + \cdots + \alpha_n x^n + \cdots = \sum_{k=0}^{\infty} \alpha_k x^k,$$

then substituting $p(x)$, $q(x)$ and y in the differential equation by the above three power series, we have

$$\sum_{k=2}^{\infty} k(k-1)\alpha_k x^{k-2} + \left(\sum_{k=0}^{\infty} a_k x^k\right)\left(\sum_{k=1}^{\infty} k\alpha_k x^{k-1}\right)$$

$$+ \left(\sum_{k=0}^{\infty} b_k x^k\right)\left(\sum_{k=0}^{\infty} \alpha_k x^k\right) = 0.$$

Comparing the coefficients of the terms, the coefficient of x^0 is

$$2 \cdot 1 \alpha_2 + a_0 \alpha_1 + b_0 \alpha_0 = 0,$$

the coefficient of x^1 is

$$3 \cdot 2 \alpha_3 + 2 a_0 \alpha_2 + a_1 \alpha_1 + b_0 \alpha_1 + b_1 \alpha_0 = 0,$$

the coefficient of x^2 is

$$4 \cdot 3 \alpha_4 + 3 a_0 \alpha_3 + 2 a_1 \alpha_2 + a_2 \alpha_1 + b_0 \alpha_2 + b_1 \alpha_1 + b_2 \alpha_0 = 0,$$

and so on.

If the values of α_0 and α_1 are given, then the values of $\alpha_2, \alpha_3, \alpha_4, \cdots$ can be found by solving above equations. Let $\alpha_0 = 1$ and $\alpha_1 = 0$, that is, $y(0) = 1$ and $y'(0) = 0$. Then we can obtain a solution y_1. Let $\alpha_0 = 0$ and $\alpha_1 = 1$, that is, $y(0) = 0$ and $y'(0) = 1$. Then we can obtain another solution y_2. Thus, the general solution of the differential equation is

$$y = C_1 y_1 + C_2 y_2,$$

where C_1 and C_2 are constants. This is the solution of the differential equation with initial conditions $y(0) = C_1$ and $y'(0) = C_2$.

If the convergence radii of the power series of $p(x)$ and $q(x)$ are the same value R, then the convergence radius of the solution is also R. We leave this proof to our readers.

Example 5. Consider the equation

$$y'' - xy = 0.$$

Suppose the solution of the equation is

$$y = \alpha_0 + \alpha_1 x + \cdots + \alpha_n x^n + \cdots = \sum_{k=0}^{\infty} \alpha_k x^k.$$

Then we have

$$(2 \cdot 1 \alpha_2 + 3 \cdot 2 \alpha_3 x + 4 \cdot 3 \alpha_4 x^2 + \cdots) - x(\alpha_0 + \alpha_1 x + \alpha_2 x^2 + \cdots) = 0.$$

Comparing the coefficients of corresponding terms, we get

$$2 \cdot 1 \alpha_2 = 0, \quad 3 \cdot 2 \alpha_3 - \alpha_0 = 0, \quad 4 \cdot 3 \alpha_4 - \alpha_1 = 0,$$

$$\cdots$$

$$(k+2)(k+1)\alpha_{k+2} - \alpha_{k-1} = 0,$$

$$\cdots .$$

Let $\alpha_0 = 1$, and $\alpha_1 = 0$. Then

$$\alpha_2 = 0, \quad \alpha_3 = \frac{1}{2 \cdot 3}, \quad \alpha_4 = \alpha_5 = 0, \quad \alpha_6 = \frac{1}{2 \cdot 3 \cdot 5 \cdot 6},$$

$$\alpha_7 = \alpha_8 = 0, \quad \alpha_9 = \frac{1}{2 \cdot 3 \cdot 5 \cdot 6 \cdot 8 \cdot 9},$$

$$\cdots .$$

In general,

$$\alpha_{3k+1} = \alpha_{3k+2} = 0, \quad \alpha_{3k} = \frac{1 \cdot 4 \cdot 7 \cdots (3k - 2)}{(3k)!}.$$

Thus, the solution is

$$y_1 = 1 + \sum_{k=1}^{\infty} \frac{1 \cdot 4 \cdot 7 \cdots (3k - 2)}{(3k)!} x^{3k}.$$

Let $\alpha_0 = 0$, and $\alpha_1 = 1$. We obtain another solution

$$y_2 = x + \sum_{k=1}^{\infty} \frac{2 \cdot 5 \cdot 8 \cdots (3k - 1)}{(3k + 1)!} x^{3k+1}.$$

By comparison criterion, y_1 and y_2 are absolutely convergent on $(-\infty, \infty)$. Since these two solutions are linearly independent, the general solution is

$$y = C_1 y_1 + C_2 y_2.$$

The same method can be applied to find the solution of the equation

$$f_0(x)y'' + f_1(x)y' + f_2(x)y = 0,$$

where $f_0(x) \neq 0$.

Example 6. Solve the equation

$$(1 - x^2)y'' - xy' + a^2 y = 0.$$

Suppose the solution of this equation is

$$y = \alpha_0 + \alpha_1 x + \cdots + \alpha_n x^n + \cdots = \sum_{k=0}^{\infty} \alpha_k x^k.$$

Then we have

$$(n + 2)(n + 1)\alpha_{n+2} - n(n - 1)\alpha_n - n\alpha_n + a^2 \alpha_n = 0.$$

That is

$$(n + 2)(n + 1)\alpha_{n+2} = (n^2 - a^2)\alpha_n.$$

Let $\alpha_0 = 1$, $\alpha_1 = 0$. We obtain the solution

$$y_1 = 1 - \frac{a^2}{2!}x^2 + \frac{a^2(a^2 - 4)}{4!}x^4 - \frac{a^2(a^2 - 4)(a^2 - 16)}{6!}x^6 + \cdots,$$

where $|x| < 1$.

Let $\alpha_0 = 0$, $\alpha_1 = 1$. We obtain the solution

$$y_2 = x - \frac{a^2 - 1}{3!}x^3 + \frac{(a^2 - 1)(a^2 - 9)}{5!}x^5$$
$$- \frac{(a^2 - 1)(a^2 - 9)(a^2 - 25)}{7!}x^7 + \cdots,$$

where $|x| < 1$.

Since these two solutions are linearly independent, the general solution of the equation is

$$y = C_1 y_1 + C_2 y_2.$$

We introduce a useful formula:

Stirling Formula:

$$n! = \left(\frac{n}{e}\right)^n \sqrt{2\pi n} e^{\frac{\theta_n}{12n}},$$

where $0 < \theta_n < 1$

This formula clarifies the order of $n!$.

First, we show the *Wallis Formula*:

$$\frac{\pi}{2} = \lim_{n\to\infty} \frac{1}{2n+1} \left[\frac{(2n)!!}{(2n-1)!!}\right]^2,$$

where

$$(2n)!! = 2n(2n - 2) \cdots 4 \cdot 2,$$
$$(2n - 1)!! = (2n - 1)(2n - 3) \cdots 3 \cdot 1.$$

In fact, for $0 < x < \frac{\pi}{2}$, we have

$$\sin^{2n+1} x < \sin^{2n} x < \sin^{2n-1} x,$$

thus

$$\int_0^{\frac{\pi}{2}} \sin^{2n+1} x \, dx < \int_0^{\frac{\pi}{2}} \sin^{2n} x \, dx < \int_0^{\frac{\pi}{2}} \sin^{2n-1} x \, dx.$$

It follows that

$$\frac{2n(2n-2)\cdots 4 \cdot 2}{(2n+1)(2n-1)\cdots 3 \cdot 1} < \frac{(2n-1)(2n-3)\cdots 3 \cdot 1}{2n(2n-2)\cdots 4 \cdot 2} \cdot \frac{\pi}{2}$$
$$< \frac{(2n-2)(2n-4)\cdots 4 \cdot 2}{(2n-1)(2n-3)\cdots 3 \cdot 1}.$$

(The calculation of the value of $\int_0^{\frac{\pi}{2}} \sin^n x \, dx$ can be found in Section 10.4.4.)

Hence,

$$\left(\frac{2n(2n-2)\cdots 4\cdot 2}{(2n+1)(2n-1)\cdots 3\cdot 1} \right)^2 \frac{1}{2n+1} < \frac{\pi}{2}$$

$$< \left(\frac{2n(2n-2)\cdots 4\cdot 2}{(2n+1)(2n-1)\cdots 3\cdot 1} \right)^2 \frac{1}{2n}.$$

Since $\lim\limits_{n\to\infty} \dfrac{2n}{2n+1} = 1$, the Wallis formula follows.

Now, we are ready to prove the Stirling formula.

Since

$$\ln(1+x) = x - \frac{x^2}{2} + \frac{x^3}{3} - \cdots + (-1)^{n-1}\frac{x^n}{n} + \cdots,$$

$$\ln(1-x) = -x - \frac{x^2}{2} - \frac{x^3}{3} - \cdots - \frac{x^n}{n} - \cdots,$$

we have

$$\ln\frac{1+x}{1-x} = 2x\left(1 + \frac{1}{3}x^2 + \frac{1}{5}x^4 + \cdots + \frac{1}{2n+1}x^{2n} + \cdots \right).$$

Let $x = \dfrac{1}{2n+1}$. Then $\dfrac{1+x}{1-x} = \dfrac{n+1}{n}$. Thus,

$$\ln\frac{n+1}{n} = \frac{2}{2n+1}\left(1 + \frac{1}{3}\frac{1}{(2n+1)^2} + \frac{1}{5}\frac{1}{(2n+1)^4} + \cdots \right).$$

That is

$$\left(n + \frac{1}{2} \right)\ln\left(1 + \frac{1}{n} \right) = 1 + \frac{1}{3}\frac{1}{(2n+1)^2} + \frac{1}{5}\frac{1}{(2n+1)^4} + \cdots.$$

The series on the right hand side is obviously greater than 1 and less than

$$1 + \frac{1}{3}\left(\frac{1}{(2n+1)^2} + \frac{1}{(2n+1)^4} + \cdots \right) = 1 + \frac{1}{12n(n+1)}.$$

Thus,

$$1 < \left(n + \frac{1}{2} \right)\ln\left(1 + \frac{1}{n} \right) < 1 + \frac{1}{12n(n+1)}.$$

Hence,

$$e < \left(1 + \frac{1}{n} \right)^{n+\frac{1}{2}} < e^{1 + \frac{1}{12n(n+1)}}.$$

Let
$$a_n = \frac{n!e^n}{n^{n+\frac{1}{2}}}.$$

Then
$$\frac{a_n}{a_{n+1}} = \frac{1}{e}\left(1 + \frac{1}{n}\right)^{n+\frac{1}{2}}.$$

Thus,
$$1 < \frac{a_n}{a_{n+1}} < e^{\frac{1}{12n(n+1)}}.$$

It follows that
$$a_n > a_{n+1}.$$

That is, a_n is a monotonically decreasing sequence and therefore has a limit a. We also have
$$a_n e^{-\frac{1}{12n}} < a_{n+1} e^{-\frac{1}{12(n+1)}}.$$

That is, $a_n e^{-\frac{1}{12n}}$ is a monotonically increasing sequence and therefore has a limit which is clearly also a. Hence,
$$a_n e^{-\frac{1}{12n}} < a < a_n.$$

Let
$$\theta_n = 12n \ln \frac{a_n}{a}.$$

Then $0 < \theta_n < 1$. Thus,
$$a_n = a e^{\frac{\theta_n}{12n}}.$$

By the definition of a_n, we have
$$n! = \left(\frac{n}{e}\right)^n \sqrt{n} a_n = \left(\frac{n}{e}\right)^n \sqrt{n} a e^{\frac{\theta_n}{12n}}.$$

Now we determine a by Wallis formula
$$\sqrt{\frac{\pi}{2}} = \lim_{n\to\infty} \frac{1}{\sqrt{2n+1}} \frac{(2n)!!}{(2n-1)!!}.$$

Since
$$\frac{(2n)!!}{(2n-1)!!} = \frac{[(2n)!!]^2}{(2n)!} = \frac{2^{2n}(n!)^2}{(2n)!},$$

and
$$n! = \left(\frac{n}{e}\right)^n \sqrt{n} a_n, \quad (2n)! = \left(\frac{2n}{e}\right)^{2n} \sqrt{2n} a_{2n},$$

we have

$$\frac{(2n)!!}{(2n-1)!!} = \frac{2^{2n}(n!)^2}{(2n)!} = \frac{2^{2n}\left(\dfrac{n}{e}\right)^{2n} na_n^2}{\left(\dfrac{2n}{e}\right)^{2n}\sqrt{2n}a_{2n}} = \sqrt{\frac{n}{2}}\frac{a_n^2}{a_{2n}}.$$

By Wallis formula,

$$\sqrt{\frac{\pi}{2}} = \lim_{n\to\infty}\frac{1}{\sqrt{2n+1}}\sqrt{\frac{n}{2}\frac{a_n^2}{a_{2n}}} = \frac{a}{2}.$$

Therefore,

$$a = \sqrt{2\pi}.$$

The Stirling formula follows.

The Stirling formula can also be written as

$$n! \sim \left(\frac{n}{e}\right)^n\sqrt{2\pi n}, (n\to\infty).$$

Example 7. Prove that

$$C_{2n}^n \sim \frac{1}{\sqrt{n\pi}}2^{2n}$$

as $n\to\infty$. Where C_n^m is a combinatorial number.

Proof: Since $C_{2n}^n = \dfrac{(2n)!}{(n!)^2}$, by Stirling formula,

$$(2n)! \sim \left(\frac{2n}{e}\right)^{2n}\sqrt{4\pi n}, \quad (n!)^2 \sim \left(\frac{n}{e}\right)^{2n}2\pi n.$$

Hence,

$$C_{2n}^n = \frac{(2n)!}{(n!)^2} \sim \frac{\left(\dfrac{2n}{e}\right)^{2n}\sqrt{4\pi n}}{\left(\dfrac{n}{e}\right)^{2n}2\pi n} = \frac{1}{\sqrt{n\pi}}2^{2n}.$$

Exercises 10.3

1. Find the radius of convergence for the following power series.

(1) $\displaystyle\sum_{n=0}^{\infty} \frac{x^n}{2^n}$;

(4) $\displaystyle\sum_{n=1}^{\infty} \frac{n^k}{n!}x^n$;

(2) $\displaystyle\sum_{n=1}^{\infty}(-1)^{n-1}\frac{x^n}{n^2}$;

(5) $\displaystyle\sum_{n=1}^{\infty} \frac{n!}{n^n}x^n$;

(3) $\displaystyle\sum_{n=1}^{\infty} \frac{x^n}{n+\sqrt{n}}$;

(6) $\displaystyle\sum_{n=1}^{\infty} \frac{(n!)^2}{(2n)!}x^n$.

2. Find the radius of convergence for the following power series, and determine the convergence or divergence of the power series at the end points of their convergence interval.

(1) $\displaystyle\sum_{n=1}^{\infty}\left(1+\frac{1}{n}\right)^{n^2}x^n$;

(4) $\displaystyle\sum_{n=1}^{\infty}\frac{3^n+(-2)^n}{n}(x+1)^n$;

(2) $\displaystyle\sum_{n=1}^{\infty} a^{n^2}x^n, \quad (0<a<1)$;

(5) $\displaystyle\sum_{n=1}^{\infty}\frac{x^n}{a^n+b^n}, \quad (a>0, b>0)$;

(3) $\displaystyle\sum_{n=1}^{\infty}\frac{n!}{a^{n^2}}x^n, \quad (a>1)$;

(6) $\displaystyle\sum_{n=1}^{\infty}\frac{x^n}{a^{\sqrt{n}}}, \quad (a>0)$.

3. Find the sum of the following series.

(1) $\displaystyle\sum_{n=0}^{\infty}(n+1)x^n$;

(3) $\displaystyle\sum_{n=1}^{\infty}nx^n$;

(2) $\displaystyle\sum_{n=1}^{\infty}(-1)^{n-1}(2n-1)x^{2n-2}$;

(4) $\displaystyle\sum_{n=1}^{\infty}\frac{x^n}{n(n+1)}$.

4. Prove the following equations for $|x|<1$.

(1) $\displaystyle\sum_{n=0}^{\infty}(n+1)(n+2)x^n = \frac{2}{(1-x)^3}$;

(2) $\displaystyle\sum_{n=1}^{\infty}n^2x^n = \frac{x+x^2}{(1-x)^3}$;

(3) $\displaystyle\sum_{n=1}^{\infty}n^3x^n = \frac{x+4x^2+x^3}{(1-x)^4}$.

5. Find the power series expansion of the following functions.

(1) e^{x^2};

(2) $\cos^2 x$;

(3) $\displaystyle\frac{x^{12}}{1-x}$;

(4) $\ln\sqrt{\dfrac{1+x}{1-x}}$;

(5) $\dfrac{x}{1+x-2x^2}$;   (6) $(1+x)e^{-x}$.

6. Find the power series expansion of the following functions by termwise differentiation and termwise integration.

(1) $\arcsin x$;   (2) $\ln(x+\sqrt{1+x^2})$;

(3) $\displaystyle\int_0^x \dfrac{\sin x}{t}\,dt$;   (4) $\displaystyle\int_1^x \dfrac{e^t}{t}\,dt$.

7. Find the power series expansion of the following functions at the given points, and determine the intervals of convergence.

(1) $x^3 - 2x^2 + 5x - 7$, $(x = 1)$;   (3) $(1+x)\ln(1+x)$, $(x = 0)$;

(2) $\dfrac{1}{x+10}$, $(x = 0)$;   (4) $\dfrac{x}{(1-x)(1-x^2)}$, $(x = 0)$;

(5) $(1+x^2)\arctan x$, $(x = 0)$;   (7) $\cosh x$, $(x = 0)$;

(6) $\ln x$, $(x = 1)$;   (8) $\cos x$, $\left(x = \dfrac{\pi}{4}\right)$.

8. Find the power series expansion of the following functions.

(1) $\ln(1 + e^x)$, (the first four terms);

(2) $\tan x$, (the first three non-zero terms).

9. Use the power series of the integrand, calculate the following integrals to the accuracy of 0.001.

(1) $\displaystyle\int_0^1 e^{-x^2}\,dx$;   (3) $\displaystyle\int_0^1 \cos x^2\,dx$.

(2) $\displaystyle\int_0^2 \dfrac{\sin x}{x}\,dx$;

10. Find the general solution of the following equations using power series.

(1) $y' = y + 2x$;   (2) $y'' + xy' + y = 0$.

11. Find the solution of the equation $y' = x + y^2$ that satisfies the initial condition $y(0) = 0$. (Find a power series solution up to the x^5 term.)

12. Find the solution of the following initial value problems.

(1) $\begin{cases} y'' - xy' + y - 1 = 0, \\ y(0) = y'(0) = 0; \end{cases}$ (2) $\begin{cases} y'' + (y')^2 = 0, \\ y(0) = 0, y'(0) = 1. \end{cases}$

13. Calculate the following values by Stirling formula.

(1) $\lg 100!$;   (2) C_{100}^{40}.

10.4 Infinite Integrals and Integrals with Parameters

10.4.1 *Convergence Criteria for Infinite Integrals*

In this section, we study theories of infinite integrals. Readers can easily find the similarities between these theories and the corresponding theories about infinite series.

First, we study the infinite integral $\int_0^{+\infty} f(x)\,dx$, which is defined in Section 9.3.3.

This integral is convergent if the limit

$$\lim_{\lambda \to +\infty} \int_a^\lambda f(x)\,dx$$

is finite.

Let

$$F(A) = \int_a^A f(x)\,dx.$$

(Notice the similarity between the function $F(A)$ and the partial sum of the infinite series.) We now prove that the function $F(x)$ has a finite limit as $x \to +\infty$ if and only if for any given $\varepsilon > 0$, there exists an $X > 0$ such that

$$|F(x) - F(x')| < \varepsilon$$

if $x > X$ and $x' > X$.

In fact, if $\lim\limits_{\lambda \to +\infty} F(x) = a$, then for any $\varepsilon > 0$, there exists an $X > 0$ such that

$$|F(x) - a| < \frac{\varepsilon}{2}, \quad |F(x') - a| < \frac{\varepsilon}{2}$$

if $x > X$, and $x' > X$. Thus,

$$|F(x) - F(x')| \leqslant |F(x) - a| + |F(x') - a| < \frac{\varepsilon}{2} + \frac{\varepsilon}{2} = \varepsilon.$$

To prove the sufficiency, we only need to show that for any sequence x_n that tends to infinity, all the limits

$$\lim_{n \to \infty} F(x_n)$$

are the same.

Since $\lim\limits_{n \to \infty} x_n = +\infty$, for $X > 0$, there exists an N, such that $x_n > X$ and $x_m > X$ if $m > N$ and $n > N$. Thus,

$$|F(x_n) - F(x_m)| < \varepsilon.$$

This implies that $F(x_n)$ is a Cauchy sequence, and the limit $\lim\limits_{n \to \infty} F(x_n)$ is finite.

Now we show that, for any sequence x_n that tends to positive infinity, all the limits of $F(x_n)$ are the same.

Indeed, suppose there are two sequences x_n and x'_n. They both tends to positive infinity as $n \to \infty$. Then the sequence

$$x_1, x'_1, x_2, x'_2, \cdots, x_n, x'_n, \cdots$$

also tends to positive infinity. Thus, the sequence

$$F(x_1), F(x'_1), F(x_2), F(x'_2), \cdots, F(x_n), F(x'_n), \cdots$$

has a finite limit. Hence,

$$\lim_{n \to \infty} F(x_n) = \lim_{n \to \infty} F(x'_n).$$

Therefore, we obtain the following results:

Cauchy Criterion: The integral $\displaystyle\int_a^{+\infty} f(x)\, dx$ is convergent if and only if for any $\varepsilon > 0$, there exists an $X > a$ such that

$$\left| \int_x^{x'} f(x)\, dx \right| < \varepsilon$$

if $x > X$ and $x' > X$.

If the integral $\displaystyle\int_a^{+\infty} |f(x)|\, dx$ is convergent, then we say that the integral $\displaystyle\int_a^{+\infty} f(x)\, dx$ is *absolutely convergent*.

Obviously, if an infinite integral is absolutely convergent, then it is convergent. However, a convergent infinite integral is not necessarily absolutely convergent.

If the infinite integral $\displaystyle\int_a^{+\infty} f(x)\, dx$ is convergent, but the infinite integral $\displaystyle\int_a^{+\infty} |f(x)|\, dx$ is divergent, then we say that $\displaystyle\int_a^{+\infty} f(x)\, dx$ is *conditionally convergent*.

By the fact that the limit of a monotonically increasing sequence exists, we have the conclusion that:

If $f(x)$ is a non-negative continuous function on $[a, +\infty)$, then the necessary and sufficient condition for the convergence of the integral $\displaystyle\int_a^{+\infty} f(x)\, dx$ is $\displaystyle\int_a^{A} f(x)\, dx$ is bounded for all $a \leqslant A < +\infty$.

Thus, we get the following criterion.

Comparison Criterion: Suppose functions $f(x)$ and $\varphi(x)$ are continuous on $[a, +\infty)$, and

$$0 \leqslant f(x) \leqslant \varphi(x)$$

for x sufficiently large. Then

(1) If $\displaystyle\int_a^{+\infty} \varphi(x)\, dx$ is convergent, then $\displaystyle\int_a^{+\infty} f(x)\, dx$ is convergent.

(2) If $\displaystyle\int_a^{+\infty} f(x)\, dx$ is divergent, then $\displaystyle\int_a^{+\infty} \varphi(x)\, dx$ is divergent.

Since $\displaystyle\int_a^{+\infty} \frac{1}{x^p}\, dx$ is convergent for $p > 1$ and is divergent for $p \leqslant 1$, the function $\dfrac{1}{x^p}$ is often used as a comparison function.

Example 1. The integrals $\displaystyle\int_1^{+\infty} \frac{\sin x}{x^2}\, dx$ and $\displaystyle\int_1^{+\infty} \frac{\cos x}{x^2}\, dx$ are absolutely convergent.

This is because

$$\left|\frac{\sin x}{x^2}\right| \leqslant \frac{1}{x^2}, \quad \left|\frac{\cos x}{x^2}\right| \leqslant \frac{1}{x^2}$$

and $\displaystyle\int_1^{+\infty} \frac{1}{x^2}$ is convergent.

The comparison criterion can also be stated in limit form:

Suppose $f(x)$ is a non-negative continuous function and $\varphi(x)$ is a positive continuous function on $[a, +\infty)$, and

$$\lim_{x \to +\infty} \frac{f(x)}{\varphi(x)} = k.$$

Then we have:

(1) If $0 < k < +\infty$, then integrals $\displaystyle\int_a^{+\infty} f(x)\, dx$ and $\displaystyle\int_a^{+\infty} \varphi(x)\, dx$ are either convergent simultaneously or divergent simultaneously.

(2) If $k = 0$, and $\displaystyle\int_a^{+\infty} \varphi(x)\, dx$ is convergent, then $\displaystyle\int_a^{+\infty} f(x)\, dx$ is also convergent.

(3) if $k = +\infty$, and $\displaystyle\int_a^{+\infty} \varphi(x)\, dx$ is divergent, then $\displaystyle\int_a^{+\infty} f(x)\, dx$ is also divergent.

The proofs of these theorems are similar to the proofs of corresponding theorems for infinite series.

Example 2. The integral $\displaystyle\int_2^{+\infty} \frac{x^2}{x^4 - x^2 - 1} \, dx$ is convergent.

In fact,

$$\frac{x^2}{x^4 - x^2 - 1} \sim \frac{1}{x^2}$$

as $x \to +\infty$.

The integral $\displaystyle\int_2^{+\infty} \frac{1}{x^2} \, dx$ is convergent. Thus, the conclusion follows.

Example 3. The integral $\displaystyle\int_1^{+\infty} \frac{\arctan x}{(1 + x^2)^{\frac{3}{2}}} \, dx$ is convergent.

In fact,

$$\frac{\arctan x}{(1 + x^2)^{\frac{3}{2}}} \sim \frac{\pi}{2x^3}$$

as $x \to +\infty$.

The integral $\displaystyle\int_1^{+\infty} \frac{\pi}{2x^3} \, dx$ is convergent. Thus, the conclusion follows.

Corresponding to Dirichlet criterion and Abel criterion for infinite series, we have:

Dirichlet Criterion: If $f(x)$ and $g(x)$ satisfy

(i) $F(A) = \displaystyle\int_a^A f(x) \, dx$ is bounded on $a < A < +\infty$,

(ii) $g(x)$ is monotonic in $[a, +\infty)$ and $\displaystyle\lim_{x \to +\infty} g(x) = 0$,

then the integral

$$\int_1^{+\infty} f(x) g(x) \, dx$$

is convergent.

Abel Criterion: If $f(x)$ and $g(x)$ satisfy

(i) $\displaystyle\int_a^{+\infty} f(x) \, dx$ is convergent,

(ii) $g(x)$ is monotonic and bounded in $[a, +\infty)$,

then the integral

$$\int_a^{+\infty} f(x) g(x) \, dx$$

is convergent.

To prove these two criteria, we need to verify a result that corresponds to the Abel Lemma for infinite series in Section 10.1.3 called the Second Mean-Value Theorem of Integrals.

The Second Mean-Value Theorem of Integrals: If a function $f(x)$ is integrable on $[a, b]$, a function $g(x) \geqslant 0$ and is monotonically decreasing, then there exists a number ξ, $(a \leqslant \xi \leqslant b)$, such that

$$\int_a^b f(x)g(x)\,dx = g(a) \int_a^\xi f(x)\,dx.$$

Another form of this theorem is:

If a function $f(x)$ is integrable on $[a, b]$, a function $g(x)$ is monotonic in $[a, b]$, then there exists a number ξ in $[a, b]$, such that

$$\int_a^b f(x)g(x)\,dx = g(a) \int_a^\xi f(x)\,dx + g(b) \int_\xi^b f(x)\,dx.$$

Proof: We first show that the product of two integrable functions $f(x)$ and $g(x)$ is also integrable.

In fact, for any partition T, let $S_1(f, T)$, $S_0(f, T)$. $S_1(g, T)$, $S_0(g, T)$, $S_1(fg, T)$, $S_0(fg, T)$ be the upper sums and the lower sums of $f(x)$, $g(x)$ and $f(x)g(x)$ respectively. Then

$$S_1(fg, T) - S_0(fg, T) = \sum_{i=1}^n \{M_i(fg) - m_i(fg)\}\Delta x_i$$

$$\leqslant \sum_{i=1}^n [M_i(f)M_i(g) - m_i(f)m_i(g)]\Delta x_i$$

$$= \sum_{i=1}^n \{[M_i(f) - m_i(f)]M_i(g) + [M_i(g) - m_i(g)]m_i(f)\}\Delta x_i$$

$$\leqslant M(g) \sum_{i=1}^n [M_i(f) - m_i(f)]\Delta x_i + M(f) \sum_{i=1}^n [M_i(g) - m_i(g)]\Delta x_i$$

$$= M(g)[S_1(f, T) - S_0(f, T)] + M(f)[S_1(g, T) - S_0(g, T)],$$

where $M_i(f)$, $m_i(f)$, $M_i(g)$, $m_i(g)$ are supremum and infimum of $f(x)$ and $g(x)$ on $[x_{i-1}, x_i]$ respectively, $M(f)$ and $M(g)$ are the supremum of $f(x)$ and $g(x)$ on $[a, b]$ respectively.

Since $f(x)$ and $g(x)$ are integrable on $[a, b]$, for any $\varepsilon > 0$, there exists a $\delta > 0$, such that

$$S_1(f, T) - S_0(f, T) < \frac{\varepsilon}{2M(g)},$$

$$S_1(g, T) - S_0(g, T) < \frac{\varepsilon}{2M(f)}$$

if $\lambda(T) < \delta$. Thus,

$$S_1(fg, T) - S_0(fg, T) < \frac{\varepsilon}{2} + \frac{\varepsilon}{2} = \varepsilon.$$

That is, the function fg is integrable on $[a, b]$.

Since $f(x)$ is integrable, and $g(x)$, being a non-negative monotonically decreasing function, is also integrable, the product $f(x)g(x)$ is integrable. The integral $\int_a^b f(x)g(x)\, dx$ is the limit of

$$I = f(a)g(a)(x_1 - a) + f(x_1)g(x_1)(x_2 - x_1)$$
$$+ \cdots + f(x_{n-1})g(x_{n-1})(b - x_{n-1}).$$

Let

$$I' = \sum_{\nu=1}^n M_\nu g(x_{\nu-1})(x_\nu - x_{\nu-1}),$$

$$I'' = \sum_{\nu=1}^n m_\nu g(x_{\nu-1})(x_\nu - x_{\nu-1}),$$

where M_ν and m_ν are supremum and infimum of $f(x)$ on $[x_{\nu-1}, x_\nu]$ respectively. Then $I' \geqslant I \geqslant I''$, and

$$I' - I'' \leqslant g(a) \sum_{\nu=1}^n (M_\nu - m_\nu)(x_\nu - x_{\nu-1}).$$

Since $f(x)$ is integrable on $[a, b]$, we have

$$\lim_{\lambda \to 0} \sum_{\nu=1}^n (M_\nu - m_\nu)(x_\nu - x_{\nu-1}) = 0.$$

Thus, for any μ_i that satisfies $m_i \leqslant \mu_i \leqslant M_i$, the limit of

$$I_1 = \sum_{\nu=1}^n \mu_\nu g(x_{\nu-1})(x_\nu - x_{\nu-1})$$

is $\int_a^b f(x)g(x)\, dx$.

By the Mean-Value Theorem for Integrals,

$$\int_{x_{\nu-1}}^{x_\nu} f(x)\, dx = \mu_\nu(x_\nu - x_{\nu-1}),$$

and we can choose such μ_ν in I_1. Thus

$$g(a) \geqslant g(x_1) \geqslant \cdots \geqslant g(x_n) \geqslant 0,$$

and

$$\sum_{\nu=1}^{n} \mu_\nu (x_\nu - x_{\nu-1}) = \int_a^b f(x)\,dx,$$

and since $f(x)$ is integrable on $[a, b]$, $\int_a^b f(x)\,dx$ is finite.

By Abel lemma in 10.1.3, we have that

$$g(a) \min_{a \leqslant c \leqslant b} \int_a^c f(x)\,dx \leqslant I_1 \leqslant g(a) \max_{a \leqslant c \leqslant b} \int_a^c f(x)\,dx.$$

The limit of I_1, $\int_a^b f(x)g(x)\,dx$ is also between

$$g(a) \min_{a \leqslant c \leqslant b} \int_a^c f(x)\,dx$$

and

$$g(a) \max_{a \leqslant c \leqslant b} \int_a^c f(x)\,dx.$$

Since $\int_a^c f(x)\,dx$ is a continuous function of c, there exists a ξ such that

$$\int_a^b f(x)g(x)\,dx = g(a) \int_a^\xi f(x)\,dx,$$

where $a \leqslant \xi \leqslant b$.

Similarly, if $g(x)$ is a non-negative monotonically increasing function, then

$$\int_a^b f(x)g(x)\,dx = g(b) \int_\eta^b f(x)\,dx,$$

where $a \leqslant \eta \leqslant b$.

It is not hard to see that

$$\int_a^b f(x)g(x)\,dx = g(a) \int_a^\xi f(x)\,dx + g(b) \int_\xi^b f(x)\,dx.$$

if $g(x)$ is monotonic but without non-negative restriction.

In fact, suppose $g(x)$ is decreasing. Then $g(x) - g(b)$ is a non-negative decreasing function. Thus

$$\int_a^b (g(x) - g(b)) f(x)\,dx = (g(a) - g(b)) \int_a^\xi f(x)\,dx.$$

Therefore,

$$\int_a^b f(x)g(x)\,dx = g(a)\int_a^\xi f(x)\,dx + g(b)\int_\xi^b f(x)\,dx.$$

We can have a similar discussion if $g(x)$ is an increasing function.

Now we prove the Dirichlet criterion and Abel criterion by the Second Mean-Value Theorem of Integrals.

We start with the proof of Dirichlet criterion. By the Second Mean-Value Theorem of Integrals,

$$\int_{A'}^{A''} f(x)g(x)\,dx = g(A')\int_{A'}^\xi f(x)\,dx + g(A'')\int_\xi^{A''} f(x)\,dx.$$

Thus,

$$\left|\int_{A'}^{A''} f(x)g(x)\,dx\right| \leqslant |g(A')|\left|\int_{A'}^\xi f(x)\,dx\right| + |g(A'')|\left|\int_\xi^{A''} f(x)\,dx\right|.$$

By the assumption,

$$|F(A)| = \left|\int_a^A f(x)\,dx\right| \leqslant M$$

where $a < A < +\infty$. Hence,

$$\left|\int_{A'}^\xi f(x)\,dx\right| = \left|\int_a^\xi f(x)\,dx - \int_a^{A'} f(x)\,dx\right| \leqslant 2M,$$

$$\left|\int_\xi^{A''} f(x)\,dx\right| = \left|\int_a^{A''} f(x)\,dx - \int_a^\xi f(x)\,dx\right| \leqslant 2M.$$

Since $\lim\limits_{x\to+\infty} g(x) = 0$, for any $\varepsilon > 0$, there exists an $A_0 > 0$ such that

$$|g(A')| < \varepsilon, \quad |g(A'')| < \varepsilon$$

if $A' > A_0$ and $A'' > A_0$. Therefore,

$$\left|\int_{A'}^{A''} f(x)g(x)\,dx\right| \leqslant 4M\varepsilon$$

if $A' > A_0$ and $A'' > A_0$.

By Cauchy criterion, the integral $\int_a^{+\infty} f(x)g(x)\,dx$ is convergent.

Next we prove Abel criterion.

Since $\displaystyle\int_a^{+\infty} f(x)\,dx$ is convergent, for any $\varepsilon > 0$, there exists an $A_0 > 0$ such that

$$\left| \int_{A'}^{A''} f(x)\,dx \right| < \varepsilon$$

if $A' > A_0$ and $A'' > A_0$.

Since $g(x)$ is monotonic, by the Second Mean-Value Theorem of Integrals,

$$\int_{A'}^{A''} f(x)g(x)\,dx = g(A') \int_{A'}^{\xi} f(x)\,dx + g(A'') \int_{\xi}^{A''} f(x)\,dx,$$

where $A' \leqslant \xi \leqslant A''$. Thus, by the fact that $|g(x)| \leqslant M$, we have

$$\left| \int_{A'}^{A''} f(x)g(x)\,dx \right| \leqslant |g(A')| \left| \int_{A'}^{\xi} f(x)\,dx \right|$$

$$+ |g(A'')| \left| \int_{\xi}^{A''} f(x)\,dx \right| \leqslant 2M\varepsilon.$$

Example 4. The integral $\displaystyle\int_0^{\infty} \frac{\sin x}{x}\,dx$ is convergent, but not absolutely convergent.

Proof: Let $g(x) = \dfrac{1}{x}$ and $f(x) = \sin x$. Since

$$\left| \int_0^A \sin x\,dx \right| = |\cos 0 - \cos A| \leqslant 2,$$

by Dirichlet criterion, the integral $\displaystyle\int_0^{\infty} \frac{\sin x}{x}\,dx$ is convergent. But it is not absolutely convergent because

$$\int_0^{\infty} \frac{|\sin x|}{x}\,dx = \sum_{n=1}^{\infty} \int_{(n-1)\pi}^{n\pi} \frac{|\sin x|}{x}\,dx$$

$$\geqslant \sum_{n=1}^{\infty} \frac{1}{n\pi} \int_{(n-1)\pi}^{n\pi} |\sin x|\,dx = \frac{2}{\pi} \sum_{n=1}^{\infty} \frac{1}{n} = \infty.$$

Similarly, we can determine the convergence of $\displaystyle\int_1^{\infty} \frac{\sin x}{x^{\alpha}}\,dx$ and $\displaystyle\int_1^{\infty} \frac{\cos x}{x^{\alpha}}\,dx$, $(\alpha > 0)$.

Example 5. The integral $\int_0^\infty \sin x^2 \, dx$ is convergent.

In fact, let $x^2 = t$, we have

$$\int_0^\infty \sin x^2 \, dx = \frac{1}{2} \int_0^\infty \frac{\sin t}{\sqrt{t}} \, dt.$$

Thus, the conclusion follows.

The integral $\int_0^\infty x \sin x^3 \, dx$ is also convergent. We leave the verification of this conclusion to our readers. The criteria for infinite integrals are parallel with the criteria for infinite series, the former is "continuous" the latter is "discrete". However, there are some differences between them. For instance, a necessary condition of convergence of a number series $\sum_{n=1}^\infty a_n$ is

$$\lim_{n \to \infty} a_n = 0,$$

but the integrand function in an infinite integral does not necessarily tend to zero as x tends to infinity. Furthermore, as we can see in the integral in Example 5,

$$\int_0^\infty x \sin x^3 \, dx,$$

it is even possible for there to exist a sequence of x values that tends to infinity while the sequence of their integrand function values also tends to infinity, yet the integral is still convergent.

In Section 9.3.3, we studied some generalized versions of Riemann integrals. One is the integral with an infinite interval as its integration interval, and is called an infinite integral; the other is the integral with an unbounded integrand function, and is called an unbounded integral. The problem of convergence of an integral of an unbounded function can be converted to the problem of convergence of an infinite integral.

Suppose $f(x)$ is a function that is defined on an interval $(a, b]$, and $f(x)$ is unbounded when $x \to a$. The integral defined as

$$\int_a^b f(x) \, dx = \lim_{\varepsilon \to 0^+} \int_{a+\varepsilon}^b f(x) \, dx.$$

Let $x = a + \dfrac{1}{y}$. Then

$$\lim_{\varepsilon \to 0^+} \int_{a+\varepsilon}^b f(x) \, dx = \lim_{\varepsilon \to 0^+} \int_{\frac{1}{\varepsilon}}^{\frac{1}{b-a}} \left[-f\left(a + \frac{1}{y}\right) \right] \frac{1}{y^2} \, dy$$

$$= \int_{\frac{1}{b-a}}^\infty f\left(a + \frac{1}{y}\right) \frac{1}{y^2} \, dy.$$

Thus, the problem becomes the convergence of the infinite integral.

If we let $y = \dfrac{1}{x}$ in the integral

$$\int_0^1 \frac{1}{\sqrt{x}}\, dx,$$

then

$$\int_0^1 \frac{1}{\sqrt{x}}\, dx = \int_1^\infty \frac{1}{\sqrt{y^3}}\, dy.$$

Therefore, it is clear to see the connection between the convergence criteria for unbounded integrals $\displaystyle\int_a^b f(x)\, dx$ ($f(x)$ is unbounded as $x \to a$) and the convergence criteria for infinite integrals.

We leave the proofs of the following conclusions to our readers (assuming that the lower limit of the integral a is an improper point).

1. *Cauchy Criterion*: The integral $\displaystyle\int_a^b f(x)\, dx$ is convergent if and only if for any $\varepsilon > 0$, there exists a $\delta > 0$ such that

$$\left| \int_{a+\delta'}^{a+\delta''} f(x)\, dx \right| < \varepsilon$$

if $0 < \delta' < \delta$ and $0 < \delta'' < \delta$.

2. If the integral $\displaystyle\int_a^b |f(x)|\, dx$ is convergent, then the integral $\displaystyle\int_a^b f(x)\, dx$ is also convergent.

3. If the inequality

$$0 \leqslant f(x) \leqslant \varphi(x)$$

is true for $x(> a)$ sufficiently close to a, then

(1) If $\displaystyle\int_a^b \varphi(x)\, dx$ is convergent, then $\displaystyle\int_a^b f(x)\, dx$ is also convergent,

(2) If $\displaystyle\int_a^b f(x)\, dx$ is divergent, then $\displaystyle\int_a^b \varphi(x)\, dx$ is also divergent.

4. Suppose $f(x)$ and $\varphi(x)$ are non-negative and positive continuous functions respectively on $(a, b]$. Suppose also

$$\lim_{x \to a} \frac{f(x)}{\varphi(x)} = k.$$

Then

(1) If $0 < k < \infty$, then the integrals $\int_a^b f(x)\,dx$ and $\int_a^b \varphi(x)\,dx$ are convergent simultaneously or divergent simultaneously.

(2) If $k = 0$, and $\int_a^b \varphi(x)\,dx$ is convergent, then $\int_a^b f(x)\,dx$ is also convergent.

(3) If $k = \infty$, and $\int_a^b \varphi(x)\,dx$ is divergent, then $\int_a^b f(x)\,dx$ is also divergent.

Example 6. The integral $\int_0^1 \dfrac{1}{\sqrt[4]{1-x^4}}\,dx$ is convergent.

Proof: The improper point is $x = 1$. When $x \to 1$,

$$\frac{1}{\sqrt[4]{1-x^4}} = \frac{1}{\sqrt[4]{1-x}\,\sqrt[4]{(1+x)(1+x^2)}} \sim \frac{1}{\sqrt[4]{4}}\frac{1}{\sqrt[4]{1-x}}.$$

Thus the integral is convergent.

Example 7. Study the integral $\int_0^1 x^{p-1}(1-x)^{q-1}\,dx$.

Solution: If $p < 1$, then $x = 0$ is an improper point. If $q < 1$, then $x = 1$ is an improper point. Rewriting the integral as:

$$\int_0^1 x^{p-1}(1-x)^{q-1}\,dx = \int_0^a x^{p-1}(1-x)^{q-1}\,dx + \int_a^1 x^{p-1}(1-x)^{q-1}\,dx,$$

where $0 < a < 1$. When $x \to 0$,

$$x^{p-1}(1-x)^{q-1} \sim x^{p-1}.$$

Thus, the first integral is convergent if $1 - p < 1$ or $p > 0$. When $x \to 1$,

$$x^{p-1}(1-x)^{q-1} \sim (1-x)^{q-1}.$$

Thus, the second integral is convergent if $1 - q < 1$ or $q > 0$. Therefore, the integral converges when $p > 0$ and $q > 0$.

The integral $\int_0^1 x^{p-1}(1-x)^{q-1}\,dx$ is called the *beta function*, and is denoted by $B(p, q)$.

Example 8. Study the integral $\int_0^\infty x^{p-1}e^{-x}\,dx$.

Solution: When $p < 1$, $x = 0$ is an improper point. It is also an infinite integral. Rewriting this integral as:

$$\int_0^\infty x^{p-1}e^{-x}\,dx = \int_0^1 x^{p-1}e^{-x}\,dx + \int_1^\infty x^{p-1}e^{-x}\,dx.$$

When $x \to 0$,

$$x^{p-1}e^{-x} \sim x^{p-1}.$$

Thus, the first integral is convergent if $p > 0$.

Since

$$x^2 x^{p-1} e^{-x} \to 0$$

as $x \to \infty$, for x sufficiently large, we have

$$x^{p-1}e^{-x} < \frac{1}{x^2}.$$

It follows that, the second integral is convergent for any value of p. Therefore, the integral is convergent if $p > 0$.

This integral is called the *gamma function*, and is denoted by $\Gamma(p)$.

10.4.2 *Integrals with Parameters*

We learned that the sum of infinitely many elementary functions may becomes a non-elementary function. In this section, we show that the integral of elementary functions may also becomes a non-elementary function. The concept of infinite integrals that we studied in Section 10.4.1 is parallel to the concept of infinite series of numbers. The concept of integrals with parameters that we study in this section is parallel to the concept of series of functions. Readers can find the similarities between the results in this section and some results on series of functions.

First, we study integrals with parameters whose limits of integration are constants.

Suppose $f(x, u)$ is continuous on D: $a \leqslant x \leqslant b$, $\alpha \leqslant u \leqslant \beta$. Then $f(x, u)$ is Riemann integrable on $[a, b]$ with respect to x. The integral

$$\int_a^b f(x, u)\, dx$$

is called an *integral with parameter u*.

Of course, the function $f(x, u)$ can be unbounded in the domain, and the limits of integral can be infinity. We study these two kinds of integrals later. For now we discuss the integral with finite limits of integration and bounded integrand $f(x, u)$.

If a function $f(x, u)$ is continuous on D: $a \leqslant x \leqslant b$, $\alpha \leqslant u \leqslant \beta$, then

$$\varphi(u) = \int_a^b f(x, u)\, dx$$

is a continuous function on the interval $[\alpha, \beta]$.

In fact, for an arbitrary point u_0 in $[\alpha, \beta]$, we have

$$\varphi(u) - \varphi(u_0) = \int_a^b [f(x, u) - f(x, u_0)]\, dx.$$

It follows that

$$|\varphi(u) - \varphi(u_0)| \leqslant \int_a^b |f(x, u) - f(x, u_0)|\, dx.$$

The function $f(x, u)$ is uniformly continuous since it is continuous on the closed region D. Thus, for any $\varepsilon > 0$, there exists a $\delta > 0$ such that

$$|f(x_1, u_1) - f(x_2, u_2)| < \varepsilon$$

for any two points (x_1, u_1) and (x_2, u_2) in D that the distance between them is less than δ.

The distance between the two points (x, u) and (x, u_0) is $|u - u_0|$. Thus

$$|f(x, u) - f(x, u_0)| < \varepsilon$$

if $|u - u_0| < \delta$. Hence

$$|\varphi(u) - \varphi(u_0)| < \varepsilon(b - a).$$

This proves that the function $\varphi(u)$ is continuous at u_0.

This result can also be written as

$$\lim_{u \to u_0} \int_a^b f(x, u)\, dx = \int_a^b \lim_{u \to u_0} f(x, u)\, dx.$$

That is, the order of the integral and the limit can be exchanged.

Since $\varphi(u)$ is continuous on $[\alpha, \beta]$, we have

$$\int_\alpha^\beta \varphi(u)\, du = \int_\alpha^\beta \int_a^b f(x, u)\, dx\, du.$$

The right side of the above equation is

$$\iint_D f(x, u)\, dx\, du,$$

and it also can be written as

$$\int_a^b \int_\alpha^\beta f(x, u)\, du\, dx.$$

That is, the order of integral signs $\displaystyle\int_\alpha^\beta$ and $\displaystyle\int_a^b$ can be exchanged.

The next result is the property of the differential of $\varphi(u)$.

If a function $f(x, u)$ and its partial derivative $\dfrac{\partial f}{\partial u}$ are continuous on D: $a \leqslant x \leqslant b$, $\alpha \leqslant u \leqslant \beta$, then

$$\varphi(u) = \int_a^b f(x, u)\, dx$$

is differentiable on $[\alpha, \beta]$, and

$$\frac{d}{du}\varphi(u) = \frac{d}{du}\int_a^b f(x, u)\, dx = \int_a^b \frac{\partial}{\partial u} f(x, u)\, dx.$$

To show this result, let $g(x, u) = \dfrac{\partial f}{\partial u}$. For an arbitrary point u_0 in $[\alpha, \beta]$, we have

$$\frac{\varphi(u_0 + k) - \varphi(u_0)}{k} = \frac{1}{k} \int_a^b \left(f(x, u_0 + k) - f(x, u_0) \right) dx$$

$$= \int_a^b g(x, u_0 + \theta k)\, dx,$$

where $0 < \theta < 1$. Since the order of the operation of limit and the operation of integration can be exchanged, we have

$$\frac{d\varphi}{du} = \lim_{k \to 0} \int_a^b g(x, u_0 + \theta k)\, dx = \int_a^b g(x, u_0)\, dx.$$

In other words, the order of the differentiation and the integration can be exchanged if $f(x, u)$ and $\dfrac{\partial f}{\partial u}$ are continuous.

Example 1. Let $f(x, u) = x^u$, $0 \leqslant x \leqslant 1$, $a \leqslant u \leqslant b$, $0 < a < b$. Then

$$\int_a^b du \int_0^1 x^u\, dx = \int_0^1 dx \int_a^b x^u\, du.$$

The left hand side is

$$\int_a^b \frac{du}{1 + u} = \ln \frac{1 + b}{1 + a}.$$

The right hand side is

$$\int_0^1 \frac{x^b - x^a}{\ln x}\, dx.$$

Therefore,

$$\int_0^1 \frac{x^b - x^a}{\ln x}\, dx = \ln \frac{1 + b}{1 + a}.$$

Next, we study integrals with parameter dependent limits of integration.

Suppose $f(x, u)$ is continuous on $a \leqslant x \leqslant b, c \leqslant u \leqslant d$. Suppose also,

$$x = \alpha(u), \quad x = \beta(u), \quad c \leqslant u \leqslant d$$

are two continuous functions of u, $a \leqslant \alpha(u) \leqslant b$, $a \leqslant \beta(u) \leqslant b$. Then the integral

$$I(u) = \int_{\alpha(u)}^{\beta(u)} f(x, u)\, dx$$

is a continuous function on $[c, d]$.

The proof is not difficult. Let u_0 be a point in $[c, d]$. Rewriting $I(u)$ as

$$I(u) = \int_{\alpha(u_0)}^{\beta(u_0)} f(x, u)\, dx + \int_{\beta(u_0)}^{\beta(u)} f(x, u)\, dx - \int_{\alpha(u_0)}^{\alpha(u)} f(x, u)\, dx.$$

The upper and lower limits of the first integral are constants, so

$$\lim_{u \to u_0} \int_{\alpha(u_0)}^{\beta(u_0)} f(x, u)\, dx = \int_{\alpha(u_0)}^{\beta(u_0)} f(x, u_0)\, dx.$$

Also, we know that

$$\left| \int_{\beta(u_0)}^{\beta(u)} f(x, u)\, dx \right| \leqslant M|\beta(u) - \beta(u_0)|,$$

$$\left| \int_{\alpha(u_0)}^{\alpha(u)} f(x, u)\, dx \right| \leqslant M|\alpha(u) - \alpha(u_0)|,$$

where $M = \max |f(x, u)|$. By the continuity of functions $\alpha(u)$ and $\beta(u)$, the two integrals above tend to zero as $u \to u_0$. Thus,

$$\lim_{u \to u_0} I(u) = I(u_0).$$

Therefore, $I(u)$ is a continuous function on $[c, d]$.

For taking the derivative of an integral with parameter in the limits, we have the following result.

Suppose $f(x, u)$ is continuous on $a \leqslant x \leqslant b, c \leqslant u \leqslant d$. Suppose also,

$$x = \alpha(u), \quad x = \beta(u)$$

are two continuous functions of u, where $c \leqslant u \leqslant d$. Suppose further that the derivatives $\alpha'(u)$ and $\beta'(u)$ exist, and that $f'_u(x, u)$ exits and is continuous. Then the derivative of $I(u)$ is

$$I'(u) = \int_{\alpha(u)}^{\beta(u)} f'_u(x, u)\, dx + \beta'(u)f(\beta(u), u) - \alpha'(u)f(\alpha(u), u).$$

The same method can be used to prove this result. We rewrite $I(u)$ as the sum of three integrals. For the first integral, we have

$$\frac{d}{du}\left(\int_{\alpha(u_0)}^{\beta(u_0)} f(x, u)\, dx\right)_{u=u_0} = \int_{\alpha(u_0)}^{\beta(u_0)} f'_u(x, u_0)\, dx.$$

For the second integral, we apply the Mean-Value Theorem, and get that

$$\frac{1}{u - u_0} \int_{\beta(u_0)}^{\beta(u)} f(x, u)\, dx = \frac{\beta(u) - \beta(u_0)}{u - u_0} f(\bar{x}, u),$$

where $\bar{x}$ is a number between $\beta(u)$ and $\beta(u_0)$. This integral tends to

$$\beta'(u_0) f(\beta(u_0), u_0)$$

as $u \to u_0$.

Readers can treat the third integral in a similar way.

Example 2. Evaluate the integral $I = \displaystyle\int_0^1 \frac{\arctan x}{x\sqrt{1 - x^2}}\, dx.$

Solution: Consider the integral with parameter u

$$I(u) = \int_0^1 \frac{\arctan ux}{x\sqrt{1 - x^2}}\, dx \qquad (I(1) = I).$$

The derivative is

$$I'(u) = \int_0^1 \frac{\partial}{\partial u}\left(\frac{\arctan ux}{x\sqrt{1 - x^2}}\right) dx = \int_0^1 \frac{1}{x\sqrt{1 - x^2}} \frac{x}{1 + u^2 x^2}\, dx$$

$$= \int_0^1 \frac{1}{(1 + u^2 x^2)\sqrt{1 - x^2}}\, dx.$$

Let $x = \cos\theta$. Then

$$I'(u) = \int_{\frac{\pi}{2}}^0 \frac{1}{1 + u^2 \cos^2\theta} \frac{1}{\sin\theta}(-\sin\theta)\, d\theta = \int_0^{\frac{\pi}{2}} \frac{1}{1 + u^2\cos^2\theta}\, d\theta$$

$$= \frac{1}{\sqrt{1 + u^2}} \arctan \frac{\tan\theta}{\sqrt{1 + u^2}}\bigg|_0^{\frac{\pi}{2}} = \frac{\pi}{2} \frac{1}{\sqrt{1 + u^2}}.$$

Integrating both sides of the above equation, we have

$$I(u) = \frac{\pi}{2} \ln(u + \sqrt{1 + u^2}) + C.$$

The constant $C = 0$ since $I(0) = 0$. Thus

$$I(u) = \frac{\pi}{2} \ln(u + \sqrt{1 + u^2}).$$

Therefore,

$$I = I(1) = \int_0^1 \frac{\arctan x}{x\sqrt{1 - x^2}}\, dx = \frac{\pi}{2} \ln(1 + \sqrt{2}).$$

10.4.3   *Infinite Integrals with Parameters*

Suppose $f(x, u)$ is continuous on the region $a \leqslant x < \infty, \alpha \leqslant u \leqslant \beta$. If for any u in $[\alpha, \beta]$, the improper integral

$$\int_a^\infty f(x, u)\, dx$$

is convergent, then it determines a function

$$\varphi(u) = \int_a^\infty f(x, u)\, dx$$

on $[\alpha, \beta]$.

In this definition, the improper integral with a parameter is parallel to

the infinite series of functions $S(u) = \sum_{n=1}^{\infty} f_n(u)$. One is a "continuous sum"

and the other is a "discrete sum". Readers can find the similarities between the results in this section and the results for infinite series of functions.

First, we introduce the concept of uniform convergence.

When we say "a integral $\int_a^\infty f(x, u)\, dx$ is convergent", we mean that, for a fixed u,

$$\lim_{A \to \infty} \int_a^A f(x, u)\, dx = \varphi(u) = \int_a^\infty f(x, u)\, dx.$$

That is, for any given $\varepsilon > 0$, there exists an $A_0(> a)$, such that

$$\left| \int_a^A f(x, u)\, dx - \int_a^\infty f(x, u)\, dx \right| = \left| \int_A^\infty f(x, u)\, dx \right| < \varepsilon$$

if $A > A_0$. The number A_0 depends on ε, and also depends on u.

If for any given ε, there exists an $A_0(> a)$ that only depends on ε, such that

$$\left| \int_A^\infty f(x, u)\, dx \right| < \varepsilon$$

for all u in $[\alpha, \beta]$, if $A > A_0$, then we say that $\int_a^\infty f(x, u)\, dx$ is *uniformly convergent* with respect to u on $[\alpha, \beta]$.

Next, we discuss uniform convergence for integrals with parameters where the integrand is unbounded at a point.

Suppose a is an improper point, for any given $\varepsilon > 0$, there exists a $\delta_0(\varepsilon) > 0$ that depends only on ε, such that

$$\left| \int_a^{a+\delta} f(x, u)\, dx \right| < \varepsilon$$

for all u in $[\alpha, \beta]$, if $0 < \delta < \delta_0$, then we say that $\int_a^b f(x, u) \, dx$ is uniformly convergent with respect to u on $[\alpha, \beta]$.

Same as for series of functions, there are criteria for uniform convergence for infinite integrals with parameters.

Cauchy Criterion: The integral $\int_a^\infty f(x, u) \, dx$ is uniformly convergent on an interval $[\alpha, \beta]$ if and only if for any given $\varepsilon > 0$, there exists an A_0 that depends only on ε, such that

$$\left| \int_{A'}^{A''} f(x, u) \, dx \right| < \varepsilon$$

for every u in $[\alpha, \beta]$, if $A' > A_0$ and $A'' > A_0$.

Weierstrass Criterion: Suppose $f(x, u)$ is continuous in x if $x \geqslant a$. Suppose also that there exists a continuous function $F(x)$ such that

$$|f(x, u)| \leqslant F(x)$$

for x sufficiently large and every u in $[\alpha, \beta]$. If $\int_a^\infty F(x) \, dx$ is convergent, then the integral $\int_a^\infty f(x, u) \, dx$ is uniformly convergent on $[\alpha, \beta]$.

The proof is not difficult: Since $\int_a^\infty F(x) \, dx$ is convergent, for any given $\varepsilon > 0$, there exists an A_0 such that

$$\left| \int_{A'}^{A''} F(x) \, dx \right| < \varepsilon$$

if $A' > A_0$ and $A'' > A_0$. Thus,

$$\left| \int_{A'}^{A''} f(x, u) \, dx \right| \leqslant \int_{A'}^{A''} |f(x, u)| \, dx \leqslant \int_{A'}^{A''} F(x) \, dx < \varepsilon.$$

Dirichlet Criterion: Suppose $\Phi(x, u)$ is monotonically decreasing in x, and is uniformly convergent to zero with respect to u as $x \to \infty$. Let

$$F(x, u) = \int_a^x f(t, u) \, dt.$$

If $|F| < M$, where M is a constant independent to x and u, then the integral

$$\int_a^\infty \Phi(x, u) f(x, u) \, dx$$

is uniformly convergent on $[\alpha, \beta]$.

In fact, since $F(x, u)$ is uniformly bounded, for any given $A' > a$ and $A'' > a$,

$$\left| \int_{A'}^{A''} f(x, u) \, dx \right| \leqslant \left| \int_{a}^{A'} f(x, u) \, dx \right| + \left| \int_{a}^{A''} f(x, u) \, dx \right| \leqslant 2M$$

for any u in $[\alpha, \beta]$.

Since $\Phi(x, u)$ tends to zero uniformly, for any $\varepsilon > 0$, there exists $A_0 > a$ such that

$$|\Phi(x, u)| < \varepsilon$$

for every u in $[\alpha, \beta]$, and $x > A_0$. By the Second Mean-Value Theorem of Integrals, we have

$$\left| \int_{A'}^{A''} f(x, u)\Phi(x, u) \, dx \right| \leqslant |\Phi(A', u)| \left| \int_{A'}^{\xi} f(x, u) \, dx \right|$$

$$+ |\Phi(A'', u)| \left| \int_{\xi}^{A''} f(x, u) \, dx \right|$$

$$\leqslant 2M\varepsilon + 2M\varepsilon = 4M\varepsilon$$

if $A' > A_0$ and $A'' > A_0$. Therefore, the integral is uniformly convergent.

Abel Criterion: Suppose $\displaystyle\int_{a}^{\infty} f(x, u) \, dx$ is uniformly convergent with respect u, $\alpha \leqslant u \leqslant \beta$. If $\Phi(x, u)$ is monotonic with respect to x, and is uniformly bounded with respect to u, then the integral

$$\int_{a}^{\infty} f(x, u)\Phi(x, u) \, dx$$

is uniformly convergent on $[\alpha, \beta]$.

We leave the proof of this result to our readers.

Example 1. The integral $\displaystyle\int_{0}^{\infty} u e^{-ux} \, dx$ is convergent for all $u \geqslant 0$, and it is uniformly convergent in $[\delta, \infty)$, $(\delta > 0)$.

In fact, for any $\varepsilon > 0$, let $A_0 = \dfrac{1}{\delta} \ln \dfrac{1}{\varepsilon}$, then we have

$$\int_{A}^{\infty} u e^{-ux} \, dx = e^{-uA} < \varepsilon, \quad (\delta \leqslant u < \infty),$$

if $A > A_0$.

The integral is not uniformly convergent in $[0, \infty)$ because

$$\int_A^\infty u e^{-ux} \, dx = e^{-uA} \to 1$$

as $u \to 0$, regardless the value of A.

Example 2. The integral $\displaystyle\int_0^\infty \frac{x \sin \beta x}{\alpha^2 + x^2} \, dx$, $(\alpha, \beta > 0)$, is uniformly convergent for $\beta \geqslant \beta_0 (> 0)$.

Proof: If $\beta \geqslant \beta_0$, then

$$\left| \int_0^A \sin \beta x \, dx \right| = \left| \frac{1 - \cos \beta A}{\beta} \right| \leqslant \frac{2}{\beta_0}.$$

The function $\dfrac{x}{\alpha^2 + x^2}$ is independent of β. It is monotonically decreasing for $x \geqslant \alpha$. Also, it tends to zero as $x \to \infty$. By the Dirichlet criterion,

$$\int_0^\infty \frac{x \sin \beta x}{\alpha^2 + x^2} \, dx$$

is uniformly convergent for $\beta \geqslant \beta_0 (> 0)$.

Example 3. The integral $\displaystyle\int_0^\infty e^{-xu} \frac{\sin x}{x} \, dx$ is uniformly convergent with respect to $u \geqslant 0$.

Proof: Since $\displaystyle\int_0^\infty \frac{\sin x}{x} \, dx$ is convergent, and e^{-xu} is a monotonically decreasing function in x, and $|e^{-xu}| \leqslant 1$ $(0 \leqslant x < \infty, 0 \leqslant u < \infty)$, by Abel criterion, the integral is uniformly convergent for $u \geqslant 0$.

Using the concept of uniform convergence, now we study the conditions of a series for termwise integrable on an infinite interval. That is, conditions for the order of the sum and the integration to be able to be exchanged.

Suppose $\displaystyle\sum_{n=1}^\infty u_n(x)$ is convergent on $[a, \infty)$, and the sum function is $S(x)$.

If for any $A > a$, the series $\displaystyle\sum_{n=1}^\infty u_n(x)$ is uniformly convergent on $[a, A)$; $\displaystyle\int_a^\infty S_n(x) \, dx$ is uniformly convergent with respect to n, where $S_n(x) = \displaystyle\sum_{k=1}^n u_k(x)$; and $\displaystyle\int_a^\infty S(x) \, dx$ is convergent, then

$$\int_a^\infty S(x) \, dx = \int_a^\infty \sum_{n=1}^\infty u_n(x) \, dx = \sum_{n=1}^\infty \int_a^\infty u_n(x) \, dx.$$

In fact, since $\displaystyle\int_a^\infty S_n(x)\,dx$ is uniformly convergent and $\displaystyle\int_a^\infty S(x)\,dx$ is convergent, for any given $\varepsilon > 0$, there exists an A_0, such that

$$\left|\int_{A_0}^\infty S(x)\,dx\right| < \frac{\varepsilon}{3},$$

$$\left|\int_{A_0}^\infty S_n(x)\,dx\right| < \frac{\varepsilon}{3}, \quad (n = 1, 2, \cdots).$$

Thus,

$$\left|\int_a^\infty S_n(x)\,dx - \int_a^\infty S(x)\,dx\right|$$

$$\leqslant \int_a^{A_0} |S_n(x) - S(x)|\,dx + \left|\int_{A_0}^\infty S_n(x)\,dx\right| + \left|\int_{A_0}^\infty S(x)\,dx\right|$$

$$< \int_a^{A_0} |S_n(x) - S(x)|\,dx + \frac{2}{3}\varepsilon.$$

Since $\displaystyle\sum_{n=1}^\infty u_n(x)$ is uniformly convergent on $[a, A_0]$, for any $\varepsilon > 0$, there exists an $N(> 0)$, such that

$$|S_n(x) - S(x)| < \frac{\varepsilon}{3(A_0 - a)}$$

for every x in $[a, A_0]$ and $n > N$. Thus, we have

$$\left|\int_a^\infty S_n(x)\,dx - \int_a^\infty S(x)\,dx\right| < \frac{\varepsilon}{3(A_0 - a)}(A_0 - a) + \frac{2}{3}\varepsilon = \varepsilon$$

if $n > N$. That is

$$\int_a^\infty \sum_{n=1}^\infty u_n(x)\,dx = \sum_{n=1}^\infty \int_a^\infty u_n(x)\,dx.$$

Example 4. Compute $\displaystyle\int_0^\infty e^{-x^2}\,dx$.

Solution: Since

$$e^{-x^2} = \lim_{n\to\infty} \left(1 + \frac{x^2}{n}\right)^{-n},$$

we have

$$\int_0^\infty e^{-x^2}\,dx = \int_0^\infty \lim_{n\to\infty}\left(1 + \frac{x^2}{n}\right)^{-n}\,dx.$$

Let

$$u_1(x) = (1 + x^2)^{-1}, \quad u_n(x) = \left(1 + \frac{x^2}{n}\right)^{-n} - \left(1 + \frac{x^2}{n-1}\right)^{-(n-1)}$$

for $n = 2, 3, \cdots$. Then

$$e^{-x^2} = \sum_{n=1}^{\infty} u_n(x).$$

It is not hard to verify that the series satisfies all the conditions of exchanging the order of the sum and the integral. Therefore

$$\int_0^\infty e^{-x^2} \, dx = \lim_{n \to \infty} \int_0^\infty \frac{1}{\left(1 + \dfrac{x^2}{n}\right)^n} \, dx.$$

Let $x = \sqrt{n}\, t$, we have

$$\int_0^\infty \frac{1}{\left(1 + \dfrac{x^2}{n}\right)^n} \, dx = \sqrt{n} \int_0^\infty \frac{1}{(1 + t^2)^n} \, dt = \sqrt{n}\, \frac{(2n-3)!!}{(2n-2)!!} \frac{\pi}{2}.$$

The value of $\displaystyle\int_0^\infty \frac{1}{(1 + t^2)^n} \, dt$ can be calculated using the beta function. See section 10.4.4. By Wallis formula (Section 10.3.4.), we have

$$\lim_{n \to \infty} \int_0^\infty \frac{1}{(1 + \dfrac{x^2}{n})^n} \, dx = \frac{1}{\sqrt{\pi}} \frac{\pi}{2} = \frac{\sqrt{\pi}}{2}.$$

Thus,

$$\int_0^\infty e^{-x^2} \, dx = \frac{\sqrt{\pi}}{2}.$$

This integral is useful in Statistics.

Similar to the series of functions, under the condition of uniform convergence, an infinite integral with parameters is continuous, differentiable and integrable if the integrand is continuous, differentiable and integrable respectively.

If a function $f(x, u)$ is continuous in $a \leqslant x < \infty, \alpha \leqslant u \leqslant \beta$, and the integral

$$\varphi(u) = \int_0^\infty f(x, u) \, dx$$

is uniformly convergent on $[\alpha, \beta]$, then $\varphi(u)$ is continuous on $[\alpha, \beta]$, and

$$\lim_{u \to u_0} \int_a^\infty f(x, u) \, dx = \int_a^\infty \lim_{u \to u_0} f(x, u) \, dx.$$

That is, the order of the limit and the integral can be exchanged.

The proof is similar to the corresponding result for series of functions:

Since $\int_a^\infty f(x, u)\, dx$ is uniformly convergent on $[\alpha, \beta]$, for any given $\varepsilon > 0$, there exists an A_0, such that

$$\left| \int_A^\infty f(x, u)\, dx \right| < \frac{\varepsilon}{3}$$

for every u in $[\alpha, \beta]$ if $A > A_0$. For an arbitrary point u_0 in $[\alpha, \beta]$, we have

$$\varphi(u) - \varphi(u_0) = \int_a^\infty f(x, u)\, dx + \int_a^\infty f(x, u_0)\, dx.$$

Thus,

$$\left| \varphi(u) - \varphi(u_0) \right| \leqslant \left| \int_a^A f(x, u)\, dx - \int_a^A f(x, u_0)\, dx \right|$$
$$+ \left| \int_A^\infty f(x, u)\, dx \right| + \left| \int_A^\infty f(x, u)\, dx \right|.$$

The value of the second and the third integrals on the right hand side is less than $\frac{\varepsilon}{3}$.

Since $f(x, u)$ is continuous on $a \leqslant x < \infty$, $\alpha \leqslant u \leqslant \beta$,

$$\int_a^A f(x, u)\, dx$$

is continuous on $[\alpha, \beta]$. Thus, for any $\varepsilon > 0$, there exists a $\delta > 0$ such that

$$\left| \int_a^A f(x, u)\, dx - \int_a^A f(x, u_0)\, dx \right| < \frac{\varepsilon}{3}$$

if $|u - u_0| < \delta$. Thus, we have

$$\left| \varphi(u) - \varphi(u_0) \right| < \varepsilon$$

if $|u - u_0| < \delta$.

If $\varphi(u) = \int_a^\infty f(x, u)\, dx$ is uniformly convergent on $[\alpha, \beta]$, then the order of the integrals $\int_\alpha^\beta$ and $\int_a^\infty$ can be exchanged. That is

$$\int_\alpha^\beta \varphi(u)\, du = \int_\alpha^\beta \left[\int_a^\infty f(x, u)\, dx \right] du = \int_a^\infty \left[\int_\alpha^\beta f(x, u)\, du \right] dx.$$

To prove this result, express $\displaystyle\int_a^\infty f(x,u)\,dx$ as a series

$$\int_a^\infty f(x,u)\,dx = \sum_{n=1}^\infty a_n(u),$$

where $\displaystyle a_n(u) = \int_{a+n-1}^{a+n} f(x,u)\,dx$.

Since $\displaystyle\int_a^\infty f(x,u)\,dx$ is uniformly convergent on $[\alpha,\beta]$, for any $\varepsilon > 0$, there exists an $A_0 > a$ such that

$$\left| \int_A^\infty f(x,u)\,dx \right| < \varepsilon$$

if $A > A_0$.

Choose $m > A_0 - a$. That is $a + m > A_0$. Then

$$\left| \sum_{n=m+1}^\infty a_n(u) \right| = \left| \sum_{n=m+1}^\infty \int_{a+n-1}^{a+n} f(x,u)\,dx \right| = \left| \int_{a+m}^\infty f(x,u)\,dx \right| \leqslant \varepsilon.$$

This implies that $\displaystyle\sum_{n=1}^\infty a_n(u)$ is uniformly convergent on $[\alpha,\beta]$.

By the corresponding theorem for series of functions (the order of the sum and the integral can be exchanged), we have

$$\int_\alpha^\beta \int_a^\infty f(x,u)\,dx\,du = \int_\alpha^\beta \sum_{n=1}^\infty a_n(u)\,du$$

$$= \sum_{n=1}^\infty \int_\alpha^\beta a_n(u)\,du = \sum_{n=1}^\infty \int_\alpha^\beta \int_{a+n-1}^{a+n} f(x,u)\,dx\,du$$

$$= \sum_{n=1}^\infty \int_{a+n-1}^{a+n} \int_\alpha^\beta f(x,u)\,du\,dx = \int_a^\infty \int_\alpha^\beta f(x,u)\,du\,dx.$$

The corresponding theorem for the derivatives is the following:

Suppose $f(x,u)$ is continuous on $a \leqslant x < \infty, \alpha \leqslant u \leqslant \beta$; the partial derivative $\dfrac{\partial f(x,u)}{\partial u}$ is also continuous on the same region; the integral $\displaystyle\int_a^\infty \dfrac{\partial f(x,u)}{\partial u}\,dx$ is uniformly convergent on $[\alpha,\beta]$. Then

$$\varphi(u) = \int_a^\infty f(x,u)\,dx$$

is differentiable on $[\alpha, \beta]$, and

$$\varphi'(u) = \int_a^\infty \frac{\partial f(x, u)}{\partial u}\, dx,$$

where $\alpha \leqslant u \leqslant \beta$.

This result can be proved by the corresponding theorem for series of functions:

Since

$$\varphi(u) = \int_a^\infty f(x, u)\, dx = \sum_{n=1}^\infty a_n(u),$$

where $a_n(u) = \int_{a+n-1}^{a+n} f(x, u)\, dx$, $\quad (n = 1, 2, \cdots)$, by the assumption, the order of the derivative and the integral can be exchanged, so we have

$$a_n'(u) = \int_{a+n-1}^{a+n} \frac{\partial f(x, u)}{\partial u}\, dx,$$

where $\alpha \leqslant u \leqslant \beta$, and the function $a_n'(u)$ is continuous on $[\alpha, \beta]$. Thus

$$\int_a^\infty \frac{\partial f(x, u)}{\partial u}\, dx = \sum_{n=1}^\infty a_n'(u).$$

Since $\displaystyle\int_a^\infty \frac{\partial f(x, u)}{\partial u}\, dx$ is uniformly convergent on $[\alpha, \beta]$, $\displaystyle\sum_{n=1}^\infty a_n'(u)$ is uniformly convergent on $[\alpha, \beta]$. Hence, $\displaystyle\sum_{n=1}^\infty a_n(u)$ satisfies the conditions of termwise differentiation. Therefore

$$\varphi'(u) = \sum_{n=1}^\infty a_n'(u) = \int_a^\infty \frac{\partial f(x, u)}{\partial u}\, dx.$$

Example 5. Compute $\displaystyle I = \int_0^{+\infty} \frac{e^{-2x} - e^{-3x}}{x}\, dx$.

Solution: Since

$$\frac{e^{-2x} - e^{-3x}}{x} = \int_2^3 e^{-xu}\, du,$$

we have

$$I = \int_0^{+\infty} \frac{e^{-2x} - e^{-3x}}{x}\, dx = \int_0^{+\infty} \left[\int_2^3 e^{-xu}\, du \right] dx.$$

Since $f(x, u) = e^{-xu}$ is continuous on $0 \leqslant x < \infty$, $2 \leqslant x \leqslant 3$, and the integral $\displaystyle\int_0^{+\infty} e^{-xu}\,dx$ is uniformly convergent on $[2, 3]$, the order of the integrals can be exchanged. That is

$$I = \int_0^\infty \frac{e^{-2x} - e^{-3x}}{x}\,dx = \int_0^\infty \left[\int_2^3 e^{-xu}\,du\right] dx$$

$$= \int_2^3 \left[\int_0^\infty e^{-xu}\,dx\right] du = \int_2^3 \frac{1}{u}\,du = \ln\frac{3}{2}.$$

Example 6. Compute

$$I(\beta) = \int_0^\infty e^{-x^2} \cos 2\beta x\,dx.$$

Solution: Since

$$|e^{-x^2} \cos 2\beta x| \leqslant e^{-x^2},$$

and the integral $\displaystyle\int_0^\infty e^{-x^2}\,dx$ is convergent, the integral $\displaystyle\int_0^\infty e^{-x^2} \cos 2\beta x\,dx$ is also convergent. Moreover

$$\frac{\partial(e^{-x^2} \cos 2\beta x)}{\partial \beta} = -2xe^{-x^2} \sin 2\beta x$$

is continuous on $0 \leqslant x < \infty$, $-\infty < \beta < \infty$, and the integral

$$\int_0^\infty \frac{\partial}{\partial \beta}(e^{-x^2} \cos 2\beta x)\,dx = -\int_0^\infty 2xe^{-x^2} \sin 2\beta x\,dx$$

is uniformly convergent on $-\infty < \beta < \infty$. Therefore the order of the derivative and the integral can be exchanged. That is

$$I'(\beta) = \frac{dI}{d\beta} = \int_0^\infty \frac{\partial}{\partial \beta}(e^{-x^2} \cos 2\beta x)\,dx$$

$$= \int_0^\infty (-2xe^{-x^2} \sin 2\beta x)\,dx$$

$$= \sin 2\beta x e^{-x^2}\Big|_0^\infty - \int_0^\infty e^{-x^2} 2\beta \cos 2\beta x\,dx = -2\beta I(\beta).$$

Since

$$I(0) = \int_0^\infty e^{-x^2}\,dx = \frac{\sqrt{\pi}}{2},$$

$I(\beta)$ is the solution of the initial value problem of the first-order differential equation

$$\begin{cases} \dfrac{dI(\beta)}{d\beta} = -2\beta I(\beta), \\[2mm] I(0) = \dfrac{\sqrt{\pi}}{2}. \end{cases}$$

Solve this equation, we obtain

$$I(\beta) = \frac{\pi}{2}e^{-\beta^2}.$$

For a function $f(x, u)$ that is defined on $a \leqslant x < \infty$, $\alpha \leqslant u < \infty$, what condition is needed to exchange the order of the integral of $f(x, u)$ with respect to x and the integral of $f(x, u)$ with respect to u?

If integrals

$$\int_a^\infty f(x, u)\, dx, \quad \int_\alpha^\infty f(x, u)\, du$$

are uniformly convergent on $[\alpha, \beta]$ with respect to u, and on $[a, b]$ with respect to x; at lease one of the following integrals exist

$$\int_a^\infty \int_\alpha^\infty |f(x, u)|\, du\, dx, \quad \int_\alpha^\infty \int_a^\infty |f(x, u)|\, dx\, du,$$

then the integrals

$$\int_a^\infty \int_\alpha^\infty f(x, u)\, du\, dx, \quad \int_\alpha^\infty \int_a^\infty f(x, u)\, dx\, du$$

exist and are equal. That is

$$\int_a^\infty \int_\alpha^\infty f(x, u)\, du\, dx = \int_\alpha^\infty \int_a^\infty f(x, u)\, dx\, du.$$

This implies that the integral $\displaystyle\int_a^\infty \int_\alpha^\infty$ can be exchanged to $\displaystyle\int_\alpha^\infty \int_a^\infty$.

To prove this result, suppose the integral $\displaystyle\int_a^\infty \int_\alpha^\infty |f(x, u)|\, du\, dx$ exists. Then

$$\int_a^\infty \int_\alpha^\infty f(x, u)\, du\, dx$$

exists. What we need to prove is

$$\lim_{\beta \to \infty} \int_\alpha^\beta \int_a^\infty f(x, u)\, dx\, du = \int_a^\infty \int_\alpha^\infty f(x, u)\, du\, dx.$$

Since $\displaystyle\int_a^\infty f(x, u)\, dx$ is uniformly convergent with respect to u on $[\alpha, \beta]$, we have

$$\int_\alpha^\beta \int_a^\infty f(x, u)\, dx\, du = \int_a^\infty \int_\alpha^\beta f(x, u)\, du\, dx.$$

Thus, we need to show that

$$\lim_{\beta \to \infty} \int_a^\infty \int_\alpha^\beta f(x, u)\, du\, dx = \int_a^\infty \int_\alpha^\infty f(x, u)\, du\, dx.$$

That is

$$\lim_{\beta \to \infty} \int_a^\infty \int_\beta^\infty f(x, u) \, du \, dx = 0.$$

Since the integral

$$\int_a^\infty \int_\alpha^\infty |f(x, u)| \, du \, dx$$

is convergent, there exists a $b > a$, such that

$$\int_b^\infty \int_\alpha^\infty |f(x, u)| \, du \, dx < \frac{\varepsilon}{2}.$$

Thus,

$$\left| \int_b^\infty \int_\beta^\infty f(x, u) \, du \, dx \right| \leqslant \int_b^\infty \int_\beta^\infty |f(x, u)| \, du \, dx$$

$$\leqslant \int_b^\infty \int_\alpha^\infty |f(x, u)| \, du \, dx < \frac{\varepsilon}{2}.$$

However,

$$\int_a^\infty \int_\beta^\infty f(x, u) \, du \, dx = \int_a^b \int_\beta^\infty f(x, u) \, du \, dx$$

$$+ \int_b^\infty \int_\beta^\infty f(x, u) \, du \, dx.$$

The absolute value of the second integral on the right side is less than $\frac{\varepsilon}{2}$.

For the first integral, since $\int_\alpha^\infty f(x, u) \, du$ is uniformly convergent on $[a, b]$ with respect to x, there exists a β_0 such that

$$\left| \int_\beta^\infty f(x, u) \, du \right| < \frac{\varepsilon}{2(b - a)}$$

for every x in $[a, b]$, if $\beta > \beta_0$. Thus,

$$\left| \int_a^b \int_\beta^\infty f(x, u) \, du \, dx \right| \leqslant \int_a^b \left| \int_\beta^\infty f(x, u) \, du \right| dx < \frac{\varepsilon}{2}$$

Therefore

$$\left| \int_a^\infty \int_\beta^\infty f(x, u) \, du \, dx \right| < \varepsilon$$

if $\beta > \beta_0$.

Example 7. Compute

$$I = \int_0^\infty \frac{\cos \alpha x}{1 + x^2}\, dx,$$

where α is a constant.

Solution: Since

$$\frac{1}{1 + x^2} = \int_0^\infty e^{-y(1+x^2)}\, dy,$$

we have

$$I = \int_0^\infty \left[\int_0^\infty e^{-y(1+x^2)}\, dy \right] \cos \alpha x\, dx$$

$$= \int_0^\infty \int_0^\infty \cos \alpha x\, e^{-y(1+x^2)}\, dy\, dx.$$

Since the integrand function $f(x, y) = \cos \alpha x\, e^{-y(1+x^2)}$ is continuous on $0 \leqslant x < \infty, 0 \leqslant y < \infty$, and the integral of the absolute value function is

$$\int_0^\infty dy \int_0^\infty |e^{-y(1+x^2)} \cos \alpha x|\, dx \leqslant \int_0^\infty e^{-y}\, dy \int_0^\infty e^{-yx^2}\, dx$$

$$= \frac{\sqrt{\pi}}{2} \int_0^\infty \frac{e^{-y}}{\sqrt{y}}\, dy = \sqrt{\pi} \int_0^\infty e^{-t^2}\, dt = \frac{\pi}{2},$$

the order of the integrals can be exchanged, and we have

$$I = \int_0^\infty e^{-y}\, dy \cdot \int_0^\infty e^{-yx^2} \cos \alpha x\, dx.$$

By Example 6, we obtain

$$\int_0^\infty e^{-yx^2} \cos \alpha x\, dx = \frac{1}{2}\sqrt{\frac{\pi}{y}}\, e^{-\frac{\alpha^2}{4y}}.$$

Therefore

$$I = \int_0^\infty e^{-y} \frac{1}{2}\sqrt{\frac{\pi}{y}}\, e^{-\frac{\alpha^2}{4y}}\, dy = \sqrt{\pi} \int_0^\infty e^{-(y + \frac{\alpha^2}{4y})}\, d\sqrt{y}.$$

Let $\sqrt{y} = t$. Then

$$I = \sqrt{\pi} \int_0^\infty e^{-(t^2 + \frac{\alpha^2}{4y})}\, dt = e^{|\alpha|}\sqrt{\pi} \int_0^\infty e^{-(t + \frac{|\alpha|}{2t})^2}\, dt.$$

Let $u = t - \dfrac{|\alpha|}{2t}$. Then $t + \dfrac{|\alpha|}{2t} = \sqrt{u^2 + 2|\alpha|}$. Since $-\infty < u < \infty$ when $0 < t < \infty$, we have

$$I = \sqrt{\pi} e^{|\alpha|} \int_0^\infty e^{-(u^2 + 2|\alpha|)}\, du = \sqrt{\pi} e^{-|\alpha|} \int_0^\infty e^{-u^2}\, du = \frac{\pi}{2} e^{-|\alpha|}.$$

10.4.4 *Several Important Infinite Integrals*

In this section, we study several useful infinite integrals.

1. *Dirichlet Integral* $\displaystyle\int_0^\infty \frac{\sin x}{x}\, dx$.

We have proved before that this integral is convergent, but not uniformly convergent.

We introduce a convergence factor $e^{-\alpha x}$, and consider the integral with a parameter

$$I(\alpha) = \int_0^\infty e^{-\alpha x} \frac{\sin x}{x}\, dx.$$

It can be proved that this integral is uniformly convergent for $\alpha \geqslant 0$. Since the integrand function is continuous in $0 \leqslant \alpha < \infty$, $0 \leqslant x < \infty$, $I(\alpha)$ is continuous in $0 \leqslant \alpha < \infty$. Especially,

$$\lim_{\alpha \to 0} I(\alpha) = I(0) = \int_0^\infty \frac{\sin x}{x}\, dx.$$

The derivative of $I(\alpha)$ with respect to α is

$$I'(\alpha) = -\int_0^\infty e^{-\alpha x} \sin x\, dx, \quad (\alpha > 0).$$

To show that the order of the differentiation and the integration can be exchanged, we observe the fact that

$$|e^{-\alpha x} \sin x| \leqslant e^{-\delta x}$$

if $\alpha \geqslant \delta > 0$, hence the integral $\displaystyle\int_0^\infty e^{-\alpha x} \sin x\, dx$ is uniformly convergent for $\alpha \geqslant \delta(> 0)$.

By using the method of integration by parts twice for the integral $\displaystyle\int_0^\infty e^{-\alpha x} \sin x\, dx$, we obtain

$$I'(\alpha) = -\frac{1}{1+\alpha^2}, \quad (\alpha > 0).$$

Thus,

$$I(\alpha) = -\arctan\alpha + C.$$

Since

$$|I(\alpha)| = \left| \int_0^\infty e^{-\alpha x} \frac{\sin x}{x}\, dx \right| \leqslant \int_0^\infty e^{-\alpha x}\, dx = \frac{1}{\alpha},$$

we have $I(\alpha) \to 0$ when $\alpha \to \infty$. It follows that $C = \dfrac{\pi}{2}$. Hence

$$I(\alpha) = \frac{\pi}{2} - \arctan \alpha, \quad (\alpha > 0).$$

Let $\alpha \to 0$. We get

$$I(0) = \int_0^\infty \frac{\sin x}{x} \, dx = \frac{\pi}{2}.$$

From this result, it is not hard to obtain that

$$\int_0^\infty \frac{\sin \alpha x}{x} \, dx = \frac{\pi}{2} \operatorname{sgn} \alpha.$$

2. Evaluate $J = \displaystyle\int_0^\infty e^{-x^2} \, dx$.

Solution: We have calculated this integral before. Here we find the value of this integral by the theorem of exchanging the operations of integrations.

Let $x = ut$, $(u > 0)$. Then

$$J = u \int_0^\infty e^{-u^2 t^2} \, dt.$$

Multiplying $e^{-u^2} \, du$ on both sides of this equation, and integrating, we obtain

$$J \int_0^\infty e^{-u^2} \, du = J^2 = \int_0^\infty e^{-u^2} u \, du \int_0^\infty e^{-u^2 t^2} \, dt.$$

Since

$$\int_0^\infty e^{-(1+t^2)u^2} u \, du = \frac{1}{2} \frac{1}{1+t^2},$$

we have

$$J^2 = \int_0^\infty dt \int_0^\infty e^{-(1+t^2)u^2} u \, du = \frac{1}{2} \int_0^\infty \frac{1}{1+t^2} \, dt = \frac{\pi}{4}.$$

Therefore

$$J = \int_0^\infty e^{-x^2} \, dx = \frac{\sqrt{\pi}}{2}.$$

3. *Fresnel Integral*

$$\int_0^\infty \sin x^2 \, dx, \quad \int_0^\infty \cos x^2 \, dx.$$

Let $t = x^2$. Then

$$\int_0^\infty \sin x^2 \, dx = \frac{1}{2} \int_0^\infty \frac{\sin t}{\sqrt{t}} \, dt.$$

Since

$$\frac{1}{\sqrt{t}} = \frac{2}{\sqrt{\pi}} \int_0^\infty e^{-tu^2} \, du,$$

we have

$$\int_0^\infty \frac{\sin t}{\sqrt{t}} e^{-kt} \, dt = \frac{2}{\sqrt{\pi}} \int_0^\infty e^{-kt} \sin t \, dt \int_0^\infty e^{-tu^2} \, du.$$

By the theorem of exchanging the order of the integrals, we have

$$\int_0^\infty \frac{\sin t}{\sqrt{t}} e^{-kt} \, dt = \frac{2}{\sqrt{\pi}} \int_0^\infty du \int_0^\infty e^{-(k+u^2)t} \sin t \, dt$$

$$= \frac{2}{\sqrt{\pi}} \int_0^\infty \frac{1}{1 + (k+u^2)^2} \, du.$$

By Abel criterion, the integral on the left side is uniformly convergent with respect to $k \geqslant 0$, so we can take the limit for $k \to 0$ inside the integral. The limit of the integral on the right hand side can also be taken under the integral sign. Thus

$$\int_0^\infty \frac{\sin t}{\sqrt{t}} \, dt = \frac{2}{\sqrt{\pi}} \int_0^\infty \frac{1}{1 + u^4} \, du = \frac{2}{\sqrt{\pi}} \frac{\pi}{2\sqrt{2}} = \sqrt{\frac{\pi}{2}}.$$

Therefore

$$\int_0^\infty \sin x^2 \, dx = \frac{1}{2} \sqrt{\frac{\pi}{2}}.$$

Similarly, we have

$$\int_0^\infty \cos x^2 \, dx = \frac{1}{2} \sqrt{\frac{\pi}{2}}.$$

From this result, it is not hard to obtain that

$$\int_0^\infty \cos \alpha x^2 \, dx = \frac{1}{2} \sqrt{\frac{\pi}{2\alpha}}.$$

4. *Gamma Function*

$$\Gamma(s) = \int_0^\infty t^{s-1} e^{-t} \, dt.$$

It was proved before that this integral is convergent for $s > 0$.

First, we show that $\Gamma(s)$ is a continuous function in $(0, \infty)$. Rewrite the integral of $\Gamma(s)$ into two parts:

$$\Gamma(s) = \int_0^1 t^{s-1} e^{-t} \, dt + \int_1^\infty t^{s-1} e^{-t} \, dt.$$

For any $\beta > \alpha > 0$, if s is in $[\alpha, \beta]$, then

$$t^{s-1}e^{-t} \leqslant t^{\alpha-1}e^{-t}$$

if $0 < t < 1$.

Since $\int_0^1 t^{\alpha-1}e^{-t}\,dt$ is convergent, the integral $\int_0^1 t^{s-1}e^{-t}\,dt$ is uniformly convergent on $[\alpha, \beta]$. Notice that $t^{s-1}e^{-t} \leqslant t^{\beta-1}e^{-t}$ if $t > 1$. Since $\int_1^\infty t^{\beta-1}e^{-t}\,dt$ is convergent, the integral $\int_1^\infty t^{s-1}e^{-t}\,dt$ is also uniformly convergent on $[\alpha, \beta]$. Thus, the integral $\int_0^\infty t^{s-1}e^{-t}\,dt$ is uniformly convergent on $[\alpha, \beta]$. It follows that $\Gamma(s)$ is continuous on $[\alpha, \beta]$. Since α and β are arbitrary $(\beta > \alpha > 0)$, $\Gamma(s)$ is continuous on $(0, \infty)$.

Next, we study the recurrent relation.

For $s > 0$, we have

$$\Gamma(s+1) = s\Gamma(s).$$

Integrating by the method of integration by parts, we have

$$\Gamma(s+1) = \int_0^\infty t^s e^{-t}\,dt = -t^s e^{-t}\Big|_0^\infty + s\int_0^\infty t^{s-1}e^{-t}\,dt = s\Gamma(s).$$

Since

$$\Gamma(1) = \int_0^\infty e^{-t}\,dt = 1,$$

we have

$$\Gamma(n+1) = n\Gamma(n) = \cdots = n(n-1)\ldots 2\cdot 1\Gamma(1) = n!,$$

where n is a positive integer. Thus, the gamma function can also be considered as a generalization of the factorial.

Especially, $\Gamma\left(\dfrac{1}{2}\right)$ is useful. Its value is

$$\Gamma\left(\frac{1}{2}\right) = \int_0^\infty t^{-\frac{1}{2}}e^{-t}\,dt.$$

Let $t = x^2$. Then

$$\Gamma\left(\frac{1}{2}\right) = 2\int_0^\infty e^{-t^2}\,dt = \sqrt{\pi}.$$

The gamma function is also called the *Euler integral of the second kind*.

5. *Beta Function*

$$B(p, q) = \int_0^1 t^{p-1}(1-t)^{q-1}\,dt.$$

It has been proved before that this integral is convergent if $p > 0$ and $q > 0$.

Now, we show that the beta function is continuous for $p > 0$ and $q > 0$. Let $p_0 > 0$ and $q_0 > 0$. Then

$$t^{p-1}(1-t)^{q-1} \leqslant t^{p_0-1}(1-t)^{q_0-1}, \quad (0 < t < 1)$$

if $p \geqslant p_0 > 0$ and $q \geqslant q_0 > 0$.

Since the integral

$$\int_0^1 t^{p_0-1}(1-t)^{q_0-1}\, dt$$

is convergent, the integral

$$\int_0^1 t^{p-1}(1-t)^{q-1}\, dt$$

is uniformly convergent with respect to p and q, for $p \geqslant p_0 > 0$ and $q \geqslant q_0 > 0$. Thus, the beta function is continuous on $p \geqslant p_0 > 0$, $q \geqslant q_0 > 0$. Since p_0 and q_0 are arbitrary, the beta function is continuous for $p > 0$ and $q > 0$.

Next, we study the recurrent relation for the beta function:

For any $p > 0$, $q > 0$ we have

$$B(p+1, q+1) = \frac{pq}{(p+q+1)(p+q)} B(p, q).$$

This can be proved by the method of integration by parts

$$\begin{aligned}
B(p+1, q+1) &= \int_0^1 t^p (1-t)^q\, dt \\
&= \left[\frac{t^{p+1}}{p+1} (1-t)^q \right]\Big|_0^1 + \frac{q}{p+1} \int_0^1 t^{p+1}(1-t)^{q-1}\, dt \\
&= \frac{q}{p+1} \int_0^1 t^{p+1}(1-t)^{q-1}\, dt \\
&= \frac{q}{p+1} \int_0^1 [t^p - t^p(1-t)](1-t)^{q-1}\, dt \\
&= \frac{q}{p+1} [B(p+1, q) - B(p+1, q+1)].
\end{aligned}$$

Thus,

$$B(p+1, q+1) = \frac{q}{p+q+1} B(p+1, q).$$

Similarly, we can obtain

$$B(p+1, q+1) = \frac{p}{p+q+1} B(p, q+1).$$

Substituting $q + 1$ by q, we have

$$B(p+1, q) = \frac{p}{p+q} a(p, q).$$

Hence

$$B(p+1, q+1) = \frac{q}{p+q+1} B(p+1, q) = \frac{q}{p+q+1} \frac{p}{p+q} B(p, q).$$

Let $p = m - 1$ and $q = n - 1$ (where m and n are positive integers). By applying the recurrent formula repeatedly we obtain that

$$
\begin{aligned}
B(m, n) &= \frac{m-1}{m+n-1} B(m-1, n) \\
&= \frac{(m-1)(m-2)}{(m+n-1)(m+n-2)} B(m-2, n) \\
&= \cdots = \frac{(m-1)!}{(m+n-1)(m+n-2)\cdots(n+1)} B(1, n) \\
&= \frac{(m-1)!}{(m+n-1)(m+n-2)\cdots(n+1)} \frac{n-1}{n} B(1, n-1) \\
&= \cdots = \frac{(m-1)!(n-1)!}{(m+n-1)!} B(1, 1).
\end{aligned}
$$

Since $B(1, 1) = 1$, we have

$$B(m, n) = \frac{(m-1)!(n-1)!}{(m+n-1)!} = \frac{\Gamma(m)\Gamma(n)}{\Gamma(m+n)}.$$

This equation is true not only for positive integers p, q, but also for positive real numbers p, q. That is, for any $p > 0$ and $q > 0$,

$$B(p, q) = \frac{\Gamma(p)\Gamma(q)}{\Gamma(p+q)}.$$

Before we prove this equation, we need to show that

$$B(p, q) = \int_0^\infty \frac{z^{q-1}}{(1+z)^{p+q}} \, dz$$

for any $p > 0$ and $q > 0$.

Let $t = \dfrac{1}{1+z}$ in the definition of the beta function

$$B(p, q) = \int_0^1 t^{p-1}(1-t)^{q-1} \, dt.$$

Then $1 - t = \dfrac{z}{1+z}$ and $dt = \dfrac{-1}{(1+z)^2}\, dz$. Thus, we have

$$B(p,q) = \int_0^1 t^{p-1}(1-t)^{q-1}\, dt$$

$$= \int_\infty^0 \frac{1}{(1+z)^{p-1}} \frac{z^{q-1}}{(1+z)^{q-1}} \frac{-1}{(1+z)^2}\, dz$$

$$= \int_0^\infty \frac{z^{q-1}}{(1+z)^{p+q}}\, dz.$$

By this equation, it is easy to prove that

$$B(p,q) = \frac{\Gamma(p)\Gamma(q)}{\Gamma(p+q)}.$$

Let $u = tv$, $(t > 0)$. Then $\Gamma(y) = \displaystyle\int_0^\infty u^{y-1}e^{-u}\, du$ becomes

$$\Gamma(y) = \int_0^\infty (tv)^{y-1}e^{-tv}t\, dv.$$

That is

$$\frac{\Gamma(y)}{t^y} = \int_0^\infty v^{y-1}e^{-tv}\, dv.$$

Substituting y by $x + y$ and substituting t by $1 + t$, we have

$$\frac{\Gamma(x+y)}{(1+t)^{x+y}} = \int_0^\infty v^{x+y-1}e^{-(1+t)v}\, dv.$$

Multiplying t^{y-1} on both side of the equation and integrating with respect to t from 0 to ∞, we obtain

$$\Gamma(x+y)\int_0^\infty \frac{t^{y-1}}{(1+t)^{x+y}}\, dt = \int_0^\infty t^{y-1}\, dt \int_0^\infty v^{x+y-1}e^{-(1+t)v}\, dv$$

The integral on the left hand side is $B(x,y)$. Exchange the order of the integrals, we get

$$\int_0^\infty v^{x+y-1}e^{-v}\, dv \int_0^\infty t^{y-1}e^{-tv}\, dt = \int_0^\infty v^{x+y-1}e^{-v}\frac{\Gamma(y)}{v^y}\, dv$$

$$= \Gamma(y)\int_0^\infty v^{x-1}e^{-v}\, dv = \Gamma(x)\Gamma(y)$$

That is

$$B(x,y) = \frac{\Gamma(x)\Gamma(y)}{\Gamma(x+y)}.$$

From this equation, we also have

$$B(x, y) = B(y, x).$$

Furthermore, using knowledge of complex analysis, we can derive the Euler's reflection formula:

$$\Gamma(p)\Gamma(1 - p) = \frac{\pi}{\sin p\pi}$$

for any $0 < p < 1$, and the Legendre duplication formula:

$$\Gamma(2p) = \frac{2^{2p-1}}{\sqrt{\pi}}\Gamma(p)\Gamma\left(p + \frac{1}{2}\right)$$

for $p > 0$.

The beta function is also called the *Euler integral of the first kind.* Together with the gamma function, they are called the *Euler integrals.*

Example 1. Evaluate $I = \displaystyle\int_0^{+\infty} \frac{1}{(1 + t^2)^n}\, dt.$

Solution: For

$$B(p, q) = \int_0^{+\infty} \frac{z^{q-1}}{(1 + z)^{p+q}}\, dz,$$

let $q = \dfrac{1}{2}$, we have

$$B\left(p, \frac{1}{2}\right) = \int_0^{+\infty} \frac{z^{-\frac{1}{2}}}{(1 + z)^{p+\frac{1}{2}}}\, dz.$$

Let $t = z^{\frac{1}{2}}$, we have

$$B\left(p, \frac{1}{2}\right) = \int_0^{+\infty} \frac{2}{(1 + t^2)^{p+\frac{1}{2}}}\, dt.$$

Let $p + \dfrac{1}{2} = n$. Then $p = \dfrac{2n - 1}{2}$. Thus

$$I = \int_0^{+\infty} \frac{1}{(1 + t^2)^n}\, dt = \frac{1}{2}B\left(\frac{2n - 1}{2}, \frac{1}{2}\right)$$

$$= \frac{1}{2}\frac{\Gamma\left(n - \frac{1}{2}\right)\Gamma\left(\frac{1}{2}\right)}{\Gamma(n)} = \frac{(2n - 3)!!}{(2n - 2)!!}\frac{\pi}{2}.$$

Example 2. Evaluate $I = \displaystyle\int_0^{\frac{\pi}{2}} \cos^m x \sin^n x\, dx$ where $m > -1, n > -1$.

Solution: Let $t = \sin^2 x$. Then $x = \arcsin \sqrt{t}$, and we have

$$\int_0^{\frac{\pi}{2}} \cos^m x \sin^n x \, dx = \frac{1}{2} \int_0^1 t^{\frac{n-1}{2}} (1-t)^{\frac{m-1}{2}} \, dt$$

$$= \frac{1}{2} B\left(\frac{n+1}{2}, \frac{m+1}{2}\right) = \frac{1}{2} \frac{\Gamma\left(\dfrac{n+1}{2}\right) \Gamma\left(\dfrac{m+1}{2}\right)}{\Gamma\left(\dfrac{m+n}{2}+1\right)}.$$

Let $m = 0$. Then

$$\int_0^{\frac{\pi}{2}} \sin^n x \, dx = \frac{1}{2} \frac{\Gamma\left(\dfrac{1}{2}\right) \Gamma\left(\dfrac{n+1}{2}\right)}{\Gamma\left(\dfrac{n}{2}+1\right)}$$

$$= \begin{cases} \dfrac{(n-1)!!}{n!!} \dfrac{\pi}{2}, & \text{if } n \text{ is an even number;} \\[2mm] \dfrac{(n-1)!!}{n!!}, & \text{if } n \text{ is an odd number.} \end{cases}$$

Let $m = -\alpha$ and $n = \alpha$. Then

$$\int_0^{\frac{\pi}{2}} (\tan x)^\alpha \, dx = \frac{1}{2} \Gamma\left(\frac{1+\alpha}{2}\right) \Gamma\left(\frac{1-\alpha}{2}\right).$$

If $p = \dfrac{1+\alpha}{2}$, then $1-p = \dfrac{1-\alpha}{2}$. By Euler's reflection formula, we have that

$$\int_0^{\frac{\pi}{2}} (\tan x)^\alpha \, dx = \frac{1}{2} \frac{\pi}{\sin \dfrac{1+\alpha}{2}\pi} = \frac{\pi}{2\cos \dfrac{\alpha\pi}{2}}.$$

Exercises 10.4

1. Determine whether the following integrals are convergent.

(1) $\displaystyle\int_0^{+\infty} \frac{x}{1+x^2} \, dx$;

(2) $\displaystyle\int_0^{+\infty} \frac{3x^2-2}{x^5-x^3+1} \, dx$;

(3) $\displaystyle\int_1^{\infty} \frac{1}{x\sqrt[3]{x^2+1}} \, dx$;

(4) $\displaystyle\int_2^{+\infty} \frac{x}{x(\ln x)^p} \, dx$;

(5) $\displaystyle\int_0^{+\infty} \frac{x^{13}}{(x^5+x^3+1)^3} \, dx$;

(6) $\displaystyle\int_0^{+\infty} \frac{x \arctan x}{\sqrt[3]{1+x^4}} \, dx$;

(7) $\displaystyle\int_0^{+\infty} x \sin x \, dx;$ (8) $\displaystyle\int_0^{+\infty} e^{-ax} \sin bx \, dx;$

(9) $\displaystyle\int_0^{+\infty} \frac{\sin^2 x}{x^2} \, dx;$ (10) $\displaystyle\int_0^{+\infty} \sqrt{x} e^{-x} \, dx;$

(11) $\displaystyle\int_0^{+\infty} \frac{x^m}{1 + x^n} \, dx, \ (n \geqslant 0);$ (12) $\displaystyle\int_{e^2}^{+\infty} \frac{1}{x \ln \ln x} \, dx;$

(13) $\displaystyle\int_0^1 \frac{\ln x}{\sqrt{1 - x^2}} \, dx;$ (14) $\displaystyle\int_0^1 \frac{1}{e^{\sqrt{x}} - 1} \, dx;$

(15) $\displaystyle\int_a^b \frac{x}{\sqrt{(x - a)(x - b)}} \, dx, \ (a < b);$

(16) $\displaystyle\int_0^1 \frac{x^n}{\sqrt{1 - x^4}} \, dx, \ (n \geqslant 0);$

(17) $\displaystyle\int_0^{\frac{\pi}{2}} \frac{\ln \sin x}{\sqrt{x}} \, dx;$ (18) $\displaystyle\int_0^1 \frac{\sqrt{x}}{e^{\sin x} - 1} \, dx;$

(19) $\displaystyle\int_0^1 \frac{1}{e^x - \cos x} \, dx;$ (20) $\displaystyle\int_0^{+\infty} \frac{\arctan x}{x^a} \, dx.$

2. Find the following limits.

(1) $\displaystyle\lim_{a \to 0} \int_{-1}^1 \sqrt{x^2 + a^2} \, dx;$ (2) $\displaystyle\lim_{a \to 0} \int_0^2 x^2 \cos ax \, dx;$

(3) $\displaystyle\lim_{a \to 0} \int_a^{1+a} \frac{1}{1 + x^2 + a^2} \, dx.$

3. Suppose that $f(x)$ is continuous on $[a, A]$. Prove that

$$\lim_{h \to 0} \frac{1}{h} \int_a^x [f(t + h) - f(t)] \, dt = f(x) - f(a).$$

4. Prove that the nth-order Bessel function

$$J_n(x) = \frac{1}{\pi} \int_0^\pi (\cos n\varphi - x \sin \varphi) \, d\varphi$$

satisfies the Bessel equation

$$x^2 J_n''(x) + x J_n'(x) + (x^2 - n^2) J_n(x) = 0.$$

5. Determine whether the following integrals are absolutely convergent or conditionally convergent.

(1) $\displaystyle\int_0^{+\infty} \frac{\sqrt{x}\cos x}{1+x}\,dx;$ (2) $\displaystyle\int_1^{+\infty} \frac{\cos(1-2x)}{\sqrt{x^3}\sqrt[3]{x^2+1}}\,dx;$

(3) $\displaystyle\int_0^{+\infty} \frac{\sin x}{\sqrt[3]{x^2+x+1}}\,dx.$

6. Suppose $f(x,y) = \dfrac{y^2 - x^2}{(x^2+y^2)^2}$. Prove that

$$\int_0^1 dy \int_0^1 f(x,y)\,dx \neq \int_0^1 dx \int_0^1 f(x,y)\,dy,$$

and interpret the reason why the inequality holds.

7. Calculate the derivative of the following functions.

(1) $\displaystyle f(x) = \int_{\sin x}^{\cos x} e^{x\sqrt{1-t^2}}\,dt;$ (2) $\displaystyle f(x) = \int_{a+x}^{b+x} \frac{\sin xt}{t}\,dt;$

(3) $\displaystyle f(x) = \int_x^{x^2} e^{-xy^2}\,dy;$ (4) $\displaystyle \varphi(\alpha) = \int_0^\alpha f(x+\alpha, x-\alpha)\,dx;$

(5) $\displaystyle f(\alpha) = \int_0^\alpha \frac{\ln(1+\alpha x)}{x}\,dx.$

8. Suppose $u(x) = \displaystyle\int_0^1 k(x,y)v(y)\,dy$, where $v(y)$ is a continuous function, and

$$k(x,y) = \begin{cases} x(1-y), & \text{if } x \leqslant y; \\ y(1-x), & \text{if } x > y. \end{cases}$$

Prove that $u''(x) = -v(x)$ for $0 \leqslant x \leqslant 1$.

9. Suppose $\varphi(x)$ has the second derivative, and $\psi(x)$ has the first derivative. Prove that

$$u(x,t) = \frac{1}{2}[\varphi(x-at) + \varphi(x+at)] + \frac{1}{2a}\int_{x-at}^{x+at} \psi(s)\,ds$$

satisfies the wave equation (equation of the vibrating string)

$$\frac{\partial^2 u}{\partial t^2} = a^2 \frac{\partial^2 u}{\partial x^2}.$$

10. Determine the convergence region of the following integrals with parameters.

(1) $\int_a^b \frac{1}{(b-x)^u} \, dx, \quad (b > a);$ (2) $\int_0^\infty x^u \, dx;$

(3) $\int_1^\infty x^u \frac{x + \sin x}{x - \sin x} \, dx;$ (4) $\int_2^\infty \frac{1}{x^u \ln x} \, dx;$

(5) $\int_0^{\frac{\pi}{2}} \frac{1 - \cos x}{x^u} \, dx;$ (6) $\int_2^\pi \frac{1}{\sin^u x} \, dx.$

11. Determine whether the following integrals are uniformly convergent on the given regions.

(1) $\int_{-\infty}^{+\infty} \frac{\cos ux}{1 + x^2} \, dx, \quad (-\infty < u < +\infty);$

(2) $\int_0^{+\infty} e^{-u^2(1+x^2)} \sin x \, dx, \quad (0 < u_0 \leqslant u < +\infty);$

(3) $\int_0^{+\infty} e^{-ax} \sin x \, dx,$ (i) $(0 < a_0 < a < +\infty),$ (ii) $(0 < a < +\infty);$

(4) $\int_1^{+\infty} e^{-ax} \frac{\cos x}{\sqrt{x}} \, dx, \quad (0 \leqslant a < +\infty);$

(5) $\int_0^{+\infty} \sqrt{a} e^{-ax^2} \, dx, \quad (0 \leqslant a < +\infty).$

12. Evaluate the following integrals.

(1) $\int_0^1 \frac{x^\beta - x^\alpha}{\ln x} \, dx, \quad (\alpha > -1, \ \beta > -1);$

(2) $\int_0^{+\infty} \frac{e^{-ax} - e^{bx}}{x} \, dx, \quad (a > 0, b > 0);$

(3) $\int_0^{+\infty} \frac{1 - e^{-ax^2}}{x^2} \, dx, \quad (a > 0);$

(4) $\int_0^{+\infty} \frac{e^{-ax^2} - e^{-\beta x^2}}{x} \, dx, \quad (\alpha > 0, \ \beta > 0);$

(5) $\int_0^{+\infty} \frac{\arctan ax}{x(1 + x^2)} \, dx;$

(6) $\int_0^1 \frac{\arctan x}{x} \frac{1}{\sqrt{1 - x^2}} \, dx.$ Hint: $\frac{\arctan x}{x} = \int_0^1 \frac{1}{1 + x^2 u^2} \, du.$

13. Using the integrals

$$\int_0^{+\infty} e^{-x^2} \, dx = \frac{\sqrt{\pi}}{2}, \quad \int_0^{+\infty} \frac{\sin x}{x} \, dx = \frac{\pi}{2}$$

to evaluate the following integrals.

(1) $\displaystyle\int_{-\infty}^{+\infty} e^{-(2x^2+x+2)}\,dx;$ (2) $\displaystyle\int_{0}^{+\infty} \frac{e^{-x}}{\sqrt{x}}\,dx;$

(3) $\displaystyle\int_{0}^{+\infty} \frac{\sin ax \cos bx}{x}\,dx,\ (a>0,\ b>0);$

(4) $\displaystyle\int_{0}^{+\infty} \frac{\sin^2 x}{x^2}\,dx;$

(5) $\displaystyle\int_{0}^{+\infty} \frac{e^{-\alpha x^2} - e^{-\beta x^2}}{x^2}\,dx,\ (\alpha>0,\ \beta>0);$

(6) $\displaystyle\int_{0}^{+\infty} x^{2n} e^{-x^2}\,dx,\ (n \text{ is a positive integer}).$

14. Evaluate the following integrals by using Euler integrals.

(1) $\displaystyle\int_{0}^{1} \sqrt{x - x^2}\,dx;$ (2) $\displaystyle\int_{0}^{+\infty} \frac{\sqrt[4]{x}}{(1+x)^2}\,dx;$

(3) $\displaystyle\int_{0}^{+\infty} \frac{x^2}{1+x^4}\,dx;$ (4) $\displaystyle\int_{0}^{a} x^2\sqrt{a^2 - x^2}\,dx;$

(5) $\displaystyle\int_{0}^{+\infty} \frac{1}{1+x^3}\,dx;$

(6) $\displaystyle\int_{0}^{+\infty} x^{2n+1} e^{-x^2}\,dx,\ (n \text{ is a positive integer});$

(7) $\displaystyle\int_{0}^{\frac{\pi}{2}} \sin^6 x \cos^4 x\,dx;$ (8) $\displaystyle\int_{0}^{\frac{\pi}{2}} \tan^n x\,dx,\ (|n|<1);$

(9) $\displaystyle\int_{0}^{1} x^{n-1}(1-x^m)^{q-1}\,dx,\ (n,q,m>0);$

(10) $\displaystyle\int_{0}^{+\infty} \frac{1}{(a^2+x^2)^n}\,dx,\ (n \text{ is a positive integer}).$

15. Prove the equation

$$\int_{0}^{+\infty} f\left(ax + \frac{b}{x}\right)\,dx = \frac{1}{a}\int_{0}^{+\infty} f(\sqrt{x^2 + 4ab})\,dx,$$

where $a>0$, $b>0$ and assume the integral on the left hand side is well defined.

Hint: Let $ax - \dfrac{b}{x} = t$, so $-\infty < t < +\infty$ when $0 < x < \infty$, and $ax + \dfrac{b}{x} = \sqrt{t^2 + 4ab}$.

Chapter 11

Fourier Series and Fourier Integrals

11.1 Fourier Series

11.1.1 *Orthogonality of the System of Trigonometric Functions*

In Chapter 10 we learned that if infinitely many arithmetic operations are applied on elementary functions, the result is not necessarily an elementary function. On the other hand, expanding a function into a series of elementary functions can provide conveniences for studying the function itself. The three basic elementary functions are: power functions, trigonometric functions and exponential functions. We focused our study on power series expansions of functions, the Taylor series, in the last chapter. In this chapter, we study trigonometric series expansions of functions, the Fourier series. Since trigonometric functions are periodic, trigonometric series expansions are useful for solving some real world problems such as frequency spectrum analysis. From Euler formula

$$e^{ix} = \cos x + i \sin x, \ (i = \sqrt{-1}),$$

the power series expansion of a function can be studied through the trigonometric series expansion of the function. (The proof of the Euler formula can be found in Complex Analysis.)

On the other hand, we studied that the infinite integral can be considered as a "continuous version" of the series, a "discrete sum". Also, we found the correspondence between the theorems for series and the theorems for infinite integrals. Similarly, the "continuous version" of Fourier series is Fourier integrals.

The functions $\sin x$ and $\cos x$ are periodic functions with period 2π. Any function $f(x)$ with period ω (that is, $f(x + \omega) = f(x)$) can be transformed into a function with period 2π by a change of variables

$$g(x) = f\left(\frac{\omega}{2\pi}x\right).$$

Therefore, we only need to study functions with period 2π.

For any integer k, the functions $\sin kx$ and $\cos kx$ are periodic functions with period 2π. The linear combination of these functions

$$\frac{1}{2}a_0 + \sum_{k=1}^{n}(a_k \cos kx + b_k \sin kx)$$

is also a periodic function with period 2π. The series

$$\frac{1}{2}a_0 + \sum_{k=1}^{\infty}(a_k \cos kx + b_k \sin kx)$$

is called a *trigonometric series*.

First, we prove that the system of trigonometric functions

$$\sin x, \ \cos x, \ \sin 2x, \ \cos 2x, \ \cdots, \ \sin nx, \ \cos nx, \cdots$$

from an *orthogonal system*.

In fact, it is obvious that

$$\int_{-\pi}^{\pi} \cos kx \, dx = 0, \quad \int_{-\pi}^{\pi} \sin kx \, dx = 0,$$

for $k = 1, 2, 3, \cdots$.

Since

$$\sin kx \cos lx = \frac{1}{2}(\sin(k+l)x + \sin(k-l)x),$$

$$\sin kx \sin lx = \frac{1}{2}(\cos(k-l)x - \cos(k+l)x),$$

$$\cos kx \cos lx = \frac{1}{2}(\cos(k+l)x + \cos(k-l)x),$$

we have

$$\int_{-\pi}^{\pi} \cos kx \sin lx \, dx = 0.$$

If $k \neq l$, then

$$\int_{-\pi}^{\pi} \cos kx \cos lx \, dx = 0, \quad \int_{-\pi}^{\pi} \sin kx \sin lx \, dx = 0.$$

On the other hand, since

$$\cos^2 kx = \frac{1 + \cos 2kx}{2}, \quad \sin^2 kx = \frac{1 - \cos 2kx}{2},$$

we have

$$\int_{-\pi}^{\pi} \cos^2 kx \, dx = \pi, \qquad \int_{-\pi}^{\pi} \sin^2 kx \, dx = \pi.$$

From the above results we conclude that the trigonometric functions

$$\frac{1}{\sqrt{2\pi}}, \quad \frac{1}{\sqrt{\pi}} \cos kx, \quad \frac{1}{\sqrt{\pi}} \sin kx, \quad k = 1, 2, \cdots$$

form an orthogonal system.

The integral from 0 to 2π of the product of any two distinct functions in the system is 0, and the integral from 0 to 2π of the square of any function in the system is 1. This is called a *system of orthonormal functions*.

Consider a function $f(x)$ on $[-\pi, \pi]$ and with period 2π. If we let

$$\int_{-\pi}^{\pi} f(x) \, dx = \pi a_0,$$

$$\int_{-\pi}^{\pi} f(x) \cos lx \, dx = \pi a_l, \qquad \int_{-\pi}^{\pi} f(x) \sin lx \, dx = \pi b_l,$$

then we get a trigonometric series

$$\frac{1}{2} a_0 + \sum_{k=1}^{\infty} (a_k \cos kx + b_0 \sin kx).$$

This is called the *Fourier series of the function* $f(x)$. The numbers a_0, a_l, b_l, $(l = 1, 2, \cdots)$ are called the *Fourier coefficients of* $f(x)$. The Fourier series obtained in this way is not necessarily convergent. Even if it is convergent, it is not necessarily convergent to $f(x)$. We denote this by

$$f(x) \sim \frac{1}{2} a_0 + \sum_{k=1}^{\infty} (a_k \cos kx + b_0 \sin kx).$$

Of course, the interval of integration is not necessarily from $-\pi$ to π. It can be any interval of length 2π, $(c, c + 2\pi)$, and especially $(0, 2\pi)$.

With Euler's formula

$$\cos kx = \frac{e^{ikx} + e^{-ikx}}{2}, \qquad \sin kx = \frac{e^{ikx} - e^{-ikx}}{2i},$$

the Fourier series of $f(x)$ can be written as

$$\frac{a_0}{2} + \sum_{k=1}^{\infty} \left(a_k \frac{e^{ikx} + e^{-ikx}}{2} + ib_k \frac{e^{-ikx} - e^{ikx}}{2} \right)$$

$$= \frac{a_0}{2} + \sum_{k=1}^{\infty} \frac{a_k - ib_k}{2} e^{ikx} + \sum_{k=1}^{\infty} \frac{a_k - ib_k}{2} e^{-ikx}$$

$$= \sum_{k=-\infty}^{\infty} F_k e^{ikx}.$$

Since

$$F_0 = \frac{a_0}{2} = \frac{1}{2\pi} \int_{-\pi}^{\pi} f(x) \, dx,$$

$$F_{\pm k} = \frac{1}{2}(a_k \mp i b_k) = \frac{1}{2\pi} \int_{-\pi}^{\pi} f(x)(\cos kx \mp i \sin kx) \, dx$$

$$= \frac{1}{2\pi} \int_{-\pi}^{\pi} f(x) e^{\mp ikx} \, dx$$

for $k = 1, 2, 3, \cdots$. The series

$$\sum_{k=-\infty}^{\infty} F_k e^{ikx}$$

is called the *complex form of the Fourier series of $f(x)$*, and F_0, $F_{\pm k}$ are the *complex Fourier coefficients of $f(x)$*. Moreover

$$F_k = \bar{F}_{-k}, \quad (k = 1, 2, \cdots)$$

are conjugate complex numbers.

The complex form of Fourier series of $f(x)$ can be derived in another way:

The following functions

$$\cdots, e^{-ikx}, \cdots, e^{-ix}, 1, e^{ix}, \cdots, e^{ikx}, \cdots$$

are pairwise orthogonal on $(-\pi, \pi)$, that is

$$\int_{-\pi}^{\pi} e^{i(m+n)x} \, dx = \begin{cases} 0, & \text{if } m + n \neq 0; \\ 2\pi, & \text{if } m + n = 0. \end{cases}$$

The complex Fourier coefficients of a given periodic function $f(x)$ are

$$F_k = \frac{1}{2\pi} \int_{-\pi}^{\pi} f(x) e^{-ikx} \, dx,$$

where $k = \cdots, -2, -1, 0, 1, 2, \cdots$. Thus, the complex Fourier series of $f(x)$ is:

$$\sum_{k=-\infty}^{\infty} F_k e^{ikx}.$$

Essentially, the two forms of Fourier series have no differences. We can choose a form that is convenient for the problem at hand.

Especially, if $f(x)$ is an even function, that is, $f(-x) = f(x)$, then $f(x) \sin kx$ is an odd function. Thus

$$\int_{-\pi}^{\pi} f(x) \sin kx \, dx = 0.$$

It follows that, if $f(x)$ is an even function, the Fourier series of $f(x)$ is

$$f(x) \sim \frac{a_0}{2} + \sum_{k=1}^{\infty} a_k \cos kx,$$

where

$$a_k = \frac{2}{\pi} \int_0^{\pi} f(x) \cos kx \, dx \qquad (k = 0, 1, 2, \cdots).$$

This series is called a *cosine series*.

Similarly, if $f(x)$ is an odd function, that is, $f(-x) = -f(x)$, then the Fourier series of $f(x)$ only has terms of sine functions:

$$f(x) \sim \sum_{k=1}^{\infty} b_k \sin kx,$$

where

$$b_k = \frac{2}{\pi} \int_0^{\pi} f(x) \sin kx \, dx \qquad (k = 1, 2, \cdots).$$

This series is called a *sine series*.

Example 1. Find the Fourier series of $f(x) = x$ on $(-\pi, \pi)$ (Figure 11.1).

Solution: Since $f(x) = x$ is an odd function, it can be expanded to a sine series

$$f(x) \sim \sum_{n=1}^{\infty} b_n \sin nx,$$

where

$$b_n = \frac{2}{\pi} \int_0^{\pi} x \sin nx \, dx = \frac{2}{n}(-1)^{n+1}$$

for $n = 1, 2, \cdots$.

Thus, the Fourier series of $f(x) = x$ is

$$2 \sum_{n=1}^{\infty} \frac{(-1)^{n+1}}{n} \sin nx.$$

Example 2. Find the Fourier series of $f(x) = |x|$ on $(-\pi, \pi)$ (Figure 11.2).

Solution: Since $f(x) = |x|$ is an even function, it can be expanded to a cosine series

$$f(x) \sim \frac{a_0}{2} + \sum_{n=1}^{\infty} a_n \cos nx,$$

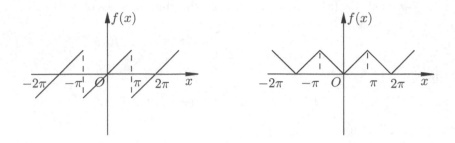

Fig. 11.1   Fig. 11.2

where

$$a_0 = \frac{2}{\pi} \int_0^\pi x \, dx = \pi,$$

$$a_n = \frac{2}{\pi} \int_0^\pi x \cos nx \, dx = \frac{2}{n\pi} \int_0^\pi x \, d\sin nx$$

$$= \frac{2}{n\pi} \left(x \sin nx \Big|_0^\pi - \int_0^\pi \sin nx \, dx \right)$$

$$= \frac{2}{n\pi} \frac{1}{n} \cos nx \Big|_0^\pi = \frac{2}{n^2\pi} [(-1)^n - 1]$$

$$= \begin{cases} 0, & n = 2k; \\ \dfrac{-4}{(2k+1)^2\pi}, & n = 2k+1. \end{cases}$$

Thus,

$$f(x) \sim \frac{\pi}{2} - \frac{4}{\pi} \sum_{k=0}^{\infty} \frac{1}{(2k+1)^2} \cos(2k+1)x.$$

Example 3. The half wave rectification eliminates the negative voltage in the alternating voltage $E(t) = E \sin t$. Find the Fourier series for the function of half wave rectification (Figure 11.3).

Solution: The function is

$$f(t) = \begin{cases} 0, & -\pi \leqslant t \leqslant 0; \\ E \sin t, & 0 \leqslant t \leqslant \pi. \end{cases}$$

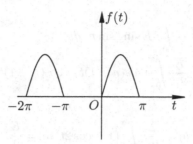

Fig. 11.3

Thus,

$$a_0 = \frac{1}{\pi} \left(\int_{-\pi}^{0} 0 \, dt + \int_{0}^{\pi} E \sin t \, dt \right) = \frac{1}{\pi} \int_{0}^{\pi} E \sin t \, dt = \frac{2E}{\pi},$$

$$a_n = \frac{1}{\pi} \int_{0}^{\pi} E \sin t \cos nt \, dt$$

$$= \frac{E}{2\pi} \int_{0}^{\pi} (\sin(n+1)t - \sin(n-1)t) \, dt.$$

For $n = 1$, we have

$$a_1 = \frac{E}{2\pi} \int_{0}^{\pi} \sin 2t \, dt = 0.$$

For $n \neq 1$, we have

$$a_n = \frac{E}{2\pi} \int_{0}^{\pi} (\sin(n+1)t - \sin(n-1)t) \, dt$$

$$= \frac{E}{2\pi} \left[-\frac{\cos(n+1)t}{n+1} + \frac{\cos(n-1)t}{n-1} \right] \Big|_{0}^{\pi}$$

$$= \frac{E}{2\pi} \left[-\frac{(-1)^{n+1}}{n+1} + \frac{1}{n+1} + \frac{(-1)^{n-1}}{n-1} - \frac{1}{n-1} \right]$$

$$= -\frac{E(1 + (-1)^n)}{(n^2 - 1)\pi}$$

$$= \begin{cases} 0, & n = 2k+1; \\ \dfrac{2E}{(1 - (2k)^2)\pi}, & n = 2k. \end{cases}$$

Also,

$$b_n = \frac{1}{\pi} \int_0^\pi E \sin t \sin nt \, dt$$

$$= \frac{E}{2\pi} \int_0^\pi (\cos(n-1)t - \cos(n+1)t) \, dt.$$

For $n = 1$, we have

$$b_1 = \frac{E}{2\pi} \int_0^\pi (1 - \cos 2t) \, dt = \frac{E}{2}.$$

For $n \neq 1$, we have

$$b_n = \frac{E}{2\pi} \int_0^\pi (\cos(n-1)t - \cos(n+1)t) \, dt$$

$$= \frac{E}{2\pi} \left[\frac{\sin(n-1)t}{n-1} - \frac{\sin(n+1)t}{n+1} \right]\Big|_0^\pi = 0.$$

Therefore, the Fourier series of $f(t)$ is

$$\frac{E}{\pi} + \frac{1}{2} E \sin t + \frac{2E}{\pi} \sum_{k=1}^\infty \frac{1}{1-(2k)^2} \cos 2kt.$$

To find the sine series expansion of a given function $f(x)$ on $[0, \pi]$, we can construct the odd extension of $f(x)$ on the interval $[-\pi, 0]$. That is, define

$$f_1(x) = \begin{cases} f(x), & 0 \leqslant x \leqslant \pi; \\ -f(-x), & -\pi \leqslant x \leqslant 0. \end{cases}$$

Then, we extend $f_1(x)$ as a periodic function to get an odd function $F(x)$ with period 2π (Figure 11. 4). Thus, the sine series expansion of $f(x)$ on $[0, \pi]$ is:

$$\sum_{n=1}^\infty b_n \sin nx,$$

where

$$b_n = \frac{2}{\pi} \int_0^\pi F(x) \sin nx \, dx = \frac{2}{\pi} \int_0^\pi f(x) \sin nx \, dx \quad (n = 1, 2, \cdots).$$

In a similar way, we can find the cosine series expansion of $f(x)$ on $[0, \pi]$:

$$\frac{a_0}{2} + \sum_{n=1}^\infty a_n \cos nx,$$

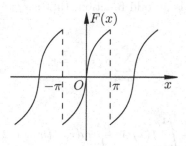

Fig. 11.4

where

$$a_n = \frac{2}{\pi} \int_0^\pi f(x) \cos nx \, dx \quad (n = 0, 1, 2, \cdots).$$

For a given function $f(x)$ on $[-l, l]$ with period $2l$, the Fourier series of $f(x)$ on $[-l, l]$ can be obtained by a variable transformation:

$$\frac{a_0}{2} + \sum_{n=1}^\infty \left(a_n \cos \frac{n\pi}{l} x + b_n \sin \frac{n\pi}{l} x \right),$$

where

$$a_n = \frac{1}{l} \int_{-l}^l f(x) \cos \frac{n\pi}{l} x \, dx, \quad (n = 0, 1, 2, \cdots),$$

$$b_n = \frac{1}{l} \int_{-l}^l f(x) \sin \frac{n\pi}{l} x \, dx, \quad (n = 1, 2, 3, \cdots).$$

Especially, if $f(x)$ is an even function, that is, $f(-x) = f(x)$, then

$$b_n = \frac{1}{l} \int_{-l}^l f(x) \sin \frac{n\pi}{l} x \, dx = 0.$$

Thus, the cosine series of $f(x)$ is

$$\frac{a_0}{2} + \sum_{n=1}^\infty a_n \cos \frac{n\pi}{l} x,$$

where

$$a_n = \frac{2}{l} \int_0^1 f(x) \cos \frac{n\pi}{l} x \, dx, \quad (n = 0, 1, 2, \cdots).$$

Similarly, if $f(x)$ is an odd function, that is, $f(-x) = -f(x)$, then the sine series of $f(x)$ is:

$$\sum_{n=1}^{\infty} b_n \sin \frac{n\pi}{l} x,$$

where

$$b_n = \frac{2}{l} \int_0^1 f(x) \sin \frac{n\pi}{l} x \, dx, \quad (n = 1, 2, 3, \cdots).$$

By the extension method above, we can find the cosine series expansion or the sine series expansion of a function $f(x)$ that is defined on $[0, l]$.

Example 4. Find the cosine series expansion of the function $f(x) = x$ on $[0, \pi]$.

Solution: We construct an even extension of $f(x) = x$ to $(-\pi, 0)$ (Figure 11.5). Then we have

$$a_0 = \frac{2}{\pi} \int_0^\pi x \, dx = \pi,$$

$$a_n = \frac{2}{\pi} \int_0^\pi x \cos nx \, dx = \frac{2}{n\pi} \left[x \sin nx \Big|_0^\pi - \int_0^\pi \sin nx \, dx \right]$$

$$= \frac{2}{n\pi} \frac{\cos nx}{n} \Big|_0^\pi = \frac{2}{n^2\pi} [(-1)^n - 1]$$

$$= \begin{cases} 0, & n = 2m; \\ \dfrac{-4}{(2m+1)^2 \pi}, & n = 2m+1. \end{cases}$$

Therefore the cosine series of $f(x) = x$ is

$$\frac{\pi}{2} - \frac{4}{\pi} \sum_{n=0}^{\infty} \frac{\cos(2n+1)x}{(2n+1)^2}.$$

Example 5. Find the sine series expansion of the following function on $[0, l]$.

$$f(x) = \begin{cases} x, & 0 \leqslant x < \dfrac{l}{2}; \\ l - x, & \dfrac{l}{2} \leqslant x \leqslant l. \end{cases}$$

Solution: We construct an odd extension of $f(x)$ to $(-l, 0)$ (Figure 11.6).

Then

$$b_n = \frac{2}{l} \int_0^l f(x) \sin \frac{n\pi}{l} x \, dx$$

$$= \frac{2}{l} \int_0^{\frac{l}{2}} x \sin \frac{n\pi}{l} x \, dx + \frac{2}{l} \int_{\frac{l}{2}}^l (l - x) \sin \frac{n\pi}{l} x \, dx$$

$$= \frac{4l}{n^2 \pi^2} \sin \frac{n\pi}{2} = \begin{cases} 0, & n = 2m; \\ \dfrac{(-1)^m 4l}{(2m+1)^2 \pi^2}, & n = 2m + 1. \end{cases}$$

Therefore the sine series of $f(x)$ is

$$\frac{4l}{\pi^2} \sum_{n=0}^{\infty} \frac{(-1)^n \sin(2n+1)x}{(2n+1)^2}.$$

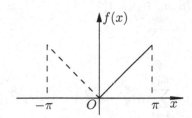

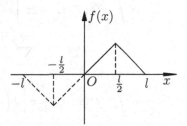

Fig. 11.5   Fig. 11.6

11.1.2   Bessel Inequality

Suppose $f(x)$ is a function on $(-\pi, \pi)$. We would like to determine a trigonometric polynomial

$$g(x) = \frac{\alpha_0}{2} + \sum_{k=1}^{n} (\alpha_k \cos kx + \beta_k \sin kx)$$

such that the value of the integral of the squared difference (mean squared difference)

$$\int_{-\pi}^{\pi} |f(x) - g(x)|^2 \, dx$$

is a minimum. That is, determine the coefficients α_k, β_k, such that the value of the integral

$$\int_{-\pi}^{\pi} |f(x) - g(x)|^2 \, dx$$

is a minimum.

Obviously,

$$\int_{-\pi}^{\pi} \left(f(x) - \frac{\alpha_0}{2} - \sum_{k=1}^{n} (\alpha_k \cos kx + \beta_k \sin kx) \right)^2 dx$$

$$= \int_{-\pi}^{\pi} f^2(x) \, dx - 2 \int_{-\pi}^{\pi} \left[\frac{\alpha_0}{2} + \sum_{k=1}^{n} (\alpha_k \cos kx + \beta_k \sin kx) \right] f(x) \, dx$$

$$+ \int_{-\pi}^{\pi} \left[\frac{\alpha_0}{2} + \sum_{k=1}^{n} (\alpha_k \cos kx + \beta_k \sin kx) \right]^2 dx$$

$$= \int_{-\pi}^{\pi} f^2(x) \, dx - 2\pi \left[\frac{\alpha_0 a_0}{2} + \sum_{k=1}^{n} (\alpha_k a_k + \beta_k b_k) \right]$$

$$+ \pi \left[2 \left(\frac{\alpha_0}{2} \right)^2 + \sum_{k=1}^{n} (\alpha_k^2 + \beta_k^2) \right]$$

$$= \int_{-\pi}^{\pi} f^2(x) \, dx + \pi \left[2 \left(\frac{\alpha_0 - a_0}{2} \right)^2 + \sum_{k=1}^{n} ((\alpha_k - a_k)^2 + (\beta_k - b_k)^2) \right]$$

$$- \pi \left[\frac{a_0^2}{2} + \sum_{k=1}^{n} (a_k^2 + b_k^2) \right]$$

$$\geqslant \int_{-\pi}^{\pi} f^2(x) \, dx - \pi \left[\frac{a_0^2}{2} + \sum_{k=1}^{n} (a_k^2 + b_k^2) \right].$$

Thus, the value of the integral of squared difference is a minimum if $\alpha_k = a_k$ and $\alpha_k = a_k$, and it is

$$\int_{-\pi}^{\pi} f^2(x) \, dx - \pi \left[\frac{a_0^2}{2} + \sum_{k=1}^{n} (a_k^2 + b_k^2) \right].$$

In other words, if the coefficients of the trigonometric polynomial are the Fourier coefficients of $f(x)$, then the value of the integral of squared difference (mean squared difference) between the given function and the trigonometric polynomial is a minimum.

Since

$$\int_{-\pi}^{\pi} |f(x) - g(x)|^2 \, dx \geqslant 0,$$

we have

$$\frac{1}{\pi} \int_{-\pi}^{\pi} f^2(x)\, dx \geqslant \frac{1}{2} a_0^2 + \sum_{k=1}^{n} (a_k^2 + b_k^2).$$

Therefore,

$$\frac{1}{\pi} \int_{-\pi}^{\pi} f^2(x)\, dx \geqslant \frac{1}{2} a_0^2 + \sum_{k=1}^{\infty} (a_k^2 + b_k^2)$$

when we let $n \to \infty$.

By the fact that the limit of the monotonically increasing and bounded sequence exists, the series on the right hand side of the inequality is convergent. This is called the *Bessel inequality*. It follows that

$$a_k = \frac{1}{\pi} \int_{-\pi}^{\pi} f(x) \cos kx\, dx \to 0, \quad b_k = \frac{1}{\pi} \int_{-\pi}^{\pi} f(x) \sin kx\, dx \to 0$$

as $k \to \infty$.

If $f(x)$ is an integrable function, then the Bessel inequality becomes an equality

$$\frac{1}{\pi} \int_{-\pi}^{\pi} f^2(x)\, dx = \frac{1}{2} a_0^2 + \sum_{k=1}^{\infty} (a_k^2 + b_k^2).$$

This is called the *Parseval equality*.

We will not give a proof of the Parseval equality here.

Moreover, there is a generalized theorem:

Riemann–Lebesgue Theorem: For any Riemann integrable function $f(x)$ on a finite interval $[a, b]$,

$$\lim_{\lambda \to \infty} \int_a^b f(x) \sin \lambda x\, dx = 0, \quad \lim_{\lambda \to \infty} \int_a^b f(x) \cos \lambda x\, dx = 0.$$

Proof: Partition the interval $[a, b]$ into n segment

$$a = x_0 < x_1 < \cdots < x_n = b.$$

Let M_i and m_i be the supremum and the infimum of $f(x)$ on $[x_i, x_{i+1}]$ respectively. Then

$$\int_a^b f(x) \sin \lambda x\, dx = \sum_{i=0}^{n-1} \int_{x_i}^{x_{i+1}} [f(x) - m_i] \sin \lambda x\, dx$$

$$+ \sum_{i=0}^{n-1} m_i \int_{x_i}^{x_{i+1}} \sin \lambda x\, dx.$$

Since

$$\left| \int_{x_i}^{x_{i+1}} [f(x) - m_i] \sin \lambda x \, dx \right| \leqslant \int_{x_i}^{x_{i+1}} (M_i - m_i) \, dx$$

$$= (M_i - m_i)(x_{i+1} - x_i),$$

and

$$\left| \int_{x_i}^{x_{i+1}} \sin \lambda x \, dx \right| \leqslant \frac{2}{\lambda},$$

we have

$$\left| \int_a^b f(x) \sin \lambda x \, dx \right| \leqslant \sum_{i=0}^{n-1} (M_i - m_i)(x_{i+1} - x_i) + \frac{2}{\lambda} \sum_{i=0}^{n-1} |m_i|.$$

Since $f(x)$ is Riemann integrable, for any given $\varepsilon > 0$, we can choose a partition of $[a, b]$ such that

$$\sum_{i=0}^{n-1} (M_i - m_i)(x_{i+1} - x_i) < \frac{\varepsilon}{2}.$$

With such a partition determined, the sum

$$\sum_{i=0}^{n-1} |m_i|$$

is a fixed number.

Let

$$\delta = \frac{4}{\varepsilon} \sum_{i=0}^{n-1} |m_i|.$$

Then

$$\left| \int_a^b f(x) \sin \lambda x \, dx \right| < \varepsilon$$

if $\lambda > \delta$. Similarly,

$$\left| \int_a^b f(x) \cos \lambda x \, dx \right| < \varepsilon$$

if $\lambda > \delta$. Therefore, the Riemann-Lebesgue theorem follows.

11.1.3 *Convergence Criterion for Fourier Series*

Consider the partial sum of the Fourier series of $f(x)$

$$S_n = S_n(x) = \frac{a_0}{2} + \sum_{k=1}^{n}(a_k \cos kx + b_k \sin kx),$$

where

$$a_k = \frac{1}{\pi}\int_{-\pi}^{\pi} f(t)\cos kt\, dt, \quad b_k = \frac{1}{\pi}\int_{-\pi}^{\pi} f(t)\sin kt\, dt.$$

Thus, S_n can also be written as

$$S_n = \frac{1}{2\pi}\int_{-\pi}^{\pi} f(t)\, dt$$

$$+ \sum_{k=1}^{n}\left(\cos kx \int_{-\pi}^{\pi} f(t)\cos kt\, dt + \sin kx \int_{-\pi}^{\pi} f(t)\sin kt\, dt\right)$$

$$= \frac{1}{\pi}\int_{-\pi}^{\pi}\left(\frac{1}{2} + \sum_{k=1}^{n}\cos k(x-t)\right) f(t)\, dt$$

$$= \frac{1}{2\pi}\int_{-\pi}^{\pi} \frac{\sin\left(n+\frac{1}{2}\right)(x-t)}{\sin\frac{1}{2}(x-t)} f(t)\, dt.$$

Here we used the fact

$$\frac{1}{2} + \sum_{k=1}^{n}\cos k\theta = \frac{\sin\left(n+\frac{1}{2}\right)\theta}{2\sin\frac{1}{2}\theta}.$$

Let $t = x + u$. Then

$$S_n = \frac{1}{2\pi}\int_{-\pi-x}^{\pi-x} \frac{\sin\left(n+\frac{1}{2}\right)u}{\sin\frac{1}{2}u} f(x+u)\, du.$$

Since the integrand function is periodic with period 2π, we have

$$S_n = \frac{1}{2\pi}\int_{-\pi}^{\pi} \frac{\sin\left(n+\frac{1}{2}\right)u}{\sin\frac{1}{2}u} f(x+u)\, du.$$

This is called the *Dirichlet integral*. Rewriting this integral as the sum of two integrals, one with the interval of integration $[0, \pi]$ and the other with the interval of integration $[-\pi, 0]$, and substituting u by $-u$ in the latter, we have that

$$S_n = \frac{1}{2\pi} \int_0^\pi \frac{\sin\left(n + \frac{1}{2}\right)u}{\sin\frac{1}{2}u} (f(x + u) + f(x - u))\, du.$$

Especially, $S_n = 1$ for $f(x) = 1$. Thus,

$$1 = \frac{1}{2\pi} \int_0^\pi \frac{\sin\left(n + \frac{1}{2}\right)u}{\sin\frac{1}{2}u}\, 2\, du.$$

Multiplying this equation by S (S is a constant), and subtract the result from the expression for S_n, we get

$$S_n - S = \frac{1}{2\pi} \int_0^\pi \frac{\sin\left(n + \frac{1}{2}\right)u}{\sin\frac{1}{2}u} (f(x + u) + f(x - u) - 2S)\, du.$$

Let

$$\Phi(u) = f(x + u) - f(x - u) - 2S.$$

Then the problem of convergence becomes the problem of whether or not the equation

$$\lim_{n \to \infty} \int_0^\pi \frac{\sin\left(n + \frac{1}{2}\right)u}{\sin\frac{1}{2}u} \Phi(u)\, du = 0.$$

is true. Since

$$\frac{\Phi(u)}{\sin\frac{u}{2}}$$

is Riemann integrable on $[\delta, \pi]$, we have

$$\lim_{n \to \infty} \int_\delta^\pi \frac{\sin\left(n + \frac{1}{2}\right)u}{\sin\frac{1}{2}u} \Phi(u)\, du = 0.$$

by Riemann-Lebesgue theorem.

Now the problem becomes whether or not the equation

$$\lim_{n\to\infty} \int_0^\delta \frac{\sin\left(n+\frac{1}{2}\right)u}{\sin\frac{1}{2}u} \Phi(u)\,du = 0$$

is true. Since

$$\left(\frac{1}{\sin\frac{u}{2}} - \frac{1}{\frac{u}{2}}\right)\Phi(u)$$

is Riemann integrable on $[0,\delta]$, by Riemann-Lebesgue theorem we have

$$\lim_{n\to\infty} \int_0^\delta \sin\left(n+\frac{1}{2}\right)u \left(\frac{1}{\sin\frac{1}{2}u} - \frac{1}{\frac{u}{2}}\right)\Phi(u)\,du = 0.$$

Thus, the condition of convergence becomes

$$\lim_{n\to\infty} \int_0^\delta \frac{\Phi(u)}{u} \sin\left(n+\frac{1}{2}\right)u\,du = 0.$$

If the limit

$$\lim_{u\to 0} \frac{\Phi(u)}{u}$$

exists, then by Riemann-Lebesgue theorem, we have

$$\lim_{n\to\infty} \int_0^\delta \frac{\Phi(u)}{u} \sin\left(n+\frac{1}{2}\right)u\,du = 0.$$

It follows that the Fourier series converges to S. Therefore, we obtain the following result:

The convergence criterion for Fourier series: If $f(x)$ is integrable on $[-\pi, \pi]$, and the limit

$$\lim_{u\to 0} \frac{\Phi(u)}{u} = \lim_{u\to 0} \frac{f(x_0+u) + f(x_0-u) - 2S}{u}$$

exists at $x = x_0$, then $S_n \to S$ as $n \to \infty$. In other words, the partial sum of the Fourier series of $f(x)$ converges to S at $x = x_0$.

If $f(x)$ is continuous at this point, then S_n tends to $S = f(x)$. If $f(x)$ is continuous on the left and also continuous on the right at $x = x_0$, that is $f(x_0 - 0)$ and $f(x_0 + 0)$ exist, but are not necessarily the same, then

$$S = \frac{1}{2}[f(x_0+0) + f(x_0-0)]$$

and

$$S_n \to \frac{1}{2}[f(x_0 + 0) + f(x_0 - 0)].$$

If $f(x)$ is a piecewise smooth function on $[-\pi, \pi]$ with period 2π, then the Fourier series of $f(x)$ converges to

$$S(x) = \frac{1}{2}[f(x + 0) + f(x - 0)]$$

at any point x in $[-\pi, \pi]$.

In fact, for a piecewise smooth function

$$\lim_{u \to 0} \frac{\Phi(x, u)}{u} = \lim_{u \to 0} \frac{f(x + u) + f(x - u) - f(x + 0) - f(x - 0)}{u}$$

$$= f'(x + 0) + f'(x - 0).$$

That is,

$$\lim_{u \to 0} \frac{\Phi(x, u)}{u}$$

exists.

By this convergence criterion, we can conclude that the Fourier series in the five examples of Section 11.1.1 are convergent in their domains. At points of discontinuity, they converges to

$$\frac{1}{2}[f(x + 0) + f(x - 0)].$$

In Example 1, let $x = \frac{\pi}{2}$. Then we get the formula

$$1 - \frac{1}{3} + \frac{1}{5} - \cdots = \frac{\pi}{4}.$$

In Example 2, let $x = \pi$. Then we get the formula

$$1 + \frac{1}{3^2} + \frac{1}{5^2} + \frac{1}{7^2} \cdots = \frac{\pi^2}{8}.$$

There are more convergence criteria for Fourier series. We encourage our readers to find them out.

Exercises 11.1

1. Show that the following systems of functions are orthogonal systems on $[0, \pi]$.

   (1) $1, \cos x, \cos 2x, \cdots, \cos nx, \cdots$;

   (2) $\sin x, \sin 2x, \sin 3x, \cdots, \sin nx, \cdots$.

2. Prove that

$$\sin \frac{\pi x}{2l}, \quad \sin \frac{3\pi x}{2l}, \quad \cdots, \quad \sin \frac{(2n+1)\pi x}{2l}, \quad \cdots$$

is an orthogonal system on $[0, l]$.

3. Prove that the Fourier series of the trigonometric polynomial

$$T_n(x) = \sum_{k=1}^{n} (\alpha_k \cos kx + \beta_k \sin kx)$$

is itself.

4. Expand the following functions to Fourier series.
   (1) $f(x) = x^2, \ (-\pi < x < \pi)$;
   (2) $f(x) = \begin{cases} -x, & -\pi \leqslant x < 0, \\ x, & 0 \leqslant x < \pi; \end{cases}$ (3) $f(x) = \begin{cases} e^x, & -\pi < x < 0, \\ 1, & 0 \leqslant x < \pi; \end{cases}$

   (4) $f(x) = \dfrac{\pi - x}{2}, \ 0 < x < 2\pi$;
   (5) $f(x) = |\cos x|, \ -\pi < x < \pi$;   (6) $f(x) = e^{a^x}, \ (-l \leqslant x \leqslant l)$.

5. Expand the following functions to sine series and cosine series.

   (1) $f(x) = \begin{cases} A, & 0 \leqslant x < \dfrac{\pi}{2}, \\ 0, & \dfrac{\pi}{2} \leqslant x \leqslant \pi; \end{cases}$ (2) $f(x) = \begin{cases} 1 - \dfrac{x}{h}, & 0 \leqslant x \leqslant h, \\ 0, & h < x \leqslant \pi. \end{cases}$

6. Expand the following functions to Fourier series, and study the convergence of each series.

   (1) $f(x) = |\sin x|, \ -\pi < x < \pi$;   (2) $f(x) = \sinh ax, \ -\pi < x < \pi$;
   (3) $f(x) = x \sin x, \ -\pi < x < \pi$;

   (4) $f(x) = \begin{cases} 1, & |x| < 1, \\ -1, & 1 \leqslant |x| \leqslant 2; \end{cases}$

   (5) $f(x) = \begin{cases} x, & 0 \leqslant x < 1, \\ 1, & 1 \leqslant x < 2, \\ 3 - x, & 2 \leqslant x \leqslant 3; \end{cases}$ (6) $f(x) = \begin{cases} 0, & -2 \leqslant x \leqslant -1, \\ E, & -1 < x < 1, \\ 0, & 1 \leqslant x \leqslant 2. \end{cases}$

7. Find the sum of the following series by Parseval equation.

(1) $\displaystyle\sum_{n=1}^{\infty} \frac{1}{n^2}$;

(using the Fourier expansion of $f(x) = x$ on $(-\pi < x < \pi)$)

(2) $\displaystyle\sum_{n=1}^{\infty} \frac{1}{(2n+1)^4}$.

(using the Fourier expansion of $f(x) = |x|$ on $(-\pi \leqslant x \leqslant \pi)$)

8. Expand the function

$$f(x) = \begin{cases} 1, & |x| < a, \\ 0, & a \leqslant |x| < \pi; \end{cases}$$

to Fourier series, and find the sum of the following series by Parseval equation.

(1) $\displaystyle\sum_{n=1}^{\infty} \frac{\sin na}{n^2}$;   (2) $\displaystyle\sum_{n=1}^{\infty} \frac{\cos na}{n^2}$.

9. Expand this function to complex Fourier series.

$$f(x) = \begin{cases} -1, & -\pi \leqslant x < 0, \\ 1, & 0 \leqslant x < \pi. \end{cases}$$

10. Suppose $|a| < 1$. Applying the equation

$$\sum_{n=0}^{\infty} a^n e^{inx} = 1 + \sum_{n=1}^{\infty}(a^n \cos nx + ia^n \sin nx)$$

to prove the following equations.

(1) $1 + \displaystyle\sum_{n=1}^{\infty} a^n \cos nx = \frac{1 - a\cos x}{1 - 2a\cos x + a^2}$;

(2) $\displaystyle\sum_{n=1}^{\infty} a^n \sin nx = \frac{a\sin x}{1 - 2a\cos x + a^2}$;

(3) $1 + 2\displaystyle\sum_{n=1}^{\infty} a^n \cos nx = \frac{1 - a^2}{1 - 2a\cos x + a^2}$.

11.2   Fourier Integrals

11.2.1   *Fourier Integrals*

Similar to the discussion in Chapter 10, the Fourier integral is a "continuous sum" corresponding to the Fourier series, a "discrete sum".

Suppose the Fourier series of $f(x)$ on $[-l, l]$ is

$$f(x) = \sum_{n=-\infty}^{\infty} F_n e^{in\omega x}, \quad \omega = \frac{\pi}{l},$$

where

$$F_n = \frac{1}{2l} \int_{-l}^{l} f(\xi) e^{-in\omega\xi} \, d\xi.$$

Substituting F_n into the series, and let

$$\lambda_n = n\omega = \frac{n\pi}{l}, \quad (n = 0, \pm 1, \pm 2, \cdots),$$

$$\Delta\lambda = \omega = \frac{\pi}{l} = \lambda_n - \lambda_{n-1}.$$

Then we have

$$f(x) = \sum_{n=-\infty}^{\infty} \frac{1}{2l} \int_{-l}^{l} f(\xi) e^{-in\omega(\xi-x)} \, d\xi$$

$$= \frac{1}{2\pi} \sum_{n=-\infty}^{\infty} \left(\int_{-l}^{l} f(\xi) e^{-i\lambda_n(\xi-x)} \, d\xi \right) \Delta\lambda.$$

Since $\Delta\lambda \to 0$ as $l \to \infty$, we can consider $\Delta\lambda$ as the infinitesimal interval of a definite integral, and λ_n as the point of division on the x-axis. The sum on the right hand side can be considered as the Riemann sum of the function of λ

$$\int_{-l}^{l} f(\xi) e^{-i\lambda_n(\xi-x)} \, d\xi$$

on $(-\infty, \infty)$. In other words, the above equation becomes

$$f(x) = \frac{1}{2\pi} \int_{-\infty}^{\infty} d\lambda \int_{-\infty}^{\infty} f(\xi) e^{-i\lambda(\xi-x)} \, d\xi$$

as $l \to \infty$.

There is no restriction that $f(x)$ must be a periodic function. The right hand side of this equation is called a *Fourier integral*. This equation is called the Fourier integral formula.

Since

$$\int_{-\infty}^{\infty} f(\xi) \sin\lambda(\xi - x) \, d\xi$$

is an odd function of λ, we have

$$\int_{-\infty}^{\infty} d\lambda \int_{-\infty}^{\infty} f(\xi) \sin\lambda(\xi - x) \, d\xi = 0.$$

The integral

$$\int_{-\infty}^{\infty} f(\xi) \cos \lambda(\xi - x) \, d\xi$$

is an even function of λ, thus

$$\int_{-\infty}^{\infty} d\lambda \int_{-\infty}^{\infty} f(\xi) \cos \lambda(\xi - x) \, d\xi$$

$$= 2 \int_{0}^{\infty} d\lambda \int_{-\infty}^{\infty} f(\xi) \cos \lambda(\xi - x) \, d\xi.$$

Hence, the Fourier integral formula can be written as

$$f(x) = \frac{1}{\pi} \int_{0}^{\infty} d\lambda \int_{-\infty}^{\infty} f(\xi) \cos \lambda(\xi - x) \, d\xi$$

$$= \int_{0}^{\infty} [A(\lambda) \cos \lambda x + B(\lambda) \sin \lambda x] \, d\lambda,$$

where

$$A(\lambda) = \frac{1}{\pi} \int_{-\infty}^{\infty} f(\xi) \cos \lambda \xi \, d\xi,$$

$$B(\lambda) = \frac{1}{\pi} \int_{-\infty}^{\infty} f(\xi) \sin \lambda \xi \, d\xi.$$

The Fourier series of $f(x)$ on $(-l, l)$ is

$$f(x) \sim \frac{a_0}{2} + \sum_{n=1}^{\infty} (a_n \cos n\omega x + b_n \sin n\omega x), \quad \omega = \frac{\pi}{l},$$

where

$$a_n = \frac{1}{l} \int_{-l}^{l} f(\xi) \cos n\omega \xi \, d\xi,$$

$$b_n = \frac{1}{l} \int_{-l}^{l} f(\xi) \sin n\omega \xi \, d\xi.$$

Comparing the two expressions, it is clear that n can only be integers in $\cos n\omega x$ and $\sin n\omega x$, the terms of Fourier series. For Fourier integral, λ can be any positive real number in $\cos \lambda x$ and $\sin \lambda x$. The number n is discrete. The number λ is continuous. We take infinite "discrete sum" to get an infinite series. And we take "continuous sum" to get an infinite integral.

The discussion above is only a formal derivation. Several conditions of convergence are needed for the Fourier integral

$$\frac{1}{\pi} \int_{0}^{\infty} d\lambda \int_{-\infty}^{\infty} f(\xi) \cos \lambda(\xi - x) \, d\xi$$

to converge to $f(x)$. For instance, if $f(x)$ is absolutely integrable on $(-\infty, \infty)$, and $f(x)$ is continuous at x, then the Fourier integral converges to $f(x)$. If the left derivative and the right derivative of $f(x)$ at x exist, then the Fourier integral of $f(x)$ converges to

$$\frac{1}{2}(f(x+0) + f(x-0)).$$

We omit the proofs of these results in this textbook.

11.2.2 *Fourier Transforms*

The Fourier integral formula can also be written as

$$f(x) = \frac{1}{2\pi} \int_{-\infty}^{\infty} d\lambda \int_{-\infty}^{\infty} f(\xi) e^{i\lambda(\xi - x)}\, d\xi$$

$$= \frac{1}{2\pi} \int_{-\infty}^{\infty} \left(\int_{-\infty}^{\infty} f(\xi) e^{i\lambda\xi}\, d\xi \right) e^{-i\lambda x}\, d\lambda.$$

Let

$$F(\lambda) = \int_{-\infty}^{\infty} f(\xi) e^{i\lambda\xi}\, d\xi,$$

we have

$$f(x) = \frac{1}{2\pi} \int_{-\infty}^{\infty} F(\lambda) e^{-i\lambda x}\, d\lambda = F^{-1}[F(\lambda)].$$

The function $F(\lambda)$ is called the *Fourier transform* of $f(x)$, and $f(x)$ or $F^{-1}[F(\lambda)]$ are called the **inverse Fourier transform** of $F(\lambda)$. The two equations above are called the Fourier transform formula and the inversion formula of Fourier transform.

If $f(x)$ is an even function, then

$$F(\lambda) = \int_{-\infty}^{\infty} f(\xi) e^{i\lambda\xi}\, d\xi = \int_{-\infty}^{\infty} f(\xi)(\cos\lambda\xi + i\sin\lambda\xi)\, d\xi$$

$$= 2 \int_{0}^{\infty} f(\xi)\cos\lambda\xi\, d\xi.$$

This is called the *Fourier cosine transform*. Since $F(\lambda)$ is an even function of λ, the inverse transform is

$$f(x) = \frac{1}{2\pi} \int_{-\infty}^{\infty} F(\lambda) e^{-i\lambda x}\, d\lambda = \frac{1}{2\pi} \int_{-\infty}^{\infty} F(\lambda)(\cos\lambda x - i\sin\lambda x)\, d\lambda$$

$$= \frac{1}{\pi} \int_{0}^{\infty} F(\lambda)\cos\lambda x\, d\lambda$$

If $f(x)$ is an odd function, then the Fourier transform is

$$F(\lambda) = 2i \int_0^\infty f(\xi) \sin \lambda\xi \, d\xi.$$

To avoid i, we define the *Fourier sine transform* as

$$G(\lambda) = 2 \int_0^\infty f(\xi) \sin \lambda\xi \, d\xi.$$

Then the corresponding inverse transform is

$$f(x) = \frac{1}{\pi} \int_0^\infty G(\lambda) \sin \lambda x \, d\lambda.$$

Example 1. Find the Fourier transform for the exponential attenuation function

$$f(x) = \begin{cases} e^{-\beta x}, & x \geqslant 0, \beta > 0, \\ 0, & x < 0. \end{cases}$$

Solution: By definition

$$F(\lambda) = \int_{-\infty}^\infty f(\xi) e^{i\lambda\xi} \, d\xi = \int_0^\infty e^{(i\lambda - \beta)\xi} \, d\xi$$

$$= \frac{e^{(i\lambda - \beta)\xi}}{(\lambda i - \beta)} \Big|_0^\infty = \frac{1}{\beta - i\lambda} = \frac{\beta + i\lambda}{\beta^2 + \lambda^2}.$$

Since $f(x)$ is discontinuous at $x = 0$, by the convergence theorem for Fourier integrals, we have

$$\frac{1}{2\pi} \int_{-\infty}^\infty \frac{\beta + i\lambda}{\beta^2 + \lambda^2} e^{-i\lambda\xi} \, d\lambda = \begin{cases} f(x), & x \neq 0; \\ \dfrac{1}{2}, & x = 0. \end{cases}$$

Example 2. Find the Fourier transform for the function

$$f(x) = \begin{cases} 1, & |x| < a; \\ \dfrac{1}{2}, & |x| \pm a; \\ 0, & |x| > a. \end{cases}$$

Solution: Since $f(x)$ is an even function, we have

$$F(\lambda) = 2 \int_0^\infty f(\xi) \cos \lambda\xi \, d\xi = 2 \int_0^a \cos \lambda\xi \, dx$$

$$= \frac{2}{\lambda} \sin \lambda\xi \Big|_0^a = \frac{2 \sin \lambda a}{\lambda}.$$

At the discontinuous points $x = a$ and $x = -a$, since $f(x)$ is defined as the mean-value of the left limit and the right limit, the inverse formula is true on the entire numerical axis:

$$f(x) = \frac{2}{\pi} \int_0^\infty \frac{\sin \lambda a}{\lambda} \cos \lambda x \, d\lambda, \quad (-\infty < x < \infty).$$

Example 3. Find the Fourier sine transform for the function

$$f(x) = \frac{1}{\sqrt{x}}, \quad (x > 0).$$

Solution: Extending the domain of $f(x)$ to $(-\infty, 0)$, such that $f(x)$ is an odd function, that is, define

$$f(x) = \frac{1}{-\sqrt{-x}}, \quad (x < 0).$$

For the extended function, the sine transform is

$$F(\lambda) = 2 \int_0^\infty \frac{1}{\sqrt{\xi}} \sin \lambda \xi \, d\xi = \frac{2}{\sqrt{\lambda}} \int_0^\infty \frac{\sin t}{\sqrt{t}} \, dt.$$

Since

$$\int_0^\infty \frac{\sin t}{\sqrt{t}} \, dt = \sqrt{\frac{\pi}{2}},$$

(Section 10.4.4), the sine transform is

$$F(\lambda) = \frac{\sqrt{2\pi}}{\sqrt{\lambda}}.$$

Similarly, the cosine transform of this function is also

$$F(\lambda) = \frac{\sqrt{2\pi}}{\sqrt{\lambda}}.$$

It is not hard to derive some of the properties of the Fourier transform: If

$$F[f] = F(\lambda) = \int_{-\infty}^\infty f(\xi) e^{i\lambda \xi} \, d\xi,$$

then we have

1. Linear relation: For Fourier transforms of $f_1(x)$ and $f_2(x)$ and constants α, β,

$$F[\alpha f_1 + \beta f_2] = \alpha F[f_1] + \beta F[f_2].$$

2. Similarity relation: For Fourier transforms of $f(x)$,

$$F[f(ax)] = \frac{1}{a} F\left(\frac{\lambda}{a}\right), \quad (a > 0).$$

3. Time-delay property:

$$F[f(x - x_0)] = e^{i\lambda x_0} F[f(x)].$$

4. Frequency-shift property:

$$F[f(x)e^{i\lambda_0 x}] = F(\lambda + \lambda_0).$$

5. Differential relation (1): Suppose the value of $f(x)$ is zero when x tends to positive or negative infinity. If the Fourier transforms of $f(x)$ and $f'(x)$ exist, then we have

$$F[f'(x)] = -i\lambda F[f(x)].$$

More generally, suppose all values of $f(x), f'(x), \cdots, f^{(k-1)}(x)$ are zero when x tends to positive or negative infinity. If $f^{(k)}(x)$ have Fourier transforms for every $k = 1, 2, \cdots, k$, then

$$F[f^{(k)}(x)] = (-i\lambda)^k F[f(x)].$$

6. Differential relation (2): If $f(x)$ and $ixf'(x)$ have Fourier transforms, then

$$F[ixf(x)] = \frac{d}{d\lambda} F(\lambda).$$

The most important property is the Fourier transform formula of convolution products.

An integral with parameters

$$\int_{-\infty}^{\infty} f_1(x - t) f_2(t) \, dt$$

is called the convolution product of $f_1(x)$ and $f_2(x)$. It is denoted as $f_1 * f_2$. Obviously,

$$f_1 * f_2 = f_2 * f_1.$$

7. The Fourier transform of convolution products: If $f_1(x)$, $f_2(x)$ and $f_1(x) * f_2(x)$ have Fourier transforms, then

$$F[f_1 * f_2] = F[f_1] \cdot F[f_2].$$

In fact, by definition

$$F[f_1 * f_2] = \int_{-\infty}^{\infty} e^{i\lambda \xi} \, d\xi \int_{-\infty}^{\infty} f_1(\xi - t) f_2(t) \, dt.$$

Since $f_1(x)$ and $f_2(x)$ are absolutely integrable on $(-\infty, \infty)$, the order of the operations of two integrals can be exchanged. Thus,

$$
\begin{aligned}
F[f_1 * f_2] &= \int_{-\infty}^{\infty} f_2(t)\, dt \int_{-\infty}^{\infty} f_1(\xi - t)e^{i\lambda\xi}\, d\xi \\
&= \int_{-\infty}^{\infty} f_2(t)\, dt \int_{-\infty}^{\infty} f_1(u)e^{i\lambda(u+t)}\, du \\
&= \int_{-\infty}^{\infty} f_2(t)e^{i\lambda t}\, dt \int_{-\infty}^{\infty} f_1(u)e^{i\lambda u}\, du \\
&= F[f_1] \cdot F[f_2] = F_1(\lambda) \cdot F_2(\lambda),
\end{aligned}
$$

and

$$
F^{-1}[F_1(\lambda) \cdot F_2(\lambda)] = f_1 * f_2 = \int_{-\infty}^{\infty} f_1(x - t)f_2(t)\, dt.
$$

Example 4. Find the Fourier transform of the sine attenuation function

$$
g(x) = \begin{cases} e^{-\beta x} \sin \omega_0 x, & x \geqslant 0; \\ 0, & x < 0, \end{cases}
$$

using the Fourier transform of the exponential attenuation function

$$
f(x) = \begin{cases} e^{-\beta x}, & x \geqslant 0, \quad (\beta > 0); \\ 0, & x < 0. \end{cases}
$$

Solution: By Example 1, we have

$$
F[f(x)] = F(\lambda) = \frac{1}{\beta - i\lambda}.
$$

Since

$$
g(x) = f(x) \sin \omega_0 x = -\frac{i}{2}(e^{i\omega_0 x} - e^{-i\omega_0 x})f(x),
$$

by the linear relation and frequency-shift property, we have

$$
\begin{aligned}
F[g(x)] &= -\frac{i}{2}[F(e^{i\omega_0 x} f(x)) - F(e^{-i\omega_0 x} f(x))] \\
&= -\frac{i}{2}[F(\lambda + \omega_0) - F(\lambda - \omega_0)] \\
&= -\frac{i}{2}\left[\frac{1}{\beta - i(\lambda + \omega_0)} - \frac{1}{\beta - i(\lambda - \omega_0)}\right] \\
&= \frac{\omega_0}{(\beta - i\lambda)^2 + \omega_0^2}.
\end{aligned}
$$

11.2.3 *Applications of Fourier Transforms*

Fourier transforms can be applied to find solutions of ordinary differential equations or partial differential equations.

Example 1. Find the solution of the differential equation $y'' = y$ that satisfies

$$\lim_{x \to \pm\infty} y(x) = \lim_{x \to \pm\infty} y'(x) = 0.$$

Solution: Suppose the solution that satisfies the condition is $y(x)$. Apply Fourier transforms to $y(x)$ and $y''(x)$, we get

$$F[y(x)] = F(\lambda), \quad F[y''(x)] = -\lambda^2 F(\lambda).$$

Under the Fourier transforms, the problems for determining solutions becomes

$$-\lambda^2 F(\lambda) = F(\lambda),$$
$$F(\lambda) = 0.$$

Thus,

$$y(x) = F^{-1}[F(\lambda)] = F^{-1}(0) = 0.$$

Example 2. Find the solution of the partial differential equation

$$\begin{cases} \dfrac{\partial u}{\partial t} = a^2 \dfrac{\partial^2 u}{\partial x^2}, & (t > 0, \ -\infty < x < \infty), \\ u|_{t=0} = \varphi(x), & \lim_{x \to \pm\infty} u(t, x) = \lim_{x \to \pm\infty} u'(t, x) = 0. \end{cases}$$

Solution: Suppose the solution is $u(t, x)$. The Fourier transform in x of $u(t, x)$, $u''_{xx}(t, x)$ and $\varphi(x)$ are

$$\bar{u}(t, \lambda) = F(u(t, x)) = \int_{-\infty}^{\infty} u(t, \xi) e^{i\lambda\xi} \, d\xi,$$

$$F[u''_{xx}(t, x)] = -\lambda^2 \bar{u}(t, \lambda), \quad F[\varphi(x)] = \bar{\varphi}(\lambda).$$

Thus, the original equation becomes

$$\begin{cases} \dfrac{d\bar{u}(t, \lambda)}{dt} = -a^2 \lambda^2 \bar{u}(t, \lambda), \\ \bar{u}(t, \lambda)_{t=0} = \bar{\varphi}(\lambda). \end{cases}$$

The general solution of this differential equation is

$$\bar{u}(t, \lambda) = C e^{-a^2\lambda^2 t}, \quad \bar{u}(t, \lambda)\Big|_{t=0} = C = \bar{\varphi}(\lambda).$$

Hence,

$$\bar{u}(t, \lambda) = \bar{\varphi}(\lambda)e^{-a^2\lambda^2 t}.$$

Therefore, the solution of the original equation is

$$u(t, x) = F^{-1}[\bar{u}(t, \lambda)] = F^{-1}[\bar{\varphi}(\lambda)e^{-a^2\lambda^2 t}]$$

$$= F^{-1}[\bar{\varphi}(\lambda)] * F^{-1}[e^{-a^2\lambda^2 t}]$$

$$= \varphi(x) * \frac{1}{2a\sqrt{\pi t}}e^{-\frac{x^2}{4a^2 t}} = \int_{-\infty}^{\infty} \varphi(\xi)\frac{1}{2a\sqrt{\pi t}}e^{-\frac{(x-\xi)^2}{4a^2 t}}\, d\xi.$$

11.2.4 Higher-Dimensional Fourier Transforms

It is not hard to generalize the results of Fourier transform of single-variable functions to multivariable functions.

The function $F(\lambda, \mu, \nu)$ defined by the parametric integral

$$F(\lambda, \mu, \nu) = \iiint\limits_{-\infty}^{\infty} f(x, y, z)e^{i(\lambda x + \mu y + \nu z)}\, dx\, dy\, dz$$

is called the *Fourier transform* of $f(x, y, z)$. The inverse transform formula is

$$f(x, y, z) = \frac{1}{(2\pi)^3} \iiint\limits_{-\infty}^{\infty} F(\lambda, \mu, \nu)e^{-i(\lambda x + \mu y + \nu z)}\, d\lambda d\mu d\nu.$$

The properties of Fourier Transforms of multivariable functions are similar to the properties of Fourier Transforms of single-variable functions. Especially

$$F\left[\frac{\partial f}{\partial x}\right] = -i\lambda F[f(x, y, z)]; \quad F\left[\frac{\partial f}{\partial y}\right] = -i\mu F[f(x, y, z)];$$

$$F\left[\frac{\partial f}{\partial z}\right] = -i\nu F[f(x, y, z)].$$

Here, we assume the Fourier transform of the partial derivatives in every equation above is well defined, and the function $f(x, y, z)$ tends to zero as x, y, z tend to positive or negative infinity. It follows that when $f(x, y, z)$ satisfies certain conditions, the Fourier transform of

$$\Delta f = \nabla \cdot (\nabla f) = \frac{\partial^2 f}{\partial x^2} + \frac{\partial^2 f}{\partial y^2} + \frac{\partial^2 f}{\partial z^2}$$

is

$$F[\Delta f] = -(\lambda^2 + \mu^2 + \nu^2)F[f].$$

This formula is very useful in Equations of Mathematical Physics.

Exercises 11.2

1. Represent the following functions as Fourier integrals.

   (1) $f(x) = \dfrac{1}{a^2 + x^2}$, $(a > 0)$;   (2) $f(x) = \dfrac{x}{a^2 + x^2}$, $(a > 0)$;

   (3) $f(x) = \begin{cases} 1, & |x| \leqslant 1, \\ 0, & |x| > 1; \end{cases}$ (4) $f(x) = \begin{cases} \sin x, & |x| \leqslant \pi, \\ 0, & |x| > \pi; \end{cases}$

   (5) $f(x) = e^{-a|x|}$, $(a > 0)$.

2. Find the Fourier transforms of the following functions.

   (1) $f(x) = x e^{-a|x|}$, $(a > 0)$;   (2) $f(x) = e^{-a|x|} \cos bx$, $(a > 0)$;

   (3) $f(x) = \begin{cases} \cos x, & |x| \leqslant \dfrac{\pi}{2}, \\ 0, & |x| > \dfrac{\pi}{2}. \end{cases}$

3. Expand the function $f(x) = e^{-x}$, $(0 \leqslant x < \infty)$, to a Fourier integral by the following requirements.
   (1) Extended as an even function;
   (2) Extended as an odd function.

4. Prove the equation by Fourier transform.

$$\int_0^\infty \frac{\sin a \cos ax}{a}\, da = \begin{cases} \dfrac{\pi}{2}, & |x| < 1, \\ \dfrac{\pi}{4}, & |x| = 1, \\ 0, & |x| > 1. \end{cases}$$

Hint: Find the Fourier transform for the function

$$f(x) = \begin{cases} 1, & |x| \leqslant 1, \\ 0, & |x| > 1. \end{cases}$$

Answers

Chapter 1. Basic Concepts

Exercises 1.1

4. $V = \dfrac{1}{3}\pi R^3 (1 - \dfrac{\theta}{2\pi})^2 \sqrt{\dfrac{\theta}{2\pi}(2 - \dfrac{\theta}{2\pi})}.$

5. $W = \begin{cases} 0, & 0 \leqslant x < 2; \\ 15(x-2), & 2 \leqslant x < 4; \\ 30, & 4 \leqslant x < 5; \\ 40, & 5 \leqslant x \leqslant 8. \end{cases}$

6. (1) a;   (2) $2x + h$,

  (3) $a^x \cdot \dfrac{a^h - 1}{h}.$

7. (4) $x \neq -2$;   (5) $x \neq 0, \pm 1, \pm 2, \ldots$;

  (7) The discontinuous point is $x = 1$.

Exercises 1.2

1. 90 m.

2. (1) $\displaystyle\int_0^{\frac{\pi}{2}} \sin^{10} x \, dx < \int_0^{\frac{\pi}{2}} \sin^2 x \, dx$;   (3) $\displaystyle\int_{\frac{1}{2}}^1 \ln x \, dx < \int_{\frac{1}{2}}^1 (\ln x)^2 \, dx.$

  (2) $\displaystyle\int_{11}^{15} \lg x \, dx < \int_{11}^{15} (\lg x)^2 \, dx$;

6. (1) $m \approx \dfrac{T_1 - T}{n} \displaystyle\sum_{i=1}^n v \left(T + i \dfrac{T_1 - T}{n} \right)$;

627

(2) $m = \int_T^{T_1} v(t)\, dt.$

8. 342.

Exercises 1.3

1. $y = 3x - 2.$

2. (1) 25; 20.5; 20.05; 20.005;   (3) 20.
   (2) 20;

3. (1) 4(g/cm);   (2) 40(g/cm);   (3) 0(g/cm);   (4) $4x$(g/cm).

4. (1) 95(g/cm);   (2) 35(g/cm);   (3) 185(g/cm).

5. The equation of the tangent line at the point $(1, 0)$ is $2x - y - 2 = 0$.
   The equation of the tangent line at the point $(-1, 0)$ is $2x - y + 2 = 0$.

6. the point $(2, 4)$.

7. points $\left(\dfrac{1}{4}, \dfrac{1}{16}\right)$ and $(-1, 1)$.

9. $\arctan 2\sqrt{2} \approx 70°30'.$ 10. $\Delta y = 0.0404$, $dy = 0.04$.

11. $\Delta y = 130; 4; 0.31; 0.0301;$ $\Delta y - dy \to 0 (\Delta x \to 0).$
    $dy = 30; 3; 0.3; 0.03;$

12. (1) $\Delta y = 17 cm^2$, $dy = 16 cm^2$;   (3) $\Delta y = 1.61 cm^2$, $dy = 1.6 cm^2$.
    (2) $\Delta y = 8.25 cm^2$, $dy = 8 cm^2$;

13. $dy = \dfrac{1}{360}\pi.$

Chapter 2. Calculations of Derivatives and Integrals

Exercises 2.1

1. (1) $3ax^2 + 2bx + c$;

   (2) $\dfrac{1}{5}x^{-\frac{4}{5}} - \dfrac{a}{3}x^{-\frac{4}{3}} - \dfrac{6}{5}x^{-\frac{6}{5}}$;   (3) $\dfrac{16}{5}x^2 \cdot \sqrt[5]{x}$;

   (4) $\dfrac{1}{(x+2)^2}$;   (5) $\dfrac{15x^2 + 48x + 82}{(5x+8)^2}$;

   (6) $\cos^3 x - \sin x \cdot \sin 2x$;   (7) $\sin x + \sin x \sec^2 x - \csc^2 x$;

   (8) $\dfrac{-2}{3\sqrt[3]{(1+x)^2(1-x)^4}}$;   (9) $\dfrac{x^4(1.5\cos x + 0.3x \sin x) + a}{(a+b)\cos^2 x}$;

(10) $\dfrac{1 - \ln x^{\alpha}}{x^{\alpha+1}}$;

(11) $a^x \left(\ln a \ln x + \dfrac{1}{x} \right)$;

(12) $-\dfrac{10^x \ln 100}{(1 + 10^x)^2}$;

(13) $t^2 e^{-t} \left[(3 - t) \arctan t + \dfrac{t}{1 + t} \right]$;

(14) $(a^2 + b^2) x^{\alpha-1} e^x \left[(x + \alpha) \arctan x + \dfrac{x}{1 + x^2} \right]$;

(15) $\dfrac{-2}{x(1 + \ln x)^2}$;

(16) $2x \log_3 x + \dfrac{x}{\ln 3}$;

(17) $\sin^{\alpha-1} x \cdot \ln^{\beta-1} x \left(\alpha \cos x \ln x + \dfrac{\beta}{x} \sin x \right)$;

(18) $\dfrac{2 \ln x}{3x \cdot \sqrt[3]{(1 + \ln^2 x)^2}}$;

(19) $\dfrac{1}{(1 + x)\sqrt{x - x^2}} - \dfrac{(1 + 2x) \arcsin x}{2(x + x^2)^{\frac{3}{2}}} - \dfrac{1}{1 - x^2}$;

(20) $\dfrac{4\sqrt{x}\sqrt{x + \sqrt{x}} + 2\sqrt{x} + 1}{8\sqrt{x} \cdot \sqrt{x + \sqrt{x}} \cdot \sqrt{x + \sqrt{x + \sqrt{x}}}}$;

(21) $\left[27\sqrt[3]{x^2} \cdot \sqrt[3]{(1 + \sqrt[3]{x})^2} \cdot \sqrt[3]{(1 + \sqrt[3]{1 + \sqrt[3]{x}})^2} \right]^{-1}$;

(22) $-\sin 2x \cos(\cos 2x)$;

(23) $\cos x \cdot \cos(\sin x) \cdot \cos[\sin(\sin x)]$;

(24) $\dfrac{x}{\sqrt{x^2 + 1}} e^{\sqrt{x^2+1}}$;

(25) $-\dfrac{15x^2}{1 + x^6} \sin(\arctan x^3) \cdot \cos^4(\arctan x^3) \cdot \cos[\cos^5(\arctan x^3)]$;

(26) $2 \left[\dfrac{1}{a^2} \sec^2 \dfrac{x}{a^2} \cdot \tan \dfrac{x}{a^2} + \dfrac{1}{b^2} \sec^2 \dfrac{x}{b^2} \cdot \tan \dfrac{x}{b^2} \right]$;

(27) $a^x e^{\arctan x} \left(\ln a + \dfrac{1}{1 + x^2} \right)$;

(28) $\dfrac{2}{x^3} \sin \dfrac{1}{x^2} e^{\cos \frac{1}{x^2}} \left(1 + \cos \dfrac{1}{x^2} \right)$;

(29) $-\dfrac{2}{1 + x^2} \dfrac{x}{|x|}$, $x \neq 0$;

(30) $\dfrac{6}{x \ln x \cdot \ln(\ln^3 x)}$;

(31) $\dfrac{1}{\cos x}$;

(32) $\dfrac{1}{1 + x^2}$;

(33) $x^{\frac{1}{x}-2}(1 - \ln x)$;

(34) $x^x (\ln x + 1)$;

(35) $e^x [1 + e^{e^x} (1 + e^{e^{e^x}})]$;

(36) $x^{x^x} \cdot x^{x-1} [x \ln x (\ln x + 1) + 1]$;

(37) $\sin x^{\cos x} [\cot x \cos x - \sin x \ln(\sin x)]$;

(38) $x^{a^x}[a^x \ln a \cdot \ln x + \dfrac{a^x}{x}]$;

(39) $(\ln x)^{x-1} x^{\ln x - 1}(x + 2\ln^2 x + x \ln x \ln \ln x)$;

(40) $-\dfrac{2x(\sin x^2 + \cos x^2)}{\sqrt{\sin(2x^2)}}$;

(41) $\dfrac{4x}{\arccos^3(x^2) \cdot \sqrt{1 - x^4}}$;

(42) -1;

(44) $\sqrt{a^2 - x^2}$;

(43) $\dfrac{1}{a + b\cos\varphi}$;

(45) $\dfrac{1}{2(1 + x^2)}$;

(46) $\dfrac{e^x}{\sqrt{1 + e^{2x}}}$;

(48) $\dfrac{12t^5}{(1 + t^{12})^2}$;

(50) $\dfrac{e^{-x} - e^x}{e^{-x} + e^x}$.

(47) $\dfrac{\sin 2x}{\sqrt{1 + \cos^4 x}}$;

(49) $\dfrac{e^x - 1}{e^{2x} + 1}$;

2. (1) $-2, 1$;   (2) $-1, 0$;   (3) $-4, 3$.

3. (1) 13.5;   (2) pe.

7. (1) $\dfrac{\varphi(x)\varphi'(x) + \psi(x)\psi'(x)}{\sqrt{\varphi^2(x) + \psi^2(x)}}$;

(2) $\dfrac{\varphi(x)\sqrt{\psi(x)}}{\varphi^2(x)\psi(x)}[\varphi(x)\psi'(x) - \varphi'(x)\psi(x) \ln \psi(x)]$;

(3) $2\left[\dfrac{\varphi(x)}{\psi(x)}\right]^{\ln \frac{\varphi(x)}{\psi(x)}} \cdot \ln \dfrac{\varphi(x)}{\psi(x)} \cdot \dfrac{\varphi'(x)\psi(x) - \varphi(x)\psi'(x)}{\varphi(x)\psi(x)}$;

(4) $\dfrac{\varphi'(x) + \varphi(x)^{\psi(x)-1}[\varphi(x)\psi'(x) \ln \varphi(x) + \varphi'(x)\psi(x)]}{1 + [1 + \varphi(x) + \varphi(x)^{\psi(x)}]^2}$.

8. (1) $x'(y) = \dfrac{e^{-x}}{x + 1}$;   (3) $x'(y) = \dfrac{-1}{2(e^{-x} - e^{-2x})}$.

(2) $x'(y) = -(1 + x^2)$;

9. (1) $2(2x^2 - 1)e^{-x^2}$;

(2) $\dfrac{1}{(1 - x^2)^2}\left[3x + \dfrac{1 + 2x^2}{\sqrt{1 - x^2}} \arcsin x\right]$;

(3) $a^x(2 + 4x \ln a + x^2 \ln^2 a)$;

(4) $\dfrac{2a\cos x}{a^2 + x^2} - \dfrac{2ax\sin x}{(a^2 + x^2)^2} - \sin x \arctan \dfrac{x}{a}$;

(5) $2\arctan x + \dfrac{2x}{1 + x^2}$;   (6) $-\dfrac{2}{x}\sin(\ln x)$.

10. (1) $\dfrac{uu'' - u'^2}{u^2} - \dfrac{vv'' - v'^2}{v^2}$; (2) $\dfrac{(u^2 + v^2)(uu'' + vv'') + (u'v - uv')^2}{(u^2 + v^2)^{\frac{3}{2}}}$.

11. (1) $y'' = 2[f'(x^2) + 2x^2 f''(x^2)]$, $y^{(3)} = 4x[3f''(x^2) + 2x^2 f^{(3)}(x^2)]$;
    (2) $y'' = (e^x + 1)^2 f''(e^x + x) + e^x f'(e^x + x)$, $y^{(3)} = (e^x + 1)^3 f^{(3)}(e^x + x) + 3e^x(e^x + 1)f''(e^x + x) + e^x f'(e^x + x)$.

12. (1) $e^x(x^2 + 100x + 2450)$; (3) $(x^2 - 379)\sin x - 40x\cos x$;
    (2) $\dfrac{28!(x + 30)}{(1 + x)^{30}}$; (4) $\dfrac{197!!(399 - x)}{2^{100}(1 - x)^{100}\sqrt{1 - x}}$;
    (5) $(-1)^n n!\left[\dfrac{1}{(x - 2)^{n+1}} - \dfrac{1}{(x - 1)^{n+1}}\right]$;
    (6) $e^x\left[\dfrac{1}{x} + \displaystyle\sum_{i=1}^{n}(-1)^i\dfrac{n(n - 1)\cdots(n - i + 1)}{x^{i+1}}\right]$;
    (7) $\dfrac{2(-1)^n n!}{(1 + x)^{n+1}}$.

16. (1) $\dfrac{1}{x - 2\pi}\,dx$; (2) $\dfrac{a^{\ln\tan x}\ln a}{\sin x\cos x}$;
    (3) $x\sin x\,dx$; (4) $\dfrac{1}{x^2 - a^2}\,dx$.

18. (1) 1.002; (3) 1.9953; (5) 0.03;
    (2) 2.7455; (4) 0.4849; (6) 1.0023.

21. $\dfrac{dy}{dx} = \dfrac{t}{2}$, $\dfrac{d^2y}{dx^2} = \dfrac{1 + t^2}{4t}$.

22. $\dfrac{dy}{dx} = \cot\dfrac{t}{2}$, $\dfrac{d^2y}{dx^2} = -\dfrac{1}{a(1 - \cos t)^2}$.

23. (1) tangent line $y = 2a$, normal line $x = a\pi$.
    (2) tangent line $4x + 2y - 3 = 0$, normal line $2x - 4y + 1 = 0$.
    (3) tangent line $4x + 3y - 12a = 0$, normal line $3x - 4y + 6a = 0$.

24. $0, 4, 8$ seconds.

25. $\dfrac{dV}{dt} = 3141.6\,\text{cm}^3/\text{second}$, $\dfrac{dS}{dt} = 2513.3\,\text{cm}^2/\text{second}$.

26. $v(t) = 3aRt^2$. 28. 0.64 cm/minute.

27. 8 degree/minute. 29. -0.875 m/second.

Exercises 2.2.1

1. (1) $2x\,dx$;   (3) $-x^{-2}\,dx$;

   (2) $\dfrac{1}{2\sqrt{x}}\,dx$;   (4) $\dfrac{x}{\sqrt{1+x^2}}\,dx$;

   (5) $\dfrac{1}{\sqrt{x^2+1}}\,dx$;

   (6) $-\sin x\,dx$;   (7) $2e^{2x}\,dx$;

   (8) $\dfrac{a}{a^2+x^2}\,dx$.

2. (1) $\dfrac{1}{2}x^2$;   (3) $-\dfrac{1}{x}$;

   (2) $\ln|x|$;   (4) $2\sqrt{x}$;

   (5) $-\cos x$;   (8) $2\sqrt{x+2}$;   (10) $\arcsin x$;

   (6) $\tan x$;   (9) $\sqrt{x^2+a^2}$;   (11) $\ln(x+\sqrt{1+x^2})$;

   (7) $-e^{-x}$;   (12) $\arctan x$.

3. (1) $\dfrac{2}{3}x^{\frac{3}{2}} + 2x^{\frac{1}{2}} + C$;   (2) $\dfrac{m}{n+m}x^{\frac{n}{m}+1} + C$;

   (3) $\dfrac{2}{5}x^{\frac{5}{2}} + x + C$;   (4) $-\dfrac{2}{3}x^{-\frac{3}{2}} - e^x + \ln|x| + C$;

   (5) $\dfrac{4}{7}x^{\frac{7}{4}} + 4x^{-\frac{1}{4}} + C$;   (6) $x - \dfrac{1}{x} - 2\ln|x| + C$;

   (7) $\dfrac{x^2}{2} - \dfrac{2}{3}x^{\frac{3}{2}} + x + C$;   (8) $\dfrac{1}{2}e^{2x} - e^x + x + C$;

   (9) $\arcsin x + \ln(x+\sqrt{1+x^2}) + C$;

   (10) $\dfrac{4^x}{\ln 4} + 2\dfrac{6^x}{\ln 6} + \dfrac{9^x}{\ln 9} + C$.

4. (1) $\dfrac{1}{2}\ln(1+x^2) + C$;

   (2) $\ln|x^2 - 3x + 8| + C$;   (3) $-\ln|\cos x| + C$;

   (4) $-\ln(1+\cos x) + C$;

                                                          (5) $\dfrac{1}{2}\ln(1+e^{2x}) + C$.

5. (1) $\dfrac{1}{202}(2x-3)^{101} + C$;   (2) $\dfrac{1}{2}(\arctan x)^2 + C$;

   (3) $-\dfrac{1}{2}\cot\left(2x + \dfrac{\pi}{4}\right) + C$;

   (4) $e^{\sin x} + C$;   (5) $-\sqrt{1-x^2} + C$;

   (6) $\dfrac{1}{4}(1+x^3)^{\frac{4}{3}} + C$;

   (7) $2\sqrt{3x^2 - 5x + 6} + C$;

(8) $\arcsin\left(\dfrac{\sin^2 x}{\sqrt{2}}\right) + C;$

(9) $\dfrac{2}{3}\sqrt{\tan^3 x} + C;$

(10) $\dfrac{1}{2}\left(\arcsin\dfrac{x}{2}\right)^2 + C;$

(12) $(b - a)\ln\left|\sin\dfrac{x}{b - a}\right| + C;$

(11) $(\arctan\sqrt{x})^2 + C;$

(13) $\ln|\ln(\ln x)| + C;$

(14) $\dfrac{1}{4}(e^x + 1)^4 + C;$

(15) $\dfrac{2}{3}\sqrt{[\ln(x + \sqrt{1 + x^2})]^3} + C.$

6. (1) $-\dfrac{1}{2}x\cos 2x + \dfrac{1}{4}\sin 2x + C;$

(3) $\dfrac{a^x}{\ln a}\left(x^2 - \dfrac{2x}{\ln a} + \dfrac{2}{\ln^2 a}\right) + C;$

(2) $-e^{-x}(x + 1) + C;$

(4) $x\arcsin x + \sqrt{1 - x^2} + C;$

(5) $\dfrac{1}{2}(x^2\arctan x + \arctan x - x) + C;$

(6) $\dfrac{1}{3}(x^3 + 1)\ln(1 + x) - \dfrac{x^3}{9} + \dfrac{x^2}{6} - \dfrac{x}{3} + C;$

(7) $\dfrac{x^{n+1}}{n + 1}\left(\ln x - \dfrac{1}{n + 1}\right) + C, \ (n \neq -1);$

(8) $x\tan x + \ln|\cos x| + C;$

(9) $\dfrac{1}{6}(1 + 2x)^{\frac{3}{2}} - \dfrac{1}{2}(1 + 2x)^{\frac{1}{2}} + C;$

(10) $-\dfrac{x^2}{4}\cos 2x + \dfrac{x}{4}\sin 2x + \dfrac{1}{8}\cos 2x + C;$

(11) $\dfrac{1}{2}(x^2 - 1)\ln\dfrac{1 + x}{1 - x} + x + C;$

(12) $x\ln(x + \sqrt{1 + x^2}) - \sqrt{1 + x^2} + C;$

(13) $\dfrac{1}{2}\left(\arctan x - \dfrac{x}{1 + x^2}\right) + C;$

(14) $\dfrac{x - 2}{x + 2}e^x + C;$

(15) $-\dfrac{1}{2}\left(\dfrac{x}{\sin^2 x} + \cot x\right) + C.$

7. (1) $\dfrac{1}{3}\ln\left|\dfrac{x - 1}{x + 2}\right| + C;$

(2) $\arctan x - \dfrac{1}{\sqrt{2}}\arctan\dfrac{x}{\sqrt{2}} + C;$

(3) $\dfrac{1}{5}x^5 - \dfrac{1}{4}x^4 + \dfrac{1}{3}x^3 - \dfrac{1}{2}x^2 + x - \ln|1 + x| + C;$

(4) $\sqrt{2}\arctan\dfrac{x}{\sqrt{2}} - \arctan x + C;$

(5) $\dfrac{1}{2\sqrt{2}}\ln\left|\dfrac{x - \sqrt{2}}{x + \sqrt{2}}\right| + \dfrac{1}{2\sqrt{3}}\ln\left|\dfrac{x - \sqrt{3}}{x + \sqrt{3}}\right| + C;$

(6) $4\ln|x-2| - \dfrac{5}{x-2} + C;$ $\qquad$ (7) $x + \dfrac{1}{x} + \ln\dfrac{(x-1)^2}{|x|} + C;$

(8) $\dfrac{1}{2}\ln|x+1| - \dfrac{1}{4}\ln(x^2+1) - \dfrac{1}{2(x+1)} + C;$

(9) $\dfrac{1}{2}\ln|(x+1)(x+3)| - \ln|x+2| + C;$

(10) $\dfrac{1}{4}\dfrac{x+2}{x^2+4x+6} + \dfrac{1}{4\sqrt{2}}\arctan\dfrac{x+2}{\sqrt{2}} + C.$

8. (1) $\dfrac{1}{\sqrt{2}}\ln\left|\tan\dfrac{x+\dfrac{\pi}{4}}{2}\right| + C$, or $\dfrac{1}{\sqrt{2}}\ln\left|\dfrac{\sqrt{2}-1+\tan\dfrac{x}{2}}{\sqrt{2}+1-\tan\dfrac{x}{2}}\right| + C;$

(2) $\tan x - \sec x + C$, or $\tan\left(\dfrac{x}{2} - \dfrac{\pi}{4}\right) + C;$

(3) $\dfrac{1}{\sqrt{a^2+b^2}}\ln\left|\tan\dfrac{x+\varphi}{2}\right| + C$, where $\varphi = \arcsin\dfrac{b}{\sqrt{a^2+b^2}};$

(4) $\dfrac{1}{\sqrt{2}}\arctan\left(\dfrac{\tan 2x}{\sqrt{2}}\right) + C;$ $\qquad$ (5) $\dfrac{1}{2}\arctan\sin^2 x + C;$

(6) $\dfrac{1}{6}\ln\dfrac{(2+\cos x)^2(1-\cos x)}{(1+\cos x)^3} + C;$

(7) $2[\sqrt{1+x} - \ln(1+\sqrt{1+x})] + C;$

(8) $\dfrac{\sqrt{2+4x}}{6}(x-1) + C;$

(9) $x + \dfrac{6}{5}x^{\frac{5}{6}} + \dfrac{3}{2}x^{\frac{2}{3}} + 2x^{\frac{1}{2}} + 3x^{\frac{1}{3}} + 6x^{\frac{1}{6}} + 6\ln|\sqrt[6]{x} - 1| + C;$

(10) $\arcsin x + \sqrt{1-x^2} + C;$

(11) $\arccos\dfrac{1}{x} + C;$ $\qquad\qquad$ (12) $\dfrac{a^2}{2}\arcsin\dfrac{x}{a} - \dfrac{x}{2}\sqrt{a^2-x^2} + C;$

(13) $\arcsin\dfrac{2x-1}{2} + \dfrac{2x-1}{4}\sqrt{3+4x-4x^2} + C;$

(14) $2(\sqrt{x} - 1)e^{\sqrt{x}} + C;$

(15) $\tan x + \dfrac{1}{3}\tan^3 x + C;$

(16) $\ln|\cos x + \sin x| + C;$ $\qquad$ (17) $\ln(e^x + e^{-x}) + C;$

(18) $2\operatorname{sgn}\left(\sin\dfrac{x}{2} + \cos\dfrac{x}{2}\right)\cdot\left(\sin\dfrac{x}{2} - \cos\dfrac{x}{2}\right) + C;$

(Note:

$$\operatorname{sgn} x = \begin{cases} 1, & x > 0, \\ 0, & x = 0, \\ -1, & x < 0; \end{cases}$$

and $|x| = x \operatorname{sgn} x$.)

(19) $\arcsin \dfrac{x+1}{\sqrt{2}} + C$;

(20)

$$F(x) = \begin{cases} \dfrac{1}{3}(x+x^2)^{\frac{3}{2}} - \dfrac{1+2x}{8}\sqrt{x+x^2}, \\ \quad + \dfrac{1}{16}\ln\left|x + \dfrac{1}{2} + \sqrt{x+x^2}\right| + C, \quad x \geqslant 0, \\ -\dfrac{1}{3}(x+x^2)^{\frac{3}{2}} + \dfrac{1+2x}{8}\sqrt{x+x^2}, \\ \quad - \dfrac{1}{16}\ln\left|x + \dfrac{1}{2} + \sqrt{x+x^2}\right| + C, \quad x \leqslant -1; \end{cases}$$

(21) $\dfrac{2}{3}\left[(x-1)^{\frac{3}{2}} + (x-2)^{\frac{3}{2}}\right] + C$;

(22) $\dfrac{1}{3}\left[x^3 + (x^2-1)^{\frac{3}{2}}\right] + C$;

(23) $\dfrac{2}{3}x\sqrt{x}\left(\ln^2 x - \dfrac{4}{3}\ln x + \dfrac{8}{9}\right) + C$;

(24) $\arctan(\ln x) + C$;

(25) $\dfrac{x}{4} + \dfrac{1}{8}\sin 2x + \dfrac{1}{16}\sin 4x + \dfrac{1}{24}\sin 6x + C$;

(26) $\dfrac{1}{3}(x^3 - 1)e^{x^3} + C$;   (27) $\dfrac{1}{3}\tan^3 x + C$;

(28) $\ln|1 + \cot x| - \cot x + C$;   (29) $\tan x - \cot x + C$;

(30) $\dfrac{x\arcsin x}{\sqrt{1-x^2}} + \ln\sqrt{1-x^2} + C$;   (31) $\dfrac{1}{\ln 4}\ln\left|\dfrac{1+2^x}{1-2^x}\right| + C$;

(32) $2\sqrt{x+1}[\ln(x+1) - 2] + C$;

(33) $x\ln(1+x^2) - 2x + 2\arctan x + C$;

(34) $\dfrac{1}{2}\arcsin x - \dfrac{x}{2}\sqrt{1-x^2} + C$;

(35) $-\dfrac{\sqrt{4-x^2}}{4x} + C$;   (36) $-\dfrac{1}{1+\tan x} + C$;

(37) $2\sqrt{x} - 2\sqrt{1-x}\arcsin\sqrt{x} + C$;

(38) $-\ln(\cos^2 x + \sqrt{1+\cos^4 x}) + C$;

(39) $x\arctan x - \dfrac{1}{2}\ln(1+x^2) - \dfrac{1}{2}(\arctan x)^2 + C$;

(40) $\dfrac{1}{2\sqrt{2}} \ln \left| \dfrac{x^2 + \sqrt{2}x + 1}{x^2 - \sqrt{2}x + 1} \right| + C.$

Exercises 2.2.2

1. (1) $\dfrac{1}{2}$;

   (2) 1;

   (3) $\dfrac{1}{2}(1 - \ln 2)$;   (4) $\dfrac{\pi}{12}(2\pi^2 - 3)$;   (5) $\dfrac{1}{5} \ln \dfrac{4}{3}$;

   (6) 1;   (7) $2(2 - \ln 3)$;

   (8) $2 - \dfrac{\pi}{2}$;   (9) $\dfrac{1}{12}$;

   (10) $\dfrac{1}{\sqrt{1 + a^2}} \arctan \dfrac{1}{\sqrt{1 + a^2}}$;   (11) $\dfrac{1}{5}(2e^\pi + 1)$;

   (12) $\ln \dfrac{e + \sqrt{1 + e^2}}{1 + \sqrt{2}}$;   (14) $\dfrac{1}{2}a^2[\sqrt{2} - \ln(\sqrt{2} + 1)]$;

   (13) $a \ln(a + \sqrt{2}a) - \sqrt{2}a + a$;   (15) $6 - 2e$;

   (16) $\dfrac{\pi}{9}$;   (17) $\dfrac{1}{6}$;

   (18) $4(\sqrt{2} - 1)$;   (20) 0.

   (19) $\dfrac{1}{2} + \dfrac{\pi}{4}$;

4. (1) $\dfrac{8}{15}$;   (2) $\dfrac{35}{128}\pi$;   (3) $\dfrac{512}{693}$;   (4) $\dfrac{8}{35}$;

   (5) $\dfrac{5}{16}\pi$;   (6) $\dfrac{3}{16}\pi$;   (7) $\dfrac{(2n)!!}{(2n+1)!!}$;   (8) $\dfrac{1}{16}\pi a^4.$

Exercises 2.2.3

1. 3.244, 3.144.   2. 1.37039.

Chapter 3. Some Applications of Differentiation and Integration

Exercises 3.1

1. (1) 4.5;

(2) $\dfrac{32}{3}\sqrt{6}$;

(3) $\dfrac{4}{3}a^3$;

(4) πa^2;

(5) $\dfrac{3}{2}\pi a^2$;

(6) $\dfrac{5}{8}\pi a^2$.

2. (1) $\dfrac{\pi^2}{2}$, $2\pi^2$;

(2) $2\pi^2 a^2 b$;

(3) 2π;

(4) $\dfrac{32}{105}\pi a^3$, $\dfrac{32}{105}\pi a^3$.

3. π.

6. (1) 2π;

(3) $6a$.

(2) $a\sinh\dfrac{b}{a}$;

8. (1) $8a$;   (2) $\pi a\sqrt{1+4\pi^2}+\dfrac{a}{2}\ln(2\pi+\sqrt{1+4\pi^2})$.

Exercises 3.2

4. (1) Concave up on $(-\infty, 1)$. Concave down on $(1, +\infty)$. The point on the curve where $x = 1$ is an inflection point.

(2) Concave down on $(-\infty, -3)$ and $(-1, +\infty)$. Concave up on $(-3, -1)$.
The points $x = -3$ and $x = -1$ are inflection points.

(3) Concave up on $(2k\pi, (2k+1)\pi)$. Concave down on $((2k+1)\pi, (2k+2)\pi)$.
The points $x = k\pi$, $(k = 0, \pm 1, \pm 2, \cdots)$ are inflection points.

(4) Concave down on $(-\infty, -1)$ and $(1, +\infty)$. Concave up on $(-1, 1)$.
The points $x = -1$ and $x = 1$ are inflection points.

5. $a = -\dfrac{3}{2}$, $b = \dfrac{9}{2}$.

6. (1) $y = 0$, $x = 1$, $x = 2$;

(2) $y = c$, $x = b$;

(3) $y = \pm\dfrac{b}{a}x$;

(4) $x = 0$, $y = x + 3$.

8. (1) $\dfrac{\sqrt{2}}{4}$;   (2) $\dfrac{\sqrt{2}}{4}$;   (3) $\dfrac{6|x|}{(1+9x^4)^{\frac{3}{2}}}$;   (4) $\dfrac{1}{6}$;

(5) $\dfrac{ab}{\left(\dfrac{a^2+b^2}{2}\right)^{\frac{3}{2}}}$.

Exercises 3.3

3. (1) 2;   (2) 2;

   (3) $\dfrac{\alpha}{\beta}$;   (4) $-\dfrac{1}{2}$;

   (5) 2;

   (6) $-\dfrac{1}{3}$;   (7) $\dfrac{1}{6}$;   (8) $\dfrac{1}{2}$;

   (9) 1;   (10) 3;   (11) 2;

   (12) $\dfrac{a}{\sqrt{b}}$;

   (13) 1;

   (14) $\dfrac{1}{a}$;   (15) $\dfrac{1}{3}$;   (16) $-\dfrac{e}{2}$;

   (17) -1;

   (18) $\dfrac{1}{2}$;

   (19) 1;   (20) e^{-1};   (21) 1;   (22) 1;

   (23) $e^{-\frac{1}{6}}$;   (24) k;   (25) 1;

   (26) $-\dfrac{1}{3}$;

   (27) $e^{-\frac{2}{\pi}}$;   (28) 0;   (29) 0;   (30) $e^{-\frac{1}{2}}$.

4. (1) maximum $y(0) = 0$, minimum $y(1) = -1$;

   (2) maximum $y(0) = 4$, minimum $y(-2) = \dfrac{8}{3}$;

   (3) maximum $y(0) = \sqrt[3]{a^4}$, minimum $y(\pm a) = 0$;

   (4) minimum $y(0) = 0$;

   (5) maximum $y(\pm 1) = e^{-1}$, minimum $y(0) = 0$;

   (6) maximum $y\left(\dfrac{\pi}{4}\right) = \sqrt{2}$;

   (7) no extremal value;

   (8) maximum $y(0) = 0$, minimum $y\left(\dfrac{2}{5}\right) = -\dfrac{108}{3125}$.

5. (1) maximal value $y = 13$, minimal value $y = 4$;

   (2) maximal value $y = 10$, minimal value $y = 6$;

   (3) maximal value $y = \dfrac{3}{5}$, minimal value $y = -1$;

   (4) maximal value $y = \dfrac{\pi}{2}$, minimal value $y = -\dfrac{\pi}{2}$.

7. side $= 1$ cm.

8. hight $= \dfrac{20}{\sqrt{3}}$ cm.

10. $\dfrac{a}{\sqrt{2}}$.

11. $1\dfrac{27}{43}$ hours.

12. $x = \dfrac{1}{n}\displaystyle\sum_{i=1}^{n} x_i$.

13. radius $= 1$ m, hight $= 1.5$ m.

14. $50 - \dfrac{100}{\sqrt{6}}$ km.

15. 1.5 km.

16. 2.4 m.

17. $(a^{\frac{2}{3}} + b^{\frac{2}{3}})^{\frac{3}{2}}$.

Exercises 3.4

1. $S(t) = \dfrac{t^4}{12} - \dfrac{t^2}{2} + t$.

2. 156.25 m.

3. $v(t_1) = \dfrac{A}{b}[\ln a - \ln(a - bt_1)]$.

$h(t_1) = \displaystyle\int_0^{t_1} v(t)\,dt = \dfrac{A}{b}\left[(\ln a + 1)t_1 + \dfrac{a - bt_1}{b}\ln(a - bt_1) - \dfrac{a}{b}\ln a\right]$.

4. 3 seconds.

5. $mg\dfrac{Rh}{R + h}$, R is the radius of the Earth.

6. $6x10^4\pi$ ton $\cdot$ m.

7. 112.7 joule.

8. $135\pi^2 \cdot 10^5$ erg.

9. $708\dfrac{1}{3}$ ton.

10. $(m_0 - 9)$ g.

11. about 3 hours.

12. $\dfrac{2Glp}{a(a + 2l)}$, (G is the universal gravitational constant).

Chapter 4. Ordinary Differential Equations

Exercises 4.1

1. (1) $y = \tan \ln|x| + C$, and $x = 0$;

(2) $y = \ln\left(\dfrac{1}{2}e^{2x} + C\right)$;

(3) $y = C|x|^{1-a}e^{-\frac{6}{2}x^2}$, and $y = 0$;

(4) $y^2 = \ln x^2 - x^2 + C$;

(5) $\dfrac{1}{\sin^2 y} = \dfrac{1}{\cos^2 x} + C$;

(6) $1 + y^2 = C(x^2 - 1)$, and $x = \pm 1$;

(7) $y^2 = 3x - 3x^2 + C$;

(8) $y = \dfrac{1}{1 - Cx}$.

2. (1) $y = 1$;

(2) $y = \dfrac{1 + x}{1 - x}$;

(3) $y = \arccos\left(\dfrac{\sqrt{2}}{2}\cos x\right)$;

(4) $y = 6 - \dfrac{35x}{1 + 6x}$.

3. $xy = 6$.

4. $y^2 = Cx$.

5. $\theta = \theta_1 + (\theta_0 - \theta_1)e^{-k_0(t + \frac{a}{2}t^2)}$.

6. $x = a(1 - e^{-kt})$.

7. $50\sqrt{29}$ cm/s.

8. (1) ≈ 0.467 km/hour;

(2) 85.2 m.

9. (1) $y = \dfrac{x(2 + Cx^3)}{1 - Cx^3}$;

(2) $y^2 - x^2 = Cy^2$;

(3) $y^2 = x^2 \ln x^2 + Cx^2$;

(4) $e^{\frac{y}{x}} = Cy$;

(5) $y = xe^{Cx+1}$;

(6) $e^{-\frac{y}{x}} = -\ln|x| + C$;

(7) $y = xe^{Cx}$.

11. $(y + x - 1)^5(y - x + 1)^2 = C$.

12. $y - 2x - \dfrac{1}{5}\ln\left|x + 2y + \dfrac{7}{5}\right| = C$.

13. $x^2 = 2Cy + C^2$.

14. $y^2 = 4C(x + C)$.

15. (1) $y = (1 + x^2)(x + C)$;

(2) $y = \begin{cases} \dfrac{1}{m + a}e^{mx} + Ce^{-ax}, & \text{if } m + a \neq 0, \\ e^{-ax}(x + C), & \text{if } m + a = 0; \end{cases}$

(3) $y = x^2(Ce^{\frac{1}{x}} + 1)$;

(4) $x = Cy^3 + \dfrac{1}{2}y^2$;

(5) $x = Ce^{2y} + \dfrac{1}{2}y^2 + \dfrac{y}{2} + \dfrac{1}{4}$;

(6) $x = y\ln y + \dfrac{C}{y}$;

(7) $y = \dfrac{1}{\pm\sqrt{Ce^{x^2} + 1}}$;

(8) $y = \dfrac{1}{\pm\sqrt{Ce^{2x^2} + x^2 + \dfrac{1}{2}}}$;

(9) $y = \dfrac{1}{Ce^{-\frac{1}{2}v^2} - y^2 + 2}$; (10) $y^3 = Ce^{ax} - \dfrac{1}{a}(x+1) - \dfrac{1}{a^2}$.

16. $y = Cx \pm \dfrac{a^2}{2x}$.

17. $v = \dfrac{k_1}{k_2}t^2 - \dfrac{2mk_1}{k_2^2} + \left(V_0 + \dfrac{2mk_1}{k_2^2}\right)e^{-\frac{k_2}{2m}t^2}$.

Exercises 4.2

1. (1) $y = \dfrac{1}{6}x^3 - \sin x + C_1 x + C_2$; (2) $y = C_1\dfrac{x^2}{2} + C_2$;

   (3) $y = C_1 e^x - \dfrac{1}{2}x^2 - x + C_2$;

   (4) $y = \dfrac{1}{3}x^2 + \dfrac{1}{2}C_1 x^3 + C_2$;

   (5) $y = (1 + C_1^2)\ln|x + C_1| - C_1 x + C_2$;

   (6) $y = (C_1 x - C_1^2)e^{\frac{x}{C_1}+1} + C_2$;

   (7) $y = \pm\dfrac{2}{3C_1}\sqrt{(C_1 x - 1)^3} + C_2$;

   (8) $y = \dfrac{C_1}{4}(x + C_2)^2 + \dfrac{1}{C_1}$; (11) $y = \dfrac{x + C_1}{x + C_2}$;

   (9) $y = C_1(x + C_2)^{\frac{2}{3}}$; (12) $y^2 = x^2 + C_1 x + C_2$;

   (10) $y = C_1 e^{\frac{x}{a}} + C_2 e^{-\frac{x}{a}}$; (13) $y = C_2 e^{C_1 x}$.

2. (1) Assume the y-axis is pointing up. Then the equations of the object going up and coming down are
$$m\dfrac{d^2 y}{dt^2} = -mg - kv^2, \quad m\dfrac{d^2 y}{dt^2} = -mg + kv^2;$$

   (2) $V = \sqrt{\dfrac{mgv_0^2}{mg + kv_0^2}}$.

3. 0.00082 second.

4. (1) $y = C_1 e^{(1+\sqrt{2})x} + C_2 e^{(1-\sqrt{2})x}$; (2) $y = C_1 e^{-\frac{1}{2}x} + C_2 e^{\frac{5}{2}x}$;

   (3) $y = e^{-\frac{x}{2}}\left(C_1 \cos\dfrac{\sqrt{3}}{2}x + C_2 \sin\dfrac{\sqrt{3}}{2}x\right)$;

   (4) $y = e^x(C_1 + C_2 x)$;

   (5) $y = e^{-x}(C_1 \cos 3x + C_2 \sin 3x)$.

5. (1) $S = e^{-t}(\cos 2t + \sin 2t)$;   (3) $y = 3e^{-2x}\sin 5x$;

   (2) $S = 4e^t + 2e^{3t}$;   (4) $y = (x+2)e^{-\frac{x}{2}}$.

6. (1) $y = C_1 + C_2 e^{-\frac{5}{2}x} + \dfrac{x}{10} + \dfrac{5}{164}\sin 2x - \dfrac{1}{41}\cos 2x$;

   (2) $y = C_1 + C_2 e^{-\frac{5}{2}x} - \left(\dfrac{26}{29} + 5x\right)\cos x + \left(\dfrac{185}{29} - 2x\right)\sin x$;

   (3) $y = C_1 e^x + C_2 e^{-2x} + (2x^2 + x)e^x$;

   (4) $y = C_1 e^x + C_2 e^{2x} - \dfrac{8}{5}\left(\cos\dfrac{x}{2} + 2\sin\dfrac{x}{2}\right)e^x$;

   (5) $y = (C_1 + C_2 x)e^x + \dfrac{1}{2}e^{-x} + \dfrac{3}{2}x^2 e^x$;

   (6) $y = C_1 e^{-x} + C_2 e^{\frac{1}{2}x}\cos\dfrac{\sqrt{3}}{2}x + C_3 e^{\frac{1}{2}x}\sin\dfrac{\sqrt{3}}{2}x + \dfrac{1}{3}xe^{-x}$;

   (7) $y = C_1 + C_2 e^x + C_3 e^{-x} + \dfrac{1}{2}\cos x$;

   (8) $y = C_1 e^x + C_2 e^{-\frac{1}{2}x}\cos\dfrac{\sqrt{3}}{2}x + C_3 e^{-\frac{1}{2}x}\sin\dfrac{\sqrt{3}}{2}x + \dfrac{1}{2}e^{-x}$;

   (9) $y = C_1 e^{2x} + C_2 e^{-x} + \left(\dfrac{1}{28}x - \dfrac{67}{784}\right)e^{5x}$

   $\qquad + \left(-\dfrac{1}{2}x^3 + \dfrac{3}{4}x^2 - \dfrac{5}{4}x - \dfrac{1}{8}\right)e^{-x}$;

   (10) $y = C_1 e^x + C_2 e^{-x} - 2\cos x - 2x\sin x$;

   (11) $y = C_1 e^{-2x} + C_2 e^{\frac{4}{3}x} - \dfrac{11}{25}\cos x + \dfrac{2}{25}\sin x$;

   (12) $y = C_1 \cos 2x + C_2 \sin 2x + \dfrac{1}{5}e^x$;

   (13) $y = (C_1 + C_2 x + C_3 x^2)e^x + \dfrac{1}{6}x^2 e^x$;

   (14) $y = C_1 \cos x + C_2 \sin x + x^2 - x$;

   (15) $y = C_1 e^{-4x} + C_2 e^{3x} - \left(\dfrac{1}{10}x^2 + \dfrac{3}{50}x + \dfrac{19}{500}\right)e^x$.

7. (1) $y = -\dfrac{2}{3}e^{2x} + \dfrac{1}{96}e^{5x} + \left(\dfrac{1}{4}x^2 + \dfrac{5}{8}x + \dfrac{21}{32}\right)e^x$;

   (2) $y = -\dfrac{4}{25}e^{3x} + \dfrac{34}{25}xe^{3x} + \dfrac{4}{25}\cos x + \dfrac{3}{25}\sin x$;

   (3) $y = \dfrac{4\sqrt{3}}{3}e^{-\frac{1}{2}x}\sin\dfrac{\sqrt{3}}{2}x + \cos x + (x-2)\sin x$;

   (4) $y = \dfrac{1}{6}e^x + \dfrac{1}{3}e^{-\frac{1}{2}x}\cos\dfrac{\sqrt{3}}{2}x + \dfrac{\sqrt{3}}{3}e^{-\frac{1}{2}x}\sin\dfrac{\sqrt{3}}{2}x - \dfrac{1}{2}\cos x - \dfrac{1}{2}\sin x$;

   (5) $\cos 3x + \dfrac{1}{8}\sin 3x + \dfrac{5}{8}\sin x$.

9. (1) $y = \dfrac{C_1}{x} + \dfrac{C_2}{x}\ln x$; (2) $y = C_1 x^2 + C_2 x^3 + \dfrac{1}{2}x$;

   (3) $y = x(C_1 + C_2 \ln x)$; (4) $R = C_1 r^{-(n+1)} + C_2 r^2$;

   (5) $y = C_1 x \cos \ln x + C_2 x \sin \ln x + x \ln x$;

   (6) $y = C_1 x^2 + C_2 x + C_3 x \ln x + \dfrac{1}{4}x^3 - \dfrac{3}{2}x(\ln x)^2$.

10. (1) $\begin{cases} x = \dfrac{1}{3}(-\cos t + 2\sin t + 1), \\ y = \dfrac{1}{3}(\cos t + \sin t + 2). \end{cases}$

   (2) $\begin{cases} x = \sin t, \\ y = \cos t. \end{cases}$ (3) $\begin{cases} x = 2e^t, \\ y = 2e^t. \end{cases}$ (4) $\begin{cases} x = \cos t, \\ y = \sin t. \end{cases}$

11. (1) $\begin{cases} x = C_1 e^t + C_2 e^{-t} + \sin t, \\ y = C_2 e^{-t} + C_1 e^t. \end{cases}$ (2) $\begin{cases} x = C_1 \cos t + C_2 \sin t + 3, \\ y = -C_1 \sin t + C_2 \cos t. \end{cases}$

   (3) $\begin{cases} x = C_1 e^{-2t} + C_2 e^t, \\ y = C_3 e^{-2t} + C_2 e^t, \\ z = -(C_1 + C_2)e^{-2t} + C_2 e^t. \end{cases}$

12. The system of differential equations of the movement of the shell is

   $\begin{cases} \dfrac{W}{g} x'' = -k\, x', \\ \dfrac{W}{g} y'' = -k\, y' - W, \end{cases}$

   $x = -\dfrac{W v_0 \cos a}{kg} e^{-\frac{kg}{W}t} + \dfrac{W v_0 \cos a}{kg}$,

   $y = \left(\dfrac{W^2}{k^2 g} + \dfrac{W v_0 \sin a}{kg}\right) - \left(\dfrac{W^2}{k^2 g} + \dfrac{W v_0 \sin a}{kg}\right) e^{-\frac{kg}{W}t} - \dfrac{W}{k}t$.

13. Choose the point where the object is thrown from the plane as the origin, the horizontal direction of the plane as the x-axis, and upward direction as the y-axis. Then the equation of the movement of the

   object is $\begin{cases} m x'' = -k\, x', & x(0) = 0,\ x'(0) = v_0. \\ m y'' = -k\, y' - mg, & y(0) = 0,\ y'(0) = 0. \end{cases}$

   $\begin{cases} x = \dfrac{m}{k}v_0 - \dfrac{m}{k}v_0 e^{-\frac{k}{m}t}, \\ y = \dfrac{m^2 g}{k^2} - \dfrac{m^2 g}{k^2} e^{-\frac{k}{m}t} - \dfrac{mg}{k}t. \end{cases}$

14. (1) $x'' + \mu x' + kx = A_0 \sin 3t$;

   (2) $x = C_1 e^{-2t} + C_2 e^{-8t} - \dfrac{30}{949} A_0 \cos 3t + \dfrac{7}{949} A_0 \sin 3t$;

(3) $x = (C_1 + C_2 t)e^{-5t} - \dfrac{30}{1156} A_0 \cos 3t + \dfrac{16}{1156} A_0 \sin 3t$;

(4) $x = C_1 e^{-4t} \cos 3t + C_2 e^{-4t} \sin 3t - \dfrac{24}{1132} A_0 \cos 3t + \dfrac{16}{1132} A_0 \sin 3t$;

(5) $x = C_1 \cos 3t + C_2 \sin 3t - \dfrac{A_0}{6} t \cos 3t$.

15. (1) linearly independent;   (3) linearly dependent;
     (2) linearly independent;   (4) linearly independent.

16. $y = C_2 + (C_1 - C_2 x) \cot x$.

17. (1) $y = C_1 x + C_2 x^2$;   (3) $y = C_1 e^x + C_2(x^2 - 1)$;
     (2) $y = C_1 e^x + C_2(x + 1)$;   (4) $y = C_1 + C_2 x^2 + x^3$.

18. $y = \pi - 1 + 4 \arctan x + x^2$.

19. $y = x^2 - e^{x-1}$.

Chapter 5. Vector Algebra and Analytic Geometry in Three-Dimensional Space

Exercises 5.1

1. (1) $5\sqrt{2}, \sqrt{34}, \sqrt{41}, 5$;   (2) $z_1 = 7, z_2 = -5$;   (3) $x = 2$.

2. $D(0, 1, -2)$.   4. 13, 13.   5. $\sqrt{129}$, 7.

6. $\sqrt{21}$.   7. 22.

9. $\overrightarrow{MA} = -\dfrac{1}{2}(\vec{a} + \vec{b})$, $\overrightarrow{MB} = \dfrac{1}{2}(\vec{a} - \vec{b})$, $\overrightarrow{MC} = \dfrac{1}{2}(\vec{a} + \vec{b})$, $\overrightarrow{MD} = \dfrac{1}{2}(\vec{b} - \vec{a})$.

12. $(a_2 - a_1, b_2 - b_1, c_2 - c_1)$, $|\overrightarrow{P_1 P_2}| = \sqrt{(a_2 - a_1)^2 + (b_2 - b_1)^2 + (c_2 - c_1)^2}$,

$\cos \alpha = \dfrac{a_2 - a_1}{|\overrightarrow{P_1 P_2}|}$, $\cos \beta = \dfrac{b_2 - b_1}{|\overrightarrow{P_1 P_2}|}$, $\cos \gamma = \dfrac{c_2 - c_1}{|\overrightarrow{P_1 P_2}|}$.

13. (1) perpendicular to x axis;   (3) perpendicular to xy plane.
     (2) parallel to y axis;

14. $(0, 0, -1)$ or $\left(\dfrac{\sqrt{2}}{2}, \dfrac{\sqrt{2}}{2}, 0 \right)$.   16. $\vec{a}_\circ = \left(\dfrac{3}{13}, \dfrac{4}{13}, -\dfrac{12}{13} \right)$.

17. $\vec{a} = (1, -1, +\sqrt{2})$.

15. $B(18, 17, -17)$.

18. $\overrightarrow{OM} = \left(\dfrac{11}{4}, -\dfrac{1}{4}, 3 \right)$.

19. $\alpha = 2, \beta = -3, \gamma = 1$.   20. $\sqrt{153}, \sqrt{93}, \sqrt{114}$.

Exercises 5.2

1. (1) -6; (2) -61.

2. (1) 38; (2) -184; (3) 9.

3. $\dfrac{3}{14}\sqrt{14}$.

4. $A = \arccos \dfrac{\sqrt{2}}{6}$, $c = \arccos \dfrac{9}{14}\sqrt{2}$.

5. $2\sqrt{19}$.

6. $\dfrac{\pi}{3}$ or $\dfrac{2\pi}{3}$.

7. (1) $(0, -8, -24)$;
   (2) $(0, -1, -1)$;
   (3) 2;
   (4) $(2, 1, 21)$.

8. (1) 24; (2) 60.

9. (1) 3; (2) 300.

10. $\dfrac{1}{2}\sqrt{1562}$.

11. 1. 12. 3.

13. (1) coplanar; (2) not coplanar.

14. $\dfrac{\pi}{3}$.

22. $\vec{x} = \dfrac{1}{\vec{a}\cdot\vec{b}}(r\vec{b} + \vec{a}\times\vec{c})$.

Exercises 5.3

2. (1) $-16(x-2) + 14y + 11(z-3) = 0$, that is $-16x + 14y + 11z - 1 = 0$;
   (2) $-5(x-1) + 2(y-2) + 11(z-1) = 0$;
   (3) $-3(x-1) + y - 3 + 4(z-2) = 0$;
   (4) $y - 2z = 0$;
   (5) $x - 3y - 2x = 0$; (6) $-x + z + 1 = 0$.

3. $\left(\dfrac{1}{3}, \dfrac{2}{3}, -\dfrac{2}{3}\right)$.

4. $\pm\sqrt{2}x - y + z - 1 = 0$.

5. (1) 1; (2) 0; (3) 3.

6. 6.5.

7. $\left(\dfrac{3-\sqrt{3}}{6}, \dfrac{3-\sqrt{3}}{6}, \dfrac{3-\sqrt{3}}{6}\right)$.

8. $(0, 7, 0)$ or $(0, -5, 0)$.

9. $2x + y + 2z - 2\sqrt[3]{3} = 0$.

10. (1) $\dfrac{x-1}{2} = \dfrac{y+2}{3} = \dfrac{z-1}{-2}$; (2) $\dfrac{x-1}{2} = \dfrac{y}{-1} = \dfrac{z+3}{3}$.

11. $x = t + 1$, $y = -7t$, $z = -19t - 2$.

12. (1) $(2, -3, 6)$;   (2) The line is on the plane.

13. (1) $79°1'$;   (2) $68°58'$.

14. $\arcsin \dfrac{3}{133}$.   15. $\dfrac{x}{-2} = \dfrac{y-2}{3} = \dfrac{z-4}{1}$.

16. $x + 2y + 3z = 0$.

17. $x + 20y + 7z - 12 = 0$ or $x - z + 4 = 0$.

18. $2x + 15y + 7z + 7 = 0$.

19. $\dfrac{x+1}{2} = \dfrac{y-2}{-3} = \dfrac{z+3}{6}$.

20. $4x + 3y - 6z + 18 = 0$.   21. $9x + 11y + 5z - 16 = 0$.

22. $13x - 14y + 11z + 51 = 0$.   23. 25.

24. $\dfrac{4}{37}\sqrt{259}$.

25. $6x + 2y + 3z \pm 42 = 0$.

Exercises 5.4

1. (1), (2), (4), (6) are surfaces of revolution.

2. (1) $4x^2 - 9y^2 - 9z^2 = 36$, $4x^2 - 9y^2 + 4z^2 = 36$;

   (2) $\sqrt{y^2 + z^2} = kx$, $y = k\sqrt{x^2 + z^2}$;

   (3) $\sqrt{y^2 + z^2} = \sin x$.

3. (1) straight line; plane.   (2) straight line; plane.

   (3) circle; cylinder with generating lines parallel to z-axis.

   (4) hyperbola; hyperbolic cylindrical surface with generating lines parallel to z-axis.

   (5) point; a straight line parallel to z-axis.

   (6) two points: $\left(\dfrac{2}{3}\sqrt{5}, 2\right)$ and $\left(-\dfrac{2}{3}\sqrt{5}, 2\right)$; two straight lines parallel to z-axis and passing through these two points respectively.

7. $x^2 + y^2 = -8z + 16$, paraboloid of revolution.

Exercises 5.5

1. $A(-5, 0, 2), B(0, -7, -4), C(4, 3, -2)$.

2. (1) $3x' - 5y' + 6z' + 10 = 0$;   (3) $x'^2 + y'^2 + z'^2 - 25 = 0$.

   (2) $\dfrac{x'}{3} = \dfrac{y'}{4} = \dfrac{z'}{2}$;

3. $\left(3 + \dfrac{\sqrt{3}}{2}, -3\sqrt{3} + \dfrac{1}{2}, -2\right).$

4. $\begin{cases} x = x_0 + x_1, \\ y = y_0 + \dfrac{\sqrt{2}}{2}y_1 - \dfrac{\sqrt{2}}{2}z_1, \\ z = z_0 + \dfrac{\sqrt{2}}{2}y_1 + \dfrac{\sqrt{2}}{2}z_1. \end{cases}$

5. The origin of $O_1x_1y_1z_1$ has coordinates (x_0, y_0, z_0) in $Oxyz$. The z-axis and z_1-axis are parallel to each other and in the same direction. Rotating the x-axis about the z-axis clockwise by an angle θ makes it to have the same direction as the x_1-axis.
$M(x_0, y_0, 1 + z_0)$, $M(x_0 + 1, y_0, z_0)$.

Chapter 6. Multiple Integrals and Partial Derivatives

Exercises 6.1

1. $\dfrac{2}{\pi}, 1, 0.$

2. $\dfrac{9}{16}.$

3. $(x + y)^{xy} + (xy)^{2x}.$

4. $\sqrt{1 + x^2}.$

6. (1) $x \geqslant 1$, $y \geqslant 0$; (3) $x > 0$, $y > 0$ or $x < 0$, $y < 0$;
 (2) $-1 < x < 1$, $y \leqslant -1$ or $y \geqslant 1$; (4) $x + y > 0$, $x - y > 0$;
 (5) $\dfrac{x^2}{a^2} + \dfrac{y^2}{b^2} \leqslant 1$;
 (6) $|y| \leqslant |x|$, $x \neq 0$; (7) $y \geqslant 0$, $x \geqslant \sqrt{y}$;
 (8) $0 < x^2 + y^2 < 1$, $y^2 \leqslant 4x$;
 (9) $2k\pi \leqslant x^2 + y^2 \leqslant (2k + 1)\pi$, $k = 0, 1, 2, \cdots$;
 (10) $x \leqslant x^2 + y^2 < 2x$.

7. $\tan 2t.$

8. $(x + y)^{x-y}$, $x^y + x - y$, $x + y - x^y.$

10. (1) $(0, 0)$; (3) points on $y^2 = 2x$.
 (2) points on $x = y$;

11. (1) $\displaystyle\int_0^1 dy \int_{-2y}^{2y} f(x, y)\, dx$; (3) $\displaystyle\int_{-\sqrt{2}}^{\sqrt{2}} dx \int_{x^2}^{4-x^2} f(x, y)\, dy$;
 (2) $\displaystyle\int_0^1 dx \int_{x-1}^{1-x} f(x, y)\, dy$; (4) $\displaystyle\int_0^1 dy \int_{-\sqrt{y-y^2}}^{\sqrt{y-y^2}} f(x, y)\, dx$;
 (5) $\displaystyle\int_0^2 dx \int_{1-\sqrt{2x-x^2}}^{1+\sqrt{2x-x^2}} f(x, y)\, dy$;

$$(6) \int_0^1 dx \int_{\frac{x}{2}}^{2x} f(x,y)\,dy + \int_1^2 dx \int_{\frac{x}{2}}^{\frac{2}{x}} f(x,y)\,dy.$$

12. (1) $\displaystyle\int_0^1 dy \int_{e^y}^e f(x,y)\,dx;$ (2) $\displaystyle\int_0^a dy \int_{a-\sqrt{a^2-y^2}}^y f(x,y)\,dx;$

(3) $\displaystyle\int_{-1}^0 dy \int_{-2\sqrt{1+y}}^{2\sqrt{1+y}} f(x,y)\,dx + \int_0^8 dy \int_{-2\sqrt{1+y}}^{2-y} f(x,y)\,dx;$

(4) $\displaystyle\int_{-1}^0 dy \int_{-\sqrt{1-y^2}}^{\sqrt{1-y^2}} f(x,y)\,dx + \int_0^1 dy \int_{-\sqrt{1-y}}^{\sqrt{1-y}} f(x,y)\,dx;$

(5)

$$\int_0^a dy \left[\int_{\frac{y^2}{2a}}^{a-\sqrt{a^2-y^2}} f(x,y)\,dx + \int_{a+\sqrt{a^2-y^2}}^{2a} f(x,y)\,dx \right]$$

$$+ \int_a^{2a} dy \int_{\frac{y^2}{2a}}^{2a} f(x,y)\,dx;$$

(6) $\displaystyle\int_0^1 dy \int_{\sqrt{y}}^{3-2y} f(x,y)\,dx.$

14. $(e-1)^2.$ 15. 9.

16. $\dfrac{1}{12}.$ 17. $\dfrac{8}{5}.$ 18. $\dfrac{20}{3}a^4.$

19. $\dfrac{9}{8}\ln 3 - \ln 2 - \dfrac{1}{2}.$ 20. $\dfrac{a^2}{3}.$

21. 0.

22. $\dfrac{a^4}{2}.$ 23. $\dfrac{1}{6}.$

24. Rewrite the inequality on the left hand side as $\displaystyle\iint_{\substack{a\leqslant x\leqslant b \\ a\leqslant y\leqslant b}} f(x)f(y)\,dx\,dy,$

then study the integral $\displaystyle\iint_{\substack{a\leqslant x\leqslant b \\ a\leqslant y\leqslant b}} [f(x)-f(y)]^2\,dx\,dy.$

25. (1) $\displaystyle\int_{-a}^a dx \int_{-b\sqrt{1-\frac{x^2}{a^2}}}^{b\sqrt{1-\frac{x^2}{a^2}}} dy \int_{-c\sqrt{1-\frac{x^2}{a^2}-\frac{y^2}{b^2}}}^{c\sqrt{1-\frac{x^2}{a^2}-\frac{y^2}{b^2}}} f(x,y,z)\,dz;$

(2) $\displaystyle\int_0^1 dz \int_{-z}^z dx \int_{-\sqrt{z^2-x^2}}^{\sqrt{z^2-x^2}} f(x,y,z)\,dy;$

(3) $\displaystyle\int_{-\frac{\sqrt{3}}{2}a}^{\frac{\sqrt{3}}{2}a} dx \int_{-\sqrt{\frac{3}{4}a^2-x^2}}^{\sqrt{\frac{3}{4}a^2-x^2}} dy \int_{a-\sqrt{a^2-x^2-y^2}}^{\sqrt{a^2-x^2-y^2}} f(x,y,z)\,dz;$

$$(4) \int_{-R}^{R} dx \int_{-\sqrt{R^2-x^2}}^{\sqrt{R^2-x^2}} dy \int_{0}^{H} f(x,y,z)\, dz.$$

26. $\dfrac{7}{2}\ln 2 - \dfrac{3}{2}\ln 5.$

27. $\dfrac{1}{720}.$

28. $0.$

29. $\dfrac{1}{48}.$

30. $\dfrac{1}{364}.$

Exercises 6.2

1. $1,\, 1+2\ln 2.$

2. $-1,\, 0.$

3. $\dfrac{3ab^2}{2\sqrt{ab^3-ba^3}},\ -\dfrac{3ba^2}{2\sqrt{ab^3-ba^3}}.$

4. $\dfrac{1}{x+y^2},\ \dfrac{2y}{x+y^2}.$

5. $y^2(1+xy)^{y-1},\ xy(1+xy)^{y-1}+(1+xy)^y\ln(1+xy).$

6. $-\dfrac{y}{x^2}\left(\dfrac{1}{3}\right)^{-\frac{y}{x}}\ln 3,\ \dfrac{1}{x}\left(\dfrac{1}{3}\right)^{-\frac{y}{x}}\ln 3.$

7. $\dfrac{2}{y\sin\dfrac{2x}{y}},\ \dfrac{-2x}{y^2\sin\dfrac{2x}{y}}.$

8. $\dfrac{y^2}{(x^2+y^2)^{\frac{3}{2}}},\ -\dfrac{xy}{(x^2+y^2)^{\frac{3}{2}}}.$

9. $-\dfrac{2x\sin x^2}{y},\ \dfrac{\cos x^2}{y^2}.$

10. $-\dfrac{y}{x^2+y^2},\ \dfrac{x}{x^2+y^2}.$

11. $-ye^{-xy}\sin\sqrt{\dfrac{y}{x}}-\dfrac{e^{-xy}}{2x}\sqrt{\dfrac{y}{x}}\cos\sqrt{\dfrac{y}{x}},\ -xe^{-xy}\sin\sqrt{\dfrac{y}{x}}+\dfrac{e^{-xy}}{2\sqrt{xy}}\cos\sqrt{\dfrac{y}{x}}.$

12. $\dfrac{z}{x}\left(\dfrac{x}{y}\right)^{z},\ -\dfrac{z}{y}\left(\dfrac{x}{y}\right)^{z},\ \left(\dfrac{x}{y}\right)^{z}\ln\dfrac{x}{y}.$

13. $\dfrac{y^z}{x}x^{yz},\ zy^{z-1}x^{yz}\ln x,\ y^z x^{yz}\ln x\ln y.$

14. $-\dfrac{x}{(x^2+y^2+z^2)^{\frac{3}{2}}},\ -\dfrac{y}{(x^2+y^2+z^2)^{\frac{3}{2}}},\ -\dfrac{z}{(x^2+y^2+z^2)^{\frac{3}{2}}}.$

15. $-\dfrac{\sqrt{y}}{\sqrt{x(1-xy)}(1+\sqrt{xy})},\ -\dfrac{\sqrt{x}}{\sqrt{y(1-xy)}(1+\sqrt{xy})}.$

16. $\dfrac{\partial z}{\partial x}=\dfrac{y(2x+y)}{\sqrt{1+(xy^2+yx^2)^2}},\ \dfrac{\partial z}{\partial y}=\dfrac{x(2y+x)}{\sqrt{1+(xy^2+yx^2)^2}}.$

20. $\dfrac{\partial^2 z}{\partial x^2}=\dfrac{xy^3}{(1-x^2y^2)^{\frac{3}{2}}},\ \dfrac{\partial^2 z}{\partial x\partial y}=\dfrac{1}{(1-x^2y^2)^{\frac{3}{2}}},\ \dfrac{\partial^2 z}{\partial y^2}=\dfrac{x^3y}{(1-x^2y^2)^{\frac{3}{2}}}.$

21. $\dfrac{\partial^2 z}{\partial x^2}=-\dfrac{4y}{(x+y)^3},\ \dfrac{\partial^2 z}{\partial x\partial y}=\dfrac{2(x-y)}{(x+y)^3},\ \dfrac{\partial^2 z}{\partial y^2}=\dfrac{4x}{(x+y)^3}.$

22. $\dfrac{\partial^2 z}{\partial x^2} = -\dfrac{x}{(x^2+y^2)^{\frac{3}{2}}}, \quad \dfrac{\partial^2 z}{\partial x \partial y} = -\dfrac{y}{(x^2+y^2)^{\frac{3}{2}}},$

$\dfrac{\partial^2 z}{\partial y^2} = \dfrac{x^3 + (x^2+y^2)\sqrt{x^2+y^2}}{(x^2+y^2)^{\frac{3}{2}}(x+\sqrt{x^2+y^2})^2}.$

23. $\dfrac{\partial^2 z}{\partial x^2} = y(y-1)x^{y-2}, \quad \dfrac{\partial^2 z}{\partial x \partial y} = x^{y-1}(1+y\ln x), \quad \dfrac{\partial^2 z}{\partial y^2} = x^y \ln^2 x.$

24. $\dfrac{\partial^2 z}{\partial x^2} = \dfrac{1}{x}, \quad \dfrac{\partial^2 z}{\partial x \partial y} = \dfrac{1}{y}, \quad \dfrac{\partial^2 z}{\partial y^2} = -\dfrac{x}{y^2}.$

28. (1) $du = \dfrac{ydx - xdy}{|y|\sqrt{y^2 - x^2}};$ (2) $dz = \dfrac{2(xdx + ydy)}{x^2 + y^2};$

    (3) $du = \dfrac{2(sdt - tds)}{(s-t)^2};$

    (4) $du = (y+z)dx + (z+x)dy + (x+y)dz;$

    (5) $du = \dfrac{(x^2+y^2)dz - 2z(xdx + ydy)}{(x^2+y^2)^2}.$

29. $dz = -0.2, \ \Delta z \sim -0.20404.$

30. $dz = 22.4, \ \Delta z = 22.75.$

31. $dz = 0.075, \ \Delta z = 0.0714.$

32. (1) 2.95;   (2) 0.502;   (3) 1.055;   (4) 0.97.

33. $\dfrac{du}{dt} = \sin 2t + 6t^5 + t^3 \cos t + 3t^2 \sin t.$

34. $\dfrac{\partial u}{\partial s} = \dfrac{2s}{1 + (1 + s^2 - t^2)^2}, \quad \dfrac{\partial u}{\partial t} = \dfrac{-2t}{1 + (1 + s^2 - t^2)^2}.$

35. $\dfrac{3 - 12t^2}{\sqrt{1 - (3t - 4t^3)^2}}.$

36. $\dfrac{\partial u}{\partial \varphi} = 2[\varphi + \theta \tan(\varphi\theta)\sec^2(\varphi\theta)], \quad \dfrac{\partial u}{\partial \theta} = 2[\theta + \varphi \tan(\varphi\theta)\sec^2(\varphi\theta)].$

37. $\dfrac{\partial u}{\partial x} = \dfrac{1 + y\cos(xy)}{\sqrt{1 - [x + y + \sin(xy)]^2}}, \quad \dfrac{\partial u}{\partial y} = \dfrac{1 + x\cos(xy)}{\sqrt{1 - [x + y + \sin(xy)]^2}}.$

38. $\dfrac{1}{1 + x^2}.$

39. $e^{ax} \sin x.$

40. $3t^2 \dfrac{\partial f}{\partial \zeta} + 4t \dfrac{\partial f}{\partial \eta}.$

41. $\dfrac{\partial u}{\partial x} = 2\dfrac{\partial f}{\partial \xi}x + 2\dfrac{\partial f}{\partial \eta}x + 2\dfrac{\partial f}{\partial \zeta}y, \quad \dfrac{\partial u}{\partial y} = 2\dfrac{\partial f}{\partial \xi}y - 2\dfrac{\partial f}{\partial \eta}y + 2\dfrac{\partial f}{\partial \zeta}x.$

42. $\dfrac{\partial f}{\partial x} + 2t\dfrac{\partial f}{\partial y} + 3t^2\dfrac{\partial f}{\partial z}$.

43. $\Delta u = 3f_{11}'' + 4(x + y + z)f_{12}'' + 4(x^2 + y^2 + z^2)f_{22}'' + 6f_2'$.

49. (1) $\dfrac{dy}{dx} = \dfrac{x + y}{y - x}$;

    (2) $\dfrac{dy}{dx} = \dfrac{2x + y}{x - 2y}$;

    (3) $\dfrac{dy}{dx} = \dfrac{y^2}{1 - xy}$;

    (4) $\dfrac{dy}{dx} = \dfrac{y^2(\ln x - 1)}{x^2(\ln y - 1)}$.

50. (1) $\dfrac{yz}{z^2 + xy}, \quad \dfrac{xz}{z^2 + xy}$;

    (2) $-\dfrac{c^2 x}{a^2 z}, \quad -\dfrac{c^2 y}{b^2 z}$;

    (3) $\dfrac{ayz - x^2}{z^2 - axy}, \quad \dfrac{axz - y^2}{z^2 - axy}$;

    (4) $\dfrac{z}{x + z}, \quad \dfrac{z^2}{y(x + z)}$.

51. $\dfrac{(2 - z)^2 + x^2}{(2 - z)^3}$.

52. $\dfrac{\partial^2 z}{\partial x^2} = \dfrac{(e^z - xy)2y^2 z - y^2 z^2 e^2}{(e^z - xy)^3} = \dfrac{2z(z - 1) - z^3}{x^2(z - 1)^3}$,

$\dfrac{\partial^2 z}{\partial x \partial y} = -\dfrac{z}{xy(z - 1)^3}, \quad \dfrac{\partial^2 z}{\partial y^2} = \dfrac{2z(z - 1) - z^3}{y^2(z - 1)^3}$.

55. $dz = -\dfrac{1}{\sin 2z}(\sin 2x\, dx + \sin 2y\, dy)$.

56. $dz = -\dfrac{z}{x}\, dx + \dfrac{(2xyz - 1)z}{(2xz - 2xyz + 1)y}\, dy$.

62. $\dfrac{\partial u}{\partial x} = -\dfrac{xu + yv}{x^2 + y^2}, \dfrac{\partial u}{\partial y} = \dfrac{xv - yu}{x^2 + y^2}, \dfrac{\partial v}{\partial x} = \dfrac{yu - xv}{x^2 + y^2}, \dfrac{\partial v}{\partial y} = -\dfrac{xu + yv}{x^2 + y^2}$.

63. $x + y - 2 = 0, \ y = x$.

64. $x + 2y - 1 = 0, \ 2x - y - 2 = 0$.

65. $\dfrac{x - 1}{16} = \dfrac{y - 1}{9} = \dfrac{z - 1}{-1}, \ 16x + 9y - z = 24$.

66. $(-1, 1, -1)$ or $\left(-\dfrac{1}{3}, \dfrac{1}{9}, -\dfrac{1}{27}\right)$.

67. $x + 2y - 4 = 0, \quad \begin{cases} \dfrac{x - 2}{1} = \dfrac{y - 1}{2}, \\ z = 0. \end{cases}$

68. $17x + 11y + 5z = 60, \quad \dfrac{x - 3}{17} = \dfrac{y - 4}{11} = \dfrac{z + 7}{5}$.

69. $x - y + 2z = \pm\sqrt{\dfrac{11}{2}}$.

70. $(-3, -1, 3), \quad \dfrac{x + 3}{1} = \dfrac{y + 1}{3} = \dfrac{z - 3}{1}$.

Exercises 6.3

1. $\dfrac{\partial u}{\partial x} = \dfrac{\dfrac{\partial y}{\partial v}}{\dfrac{\partial(x,y)}{\partial(u,v)}}, \quad \dfrac{\partial u}{\partial y} = -\dfrac{\dfrac{\partial x}{\partial v}}{\dfrac{\partial(x,y)}{\partial(u,v)}}, \quad \dfrac{\partial v}{\partial x} = -\dfrac{\dfrac{\partial y}{\partial u}}{\dfrac{\partial(x,y)}{\partial(u,v)}}, \quad \dfrac{\partial v}{\partial y} = \dfrac{\dfrac{\partial x}{\partial u}}{\dfrac{\partial(x,y)}{\partial(u,v)}}.$

2. $\dfrac{\partial z}{\partial x} = \dfrac{\dfrac{\partial(z,y)}{\partial(u,v)}}{\dfrac{\partial(x,y)}{\partial(u,v)}}, \quad \dfrac{\partial z}{\partial y} = -\dfrac{\dfrac{\partial(x,z)}{\partial(u,v)}}{\dfrac{\partial(x,y)}{\partial(u,v)}}.$

5. $\dfrac{2}{3}\pi[(1+R^2)^{\frac{3}{2}} - 1].$ 6. $\dfrac{\pi}{3} - \dfrac{4}{9}.$

7. $\dfrac{\pi}{4}(2\ln 2 - 1).$

8. $\dfrac{3}{64}\pi^2.$

9. $\dfrac{8}{9}\sqrt{2}.$

10. $-6\pi^2.$

11. $18\pi.$

12. $\dfrac{8}{3}ab\arctan\dfrac{a}{b}.$

13. $\dfrac{15}{4}\pi b(a^2 + b^2).$

14. $\dfrac{2}{3}\pi ab.$

15. $\dfrac{16}{3}\pi.$

16. $\dfrac{1}{48}.$

17. $\dfrac{\pi}{10}.$

18. $\dfrac{\pi^2}{4}abc.$

19. $\dfrac{4}{15}\pi(R^5 - r^5).$

20. (1) πab; (3) $2a^2$;

(2) $\dfrac{5}{8}\pi a^2$; (4) $\dfrac{a^2 b^2}{2c^2}.$

21. $\pi a^3.$

22. $8\pi.$

23. $\dfrac{\pi}{3}(2 - \sqrt{2})(b^2 - a^2).$

24. $\dfrac{2\pi}{3}(2 - \sqrt{2})abc.$

25. $\dfrac{\pi}{3}a^3.$

26. $\dfrac{\pi}{3}\dfrac{a^2 bc}{h}.$

27. $\dfrac{3}{4}\ln 2.$

28. $\dfrac{1}{3}(b - a)(q - p).$

29. $13\dfrac{1}{3}.$

30. $\sqrt{2}\pi.$

31. $2a^2.$

32. $\dfrac{2}{3}\pi ab[(1 + c^2)^{\frac{3}{2}} - 1].$

33. $\dfrac{16}{3}\pi a^2.$

34. $\pi\left[a\sqrt{a^2 + h^2} + h^2\ln\dfrac{a + \sqrt{a^2 + h^2}}{h}\right].$

35. $a(\varphi_2 - \varphi_1)[b(\psi_2 - \psi_1) + a(\sin\psi_2 - \sin\psi_1)], 4\pi^2 ab.$

36. $8a^2.$ 37. $S\cos\gamma, S\cos\alpha, S\cos\beta.$

Chapter 7. Line Integrals, Surface Integrals and Exterior Differential Forms

Exercises 7.1

1. (1) $x + y + z = C;$ (2) $x^2 + y^2 + z^2 = R^2, (R \neq 0);$

   (3) hyperboloid of revolution of one sheet, or hyperboloid of revolution of two sheets, or circular conical surface;

   (4) $z = C\sqrt{x^2 + y^2};$

   (5) $z = \sqrt{\dfrac{x^2 + y^2}{2}};$ (6) $x^2 + y^2 + z^2 = R^2, (R \neq 0).$

2. (1) $y = Cx;$ (3) $x^2 + 2y^2 = C, (C > 0);$

   (2) $x^2 - y^2 = C;$ (4) $\dfrac{x}{x^2 + y^2} = C.$

3. $0.$

4. $\dfrac{\sqrt{2}}{2}.$

5. $5.$

6. $\dfrac{98}{13}.$ 8. $\dfrac{\sqrt{2}}{3}$ or $-\dfrac{\sqrt{2}}{3}.$ 10. $\dfrac{x_0 + y_0 + z_0}{\sqrt{x_0^2 + y_0^2 + z_0^2}}.$

7. $\dfrac{1}{2}(\cos\alpha, \sin\alpha).$ 9. $\dfrac{1}{2}.$

11. $\dfrac{2\left(\dfrac{x_0^2}{a^2} + \dfrac{y_0^2}{b^2} + \dfrac{z_0^2}{c^2}\right)}{\sqrt{\dfrac{x_0^2}{a^4} + \dfrac{y_0^2}{b^4} + \dfrac{z_0^2}{c^4}}}.$

12. $\sqrt{3} + 5, -\sqrt{3} - 5.$ 13. $\dfrac{\sqrt{3}}{9}, -\left(\dfrac{1}{21}, \dfrac{2}{21}, \dfrac{4}{21}\right).$

14. $x^2 + y^2 + z^2 = R, -e\left(\dfrac{x}{r^3}, \dfrac{y}{r^3}, \dfrac{z}{r^3}\right), e\left(\dfrac{x}{r^3}, \dfrac{y}{r^3}, \dfrac{z}{r^3}\right).$

15. $\dfrac{\partial u}{\partial r} = \dfrac{2u}{r}, \quad r = \sqrt{x^2 + y^2 + z^2}; \quad \dfrac{\partial u}{\partial r} = |\operatorname{grad} u|,$ if $a = b = c.$

16. $\left(-\dfrac{1}{3}, \dfrac{3}{4}\right)$ or $\left(\dfrac{7}{3}, -\dfrac{3}{4}\right).$

17. $\cos\varphi = -\dfrac{12}{5\sqrt{145}}$.

18. $\cos\varphi = -\dfrac{4}{405}$.

19. $\dfrac{\partial u}{\partial l} = \dfrac{\mathbf{grad}\,u \cdot \mathbf{grad}\,v}{|\mathbf{grad}\,v|}$; $\quad \dfrac{\partial u}{\partial l} = 0$, when $\mathbf{grad}\,u \perp \mathbf{grad}\,v$.

20. the point where $r = 1$.

21. (1) $\left(\dfrac{x}{r}, \dfrac{y}{r}, \dfrac{z}{r}\right)$;

    (2) $(2x, 2y, 2z)$;

    (3) $-\left(\dfrac{x}{r^3}, \dfrac{y}{r^3}, \dfrac{z}{r^3}\right)$;

    (4) $\left(\dfrac{x}{r^2}, \dfrac{y}{r^2}, \dfrac{z}{r^2}\right)$;

    (5) $f'(r)\left(\dfrac{x}{r}, \dfrac{y}{r}, \dfrac{z}{r}\right)$;

    (6) $2f'(r^2)(x, y, z)$.

23. $\dfrac{x^2 + y^2 - z(x + y)}{(x^2 + y^2 + z^2)\sqrt{x^2 + y^2}}$.

24. (1) nr^n;

    (2) (x, y, z);

    (3) $(0, 0, 0)$.

26. $(\vec{\omega} \cdot \vec{\omega})\vec{r} - (\vec{\omega} \cdot \vec{r})\vec{\omega}$.

28. (2) $(\vec{r} \cdot \vec{a})\vec{b} + (\vec{r} \cdot \vec{b})\vec{a}$;

    (3) $\vec{c} + \dfrac{\vec{c}}{2\vec{c} \cdot \vec{r}}$.

29. $(x + t)(t - y) = C$.

30. The electric field lines are rays from the origin
$$\begin{cases} x = at, \\ y = bt, \\ z = ct, \end{cases}$$
$0 < t < +\infty$, c is an constant.

31. $x^2 + y^2 = R^2$, $(R > 0)$.

Exercises 7.2

1. $1 + \sqrt{2}$.

2. $\dfrac{256}{15}a^2$.

3. $2\pi^2 a^3(1 + 2\pi^2)$.

4. 0.

5. $2a^2$.

6. $\dfrac{a^3}{6}(\cosh^{\frac{3}{2}} 2t_0 - 1)$.

7. $\dfrac{8\sqrt{2}}{3}a\pi^3$.

8. $\dfrac{a^2}{256\sqrt{2}}\left[100\sqrt{38} - 72 - 17\ln\dfrac{25 + 4\sqrt{38}}{17}\right]$.

9. $a^{\frac{7}{8}}$.

11. $2\pi a^{2n+1}$.

10. $\dfrac{\pi a}{4}e^a + 2(e^a - 1)$.

12. $\dfrac{1}{3p}[(p^2 + y_0^2)^{\frac{3}{2}} - p^3]$.

13. $\dfrac{2}{3}\pi a^3$.

14. 5.

15. $4\pi R^2$.

16. $-\dfrac{56}{15}$.

17. 3.

18. -8.

19. $\dfrac{4}{5}$.

20. $-\dfrac{14}{15}$.

21. 1 for all paths.

22. (1) 0;

(2) $\dfrac{1}{3}$;

(3) 1.

23. 0.

24. 0.

25. 6.

26. $\dfrac{1}{35}$.

27. 13.

28. -2π.

29. (1) $\dfrac{4}{3}$;

(2) 0;

(3) $\dfrac{12}{5}$;

(4) -4;

(5) 4.

30. mab.

31. See problem 32 in Section 7.1. The circulation is $4\pi l$.

32. The work done is $\dfrac{1}{r_2} - \dfrac{1}{r_1}$. The circulation is 0 for any closed curve that does not pass the origin.

Exercises 7.3

1. 9.

2. $4\sqrt{61}$.

3. $\dfrac{3 - \sqrt{3}}{2} + (\sqrt{3} - 1)\ln 2$.

4. πa^3.

5. πR^3.

6. $\dfrac{\pi}{2}(1 + \sqrt{2})$.

7. $\dfrac{64}{15}\sqrt{2}a^4$.

8. $\dfrac{125\sqrt{5} - 1}{420}$.

9. $2\pi R\left(1 - \dfrac{\sqrt{R^2 - \rho^2}}{R}\right)$.

10. 1.

11. $\dfrac{1}{8}$.

12. $\dfrac{8}{3}\pi(a + b + c)R^2$.

13. 6π.   14. 0.   15. 3.   16. 0   17. 0.

18. $-\dfrac{\pi}{2}h^4$.

19. $\dfrac{1}{2}$.

20. 0.

22. $2\pi R^2 h$.

21. $\dfrac{3\pi}{16}$.

Exercises 7.4

1. (1) $\dfrac{\pi a^4}{2}$;

   (2) $-2\pi ab$;

   (3) $-\dfrac{1}{5}(e^\pi - 1)$;   (4) $\dfrac{\pi m a^2}{8}$;   (5) (i) 0;   (ii) $\dfrac{\pi a^3}{8}$.

3. $I = \begin{cases} 2\pi, & \text{if the origin is inside of } C, \\ 0, & \text{if the origin is outside of } C. \end{cases}$

4. (1) $3a^4$;

   (2) $\dfrac{12}{5}\pi a^5$;   (3) $R^2 H\left(\dfrac{2}{3}R + \dfrac{\pi H}{8}\right)$; (4) $\dfrac{1}{8}$;

   (5) 0.

5. (1) $-\sqrt{3}\pi a^2$;   (3) $-2\pi a(a + b)$;

   (2) 0;   (4) $-\dfrac{3\pi a^2}{4}$.

6. (1) div $A = 0$, **rot** $A = 2(y - z, z - x, x - y)$;

   (2) div $\vec{c} = 6xyz$, **rot** $\vec{c} = (x(z^2 - y^2), y(x^2 - z^2), z(y^2 - x^2))$.

7. (1) 6;

   (2) $f''(r) + \dfrac{2}{r}f'(r)$.

8. (1) $\dfrac{f'(r)}{r}(\vec{r} \times \vec{c})$;

   (2) 0;   (3) 0;   (4) $\vec{a} \times \vec{b}$;   (5) $2\vec{c}$;

   (6) $2f(r)\vec{c} + \dfrac{f'(r)}{r}[\vec{c}(\vec{r} \cdot \vec{r}) - \vec{r}(\vec{c} \cdot \vec{r})]$.

9. (1) 3;

   (2) $\dfrac{2}{r}$;

   (5) $\dfrac{f'(r)}{r}(\vec{c} \cdot \vec{r})$;

   (3) 0;

   (4) 0;

   (6) $f''(r) + \dfrac{2}{r}f'(r)$.

10. (1) 0;

   (2) For a point in the rigid body, $\mathbf{rot}\,\vec{v} = 2\vec{\omega}$, where $\vec{\omega}$ is the angular velocity.

11. (1) πab;

   (2) $\dfrac{3\pi}{8}a^2$;

   (3) $3\pi a^2$.

Exercises 7.5

1. (1) 4;

   (2) $-\dfrac{3}{2}$;

   (3) 9;   (4) 62;   (5) 2;

   (6) $\displaystyle\int_{r_1}^{r_2} f(r)r\,dr$, where $r_1 = \sqrt{x_1^2 + y_1^2 + z_1^2}$, $r_2 = \sqrt{x_2^2 + y_2^2 + z_2^2}$.

2. (1) $xy = C$;   (2) $x^2 y = C$;   (3) $xe^y - y^2 = C$;

   (4) $x - \dfrac{y^2}{2} + \dfrac{1}{3}(x^2 + y^2)^{\frac{3}{2}} + C$.

3. $x^2 yz + y^2 zx + z^2 xy$.

4. $\displaystyle\int_{r_0}^{r} f(r)\,dr$, where r_0 can be chosen arbitrarily.

5. (1) not a potential field, not a source-free field;

   (2) harmonic field;   (4) source-free field;

   (3) harmonic field;   (5) harmonic field.

Exercises 7.6

1. (1) $7z^2\,dy \wedge dz - x\,dz \wedge dx - (x\sin 3x + 7yz^2)\,dx \wedge dy$;

   (2) $dx \wedge dy$;   (3) $-21\,dx \wedge dy \wedge dz$.

2. (1) $(-\cos x + \sin y)\,dx \wedge dy$;   (3) $(x + 6)\,dx \wedge dy \wedge dz$.

   (2) 0;

Chapter 8. Some Applications of Calculus in Several Variables

Exercises 8.1

1. $5 + 2(x-1)^2 - (x-1)(y+2) - (y+2)^2$.

3. $1 + (x-1) + (x-1)(y-1) + R_2$.

4. $1 - \dfrac{1}{2}(x^2 + y^2) - \dfrac{1}{8}(x^2 + y^2)^2 + R_4$.

5. $\displaystyle\sum_{n=0}^{\infty} (-1)^n \dfrac{(x^2+y^2)^{2n+1}}{(2n+1)!}$, $(x^2 + y^2 < +\infty)$.

6. (1) the minimum value of $z = 0$ at the point $(0, 1)$;
   (2) the minimum value of $z = -1$ at the point $(1, 0)$;
   (3) the minimum value of $z = -1$ at the point $(1, 1)$;
   (4) the minimum value of $z = 30$ at the point $(5, 2)$.

7. (1) maximum $z(1, -1) = 6$,   minimum $z(1, -1) = -2$;
   (2) the minimum value of $z = -(4 + 2\sqrt{6})$ when $x = y = -(3 + \sqrt{6})$;
   the maximum value of $z = 2\sqrt{6} - 4$ when $x = y = -(3 - \sqrt{6})$.

8. (1) maximum $z\left(\dfrac{1}{2}, \dfrac{1}{2}\right) = \dfrac{1}{4}$;

   (2) minimum $u\left(\dfrac{ab^2}{a^2+b^2}, \dfrac{a^2b}{a^2+b^2}\right) = \dfrac{a^2b^2}{a^2+b^2}$;

   (3) minimum $u(3, 3, 3) = 9$;

   (4) minimum $z = -\dfrac{1}{3\sqrt{6}}$ when one variable is $-\dfrac{2}{\sqrt{6}}$ and the rest are $\dfrac{1}{\sqrt{6}}$,
   maximum $z = \dfrac{1}{3\sqrt{6}}$ when one variable is $\dfrac{2}{\sqrt{6}}$ and the rest are $-\dfrac{1}{\sqrt{6}}$.

9. the inscribed isosceles triangle's base line is $y = -1$.

10. $\sqrt{9 \pm 5\sqrt{3}}$.

11. $\left(\dfrac{21}{13}, 2, \dfrac{63}{26}\right)$.

12. the perimeter of the right triangle has maximum when both legs are $\dfrac{l}{\sqrt{2}}$.

13. $H = 2R = 2\sqrt{\dfrac{S}{3\pi}}$.

14. $\dfrac{7}{4\sqrt{2}}$.

15. the rectangle with sides $\dfrac{p}{3}$ and $\dfrac{2}{3}p$.

17. a^2.

18. $\left(\dfrac{8}{5}, \dfrac{16}{5}\right)$.

Exercises 8.2

1. $\frac{\pi}{2}ab$.

2. Choose the coordinate system such that the center of the larger disk is the origin, and the positive x-axis passes through the center of the smaller disk, then $x_0 = \dfrac{-6a}{5(3\pi - 2)}$, $y_0 = 0$.

3. $2\pi r(R - r)$. 4. $2k\pi(R^2 - r^2)$.

5. $x_0 = y_0 = 0$, $z_0 = \dfrac{3}{4}c$. 6. $\left(0, 0, \dfrac{2}{3}\right)$.

7. $\left(0, 0, \dfrac{4}{5}a\right)$. 8. $\dfrac{8}{15}\pi a^5$.

9. (1) (i) $\dfrac{1}{12}ml^2$, (ii) $\dfrac{1}{3}ml^2$; (3) (i) $\dfrac{1}{2}mR^2$, (ii) $\dfrac{1}{4}mR^2$;

   (2) (i) mR^2, (ii) $\dfrac{1}{2}mR^2$; (4) $\dfrac{7}{5}mR^2$.

10. $F_x = F_y = 0$, $F_z = k\rho m\pi \ln\dfrac{b}{a}$.

11. The gravitational force is 0 if the particle is inside of the shell. The gravitational force is equal to that of a particle at the center of the shell with a mass that is equal to the mass of the shell, if the particle is outside of the shell.

12. angular momentum $I = \dfrac{8}{3}\pi\rho R^4$, kinetic energy $T = \dfrac{4}{3}\pi\rho R^4\omega^2$.

Chapter 9. The ε-δ Definitions of Limits

Exercises 9.1

2. (1) 0; (3) 0;

   (2) $\dfrac{2}{3}$; (4) 0;

   (5) $\dfrac{1}{3}$; (6) $-\dfrac{1}{2}$; (7) $\dfrac{1}{3}$; (8) $\dfrac{1}{2}$;

   (9) $\dfrac{1 - b}{1 - a}$;

   (10) 0; (12) 0;

   (11) $\dfrac{a + b}{2}$; (13) $\dfrac{1}{3}$;

   (14) 0; (15) $\ln(1 + \sqrt{2})$; (16) does not exist;

   (17) 1; (18) 3; (19) 2.

20. (1) convergent;   (3) convergent;
    (2) convergent;   (4) divergent.

Exercises 9.2

2. $f(0+0) = 1$, $f(0-0) = 1$.

3. (1) $f(2+0) = 4$, $f(2-0) = -2a$; (2) $a = -2$.

6. (1) 1;

    (2) $\dfrac{2}{3}$;   (3) $\dfrac{1}{2}$;

    (4) 6;   (5) 10;

    (6) $\dfrac{1}{2}nm(n-m)$;   (7) $\dfrac{m}{n}$;   (8) $\dfrac{n(n+1)}{2}$;

    (9) 0;   (10) 1;   (11) -1;   (12) 0;

    (13) $\dfrac{1}{2}(a+b)$;   (14) $\dfrac{3}{2}$;   (15) $\dfrac{\alpha}{m} - \dfrac{\beta}{n}$;   (16) $\dfrac{\alpha}{m} + \dfrac{\beta}{n}$.

8. (1) $a = 1$, $b = -1$;

    (2) $a = 1$, $b = -\dfrac{1}{2}$;   (3) $a = -1$, $b = \dfrac{1}{2}$.

10. (1) e^{-1};   (2) e^{-2};   (3) e^2;   (4) e^{-1};   (5) 5;

    (6) $(-1)^{m-n}\dfrac{m}{n}$;   (7) $\dfrac{1}{2}$;

    (8) 1;   (9) $-\sin a$;

    (10) $\begin{cases} 1, & \text{if } m = n, \\ 0, & \text{if } m > n, \\ \infty. & \text{if } m < n. \end{cases}$

13. (1) first order;   (2) first order;   (3) second order; (4) first order.

14. (1) 3^x is of higher order than 2^x;
    (2) $\sqrt{1+x^2}$ and $x - 1$ arc of the same order;
    (3) $\sqrt{x^3 + x + 1}$ is of higher order than $\sqrt{x + \sin^2 x}$;
    (4) $\sqrt{x^2 + x + 1}$ is of higher order than $\ln(x + \sqrt{x^2 + 1})$.

15. (3) $A = \dfrac{\pi}{2}$.

17. (1) $\dfrac{m}{n}$; 　　　　　　　　　　(2) $\dfrac{1}{4}$;

　　(3) 1; 　　　　　　　　　　　　(4) 1.

19. (1) $x = \pm 1$ are discontinuity points of the first kind;

　　(2) continuous;

　　(3) $x = 1$ is a discontinuity point of the first kind

$$(4)\ f(x) = \begin{cases} x, & \text{if } x < 0, \\ 1, & \text{if } x = 0, \\ \dfrac{1}{x^2}. & \text{if } x > 0, \end{cases}$$

　　$x = 0$ is discontinuity point of the second kind;

　　(5) $f(x) = \operatorname{sgn} x$, $x = 0$ is a discontinuity point of the first kind.

20. $a = 0$, $b = 1$, $c = 0$.

22. (1) The function $f(x) + g(x)$ is not continuous at the point x_0. If $f(x) = 0$, then $f(x)g(x)$ is continuous at the point x_0. If $f(x) \neq 0$, then $f(x)g(x)$ is not continuous at the point x_0.

　　(2) Both $f(x) + g(x)$ and $f(x)g(x)$ are possibly continuous at the point x_0.

26. (1) uniformly continuous; 　　　(5) uniformly continuous;

　　(2) not uniformly continuous; 　　(6) not uniformly continuous;

　　(3) uniformly continuous; 　　　(7) uniformly continuous.

　　(4) uniformly continuous;

Exercises 9.3

1. (1) $\inf(A) = \min\{a_1, a_2, \cdots, a_n\}$, $\sup(A) = \max\{a_1, a_2, \cdots, a_n\}$;

　　(2) $\inf(A) = 2$, $\sup(A) = e$; 　　(3) $\inf(A) = 0$, $\sup(A) = 1$;

　　(4) $\inf(A) = -1$, $\sup(A) = \dfrac{1}{2}$;

　　(5) $\inf(A) = -1$, $\sup(A) = 1$; 　　(7) $\inf(A) = e$, $\sup(A) = +\infty$;

　　(6) $\inf(A) = 0$, $\sup(A) = 1$; 　　(8) $\inf(A) = -\infty$, $\sup(A) = 1$.

2. (1) $\inf(f(x)) = 0$, $\sup(f(x)) = 25$; (3) $\inf(f(x)) = 0$, $\sup(f(x)) = 1$;

　　(2) $\inf(f(x)) = 0$, $\sup(f(x)) = 1$; (4) $\inf(f(x)) = -1$, $\sup(f(x)) = 1$;

　　(5) $\inf(f(x)) = \dfrac{1}{2}$, $\sup(f(x)) = 4$;

　　(6) $\inf(f(x)) = -\sqrt{2}$, $\sup(f(x)) = \sqrt{2}$.

3. $\sigma = \displaystyle\sum_{i=1}^{n-1} \left(i + \frac{1}{2} \right) \frac{25}{n^2} = 12\frac{1}{2}$.

4. (1) $S_1(f,T) = \dfrac{65}{4} + \dfrac{175}{2n} + \dfrac{125}{4n^2}$, $S_0(f,T) = \dfrac{65}{4} - \dfrac{175}{2n} + \dfrac{125}{4n^2}$;

   (2) $S_1(f,T) = \dfrac{1}{n} \displaystyle\sum_{i=1}^{n} \sqrt{\dfrac{i}{n}}$, $S_0(f,T) = \dfrac{1}{n} \displaystyle\sum_{i=0}^{n-1} \sqrt{\dfrac{i}{n}}$.

5. $V_0 T + \dfrac{1}{2} g T^2$.

6. (1) $\dfrac{2}{3}(2^{\frac{3}{2}} - 1)$; 　　　　　　　(3) $\dfrac{\pi}{4}$.

   (2) $\ln 2$;

7. (1) $\dfrac{1}{3}$; 　　　　　　　　　　　(2) $\dfrac{1}{a}$;

   (3) π;

   (4) $\dfrac{2}{3} \ln 2$; 　　　　　　　　　(5) $\dfrac{1}{2}$;

   (6) $\dfrac{1}{2}$; 　　　　　　　　　　(8) $\dfrac{(2n-3)!!}{(2n-2)!!} \dfrac{\pi}{2a^{2n-1}}$;

   (7) $n!$; 　　　　　　　　　　　(9) -1;

   (10) $\dfrac{\pi}{2}$; 　　(11) $-\dfrac{1}{4}$; 　　(12) $\dfrac{\pi}{2}$; 　　(13) $\dfrac{8}{3}$;

   (14) $\begin{cases} \dfrac{(n-1)!!}{n!!} \dfrac{\pi}{2}, & \text{if } n \text{ is an even number,} \\[2mm] \dfrac{(n-1)!!}{n!!}, & \text{if } n \text{ is an odd number;} \end{cases}$

   (15) $2\dfrac{(2n)!!}{(2n+1)!!}$; 　　　　(16) $(-1)^n n!$.

Chapter 10. Infinite Series and Infinite Integrals

Exercises 10.1

2. (1) $\dfrac{7}{2}$; 　　　　　　　　　(2) $\dfrac{55}{24}$.

3. (1) convergent; 　(5) divergent; 　(9) divergent; 　(13) convergent;
   (2) divergent; 　(6) convergent; 　(10) convergent; (14) convergent;
   (3) convergent; 　(7) convergent; 　(11) convergent; (15) convergent;
   (4) divergent; 　(8) divergent; 　(12) divergent; 　(16) convergent;

(17) convergent if $a > 1$, divergent if $0 < a < 1$.

9. (1) convergent if $p > 1$, divergent if $p \leqslant 1$;

(2) divergent;

(3) convergent if $p > 1$, and q is an arbitrary number, or if $p = 1$, and $q > 1$.

12. (1) convergent; (3) convergent; (5) divergent; (7) convergent.

(2) convergent; (4) convergent; (6) convergent;

13. (1) absolutely convergent if $p > 1$, conditionally convergent if $0 < p \leqslant 1$;

(2) absolutely convergent if $|a| > 1$, conditionally convergent if $|a| = 1$, divergent if $|a| < 1$;

(3) absolutely convergent if $p > 1$, conditionally convergent if $0 < p \leqslant 1$;

(4) conditionally convergent for any x that is not a negative integer.

Exercises 10.2

1. (1) absolutely convergent if $|x| > 1$; (2) absolutely convergent if $x > 0$;

(3) absolutely convergent if $x > -\dfrac{1}{4}$ and $x < -\dfrac{1}{2}$;

(4) absolutely convergent if $|x| < 1$; (6) absolutely convergent if $x > 1$;

(5) absolutely convergent if $|x| \neq 1$;

(7) absolutely convergent if $|x| < 1, 0 \leqslant y \leqslant +\infty$ or if $|x| > 1, y > |x|$, conditionally convergent if $x = -1, 0 \leqslant y \leqslant 1$;

(8) absolutely convergent if $0 < \min(x, y) < 1$.

2. (1) (i) uniformly convergent, (ii) not uniformly convergent;

(2) uniformly convergent; (3) uniformly convergent;

(4) (i) not uniformly convergent, (ii) uniformly convergent.

3. (1) uniformly convergent; (2) uniformly convergent;

(3) (i) uniformly convergent, (ii) not uniformly convergent;

(4) uniformly convergent; (9) not uniformly convergent;

(5) uniformly convergent; (10) uniformly convergent;

(6) uniformly convergent; (11) uniformly convergent;

(7) uniformly convergent; (12) uniformly convergent.

(8) uniformly convergent;

4. (1) $f(x) = \begin{cases} 0, & \text{if } x = 0, \\ \dfrac{1}{x}, & \text{if } x \neq 0. \end{cases}$

The function $f(x)$ is not continuous at the point $x = 0$, and is continuous at points $x \neq 0$;

(2) The function $f(x)$ is continuous on the real number axis.

5. $\lim\limits_{x \to 1} f(x) = \dfrac{-1}{4}$, $\lim\limits_{x \to +\infty} f(x) = 1$.

7. $f''(x) = -\sum\limits_{n=1}^{\infty} \dfrac{\sin nx}{n^2}$. $\qquad$ 8. $\displaystyle\int_{\ln 2}^{\ln 3} f(x)\, dx = \dfrac{1}{2}$.

9. $\displaystyle\int_{0}^{\pi} f(x)\, dx = 0$.

Exercises 10.3

1. (1) $R = 2$; $\qquad$ (3) $R = 1$; $\qquad$ (5) $R = e$;
   (2) $R = 1$; $\qquad$ (4) $R = +\infty$; $\qquad$ (6) $R = 4$.

2. (1) $R = \dfrac{1}{e}$, $\left(-\dfrac{1}{e}, \dfrac{1}{e}\right)$, divergent if $x = \pm\dfrac{1}{e}$;
   (2) $R = +\infty$, $(-\infty, +\infty)$; $\qquad$ (3) $R = +\infty$, $(-\infty, +\infty)$;
   (4) $R = \dfrac{1}{3}$, $\left(-\dfrac{4}{3}, -\dfrac{2}{3}\right)$, conditionally convergent at $x = -\dfrac{4}{3}$, divergent at $x = -\dfrac{2}{3}$;
   (5) $R = \max(a, b)$, $(-R, R)$, divergent at $x = \pm R$;
   (6) $R = 1$, $(-1, 1)$. For $x = \pm 1$, if $a > 1$, then it is absolutely convergent. If $a \leqslant 1$, then it is divergent.

3. (1) $s(x) = \dfrac{x}{1 - x}$, $(|x| < 1)$;

   (2) $s(x) = \dfrac{1 - x^2}{(1 + x^2)^2}$, $(|x| < 1)$;

   (3) $s(x) = \dfrac{x}{(1 - x)^2}$, $(|x| < 1)$;

   (4) $s(x) = \begin{cases} 1 + \dfrac{1 - x}{x} \ln(1 - x), & \text{if } 0 < |x| < 1, \\ 0, & \text{if } x = 0, \\ 1 - 2\ln 2, & \text{if } x = -1, \\ 1, & \text{if } x = 1. \end{cases}$

5. (1) $\sum\limits_{n=0}^{+\infty} x^{2n}$, $(|x| < 1)$;

(2) $1 + \sum\limits_{n=1}^{+\infty} (-1)^n \dfrac{2^{2n-1}}{(2n)!} x^{2n}$, $(|x| < +\infty)$;

(3) $\sum\limits_{n=12}^{+\infty} x^n$, $(|x| < 1)$;

(4) $\sum\limits_{n=0}^{+\infty} \dfrac{x^{2n+1}}{2n+1}$, $(|x| < 1)$;

(5) $\dfrac{1}{3} \sum\limits_{n=0}^{+\infty} [1 - (-2)^n] x^n$, $\left(|x| < \dfrac{1}{2} \right)$;

(6) $1 + \sum\limits_{n=1}^{+\infty} (-1)^n \left[\dfrac{1}{n!} - \dfrac{1}{(n-1)!} \right] x^n$, $(|x| < +\infty)$.

6. (1) $x + \sum\limits_{n=1}^{+\infty} \dfrac{(2n-1)!!}{(2n)!!} \dfrac{x^{2n+1}}{2n+1}$, $(|x| \leqslant 1)$;

(2) $x + \sum\limits_{n=1}^{+\infty} (-1)^n \dfrac{(2n-1)!!}{(2n)!!} \dfrac{x^{2n+1}}{2n+1}$, $(|x| \leqslant 1)$;

(3) $\sum\limits_{n=0}^{+\infty} (-1)^n \dfrac{x^{2n+1}}{(2n+1)(2n+1)!}$, $(|x| < +\infty)$;

(4) $\ln x + \sum\limits_{n=1}^{+\infty} \dfrac{x^n}{nn!} - \sum\limits_{n=1}^{+\infty} \dfrac{1}{nn!}$.

7. (1) $(x-1)^3 + (x-1)^2 + 4(x-1) - 3$, $(|x| < +\infty)$;

(2) $\sum\limits_{n=0}^{+\infty} (-1)^n \dfrac{x^n}{10^{n+1}}$, $(|x| < 10)$;

(3) $x + \sum\limits_{n=1}^{+\infty} (-1)^{n-1} \dfrac{x^{n+1}}{n(n+1)}$, $(|x| < 1)$. The region of convergence is $|x| \leqslant 1$;

(4) $\dfrac{1}{2} \sum\limits_{n=1}^{+\infty} \left[(n+1) - \dfrac{1 + (-1)^n}{2} \right] x^n$, $(|x| < 1)$;

(5) $x + 2 \sum\limits_{n=1}^{+\infty} \dfrac{(-1)^{n+1}}{4n^2 - 1} x^{2n+1}$, $(|x| \leqslant 1)$;

(6) $\sum\limits_{n=0}^{+\infty} (-1)^n \dfrac{(x-1)^{n+1}}{n+1}$, $(0 < x \leqslant 2)$;

(7) $\sum\limits_{n=0}^{+\infty} \dfrac{x^{2n}}{(2n)!}$, $(|x| < +\infty)$;

(8) $\dfrac{\sqrt{2}}{2}\displaystyle\sum_{n=0}^{+\infty}(-1)^{\frac{n(n+1)}{2}}\dfrac{\left(x-\frac{\pi}{4}\right)^n}{n!}$, $(|x| < +\infty)$.

8. (1) $\ln 2 + \dfrac{1}{2}x + \dfrac{1}{8}x^2 - \dfrac{1}{192}x^4 + \cdots$;

   (2) $x + \dfrac{x^3}{3} + \dfrac{2}{15}x^5 + \cdots$.

9. (1) 0.747;   (2) 1.605;   (3) 0.905.

10. (1) $y = Ce^x - 2x - 2$, where C is a constant;

   (2) $y = C_1 \displaystyle\sum_{n=0}^{+\infty}(-1)^n\dfrac{x^{2n+1}}{(2n+1)!!} + C_2\sum_{n=0}^{+\infty}(-1)^n\dfrac{x^{2n}}{(2n)!!}$, where C_1, C_2 are constants.

11. $y = \dfrac{x^2}{2} + \dfrac{1}{20}x^5 + \cdots$.

12. (1) $y = \dfrac{x^2}{2} + \displaystyle\sum_{n=2}^{+\infty}\dfrac{(2n-3)!!}{(2n)!}x^{2n}$;

   (2) $y = \ln(1+x)$.

13. (1) $157.970 + 0.0004\theta$, $(0 < \theta < 1)$;

   (2) $10^{28} \times 1.378\left(1 + \dfrac{\theta}{288}\right)$, $(|\theta| < 1)$.

Exercises 10.4

1. (1) divergent;   (2) convergent;   (3) convergent;

   (4) convergent if $p > 1$, divergent if $p \leqslant 1$;

   (5) convergent;   (6) divergent;   (7) divergent;

   (8) convergent if $a > 0$, divergent if $a \leqslant 0$;

   (9) convergent;   (10) convergent;

   (11) convergent if $m > -1$ and $n - m > 1$;

   (12) divergent;   (15) convergent;   (18) convergent;
   (13) convergent;   (16) convergent;   (19) divergent;
   (14) convergent;   (17) convergent;

   (20) convergent if $1 < a < 2$.

2. (1) 1;

(2) $\dfrac{8}{3}$;

(3) $\dfrac{\pi}{4}$.

5. (1) conditionally convergent;   (3) conditionally convergent.
   (2) absolutely convergent;

7. (1) $-(e^{x|\sin x|}\sin x + e^{x|\cos x|}\cos x) + \displaystyle\int_{\sin x}^{\cos x}\sqrt{1-t^2}\,e^{x\sqrt{1-t^2}}\,dt$;

   (2) $\left(\dfrac{1}{x}+\dfrac{1}{b+x}\right)\sin x(b+x) - \left(\dfrac{1}{x}+\dfrac{1}{a+x}\right)\sin x(a+x)$;

   (3) $-\displaystyle\int_{x}^{x^2} y^2 e^{-xy^2}\,dy + 2xe^{-x^5} - e^{-x^3}$;

   (4) $f(a,-a) + 2\displaystyle\int_{0}^{a} f'_u(u,v)\,dx$, where $u = x+a$, $v = x-a$;

   (5) $\dfrac{2}{a}\ln(1+a^2)$.

10. (1) $u < 1$;   (4) $u > 1$;
    (2) divergent everywhere;   (5) $u < 3$;
    (3) $u < -1$;   (6) $u < 1$.

11. (1) uniformly convergent;   (2) uniformly convergent;
    (3) (i) uniformly convergent, (ii) not uniformly convergent;
    (4) uniformly convergent;   (5) not uniformly convergent.

12. (1) $\ln\dfrac{1+\beta}{1+\alpha}$;   (2) $\ln\dfrac{b}{a}$;

    (3) $\sqrt{\pi a}$;

    (4) $\dfrac{1}{2}\ln\dfrac{\beta}{\alpha}$;   (5) $\dfrac{\pi}{2}\ln(1+a)$;   (6) $\dfrac{\pi}{2}\ln(1+\sqrt{2})$.

13. (1) $\sqrt{\dfrac{\pi}{2}}e^{-\frac{15}{8}}$;   (2) $\sqrt{\pi}$;

    (3) 0 if $a < b$, $\dfrac{\pi}{4}$ if $a = b$, $\dfrac{\pi}{2}$ if $a > b$;

    (4) $\dfrac{\pi}{2}$;   (6) $\dfrac{(2n-1)!!}{2^{n+1}}\sqrt{\pi}$.

    (5) $\sqrt{\pi}(\sqrt{\beta}-\sqrt{\alpha})$;

14. (1) $\dfrac{\pi}{3}$;   (2) $\dfrac{\pi}{2\sqrt{2}}$;   (3) $\dfrac{\pi}{2\sqrt{2}}$;   (4) $\dfrac{\pi a^4}{16}$;

(5) $\dfrac{2\pi}{3\sqrt{3}}$;   (6) $\dfrac{n!}{2}$;   (7) $\dfrac{3}{512}\pi$;   (8) $\dfrac{\pi}{2\cos\dfrac{n\pi}{2}}$;

(9) $\dfrac{1}{m}B\left(\dfrac{n}{m},q\right)$;   (10) $\dfrac{(2n-3)!!}{(2n-2)!!}\dfrac{\pi}{2a^{2n-1}}$.

Chapter 11. Fourier Series and Fourier Integrals

Exercises 11.1

4. (1) $f(x) \sim \dfrac{\pi^2}{3} + 4\displaystyle\sum_{n=1}^{+\infty} \dfrac{(-1)^n}{n^2} \cos nx$;

(2) $f(x) \sim -\dfrac{\pi}{4} + \displaystyle\sum_{n=1}^{+\infty} \left[\dfrac{(-1)^n - 1}{n^2\pi} \cos nx + \dfrac{1 - 2(-1)^n}{n} \sin nx \right]$;

(3)
$$f(x) \sim \dfrac{1 + \pi - e^{-\pi}}{2\pi} + \dfrac{1}{\pi}\sum_{n=1}^{+\infty} \left\{ \dfrac{1 - (-1)^n e^{-\pi}}{1 + n^2} \cos nx \right.$$
$$\left. + \left[\dfrac{-n + (-1)^n n e^{-x}}{1 + n^2} + \dfrac{1}{n}(1 - (-1)^n) \right] \sin nx \right\};$$

(4) $f(x) \sim \displaystyle\sum_{n=1}^{+\infty} \dfrac{\sin nx}{n}$;

(5) $f(x) \sim \dfrac{2}{\pi} + \dfrac{4}{\pi}\displaystyle\sum_{n=1}^{+\infty} \dfrac{(-1)^{n+1}}{4n^2 - 1} \cos 2nx$;

(6) $f(x) \sim 2\sinh al \left[\dfrac{1}{2al} + \displaystyle\sum_{n=1}^{+\infty} (-1)^n \dfrac{al\cos\dfrac{nx}{l}x - n\pi\sin\dfrac{n\pi}{l}x}{(al)^2 + (n\pi)^2} \right]$.

5. (1) $f(x) = \dfrac{A}{2} + \dfrac{2A}{\pi}\displaystyle\sum_{n=0}^{+\infty} \dfrac{(-1)^n}{2n+1} \cos(2n+1)x$, for $0 \leqslant x < \dfrac{\pi}{2}$ and $\dfrac{\pi}{2} < x \leqslant \pi$.

The series converges to $\dfrac{A}{2}$ at $x = \dfrac{\pi}{2}$.

$$f(x) = \dfrac{2A}{\pi}\sum_{n=0}^{+\infty} \dfrac{1 - \cos\dfrac{n\pi}{2}}{n} \sin nx,$$ for $0 \leqslant x < \dfrac{\pi}{2}$ and $\dfrac{\pi}{2} < x \leqslant \pi$.

The series converges to $\dfrac{A}{2}$ at $x = \dfrac{\pi}{2}$;

(2) $f(x) = \dfrac{h}{2\pi} + \dfrac{2}{\pi h} \displaystyle\sum_{n=1}^{+\infty} \dfrac{1 - \cos nh}{n^2} \cos nx$ for $0 \leqslant x \leqslant \pi$,

$f(x) = \dfrac{2}{\pi h} \displaystyle\sum_{n=1}^{+\infty} \dfrac{nh - \sin nh}{n^2} \sin nx$ for $0 < x \leqslant \pi$,

The series converges to 0 at $x = 0$.

6. (1) $|\sin x| = \dfrac{2}{\pi} - \dfrac{4}{\pi} \displaystyle\sum_{n=1}^{+\infty} \dfrac{\cos 2nx}{4n^2 - 1}$, $(-\infty < x < +\infty)$;

(2) $\sinh ax = \dfrac{2 \sinh a\pi}{\pi} \displaystyle\sum_{n=1}^{+\infty} (-1)^{n+1} \dfrac{n \sin nx}{n^2 + a^2}$, $(-\pi < x < \pi)$;

(3) $x \sin x = 1 - \dfrac{1}{2} \cos x + 2 \displaystyle\sum_{n=1}^{+\infty} \dfrac{(-1)^{n+1}}{n^2 - 1} \cos nx$, $(-\pi < x < \pi)$;

(4) $f(x) = \dfrac{4}{\pi} \displaystyle\sum_{n=1}^{+\infty} \dfrac{(-1)^{n+1}}{2n - 1} \cos \dfrac{(2n - 1)\pi}{2} x$, $(|x| < 1, 1 < |x| \leqslant 2)$.

The series converges to 0 at $x = \pm 1$;

(5) $f(x) = \dfrac{2}{3} - \dfrac{3}{\pi^2} \displaystyle\sum_{n=1}^{+\infty} \dfrac{1 + (-1)^{n+1} \cos \dfrac{n\pi}{3}}{n^2} \cos \dfrac{2n\pi}{3} x$, $(0 \leqslant x \leqslant 3)$;

(6) $f(x) = \dfrac{E}{2} + \displaystyle\sum_{n=1}^{+\infty} \dfrac{2}{\pi} \dfrac{\sin \dfrac{n\pi}{2}}{n} \cos \dfrac{nx}{2} x$, $(|x| < 1, 1 < |x| \leqslant 2)$.

The series converges to $\dfrac{E}{2}$ at $x = \pm 1$.

7. (1) $\dfrac{\pi^2}{6}$; (2) $\dfrac{\pi^4}{96}$.

8. $f(x) = \dfrac{a}{\pi} + \displaystyle\sum_{n=1}^{+\infty} \dfrac{2}{n\pi} \sin na \cos nx$

for $|x| < a, a < |x| \leqslant \pi$.

The series converges to $\dfrac{1}{2}$ at $x = \pm a$.

(1) $\dfrac{a(\pi - a)}{2}$; (2) $\dfrac{\pi^2 - 3\pi a + 3a^2}{6}$.

9. $f(x) = \displaystyle\sum_{n=-\infty}^{-1} \left(\dfrac{\cos n\pi - 1}{n\pi} \right) i e^{-inx} + \displaystyle\sum_{n=1}^{+\infty} \left(\dfrac{\cos n\pi - 1}{n\pi} \right) i e^{-inx}$

for $0 < |x| < \pi$.

The series converges to 0 at points $x = 0$ and $x = \pm \pi$.

Exercises 11.2

1. (1) $\dfrac{1}{a}\displaystyle\int_0^{+\infty} e^{-a\lambda}\cos\lambda x\,d\lambda,\quad (-\infty < x < \infty);$

   (2) $\displaystyle\int_0^{+\infty} e^{-a\lambda}\sin\lambda x\,d\lambda,\quad (-\infty < x < \infty);$

   (3) $f(x) = \dfrac{2}{\pi}\displaystyle\int_0^{+\infty} \dfrac{\sin\lambda}{\lambda}\cos\lambda x\,d\lambda,\quad (|x| < 1, |x| > 1),$

   $\dfrac{2}{\pi}\displaystyle\int_0^{+\infty} \dfrac{\sin\lambda}{\lambda}\cos\lambda x\,d\lambda = \dfrac{1}{2},\quad$ at points $x = \pm 1;$

   (4) $f(x) = -2\displaystyle\int_0^{+\infty} \dfrac{\sin\lambda\pi}{1-\lambda^2}\sin\lambda x\,d\lambda;$

   (5) $f(x) = \dfrac{2}{\pi}\displaystyle\int_0^{+\infty} \dfrac{1}{a^2+\lambda^2}\cos\lambda x\,d\lambda.$

2. (1) $F(\lambda) = \dfrac{4a\lambda i}{(a^2+\lambda^2)^2};$

   (2) $F(\lambda) = \dfrac{a}{(\lambda-b)^2+a^2} + \dfrac{a}{(\lambda+b)^2+a^2};$

   (3) $F(\lambda) = \dfrac{2\cos\dfrac{\lambda\pi}{2}}{1-\lambda^2}.$

3. (1) $f(x) = \dfrac{2}{\pi}\displaystyle\int_0^{+\infty} \dfrac{\cos\lambda x}{1+\lambda^2}\,d\lambda;$

   (2) $f(x) = \dfrac{2}{\pi}\displaystyle\int_0^{+\infty} \dfrac{\lambda\sin\lambda x}{1+\lambda^2}\,d\lambda,\quad (x > 0).$

   The integral is 0 at the point $x = 0.$

Index

ε-N definition of limits, 433
ε-δ definition of limits, 448

Abel criterion, 495, 511, 556, 572
Abel lemma, 494
Abel theorem, 524, 531
absolute convergence, 488, 554
alternating series, 493
antiderivative, 41
area element, 254, 296, 297, 299
asymptote, 116
at most countable set, 473
average curvature, 122
axis of revolution, 236

Bernoulli equation, 163
Bessel inequality, 609
beta function, 564, 586
Bolzano–Weierstrass criterion, 441
bounded sequence, 436

catenary, 171
Cauchy criterion, 442, 449, 487, 504,
    508, 554, 563, 571
Cauchy's Mean-Value Theorem, 36
center of curvature, 124
chain rule, 46
characteristic equation, 181
characteristic polynomial, 181
characteristic roots, 181
circle of curvature, 124
circulation, 339

closed interval, 3
comparison criterion, 488, 555
complete system of equations of fluid
    dynamics, 426
concave, 114
concave down, 114
concave up, 114
conditional convergence, 493, 554
conical surface, 238
conjugate axis, 241
continuity, 5, 454
continuous function, 5
contour surface, 317
convergence, 476, 479, 486
convergence criterion for Fourier
    series, 613
convergence criterion for series of
    positive terms, 487
convergent, 434
convex, 114
cosine series, 601
countable set, 473
cross product, 220
curl field, 383
curvature, 121
curvilinear coordinates, 297
cylindrical surface, 234

definite integral, 15, 467
dependent variable, 4
derivative, 27
differential, 31

differential equation, 151
direction cosines, 212
direction numbers, 230
direction vector, 230
directional derivative, 318
directrix, 234, 238
Dirichlet criterion, 494, 510, 556, 571
Dirichlet integral, 583, 611
discontinuity, 5
discontinuous point, 454
divergence, 363, 476, 479, 486
divergent, 434
divergent improper integral, 160
domain, 4
dot product, 217
double integral, 254

elementary function, 50, 455
elimination method, 200
ellipsoid, 239
ellipsoid of revolution, 237
elliptic cylindrical surfaces, 235
elliptic paraboloid, 242
equation of the plane, 226
equivalent infinites, 452
equivalent infinitesimals, 451
Euler integral of the first kind, 590
Euler integral of the second kind, 586
Euler integrals, 590
Euler's equation, 425
Euler's formula, 182
exact differential condition, 377
existence and uniqueness theorem,
    521
existence theorem of implicit
    functions, 512
explicit function, 278
exterior differential form, 391
exterior product of differentials, 391
extremum, 34
extremum point, 34

Fermat theorem, 35
field, 317
flux, 349
Fourier coefficient, 599

Fourier coefficients, 600
Fourier cosine transform, 619
Fourier integral, 617
Fourier series, 599, 600
Fourier sine transform, 620
Fourier transform, 619
free vector, 211
Fresnel integral, 584
function, 3
function series, 502
Fundamental Theorem of Calculus
    (Differential Form), 39
Fundamental Theorem of Calculus
    (Integral Form), 40

gamma function, 565, 585
Gauss's Theorem, 361
generating curve, 236
generating line, 234, 238
generatrix, 234, 236
gradient, 319
greatest lower bound, 459, 469
Green's Theorem, 357

harmonic function, 276
heat equation, 277, 429
higher order infinite, 452
higher order infinitesimal, 451
homogeneous equation, 162
homogeneous function, 239
hyperbolic cylindrical surface, 235
hyperbolic paraboloid, 241
hyperboloid of one sheet, 242
hyperboloid of revolution of one
    sheet, 238
hyperboloid of revolution of two
    sheets, 238
hyperboloid of two sheets, 242

implicit function, 278
improper point, 479
indefinite integral, 41
independence of the path, line
    integral, 377
independent variable, 4
infimum, 459, 469

infinite, 451
infinites of the same order, 451
infinitesimal, 451
infinitesimals of the same order, 451
inflection point, 115
initial condition, 154
initial value problem, 154
inner product, 217
integral sum, 466
integral variable, 15
integral with parameters, 565
integrand, 15, 254
integration by parts, 73, 89
integration by substitution, 69, 87
invariance of differential form, 54
invariance of the differential, 272
inverse Fourier transform, 619
irrational number, 439

Jacobian determinant, 272, 295
Jacobian matrix, 272, 295

l'Hôpital's rule, 130
Lagrange multiplier method, 412
Lagrange series, 544
Lagrange's Mean-Value Theorem, 36
Laplace's equation, 276
least upper bound, 459, 468
Lebesgue theorem, 474
left continuity, 454
left limit, 447
Leibniz criterion, 493
Leibniz formula, 57
level surface, 317
limit, 1, 447
line integral of the first kind, 327
line integral of the second kind, 332
Liouville formula, 175
lower Darboux sum, 468
lower limit, 15

Maclaurin series, 534
magnitude, 212
maximum value, 34
Mean-Value Theorem for Derivatives, 33

Mean-Value Theorem for Integrals, 17
Mean-Value Theorem of Double Integrals, 256
method of least squares, 410
method of undetermined coefficients, 542
method of variation of constants, 164
minimum value, 34
monotonically decreasing, 438
monotonically increasing, 438

natural logarithmic function, 23
negative vector, 213
Newton–Leibniz formula, 41
normal vector of the plane, 226

oblique asymptote, 117
open interval, 3
orthogonal system, 598

parabolic cylindrical surface, 235
parabolic method, 96
paraboloid of revolution, 237
parametric equation of the line, 230
Parseval equality, 609
partial derivative, 270
partial differential, 270
partial sum, 14, 486
Poincaré lemma, 396
potential field, 381
potential function, 381

quadric surface, 234

radius of convergence, 525
radius of curvature, 124
range, 4
ratio test (d'Alembert criterion), 489
rational number, 438
rectangular coordinate system, 210
region of integration, 254
remainder, 127
remainder of the Taylor expansion, 532
Riemann integrable, 467
Riemann integral, 467

Riemann sum, 467
Riemann–Lebesgue theorem, 609
right continuity, 454
right limit, 447
Rolle's Theorem, 35
Root Test, 488
rotation of axes, 248
rotor, 372
rotor field, 383

saddle surface, 241
scalar field, 317
scalar product, 213
scalar triple product, 222
Second Mean-Value Theorem of
    Integrals, 557
second order derivative, 54
separation of variables, 153
set of measure zero, 473
sine series, 601
singular point, 282
sliding vector, 212
smooth surface, 287
solenoidal vector field, 386
source-free field, 383
Stirling formula, 547
Stokes' Theorem, 367, 403
streamline, 321
summation by parts, 493
supremum, 459, 468
surface integral of the first kind, 346
surface integral of the second kind,
    350
surface of revolution, 236, 330

symmetric equations of the line, 230
system of orthonormal functions, 599

tangent line, 25
tangent method, 61
Taylor expansion, 127, 405, 532
Taylor series, 533
total differential, 270
translation of axes, 245
transverse axis, 241
trapezoidal method, 94
trapezoidal rule, 95
trigonometric series, 598
triple integral, 255

uncountable set, 473
undetermined coefficients, 190
uniform continuity, 460
uniform convergence, 504, 570
unit vector, 213
upper Darboux sum, 468
upper limit, 15

vector, 211
vector field, 317
vector potential, 383
vertex, 238
volume element, 255, 301, 302

Wallis formula, 547
wave equation, 276, 427
Weierstrass criterion, 509, 571
Wronskian determinant, 173

zero vector, 213

Printed in the United States
By Bookmasters